Jebel Bishri in Focus

Remote sensing, archaeological surveying, mapping and GIS studies of Jebel Bishri in central Syria by the Finnish project SYGIS

Minna Lönnqvist, Markus Törmä, Kenneth Lönnqvist and Milton Nuñez

with contributions by

Sanna Aro-Valjus, Minna Falck, Michael Herles, Merja Kaario, Markus Königsdörfer, Donald Lillqvist, Kirsi Lorentz, Martti Nissinen, Jari Okkonen, Juha Pakkala, Anniina Pietilä, Helena Riihiaho, Juhana Saukkonen, Taija Turunen, Arto Vuorela, Margot Stout Whiting

Edited by

Minna Lönnqvist and Kenneth Lönnqvist

BAR International Series 2230
2011

Published by

Archaeopress
Publishers of British Archaeological Reports
Gordon House
276 Banbury Road
Oxford OX2 7ED
England
bar@archaeopress.com
www.archaeopress.com

BAR S2230

Jebel Bishri in Focus: Remote sensing, archaeological surveying, mapping and GIS studies of Jebel Bishri in central Syria by the Finnish project SYGIS

ISBN 978 1 4073 0792 3

Printed in England by 4edge, Hockley

All BAR titles are available from:

Hadrian Books Ltd
122 Banbury Road
Oxford
OX2 7BP
England
www.hadrianbooks.co.uk

The current BAR catalogue with details of all titles in print, prices and means of payment is available free from Hadrian Books or may be downloaded from www.archaeopress.com

PREFACE AND ACKNOWLEDGEMENTS

Coordinates are the basic parameters of the Finnish project SYGIS (abbreviation of the "Syrian GIS") studying Jebel Bishri, a mountainous region in Central Syria. The major focus of this archaeological project has been on locating discovered sites unambiguously on Earth's surface. This is the beginning and the focal point from where to start with recording and creating data for the cultural heritage management of Syria to prevent looting and preserve cultural remains in this vulnerable area. The rest may be more or less a matter of conjecture in humanistic interpretation and relative dating in such a surface survey in which absolute dating methods have not been used; the reasons for not using absolute dating methods are obvious in the absence of stratigraphical contexts. For instance, there is no use to collect samples for radiocarbon dating without a clear provenience. However, based on the typological classification and terminus dating of the associated finds we can say that the project has covered a time span of nearly 0.5 million years of hominid and human activities in the Jebel Bishri region. In addition, it has been our aim to understand the development of human cultures from a broader perspective and gather aspects of the cultural *longue durée* in different environmental zones and the interaction of people between them.

A visit to Syria by Minna Lönnqvist, Ph.D., (University of Helsinki, Finland) and Helena Riihiaho, Theol. Cand., (University of Uppsala, Sweden) in 1997 to study the archaeological remains of the Amorites initiated the project through talks with Prof. Sultan Muhesen, then Director of the General Directory of the Antiquities and Museums (DGAM) of the country. Prof. Muhesen was keen to start a GIS-based (GIS=Geographic Information Systems) project in Syria as he was familiar with the Finnish contribution to such a developing project in Egypt. In 1999 correspondence between the Syrians from DGAM and the Finns from the University of Helsinki led to ten years of co-operation in archaeologically surveying and mapping areas of Jebel Bishri under the Finnish project SYGIS, using remote sensing methods and GIS. We are most grateful to Prof. Muhesen and Dr. Ammar Abdul-Rahman for the valuable support which we received from them at the beginning. The Finnish project was accepted to NASA's world monitoring program through German Aero-Space Center in 1999, and the Academy of Finland supported the initial phase of the project, which we acknowledge with gratitude. The first field season took place in 2000.

Subsequently NorFA (currently NordForsk), and the Finnish mobile phone company Nokia Co. continued to finance the project for which we are very grateful. The home institution of the project has been the Institute for Cultural Research at the University of Helsinki, Finland. Prof. Juhani U.E. Lehtonen, the former director of the institute, deserves our thanks in providing his support all the way through the project. The following Directors of the DGAM, such as A. Moaz, Tammam Fakouch and Bassam Jamous, are remembered with gratitude for their kind co-operation throughout the years. We are also grateful to Dr. Michel al-Maqdissi, the Director of Archaeological Excavations in Syria, who has understood the scientific importance of the project and its benefits to the cultural heritage management of Syria. The Palmyra Museum in Syria and its emeritus director Khaled Al-As'ad, present director Walid Al-As'ad, as well as emeritus deputy director Ali Taha, and present deputy director Ahmad Taha have been kindly co-operating with our project for all the years. Archaeologists Muhammed Taha, Muhammed Al-As'ad and the cooks Feisal and Hussein were provided by the Museum of Palmyra. Mr. Muhammed Turki from Palmyra/Homs has been a kind friend and has been helping us in various practical matters, which we acknowledge with gratitude. Directors Khalil Ajaz and Yasser Showhan from the Deir ez-Zor Museum have been important in supporting the survey on the Euphrates side, in which Mr. Showhan himself participated in 2006. Mr. Showhan's uncle provided topographical information of Roman sites on the southeastern piedmont area of Jebel Bishri. Muhammed Nazi from the Deir ez-Zor Museum was of valuable assistance in the field work in the area.

Initially Prof. Marjatta Hietala from the Academy of Finland supported the project, and GIS-Prof. Kirsi Virrantaus from the Helsinki University of Technology (HUT), Finland, has been supporting the project further with her understanding and mentoring for which we are most grateful. HUT has provided the invaluable skills of research scientist Markus Törmä, Tech.Lic., geodets Mervi Saario, MSci, and Jaakko Latikka, MSci. Prof. Gullög Nordquist from the University of Uppsala, Sweden, has been mentoring the project and offering valuable views, and with her kind help and support the Nordic co-operation flourished. Archaeologist Niina Turri from the University of Helsinki, Finland, has been a great contributor to planning the fund-raising and managing some public relations works, for which we all are indebted to her. Prof. Milton Nuñez from the University of Oulu, Finland, was responsible for the geological studies, and analyses of some of the lithic material, the work which the project acknowledges with gratitude. Dr. Jari Okkonen, also from the University of Oulu, has been skilfully mapping sites with GIS programs together with Dr. Tuula Okkonen. Archaeologists Dr. Kirsi Lorentz, Donald Lillqvist, BA, and Minna Falck, MA, were valuable aids in recording, documenting and mapping as well. Josep Pedrez Rodés, MSci, and Elena Garcia-Guizé, MSci, from the Universities of Oulu and Barcelona, were responsible for GPR studies. My husband, Dr. Kenneth Lönnqvist, has been invaluable in the later fieldwork and a great support in the reporting.

Prof. Christian Meyer from University of Bergen, Norway, and Prof. Ingolf Thuesen, from the University of Copenhagen, Denmark, kindly supported the project in 2003-2004, and Prof. Felix Blocher from the Martin Luther University of Halle-Wittenberg, Germany, deserves special thanks for his support in the field work of 2005-2006. Assyriologists-archaeologists Dr. Michael Herles and Markus Königsdörfer, MA, were good partners in recording and documenting in the field. Art historian Dr. Hanna-Riitta Toivanen, Prof. Martti Nissinen, and Helena Riihiaho, Theol. Cand. have made some architectural studies and plan drawing of tombs and roads, which we acknowledge with gratitude. We are also grateful to archaeologists Charlotte Stuhr-Børlit, MA, Sermin Anadol, MA, Bonnie Nilhamn, MA, Kim Darmark, MA, Dr. Kristina Josephson-Hesse, Dr. Nils Anfinset, historian Dr. Eivind Seland, students Annina Pietilä, Taija Turunen and Marlena Whiting, MA. for their contributions to the project. Researchers from different fields such as theologians Dr. Juha Pakkala, Merja Kaario, Theol.Cand., as well as Dr. Juhana Saukkonen, with assyriologists-archaeologists Margot Stout Whiting M.Phil. and assyriologist-archaeologist Dr. Sanna Aro are also to be mentioned. They all assisted in the fieldwork. The Askola High School, Finland, contributed by supplying assistance in the fieldwork as well.

Assyriologists Prof. Jean-Marie Durand, Prof. Dominique Charpin, Prof. Gianni Lanfranchi, Prof. Mario Falez, Dr. Mikko Luukko, Dr. Pirjo Lapinkivi and Dr. Jesper Eidem, are to be mentioned with special thanks for their contribution in terms of time and support in specific questions which have helped the publication process. Readers of different parts of the manuscript, Prof. Iisac Gilead and Dr. Andrew Kandel, specialists in prehistoric sites and lithics, Prof. Edward Dabrowa, specialist in the Roman history of the eastern frontier, and Prof. Andreas Kaplony, historian and arabist, have given valuable comments of which we are most indebted. They are not, however, responsible for opinions presented in the present work. Dr. Peter McNiven and Brian Roberts, BA, have made corrections to the English language of the manuscript for which we express our gratitude. However, the authors assume full responsibility for any factual or linguistic errors that have been introduced later to the original manuscript.

This report is completed exactly ten years from the first field season.

Emmanuel Gardens, Swansea, UK
28th July, 2010

Adj. Prof. Minna Lönnqvist

TABLE OF CONTENTS

1. INTRODUCTION

Minna Lönnqvist

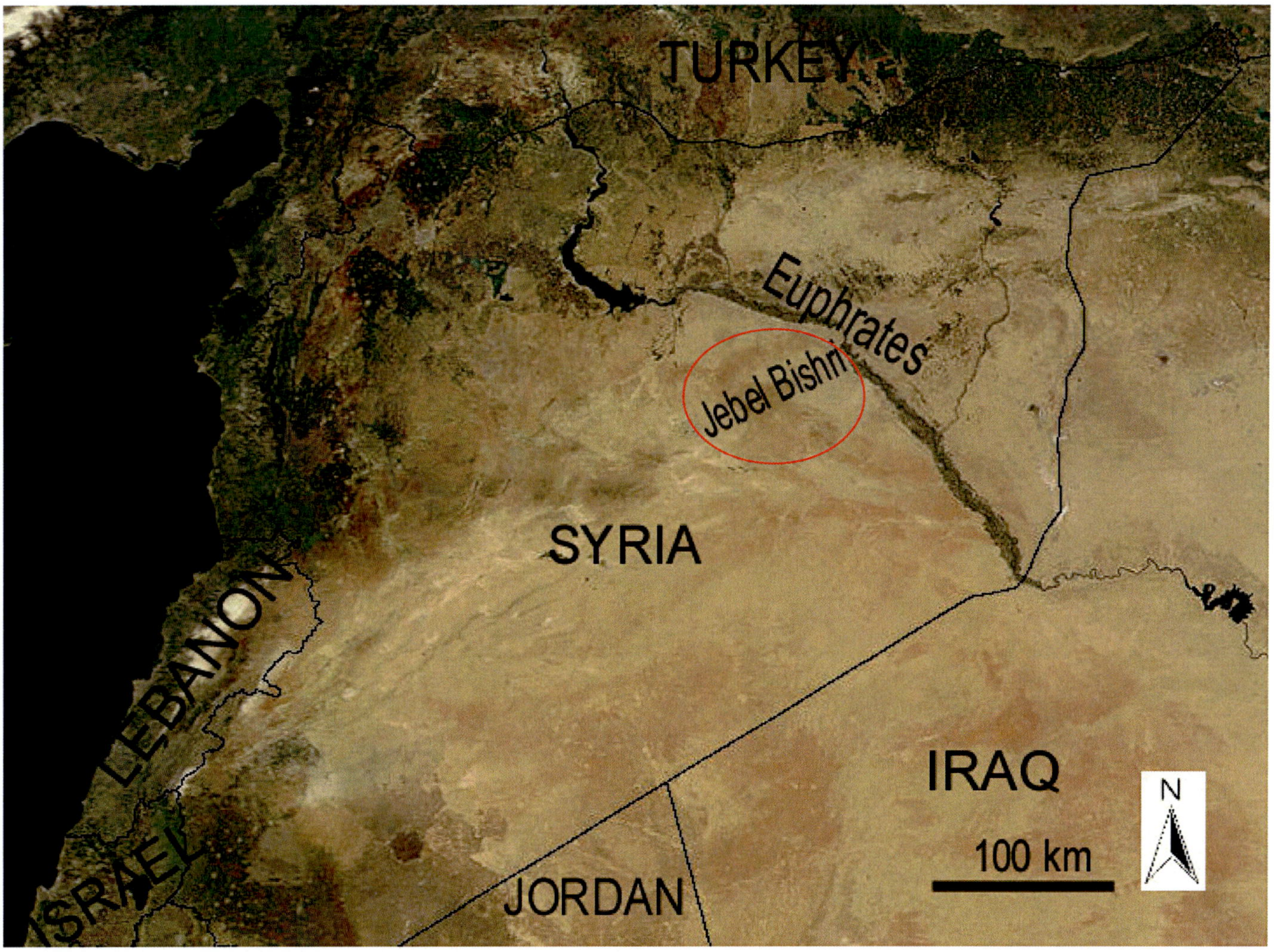

Fig. 1. A satellite map displaying the location of Jebel Bishri in Central Syria.

1.1. Zooming in on the Marginal Habitat of Jebel Bishri

As recently as fifty, even twenty five, years ago most mountain regions were at the far periphery of mainstream societal concerns. Remote, relatively inaccessible, they were generally pictured as difficult, unyielding, and unprofitable environments – amongst world's last remaining arenas for exploration and mapping, and privileged enclaves for mountaineering and adventure... Mountains constitute the most diverse landscapes on earth and thus hold a very large share of the world's biodiversity and cultural diversity...It is this biological and cultural endemism that is one of the reasons for much of the current growth of lowland interest in mountain areas. (Ives 2001).

The Oman Mountains and the Palmyrides are considered to form the largest arid mountain areas in Arabia (Edgell 2006: 447-463). The northern Palmyrides structurally consist of the Bilas block in the west, and the N-E inclined Bishri block to the east (Katzmin 2002: 124). The Bishri block, or arch in Central Syria, better known as Jebel Bishri, is located in the area between 34-36° N latitude, 39-40° E longitude (see the location in Fig. 1.). This table-mountain area is spatially limited to the Syrian Desert - a *Hamada* - in the south, and to the north and to the east it is bordered by the green Euphrates river valley, the region of the earliest human civilization. In between these lower areas there rises the peak of higher altitudes known as Jebel al-Bishri reaching ca. 867 m

(Wirth 1971: 55). With its piedmonts the mountain covers ca. 1 million ha (ACSAD = The Arab Centre for the Studies of Arid Zones and Drylands). Being like a bastion between the Upper and Lower Euphrates the higher ridges form a natural boundary following the river belt. From the earliest times, this region became a zone of interaction between different cultures and empires.

Arid lands, a number of geological structures that include mountains, were formed since the Jurassic period, broadening into arable lands and plateaus, which are remnants of Gondwana Shields, formed hundreds of millions years back in time, as well as lowlands of relatively recent sedimentary rocks which are a couple of million years old (Heathcote 1983: 43). Climatic conditions in the neighbourhood of Jebel Bishri are harsh nowadays. In UNESCO global terms the area belongs to the arid zones falling between 100 mm and 250 mm isohyets that mark the annual precipitation. Only once a decade is the rainfall in the area likely to reach 200 mm. According to the UNESCO (1963) definition, the area belongs to a sub-desert attenuated climate differing from the Mediterranean climate zone.

On Jebel Bishri winter can be very cold and summer very hot with heavy sand storms often blowing, which even prevent short-distance visibility. The region is currently covered by Irano-Turanian vegetation (see, e.g., Perrot 1993: Fig. 3). In the classification by ICARDA (The Center for Studies of Dry Land Agriculture situated near Aleppo) this area in Syria falls into Zone 5 used primarily as rangeland by nomadic and semi-nomadic pastoralists. The western parts of the mountain are now best defined as being desert-like in character, unlike the eastern part. The western ridge offers a landscape that is reminiscent of the Grand Canyon, while the greener steppe and *wadi* areas of the southeastern plateaus as well as the fringes near the Euphrates offer more steppe-like vegetation cover. The northeastern piedmonts of the mountain practically start from the outskirts of the city of Deir ez-Zor, and they continue over 100 km along the Middle Euphrates to Raqqa and beyond. The fertile fields of the Euphrates valley have been under artificial irrigation for thousands of years, due to the low precipitation which hinders rain-fed cultivation in the region.

Ives *et al.* (1997) define "steep slopes" and "altitude" as facets of mountain landscapes. In 1992 UNESCO defined the value of *cultural landscapes*, which allowed the inclusion of sites without monumental or urban nature, promoting interest in their protection and preservation (see Skeates 2000: 16). Jebel Bishri provides a mountain landscape and is a natural monument in itself having been a habitat for small-scale societies such as those of hunter-gatherers and pastoral nomads for several millennia. While approached along the Euphrates from Lower Mesopotamia Jebel Bishri forms the first imposing massif. It is especially famous for special references occurring in ancient cuneiform sources dating from the third millennium to the first millennium B.C. In Sumerian and Babylonian myths and folklore the mountain became deified, and it is described as a dwelling place of gods. It was called the mountain of the Amorites by the Akkadians and Sumerians (see RIME 2: 90-94; Buccellati 1966; Streck 2000; Nilhamn 2008) and later on, it became to be known as the mountain of the Ahlâmu-Aramaeans in the Assyrian sources (see, e.g., Kupper 1957; Darmark 2008). The mental image of Jebel Bishri and its conquest seems to have become a literary *topos* in cuneiform sources (see Pappi 2006: 241-256) to prove power over nomadic West Semites. Thus the mountain became a geopolitically central area on the Middle Euphrates, where urban powers and mountain people encountered. It was in this place that the drama of a constant dilemma with a "happy marriage" between the cultivating groups and the steppe people was also played out, for the couple cannot live without each other, despite the fact that clashes have occurred (cf. Sahlins 1968: 35). This relationship is already illustrated in the Sumerian myth of nomadic Martu (= god *Amurru* in the Akkadian sources, the personification of the Amorites) and his intention to marry a Sumerian lady (CBS 14061 in Chiera 1924). The identified dichotomy is also illustrated in the Mari cuneiform documents (ARM) dating from the eighteenth century B.C. There was a long tradition in Mesopotamian culture of seeing others as enemies, especially peoples of the mountains, marshes and deserts (Zaccagnini 1982 and Pongratz-Leisten 2001).

Already Father A. Poidebard pioneered the development of remote sensing methods in aerial archaeology in the neighbourhood of Jebel Bishri in the 1920s and 1930s (Poidebard 1934; Mouterde and Poidebard 1945; Kennedy 2002: 33-34). He used different seasons and daylight conditions to trace ancient remains left by people in the region (Brooks and Johannes 1990: 167). However, Poidebard was concentrating on the monumental Roman remains of the so-called Eastern *Limes*, the Roman imperial frontier continuing to the Euphrates and beyond. Nowadays new tools and techniques, such as satellite platforms, geophysical methods and GIS (Geographic Information Systems) are increasingly used by archaeologists to find traces left by people on the Earth's surface and beneath (see Lönnqvist and Stefanakis 2009). These tools and techniques complement the naked eye, and such traditional companions of the archaeologist as a spade and a brush. Desert and steppes provide the most favourable areas for using satellite data sources in archaeological prospecting, because of the lack of dense vegetation cover and long periods of cloudy conditions that prevent visibility (Scollar 1990: 1). In less than a decade the possibility of using satellite imagery in approaching remote areas such as Jebel Bishri has improved considerably, and nowadays anyone using GoogleEarth can zoom in on the mountain.

Since the year 1999/2000 the primary aim of the Finnish project SYGIS[1] (an anagram from the Syrian Geographic Information Systems) has been to carry out preliminary regional GIS-based archaeological prospecting, surveying and mapping of Jebel Bishri by using satellite data sources, and to exercise intensive ground studies in chosen plots, for the purpose of the cultural heritage management of Syria, and thus offer support for a developing country. Therefore empirical field work has been necessary in order to study the traces of human life in this marginal zone. The identification of an archaeological site and the recording of its location information provide the first step for its protection and preservation. Looting of archaeological sites is an especially acute problem in the neighbourhood.

Desertification is also a major environmental hazard that affects the life of nomads and agricultural endeavours in the region of Jebel Bishri today. It also affects the visibility and preservation of the archaeological sites. Already in the 1980s R.W. Dennell (1982) discussed the lack of long-term time-scale and the usability of the role of archaeology in studying the causes, processes and even diagnosis of desertification. He emphasized that the more we gather information about the past ways of using deserts by the early man, we might be better off to evaluate the impact of desertification in the present. So, archaeology can contribute to current debates on climate change and environmental change like the effects of geohazards on human populations by studying, how successive generations have exploited and altered the environment (Barker and Gilbertson 2000: 7-18; Lönnqvist and Törmä 2003; 2004; Barker *et al.* 2007; Lönnqvist *et al.* 2008; McLaren and Reynolds 2009). Using remote sensing and GIS our Finnish project has approached the long-term and short-term changes in human environment of the Jebel Bishri region. Both before and simultaneously with our project ACSAD and its co-operation of German parties have been using remote sensing methods with GIS while carrying out an environmental project to counteract desertification on the mountain.

It is a well-known fact that annual rainfall reflected in the isohyets is not static but constantly changing its spaces and boundaries. The dynamics of the arid lands can be seen in the responses of people to the surrounding environmental conditions, but we need to hold back from environmental determinism and take into account social and political factors as well, as people also affect the conditions of life in the marginal zones. Contemporaneously with our project archaeological research to the arid zone in the Levant has been applied by the French in their project *Marges arides de la Syrie du Nord* in 1995-2002 (see, e.g., Geyer *et al.* 2004-2005) and the British in the Wadi Faynan landscape project in Southern Jordan in 1997-2000 (see, e.g., Barker *et al.* 2007). Both of these projects have paid special attention to geomorphology, settlement development and land use. Our project which started in 1999/2000 resembles these endeavours in some ways but differs in the specific use of macro-scale in remote sensing and GIS, which have also been utilized, for instance, in the landscape project in the Homs Region in Syria since 2001 (Philip *et al.* 2005).

1.2. Previous and Contemporaneous Archaeological Studies in the Neighbourhood

Early Arab sources and the nineteenth to twentieth century European travellers, such as E. Sachau (1900), V. Chapot (1907), F. Sarre with E. Herzfeld (1911), R. Dussaud (1927) and A. Musil (1927, 1928) report visits to the ridges of Jebel Bishri but their observations concentrate on monumental remains.

G. Buccellati and M. Kelly-Buccellati (1967) also carried out a short survey in the neighbourhood of the Palmyrene and Jebel Bishri in the 1960s, and they published a half a page article mentioning such archaeological remains as stone circles. Apart from the early itineraries, A. Poidebard's (1934) aerial studies and archaeological visits, the mountain proper had largely remained archaeologically *terra incognita* before our Finnish Project SYGIS started its prospecting, surveying and mapping in 1999/2000 (Lönnqvist and Törmä 2003).

Observations of the earliest Palaeolithic activities on the right bank of the Euphrates following the edge of Jebel Bishri have been carried out by Syrians such as S. Muhesen (1985), by French such as J. Besançon and P.

---

[1] See SYGIS, the Archaeological Survey and Mapping Project of Jebel Bishri, in Lönnqvist and Törmä 2003, also home site of the project www.helsinki.fi/hum/arla/sygis.

Sanlaville (1981) and some British like L. Copeland (2004) as well as Westaway *et al.* (2005 and 2007) before and contemporaneously with us. The oasis basin of El Kowm on the western piedmont with its prehistoric sites ranging from the Lower Palaeolithic to the Chalcolithic period has been studied for decades by Syrians, Germans, French and Swiss researchers (e.g., Dornemann 1986; Cauvin and Sanlaville 1981; Le Tensorer *et al.* 2006). On the other hand, J. Zarins' (1989a, 1989b, 1990, 1992) interest in archaeological study concerning the process of animal domestication and the earliest signs of nomadism in the Near East had resulted in a series of short publications in the late 1980s and early 1990s that generally dealt with the district of Jebel Bishri. In Zarins' work the mountain became adjacent to the Greater Southwest Arid Zone and its nomadic technocomplex. The archaeological material which Zarins' article under the heading of Jebel Bishri (1989a) in specific deals with, however, concerns the oasis of El Kowm, which lies on the western piedmont and not in the mountain proper. In any event, thanks to the studies by J. Zarins and D. Stordeur (1993) the region has remained significant in the study of the emergence of pastoral nomadism. (See also, e.g. Anfinset 2008).

Epipalaeolithic and Pre-Pottery Neolithic sites on the Middle Euphrates testify to the first steps to sedentarism: El Kowm, Mureybet, Abu Hureyra (Moore *et al.* 2000) and Bouqras (e.g. Akkermans *et al.* 1983) are the most famous sites reflecting the transition on the Middle Euphrates. The modern study of nomadic life has focused on the need of pastoralists to have extra supply of agricultural products from sedentary people. But was this the situation throughout the past? A possible shift from a mobile hunter-gatherer lifestyle directly to other modes of mobility is attested to in marginal areas such as the Negev and Sinai and this leaves the question as to the role of agriculture and the degree of dependence upon it in the arid zones still to be a challenging problem for analysis. From the Pre-Pottery Neolithic period small livestock rearing gradually became an established subsistence economy in the neighbourhood, but only the cuneiform sources from the third millennium B.C. provide actual historical evidence of the special connection that the area of Jebel Bishri had with pastoral nomads and their economy.

Previously the more sedentary areas on the right bank of the Middle Euphrates below Jebel Bishri have been identified by A.M.T. Moore (Sanlaville 1985, and personal communication) and K. Kohlmeyer (1984; 1986) in their surveys, reaching from the Tabqa dam or Tall Bi´a (Tuttul) as far as the peninsula of Halabiya. A Bronze Age cemetery of Abu Hamed was excavated on the northwestern piedmont of Jebel Bishri ca. 40 km east of Raqqa by a German team in 1990 (Falb *et al.* 2005). In addition, a Syrian-Spanish team has been studying the surroundings of Halabiya contemporaneously with us (Montero-Fenollós 2006), and a Syrian-Japanese team has been since 2007 following in our footsteps on the northwestern slopes of Jebel Bishri and the Euphrates valley (Al-Maqdissi and Ohnuma 2008).

The Roman fortified line that runs through the oases on the western piedmont of Jebel Bishri and at the edge of the Syrian Desert consists of *castra* and *castella* and is best known as the *Strata Diocletiana* dating from ca. A.D. 300. But as several sites associated with the *Strata* are already listed in Ptolemy's *Geographia* (the second century A.D.), it appears that the Emperor Diocletian fortified and refortified many Roman and Palmyrene sites that already existed in the region. Keeping this fact in mind our archaeological study elucidates the development of the desert frontier and its relation to the Euphrates *Limes*. M. Sartre (2005: 143) points out that, although Poidebard's aerial studies in the neighbourhood were innovative, there have been little systematic studies to verify his observations on the ground. Recent German and French studies have, however, changed the situation, and our survey further provides ground observations of several sites mentioned by Poidebard, even adding unknown sites to the Roman Eastern frontier zone.

After the studies of A. Musil, F. Sarre, E. Herzfeld and A. Poidebard, Germans, for instance, have been excavating the Byzantine fortress of Resafa-Sergiopolis (see, e.g., Ulbert 1986) at the northwestern foot, and they have studied the Roman forts of Cholle, Tetrapyrgium and Sura on the *Strata Diocletiana* and the Euphrates *Limes* (Konrad 2001). Americans and Swiss have studied the remains at the Umayyad castle of Qasr al-Hayr ash-Sharqi associated with the *Strata Diocletiana* district (Grabar *et al.* 1978; Genequand 2005). Further on the Euphrates *Limes* J. Napoli (2000) has studied the Greco-Roman site of Djazla hanging on the ridge of Jebel Bishri. The peninsula of Halabiya that is protruding to the Euphrates forms part of Jebel Bishri and offers the fortress of Zenobia which the Frenchmen first led by J. Lauffray (1983; 1991) have been surveying, excavating and restroring since the 1940s. M. Gschwind (2005) has recently conducted a joint Syro-German excavation of the Roman fort of Qreiye and a tell situated next to the fort on the basalt peninsula of 'Ayyāsh on the Euphrates.

1.3. Catching Sight of Ancient Human Traces on Jebel Bishri

Visibility and preservation are factors that affect our potential to reveal ancient remains and traces left by humans (Evans and O'Connor 1999: 78). The scale on its part determines visibility. Apart from good surface visibility of deserts and steppes, obtaining visibility of mobile or ephemeral sites such as those of hunter-gatherers and nomads has often created a problem for the remote sensing carried out by our project using satellite image data sources. This is generally due to the low spatial resolution of the data, but it is also due to the types of sites involved. Therefore experimental fieldwork is needed to complement the remote sensing.

In our project no preference has been given to any archaeological period in order to provide basic mapping and understanding of the development of human life in the marginal region (see Lönnqvist *et al.* 2006). However, this report has been written in a view of providing an easy access to specialists of different periods whether those of prehistory or history so that a person does not have to go through the whole report site by site to pick up Palaeolithic, Bronze Age or Roman evidence. Each time scale or subsistence economy is given a space of its own in separate chapters providing an easy access to specialists of particular eras or anthropological interests.

Jebel Bishri provides a niche where the life of human societies has been carried on (see Evans and O'Connor 1999: 94). K. Butzer's (1982: 7) themes of contextual archaeology have been utilized in this study, and they consist of such basic components as 1) space, 2) scale, 3) complexity, 4) interaction and 5) equilibrium state. B. Winterhalder (1980) on his part has defined the dimensions of human ecology in 1) foraging range, 2) migratory range and 3) extended interaction through trade. All of these aspects clearly materialise through the study of Jebel Bishri. The area forms an ecologically important spatial boundary between the desert and the sown that affects the nomadic and sedentary life (Lönnqvist and Törmä 2004). High- and lowland interaction provides a special avenue for studying mobility and sedentism in this region.

From the captured data the mapping of basic archaeological sites was carried out from space, ground and subterranean land. From this archaeological data our special interest was to study the relationship between the pastoral nomadic life of the mountain and the sedentary sites of agricultural villagers in the Euphrates valley. However, this research problem came to encompass hunter-gatherer mobility alongside with the pastoral nomadic way of life in order to understand the continuity of mobile life in this arid zone on the desert fringe. The degree and nature of mobility and the process of sedentarization are central questions when studying the choices people take in this area so closely located to the Levantine corridor. Ethnoarchaeological approaches have also been integrated into our research design in order to understand the nomadic life and the processes of sedentarization and nomadization. In our ethnoarchaeological studies the process of sedentarization and differences in site abandonment processes are observed in order to understand the effects of the environmental processes and other variables on the life of the nomads.

Naturally the development of new methods in studying ancient nomads, many based on ethnoarchaeological approaches (e.g. Hole 1979; La Bianca 1990; Cribb 1991; Finkelstein 1995), has been vital to our project in tracing archaeological remains on the ground and making the remains of nomadic sites visible. S. Kent (1989: 2) has clarified the general definitions of mobility, nomadism and sedentism in a way that has been beneficial for our project. She regards all of these terms as conditions of group movement. Nomadism involves movement on a landscape, whereas sedentism means lack of movement, and 'mobility' simply denotes group movement in space. Both nomadism and sedentism are at the extreme ends of a mobility continuum. The degrees of nomadism and sedentism are reflected in site structure, site furniture, small finds, and the degrees between the extreme ends are expressed in semi-nomadism and semi-sedentism. (See also Salzman 1980: 11). However, differing from some definitions or the use of the word "nomadism" we exclude hunter-gatherers from the concept of nomadism and use the word "mobility" instead as our view is based on the original Greek word *nom'as* which means "to be grazing (pasturing)" and thus associated with pastoral economy (cf. also Khazanov 1994: 15; Lönnqvist 2000: 60).

1.4. Resources, Data Sources and Strategies used by SYGIS

1.4.1 Resources and Co-operation

The Finnish project was first financed by the Council of Developing Studies in the Academy of Finland and therefore the idea in supporting the cultural heritage management of Syria was also to offer information of the past changes for the benefit of present-day sustainable development in the region. Sustainability concerns long-term perspectivism, and archaeology provides this perspective that can be applied to developing studies.

The home institution of the project has been the Institute for Cultural Research, Department of Archaeology at the University of Helsinki, Finland, and remote sensing has chiefly been carried out at the Helsinki University of Technology, Finland. Apart from the Academy of Finland, NorFA (NordForsk under the Nordic Ministers) and Nokia Co. as well as Minna and Kenneth Lönnqvist have been financing the project. The General Directory of the Antiquities and Museums in Syria has kindly provided research permits, assistance and staff for the project. Acceptance into NASA's world monitoring program through DLR (the German Aerospace Center) and co-operation with ISPRS (the International Society of Photogrammetry, Remote Sensing and Spatial Information Sciences), has also been invaluable for the project. ICOMOS (International Council of Monuments and Sites) under UNESCO and its scientific committee CIPA have been an important forum for partnership, discussion and project development.

Fig. 2. The Guest House of the Syrian General Directory of the Antiquities and Museum, our headquarters in 2000, near Qasr al-Hayr ash-Sharqi. Photo: SYGIS.

In the year 2000 the headquarters of the project situated at the Guest House of the General Directory of the Antiquities and Museums next to Qasr al-Hayr ash-Sharqi (Fig. 2.) on the western piedmont, and a camp was established near the western edge of the mountain at Tar al-Sbai. During the field seasons 2003-2006 the headquarters of the project was at the Badiyah Cham Hotel (Fig. 3.) in Deir ez-Zor, and a camp situated at Nadra on the eastern piedmont area in 2004 (see Fig. 4.).

Fig. 3. Hotel Badiyah Cham in Deir ez-Zor served as the project headquarters in 2003-2006. Photo: Kenneth Lönnqvist 2003.

Fig. 4. The tent camp of the project on the mountain in 2004. Photo: Gullög Nordquist.

The hotel Badiyah Cham continued to serve as our base in subsequent years as well. The field staff has been multinational, even though the majority of its members has consisted of Finns. Beside the Finns and Syrians the participants have consisted of Swedes, Norwegians, Danes, Germans, Spaniards and US citizens. The largest groups joined the field work during the seasons of 2000 and 2004 (see Figs. 6. and 7.). The staff has consisted of archaeologists, GIS-specialists, photogrammetrists, a geologist, a geodet, historians, a bioanthropologist, a topographer, an art historian, assyriologists and theologians. The members of the staff have belonged to the Universities of Helsinki and Oulu as well as the Helsinki University of Technology in Finland, the Universities of Uppsala and Umeå in Sweden, the Universities of Oslo and Bergen in Norway, the University of Copenhagen in Denmark, the University of Cambridge in Britain, the University of Halle-Wittenberg in Germany and the University of Barcelona in Spain.

Assistantship has also been received from the Askola High School in Finland.

Fig. 5. The field team of 2000 on Jebel Bishri. From right, upper row: Juha Pakkala, Mervi Saario, Merja Kaario, Martti Nissinen, Minna Falck, Margot Stout Whiting, Donald Lillqvist, Mohammed Taha, Said Abdullah, Juhana Saukkonen, Minna Lönnqvist, Marlena Whiting, Kirsi Lorentz. In addition, not present in the picture: Sanna Aro-Valjus. Anniina Pietilä, Taija Turunen and Helena Riihiaho were visiting assistants while Lebanese Ghadi Boustani with Jean-Pierre Matar were documenting by photography and video-filming.

Fig. 6. The field team of 2004 on Jebel Bishri. From right: Kenneth Lönnqvist, Milton Nunez, Josep Pedrez Rodes, Jari Okkonen, Margot Stout-Whiting, Eivind Seland, Sermin Anadol, Nils Anfinset, Christian Meyer, Helena Riihiaho, Kristina Josephson-Hesse, Bonnie Nilhamn, Hanna-Riitta Toivanen, Elena Garcia-Guize, Charlotte Stuhr-Børlit, Kim Darmark. In addition, not present in the picture, Gullög Nordquist.

Fig. 7. Field teaching for a Nordic research training group at the foot of Jebel Bishri. Photo: Gullög Nordquist 2004.

Fig. 8. NorFA (NordForsk) seminar under the auspices of the project at the hotel Badiyah Cham in Deir ez-Zor 2004. Photo: Gullög Nordquist 2004.

The work of the project has included course planning and fund raising from the developing funds available at the Finnish Foreign Ministry, for the education of the staff of the Syrian General Directory of the Antiquities and Museums in the field of GIS in 2003-2004. The course was implemented by ACSAD. Furthermore the aspects of education and research training have been included in a Nordic PhD course financed by NorFA (NordForsk) (Figs. 6., 7. and 8.), the published papers of which form the introductory volume *Jebel Bishri in Context* (Lönnqvist ed. 2008) for the present report, also published in the BAR International series.

1.4.2 Background Data Sources and Strategies for Archaeological Prospecting, Surveying and Mapping

Different data sources, techniques and methods have been utilized to gather traces about the presence of people in ancient times. Apart from A. Poidebard's studies, aerial photographs are not generally available for civil use in Syria by comparison to many neighbouring countries. The mosaic of the physiography of Jebel Bishri is visible in tessellated satellite images nowadays, showing different components of environment and landscape. The benefits in using satellite data sources in prospecting and surveying include the ability to visualise large areas, and the low cost of the data by comparison with the lengthy work of expedition groups on the ground. However, as previously mentioned, the ground surveying has been a vital strategy to collect experimental data for identifying different types of remains as well as traces of activities left by humans, capture location data and reference data for dating the remains and traces.

The primary textual sources pertaining to the region consist of ancient cuneiform texts, Roman-Byzantine sources, old itineraries and ethnographic accounts. The main cuneiform texts comprise the mentioned accounts of Akkadian victories, Sumerian Gudea statues (Statue B, Louvre AO2), Mari archives (ARM) as well as the accounts of the victories of the Assyrian kings over Aramaeans at the foot of the mountain. Roman and Byzantine sources consist of military tabulas and lists, e.g., *Notitia Dignitatum* and authors such as ancient historians Pliny and Procopius as well as Arabic sources. Ancient maps from the Roman period such as the previously mentioned Ptolemy's *Geographia* and *Tabula Peutingeriana* partly cover the surroundings of Jebel Bishri. The Silk Road was a major link between China and the Roman Empire that reached through the neighbouring southern desert to a caravan station in Palmyra. The line that it took may have followed an earlier Bronze Age caravan route leading from Tadmor (Palmyra) over the desert towards Dûr Yahdun-Lim (Deir ez-Zor?) and Mari (Joannès 1997).

More recent ethnographic accounts and tribal studies consist of various materials dealing with the tribal groups in the area. The local tribes are described, for example, by C. Niebuhr (1792), J.L. Burckhardt (1831), M. von Oppenheim (1939), E. Wirth (1971), N.N. Lewis (1987) and J. Jabbur (1995) in their general studies. However, interviews among the local nomads, who have served as ethnographic informants, formed a vital source with regard to the current social and economic life in the region. Our interviews especially have provided information about the semi-sedentary life of transhumant pastoralists. From the locals we gathered knowledge of the terrain and the existence of some of the archaeological remains that were discovered. Bedouins are valuable guides when navigating on the ground in the desert even if modern technology with GPS and satellite images are used. Bedouins are also quite well aware of the Roman remains in their territories. However, the antiquity and significance of other remains, such as prehistoric and later historic rock shelters, workshops, hut-bases and animal pens known as corrals remain generally unnoticed by them.

The strategic outline of SYGIS can be outlined here in a nutshell:

Prospecting and mapping

- overall environmental understanding of the area through satellite images such as those provided by Landsat and SPOT
- Landsat images and CORONA satellite photographs (highest resolution 1.8 m) have been used for prospecting and mapping the archaeological sites (Landsat-7 ETM panchromatic channel with 15 m resolution have also been used for field mapping)
- high spatial resolution satellite images such as QuickBird (0.6 m) offer data source for tracing and mapping different kinds of sites and structures, including mobile camp sites, corrals and tombs
- for mapping, the images have to be orthorectified for UTM coordinate system (rectification is carried out with maps and collecting reference GPS control points on the ground)
- for mapping scanned Russian topographic maps
- background data: previous archaeological studies in the neighbourhood, historical, cartographic and ethnographic sources

Choosing areas for empirical field survey

- different environmental zones of Jebel Bishri have been targeted for tracing possible differences in the remains according to environmental zones
- desert-steppe in the west, e.g. Tar al-Sbai
- steppe in the central eastern area at Nadra
- alluvial terraces facing the irrigated agricultural river valley of the Euphrates
- tracing possible differences in the remains according to environmental zones
- mobile/nomadic sites *versus* sedentary sites
- tracing possible differences in monument and non-monument areas

Field surveying

- defining transects
- pedestrian survey at 15 m interval in transects
- defining sites
- recording sites with GPS (acquiring UTM coordinates)
- filling field forms for recording and documenting the site
- photographing
- recording site measurements, structure and find distribution with EDM at sample sites
- measuring sites by tape measure
- drawing structures by hand

Mapping and modelling

- GPS control point capturing
- UK military aviation maps 1:100 000 are used as a base
- Russian topographic maps 1: 50 000
- Technische Fachhohschule, Berlin, satellite maps edited by Meissner and Ripke
- Rectified satellite images (see above)
- Topographic SRTM radar shuttle mission 2000 DEM tiles, and Aster-DEM
- Software used in mapping and modelling: ERDAS MapSheets, ArcView, ImageDrape

Bibliography

Akkermans, P.A. *et al.* (1983) Bouqras Revisited: Preliminary Report on a Project in Eastern Syria, in *Proceedings of the Prehistoric Society*, Vol. 49, pp. 335-372.

Anfinset, N. (2008) Towards a Specialised Food Production – A Look at Jebel Bishri, in *Jebel Bishri in Context, Introduction to the Archaeological Studies and the Neighbourhood of Jebel Bishri in Central Syria,* Proceedings of a Nordic Research Training Seminar in Syria, May 2004, ed. Lönnqvist, M., BAR International Series 1817, Oxford: Archaeopress, pp. 1-13.

Al-Maqdissi, M. and Ohnuma, K. (2008) Preliminary Reports of the Syria-Japan Archaeological Joint Research in the Region of Ar-Raqqa, Syria, 2007, in *Al-Rāfidān*, Vol. XXIX, pp. 117-196.

Barker, G. and Gilbertson, D. (2000) *The Archaeology of Dry Land: Living at the margin*, One World Archaeology 39. UK: Routledge, Taylor and Francis.

Barker, G., Gilbertson, D. and Mattingly, D. (2007) *Archaeology and Desertification: The Wadi Faynan Landscape Survey, Southern Jordan.* Wadi Faynan Series 2. Levant Supplementary Series 4. Oxford: Oxbow Books.

Besançon, J. and Sanlaville, P. (1981) Aperçu Géomorphologique sur la Vallée de l' Euphrate Syrien, in *Paléorient,* vol. 7/2, pp. 5-18.

Brooks, R.R. and Johannes, D. (1990) *Phytoarchaeology.* Historical, Ethno- & Economic Botany Series. Vol. 3. T. R. Dudley general ed. Portland, Oregon: Dioscorides Press.

Buccellati, G. (1966) *The Amorites of the Ur III Period.* Pubblicazioni del seminario di semistica a cura di Giovanni Garbini. Richerche I. Naples: Istituto Orientale di Napoli.

Buccellati, G. and Kelly-Buccellati, M. (1967) Annual Meeting of the Archaeological Institute of America (Archaeological Survey of the Palmyrene and the Jebel Bishri), in *Archaeology,* Vol. 20, pp. 305-306.

Burchardt, J.L. (1831) *Notes on the Bedouins and Wahabys collected during his travels in the East* by the late John Lewis Burchardt. Vol. 1. London: Henry Colburn and Richard Bentley.

Butzer, K. (1982) *Archaeology as Human Ecology: Method and Theory for a Contextual Approach.* Cambridge: Cambridge University Press.

Cauvin, J. and Sanlaville, P. (eds.) (1981) *Préhistoire du Levant: Chronologie et organisation de l'Espace depuis les Origines jusqu'au VIe millénaire.* Colloques Internationaux du Centre National de la Recherche Scientifique No. 598. Paris: Editions du CNRS.

Chapot, V. (1907) *La Frontiére de l'Euphrate de Pompée a la conquête arabe.* Paris: A. Fontemoing.

Chiera, E. (1924) *Sumerian Religious Texts.* Crozer Theological Seminary Babylonian Publications. Vol. I. Upland. P.A.

Copeland, L. (2004) The Paleolithic of the Euphrates Valley in Syria, in *From the River to the Sea, The Palaeolithic and the Neolithic on the Euphrates and in the Northern Levant,* Studies in honour of L. Copeland, ed. by Aurenche, O., La Mière, M. and Sanlaville, P., BAR International Series, No. 1263, Oxford: Archaeopress, pp. 19-114.

Cribb, R. (1991) *Nomads in Archaeology.* New Studies in Archaeology. Cambridge: Cambridge University Press.

Darmark, K. (2008) The Archaeological Potential of the Assyro-Aramaean Hostility on the Euphrates Side of Jebel Bishri, Implications of Battlefield Archaeology, in *Jebel Bishri in Context, Introduction to the Archaeological Studies and the Neighbourhood of Jebel Bishri in Central Syria,* Proceedings of a Nordic Research Training Seminar in Syria, May 2004, ed. Lönnqvist, M., BAR International Series, No. 1817, Oxford: Archaeopress, pp. 49-58.

Dennell, R.W. (1982) Archaeology and the Study of Desertification, in *Desertification and Development: Dryland Ecology in Social Perspective,* ed. by Spooner, B. and Mann, H.S., London: Academic Press, pp. 43-60.

Dornemann, R.H. (1986) *A Neolithic Village at Tell El Kowm in the Syrian Desert,* Studies in Ancient Oriental Civilization No. 43, Chicago: Chicago University Press.

Dussaud, R. (1927) *Topographie Historique de la Syrie Antique et Médiévale.* Paris: Geuthner.

Edgell, H.S. (2006) *Arabian Deserts: Nature, Origin and Evolution.* Netherlands: Springer.

Evans, J. and O'Connor, T. (1999) *Environmental Archaeology, Principles and Methods.* Midsomer Norton, Somerset: Sutton Publishing.

Falb, C., Krasnik, K., Meyer, J.-W. and Vila, E. (2005) *Gräber des 3. Jahrtausends v. Chr. Im syrischen Euphrattal. 4. Der Friedhof von Abu Hamed.* Schriften zur vorderasiatischen Archäologie. Herausgegeben von Winfried Orthmann. Band 8. Saarwellingen: Saarländische Druckerei & Verlag.

Finkelstein, I. (1995) *Living on the Fringe: The Archaeology and History of the Negev, Sinai and Neighbouring Regions in the Bronze and Iron Ages.* Monographs in Mediterranean Archaeology 6. Midsomer Norton, Somerset: Sheffield Academic Press.

Genequand, D. (2005) From 'desert castle' to medieval town: Qasr al-Hayr al-Sharqi (Syria), in *Antiquity,* Vol. 79, pp. 350-361.

Geyer, B., Rousset, M.-O., Al-Dbiyat, M., Awad, N., Barge, O., Besançon, J., Calvet, Y., Gatier, P.-L. and Jaubert, R. (2004-2005) Les Marges Arides de la Syrie du Nord: Première Synthèse d'une Prospection Géoarchaeologique, in *Annales Archéologiques Arabes Syriennes,* Vols. XLVII-XLVIII, pp. 17-34.

Grabar, O., Holod, H., Knustad, J. and Trousdale, W., *et al.* (1978) *City in the Desert: Qasr al-Hayr East.* An Account of the Excavations Carried out at Qasr al-Hayr East on Behalf of the Kelsey Museum of Archaeology at the University of Michigan, with the Help of Harvard University and the Oriental Institute of Chicago. Harvard Middle Eastern Monographs 23-24. Cambridge, Mass: Harvard University Press.

Gschwind, M. (2005) Qreiye/'Ayyāš, in *Archäologischer Anzeiger,* 2. Halbband, 2005, Deutsches Archäologisches Institut, pp. 255-257.

Heathcote, R.L. (1983) *The Arid Lands: Their Use and Abuse.* Themes in resource management. London and New York: Longman.

Hole, F. (1979) Rediscovering the Past in the Present: Ethnoarchaeology in Luristan, Iran, in *Ethnoarchaeology, Implications of Ethnography for Archaeology*, ed. by Kramer, Carol, New York: Columbia University Press, pp. 192-218.

Ives, J.D. (2001) *Highland - Lowland Interactive Systems* (draft of document for FAO - FORC/IYM 2002), Ottawa, Canada. 15th March, 2001.

Ives, J.D., Messerli, B. and Spiess, E. (1997) Mountains of the World: A Global Priority, in *Mountains of the World: A Global Priority*, London & New York: Parthenon, pp. 1-16.

Jabbur, J. S. (1995) *The Bedouins and the Desert: Aspects of Nomadic Life in the Arab East.* SUNY Series in Near Eastern Studies. Buffalo: State University of New York Press.

Joannès, F. (1997) Palmyre et les Routes du Désert au Début du Deuxième Millénaire av. J.-C., in *MARI Annales de Recherches Interdisciplinaires*, Vol. 8, pp. 393-415.

Katzmin, V.G. (2002) The Late Paleozoic to Cainozoic Intraplate Deformation in North Arabia: A Response to Plate Boundary-forces, in *EGU Stephan Mueller Special Publication Series*, 2, pp. 123-138.

Kennedy, D. (2002) Aerial Archaeology in the Middle East: The Role of the Military Past, Present ... and Future?, in *Aerial Archaeology, Developing Future Practice*, ed. by Bewley, R. H., & Rączkowski, W., Amsterdam – Berlin – Oxford – Tokyo – Washington D.C.: IOS Press, pp. 33-48.

Kent, S. (1989) Cross-cultural perceptions of farmers as hunters and the value of meat, in *Farmers as Hunters*, ed. by Kent, S., New Directions in Archaeology, Cambridge: Cambridge University Press, pp. 1-17.

Khazanov, A.M. (1994) *Nomads and the Outside World.* Madison, Wisconsin: University of Wisconsin Press.

Kohlmeyer, K. (1984) Euphrat-Survey, Sie mit Mitteln der Gerda Henkel Stiftung durchgeführte archäologische Geländebegehung im syrischen Euphrattal, in *Mitteilungen der Deutschen Orient-Gesellschaft,* Vol. 116, pp. 95-118.

Kohlmeyer, K. (1986) Euphrat-Survey 1984, Zweiter Vorbericht über die Mitteln der Gerda Henkel Stiftung duchgeführte archäologische Geländebegehung im syrischen Euphrattal, in *Mitteilungen der Deutschen Orient-Gesellschaft,* Vol. 118, pp. 51-64.

Konrad, M. (2001) (with the contributions by Baldus, H.R. and Ulbert, T.) *Der Spätrömische Limes in Syrien: Archäologische Untersuchungen an den Grenzkastellen von Sura, Tetrapyrgium, Cholle und in Resafa.* Resafa 5. Mainz: Philipp von Zabern.

Kupper, J.-R. (1957) *Les nomades en Mésopotamie au temps des rois de Mari.* Bibliothèque de la Faculté de Philosophie et Lettres de l'Université de Liège. Fascicule CXLII. Paris: Les Belles Lettres.

Lauffray, J. (1983) *Halabiyya – Zenobia: Place Forte du Limes Oriental et La Haute-Mésopotamie au VIe Siècle.* Tome I: Les duchés frontaliers de Mésopotamie et les fortifications de Zenobia. Bibliothèque Archéologique et Historique T. CXIX. Institut français d'archéologie du Proche-Orient, Beyrouth – Damas – Amman. Paris: Paul Geuthner.

Lauffray, J. (1991) *Halabiyya – Zenobia: Place Forte du Limes Oriental et La Haute-Mésopotamie au VIe Siècle.* Tome II: L'architecture publique, religieuse, privée et funéraire. Bibliothèque Archéologique et Historique T. CXXXVIII. Institut français d'archéologie du Proche-Orient, Beyrouth – Damas – Amman. Paris: Paul Geuthner.

La Bianca, Ø.S. (1990) *Sedentarization and Nomadization: Food System Cycles at Hesban and Vicinity in Transjordan. Hesban 1.* Berrien Springs, MI: The Institute of Archaeology, St. Andrews University Press.

Le Tensorer, J.-M., Muhesen, S., Schmid, P. with the scientific collaborators Hack, Th., Jagher, R., Le Tensorer, H., Richter D. and Wojtczak, D. (2006) *Research on the Palaeolithic of the El Kowm Area (Syria).* Abstract.

Lewis, N.N. (1987) *Nomad and Settlers in Syria and Jordan, 1800-1980.* Cambridge: Cambridge University Press.

Lönnqvist, M.A. (2000) *Between Nomadism and Sedentism, Amorites from the Perspective of Contextual Archaeology.* Ph.D. dissertation. Helsinki: Juutiprint.

Lönnqvist, M. (ed.) (2008) *Jebel Bishri in Context: Introduction to the Archaeological Studies and the Neighbourhood of Jebel Bishri in Central Syria.* Proceedings of a Nordic Research Training Seminar in Syria, May 2004. BAR International Series 1817. Oxford: Archaeopress.

Lönnqvist, M. and Törmä, M. (2003) SYGIS – The Finnish Archaeological Project in Syria, in Proceedings of the XIXth International Symposium CIPA 2003 (The ICOMOS & ISPRS Committee for Documentation of Cultural Heritage): New Perspectives to Save Cultural Heritage, Antalya, Turkey, ed. by Altan, M.O., *The ISPRS International Archives of the Photogrammetry, Remote Sensing and Spatial Information Sciences,* Vol. XXXIV-5/C15, Istanbul, pp. 609-614.

Lönnqvist, M. and Törmä, M. (2004) Different Implications of a Spatial Boundary, Jebel Bishri between the Desert and the Sown in Syria, The ISPRS XXth Congress Proceedings, Edited by M. Orhan Altan, *The ISPRS International Archives of the Photogrammetry, Remote Sensing and Spatial Information Sciences,* Vol. XXXV. Part B, Istanbul, pp. 897-902.

Lönnqvist, M., Törmä, M., Lönnqvist, K., Stout Whiting, M., Nuñez, M., Okkonen, J. (2006) An Ethical Overview of Long-term Perspectivism: Experiences from an Archaeological Survey on Jebel Bishri in Syria, in *The e-volution of Information Communication Technology in Cultural Heritage, Where Hi-Tech Touches the Past: Risks and Challenges for the 21st Century,* Short papers from the joint event CIPA/VAST/EG/EuroMed 2006, 30 October – 4 November 2006, Nicosia, Cyprus, ed. by Ioannides, M., Arnold, D., Niccolucci, F., Mania, K, Budapest: EPOCH Publication, pp. 108-113.

Lönnqvist, M., Törmä, M., Nuñez, M., Okkonen, J. (2008) Desertification monitoring and archaeological perspectives from Jebel Bishri, Syria, in *Remote sensing and GIS technologies for geohazard monitoring symposium,* International Geological Congress, Oslo, August 6-14th, 2008, abstract on http://www.cprm.gov.br /33IGC/1287208.html.

Lönnqvist, M. and Stefanakis, E. (2009) GIScience in Archaeology: Ancient Human Traces in Automated Space, in *Manual of Geographic Information Systems,* ed. by Madden, M., Bethesda, MD: ASPRS, pp. 1221-1259.

McLaren, S.J. and Reynolds, T. (2009) Early Humans in Dryland Environments: A Geoarchaeological Perspective, in *Geomorphology of Desert Environments,* ed. by Parsons, A.J. and Abrahams, A.D., Second edition, Netherlands: Springer, pp. 773-797.

Montero Fenollós, J.L., Chebibe, S., Márquez, I., Caramelo, F. and Vidal, J. (2006) *5th International Congress on the Archaeology of the Ancient Near East (5ICAANE).* Abstracts. Madrid.

Moore, A.M.T., Hillman, G.C. and Legge, A.J. (2000) *Village on the Euphrates: From Foraging to Farming at Abu Hureyra.* Oxford: Oxford University Press.

Mouterde, R. and Poidebard, A. (1945) *Le Limes de Chalcis: Organisation de la steppe en Haute Syrie Romaine.* Bibliothèque archèologique et historique. Documents aériens et épigraphiques. Texte and Atlas. Paris: Paul Geuthner.

Muhesen, S. (1985) *L'Acheuléen récent évolué de Syrie.* BAR International Series. No 248. Oxford: BAR International Series.

Musil, A. (1927) *The Middle Euphrates: A Topographical Itinerary.* American Geographical Society. Oriental Explorations and Studies No 3. New York: American Geographical Society.

Musil, A. (1928) *Palmyrena: A Topographical Itinerary.* American Geographical Society. Oriental Explorations and Studies No 4. New York: American Geographical Society.

Napoli, J. (2000) Les Remparts de la Fortresse de Djazla sur le Moyen-Euphrate, in *Syria,* Vol. 77, pp. 117-136.

Niebuhr, C. (1792) *Travels through Arabia and other Countries in the East.* Performed by M. Niebuhr, translated into English by Robert Heron. Vol. II. Edinburgh: R. Morrison.

Nilhamn, B. (2008) Nomadic Life in Central and Eastern Syria, A Perspective from the Present Life on Badiyah to the Amorite Nomadism in the Bronze Age, in *Jebel Bishri in Context, Introduction to the Archaeological Studies and the Neighbourhood of Jebel Bishri in Central Syria,* Proceedings of a Nordic Research Training Seminar in Syria, May 2004, ed. by Lönnqvist, M., BAR International Series, No. 1817, pp. 15-30.

Notitia Dignitatum (Otto Seeck edidit 1962). Frankfurt am Main: Minerva.

Oppenheim, M. von (1939) *Die Beduinen.* Band 1: Die Beduinenstämme in Mesopotamien und. Unter Mitbearbeitung von Erich Bräunlich und Werner Caskel Leipzig: Otto Harrassowitz.

Pappi, C. (2006) The Jebel Bišhri in the Physical and Cultural Landscape of the Ancient Near east, in *KASKAL*, Vol. 3, pp. 241-256.

Perrot, J. (1993) Remarques Introductives, in *Paléorient*, Vol. 19, pp. 9-21.

Philip, G., Abdulkarim, M., Newsom, P., Beck, A., Bridgland, D., Bshesh, M., Shaw, A., Westaway, R. and Wilkinson, K. (2005) Settlement and Landscape Development in the Homs Region, Syria, Report on Work Undertaken 2001-2003, in *Levant*, Vol. 37, pp. 21-42.

Poidebard, A. (1934) *La Trace de Rome dans le Désert de Syrie*: Le Limes de Trajan a la conquête arabe. Recherches aériennes (1925-1932). Texte. Atlas. Bibliothèque Archéologique et Historique. Tome XVIII. Paris: Paul Geuthner.

Pongratz-Leisten, B. (2001) The Other and the Enemy in the Mesopotamian Conception of the World, in *Mythology and Mythologies, Methodological Approaches to Intercultural Influences, MELAMMU Symposia,* Vol. 2, Proceedings of the Second Annual Symposium of the Assyrian and Babylonian Intellectual Heritage Project, Held in Paris, France, October 4-7, 1999, ed. by Whiting, R.M., Helsinki: The Neo-Assyrian Text Corpus Project, pp. 195-231.

Ptolemy = Claudius Ptolemy, *The Geography*, translated and edited by Stevenson. E.L. New York: Dover Publications.

RIME 2 = Grayson, A.K. (1991*) Assyrian Rulers of the Early First Millennium BC I (1114-859 BC). The Royal Inscriptions of Mesopotamia. Assyrian Periods.* Vol. 2. Toronto: University of Toronto Press.

Sachau, E. (1900) *Am Euphrat und Tigris: Reisenotizen aus dem Winter 1897-1898.* Leipzig: Hindrichs.

Sahlins, M.D. (1968) *Tribesmen.* Foundations of Modern Anthropological Series. Englewood Cliffs, New Jersey: Prentice-Hall, Inc.

Salzman, P.C. (1980) Introduction: Processes of Sedentarization as Adaptation and Response, in *When Nomads Settle: Processes of Sedentarization as Adaptation and Response,* ed. by Salzman, P. with the assistance of Sadala, E., Praeger Special Studies, New York: Praeger Publishers, pp. 1-20.

Sanlaville, P. (ed.) (1985) *Holocene Settlement in North Syria*: résultats de deux prospections archéologiques effectuées dans la région du nahr Sajour et sur le haut Euphrate syrien. Maison de L´Orient Mediterraneen (C.N.R.S. – Université Lyon 2), Lyon France. Archaeological Series No. 1. BAR International Series. No. 238. Oxford: BAR International Series.

Sarre, F. and Herzfeld, E. (1911) *Archäologische Reise im Euphrat- und Tigris-Gebiet.* Band I. Forschungen zur islamischen Kunst I. Berlin: Dietrich Reimer (Ernst Vohsen).

Sarre, F. and Herzfeld, E. (1920) *Archäologische Reise im Euphrat- und Tigris-Gebiet.* Band II. Forschungen zur islamischen Kunst I. Berlin: Dietrich Reimer (Ernst Vohsen).

Sartre, M. (2005) *The Middle East under Rome.* USA: Harvard University Press.

Scollar, I. (1990) *Archaeological Prospecting and Remote Sensing.* Topics in Remote Sensing 2. Cambridge: Cambridge University Press.

Skeates, R. (2000) *Debating the Archaeological Heritage.* Duckworth Debates in Archaeology. London: Duckworth.

Stordeur, D. (1993) Sédentaires et nomads du PPNB final dans le desert de Palmyre (Syrie), in *Paléorient,* Vol. 19, pp. 187-204.

Streck, M.P. (2000) *Das amurritische Onomastikon der altbabylonischen Zeit.* Bd. 1: Die Amurriter, die onomastiche Forschung, Ortographie und Phonologie, Nominalmorphologie. Alter Orient und Altes Testament. Band 271/1. Münster: Ugarit-Verlag.

Tabula Peutingeriana (see, e.g., Itineraria Picta, Contributo allo Studio della Tabula Peutingeriana, by Levi, A. and Levi, M. 1967. Roma: L'Erma).

Ulbert, T. (1986) *Resafa.* Teil II. Die Basilika des Heiligen Kreuzes in Resafa – Sergiupolis, Mit Beiträgen von Bayer, I., Gatier, P.-L., Dietger Grosser, D. Raif Khoury, R.G., Mackensen M. und Römer, C. Mainz am Rhein: Deutsches Archäologisches Institut.

Westaway, R., Bridgland, D., Challis, K. and White, M. (2005) A Dating Framework for the Lower and Middle Palaeolithic in Levantine Rivers, in *CBRL 2005, Newsletter of the Council for the British Research in the Levant,* pp. 31-33.

Westaway, R., Bridgland, D., Daoud, M., Radwan, Y., Abu Romieh, M., Yassminh, R. and Al Ashkar, A. (2007) Investigations of Euphrates Terraces and Related Archaeology, NE Syria, 2006-2007, in *CBRL 2007, Newsletter of the Council for the British Research in the Levant,* Vol. 2, pp. 82-89.

Winterhalder, B. (1980) Environmental Analysis in Human Evolution and Adaptation Research, in *Human Ecology,* Vol. 8, pp. 135-155.

Wirth, E. (1971) *Syrien: Eine Geographische Landeskunde.* Wissenschaftliche Länderkunden. Band 4/5. Darmstad: Wissenschaftliche Buchgesellschaft.

Zaccagnini, C. (1982) The Enemy in the Neo-Assyrian Royal Inscriptions: the "Ethnographic" Description, in *Mesopotamie und seine Nachbarn,* Vol. 2, ed. by Nissen, H.-J. and Renger, J., 25th RAI, Berlin, pp. 409-424.

Zarins, J. (1989a) Jebel Bishri and the Amorite Homeland: the PPNB Phase, in *To the Euphrates and Beyond.* Archaeological Studies in Honour of Maurits N. Van Loon, ed. by Chaex, O.M. *et al.,* Rotterdam: Balkema, pp. 29-51.

Zarins, J. (1989b) Pastoralism in southwest Asia: the second millennium BC, in *The Walking Larder, Patterns of Domestication, Pastoralism, and Predation,* ed. by Clutton-Brock, J., One World Archaeology 2, Cambridge: Cambridge University Press, pp. 127-155.

Zarins, J. (1990) Early Pastoral Nomadism and the Settlement of Lower Mesopotamia, in *Bulletin of the American Society of Oriental Research,* No. 280, pp. 31-65.

Zarins, J. (1992) Archaeological and Chronological Problems within the Great Southwest Asian Arid Zone, 8500-1850 B.C., in *Chronologies in Old World Archaeology,* Third Edition. 2 Vols., ed. by Ehrich, R.W., Chicago: The University of Chicago Press, Vol. 1, pp. 42-62.

2. THE COVERAGE IN MACRO-SCALE: SATELLITE DATA SOURCES AND THEIR APPLICATION

Markus Törmä and Minna Lönnqvist

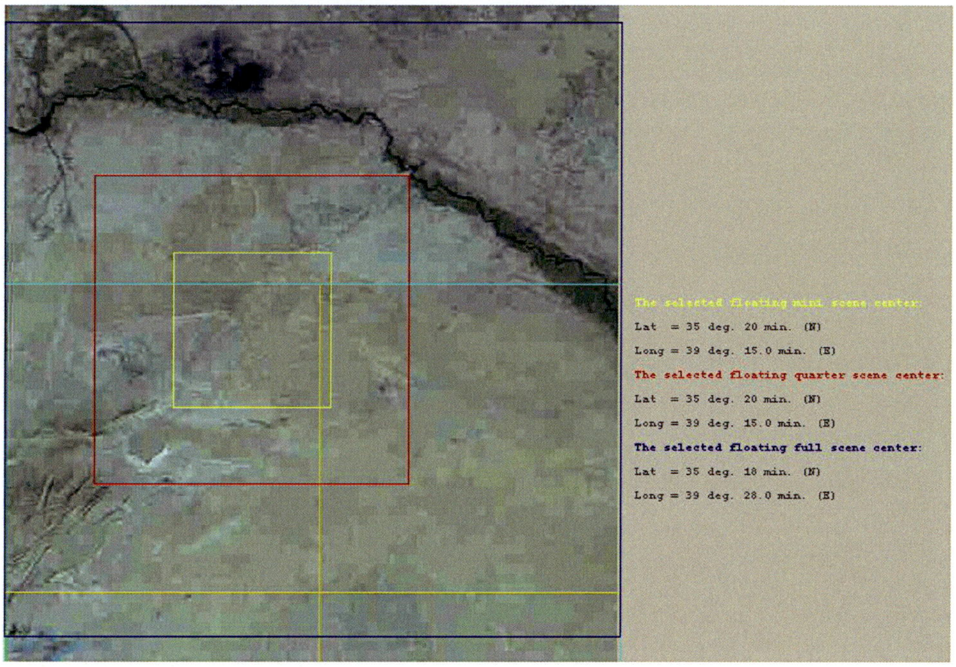

Fig. 1. Landsat-7 ETM zooming in on Jebel Bishri.

2.1. Macro-scale in Prospecting

There is a spot on the Earth where an archaeological find evolves; it is a location that can be studied in different scales and from various perspectives. The scale is a major key to attain visibility of the remains of past human activities and their contexts. Scale also determines possibilities for analysis and interpretation. The larger the scales the more ways there are to enlarge our perspectives of archaeological contexts. Space and airborne platforms enable us to see things in a *macro-scale*, even underground features, which one is sometimes unable to see or grasp from the ground level by the naked eye.

The data sources in *macro-scale* used by the Finnish project SYGIS consist of satellite images and space-borne radar data with different resolutions, generally opening vistas to environments, topography, sites, structures, trails and tracks. Using remote sensing for prospecting amounts to passive prospecting that does not disturb or alter the state of archaeological or environmental record on the ground (Scollar 1990: 1-4; Banning 2002: 41-42). In our investigations ground studies were carried out in order to locate and collect much needed experimental field data.

Geographic Information Systems (GIS) have provided the means to handle large amounts of location and spatial information embedded in remote sensed data, GIS is also a tool for combining field captured data whether using GPS, a total station/EDM, laser-scanner or digital cameras. It helps to classify the data, produce maps, design environmental models and construct landscape as well as visibility models (see, e.g., Aldenderfer 1996).

2.2. The Satellite Images Acquired

In order to determine the land cover of the study area and its changes, as well as providing data for visualization purposes, the following remote sensed images were acquired:

Landsat Multi-Spectral Scanner (MSS) images downloaded from Global Land Cover Facility (http://glcf.umiacs.umd.edu/):

- Landsat-1 MSS-image 186/36 (WRS-1) taken 14.9.1972 (Fig. 2.)
- Landsat-2 MSS-image 185/35 (WRS-1) taken 27.6.1975 (Fig. 3.)
- Landsat-2 MSS-image 185/36 (WRS-1) taken 27.6.1975 (Fig. 4.)
- Landsat-4 MSS-image 172/35 (WRS-2) taken 12.7.1983 (Fig. 5.)
- Landsat-4 MSS-image 172/36 (WRS-2) taken 12.7.1983 (Fig. 6.)

Landsat Thematic Mapper (TM) images downloaded from Global Land Cover Facility:

- Landsat-4 TM-image 172/35 (WRS-2) taken 1.9.1990 (Fig. 7.)
- Landsat-4 TM-image 172/36 (WRS-2) taken 1.9.1990 (Fig. 8.)

Landsat Enhanced Thematic Mapper (ETM) images downloaded from Global Land Cover Facility:

- Landsat-7 ETM-image 172/35 (WRS-2) taken 11.8.2000 (Figs. 9. and 10.)
- Landsat-7 ETM-image 172/36 (WRS-2) taken 26.5.2001 (Figs. 11. and 12.)

Landsat Enhanced Thematic Mapper image acquired from Eurimage via Novosat Ltd.:

- Landsat-7 ETM-image 172/halfway 35 and 36 (WRS-2) taken 29.1.1999 (Figs. 13. and 14.)

SPOT images were acquired as part of FP6/OASISproject. The Following SPOT-5 images were acquired (PAN: panchromatic, XS: multispectral image):

- PAN place 126-278 date 11.8.2006 (Fig. 15)
- PAN place 126-278 date 20.6.2006 (Fig. 16)
- XS place 127-279 date 17.8.2006
- XS place 126-278 date 11.6.2006

QuickBird pan-sharpened multispectral image acquired from Eurimage via Novosat Ltd.:

- Centre coordinates lat. 35. 422, long. 39. 4788, taken 15.7.2003 (Fig. 19.)

Several CORONA photographs were acquired from USGS:

- For example photographs dated to 28.6.1966 (figure x.15), 22.1.1967 (figure x.16) and dated to 22.1.1967 (figure x.17), places S FWD 1034-2, S FWD 1038-2 and S FWD 1038-2.

2.2.1 LANDSAT

'Landsat' is a series of polar orbiting remote sensing satellites. The first Landsat satellite was launched in 1972, and the last one, the seventh, in 1998. The most important instruments onboard the Landsat satellites were the Multi-Spectral Scanner (MSS), the Thematic Mapper (TM) and the Enhanced Thematic Mapper (ETM). These instruments are electro-optical scanners which produce digital data (Kramer 1996). The main differences between them are the number of channels that each system operates, their wavelengths, and their spatial resolution (see Table 1).

The Landsat images downloaded from Global Land Cover Facility are part of the NASA-sponsored global Landsat data set known as 'GeoCover'. The pixel sizes of GeoCover images differ from standard Landsat sizes and are 57 m for MSS, 28.5 for TM and 28.5 and 14.25 m (PAN) for ETM-images.

The GeoCover Landsat images were selected according to the year of acquisition (one epoch of the MSS images generally covers the years 1972–1980; another of the TM mages generally covers the years 1987–1993; and the third of the ETM images covers the year 1999 and 2000. There were no instrument malfunctions or errors during ground processing that took into account the cloud cover with regard to the quality of data, and phenology (maximum greenness). Single data runs, which produced consecutive images of a given path, were collected on the same date (Tucker *et al.* 2004).

2.2.2 SPOT

SPOT images are similar to the Landsat ETM images, but as previously mentioned, their spatial resolution is higher, and the number of channels and image sizes are smaller. Channels of multi-spectral SPOT images correspond to channels 2 (Green), 3 (Red), 4 (NIR) and 5 (MIR) of the Landsat ETM image. The image size is much smaller, at 60 km x 60 km, but the spatial resolution is higher.

2.2.3 CORONA

The CORONA satellites, used in 1960–1972, were the first generation of US reconnaissance satellites, and they took stereo photos using two oblique-viewing panoramic cameras. One photo is a panchromatic film strip of 7 x 90 cm. The best spatial resolution is 1.83 m for the KH-4B mission at a flight height of 150 km. The SYGIS project has mainly utilized photographs from the KH-4A mission, the size of the photos being 2.25 x 29.8 inches and the resolution 2.7 m (www.earthexplorer.usgs.gov; see, e.g., Figs. 17., 18.). Geometric correction of the CORONA images is very difficult due to the lack of information about the principal point and fiducially marks, and also because of panoramic distortion (Altmaier and Kany 2002).

2.2.4 QuickBird

QuickBird is a modern digital imager with very high spatial resolution, 0.6 m on the panchromatic channel. The images also provide four multispectral channels (Table 1). Due to the high spatial resolution, the image size is considerably smaller than that of the Landsat images. Also the radiometric resolution is higher than in previous images taken before (DigitalGlobe 2004). The SYGIS project acquired an image covering 8 km x 8 km over the central areas of Jebel Bishri (Fig. 19.).

Instrument	Landsat MSS	Landsat TM	Landsat ETM	SPOT	QuickBird	CORONA
Channels	4: 0.5–0.6 μm 5: 0.6–0.7 μm 6: 0.7–0.8 μm 7: 0.8–1.1 μm	1: 0.45–0.52 μm 2: 0.52–0.60 μm 3: 0.63–0.69 μm 4: 0.76–0.90 μm 5: 1.55–1.75 μm 6: 10.4–12.5 μm 7: 2.08–2.35 μm	1: 0.45–0.52 μm 2: 0.52–0.60 μm 3: 0.63–0.69 μm 4: 0.76–0.90 μm 5: 1.55–1.75 μm 6: 10.4–12.5 μm 7: 2.08–2.35 μm PAN: 0.50 – 0.90 μm	B1: 0.50–0.59 B2: 0.61–0.68 B3: 0.78–0.89 B4: 1.58–1.75 PAN: 0.48-0.71	1: 0.45–0.52 μm 2: 0.52–0.60 μm 3: 0.63–0.69 μm 4: 0.76–0.90 μm PAN: 0.45–0.90 μm	Panchromatic Film
Spatial resolution	79 m	30 m Ch. 6 120 m	30m PAN: 15 m Ch. 6 60 m	B1-B3: 10 m B4: 20 m PAN: 5 m	2.44 m PAN: 0.61 m at nadir	Ca. 1.8 m in the best case
Image size	185 x 185 km^2	185 x 185 km^2	185 x 185 km^2	60 x 60 km^2	16.5 x 16.5 km^2	188 x 14 km^2
Radiometric Quantization	7 bit	8 bit	8 bit	8 bit	11 bit	Scanned film

Table 1. The spatial and spectral properties of the used satellite images.

Fig. 2. Landsat-1 MSS-image 186/36 (WRS-1), date 14.09.1972, channel 574.

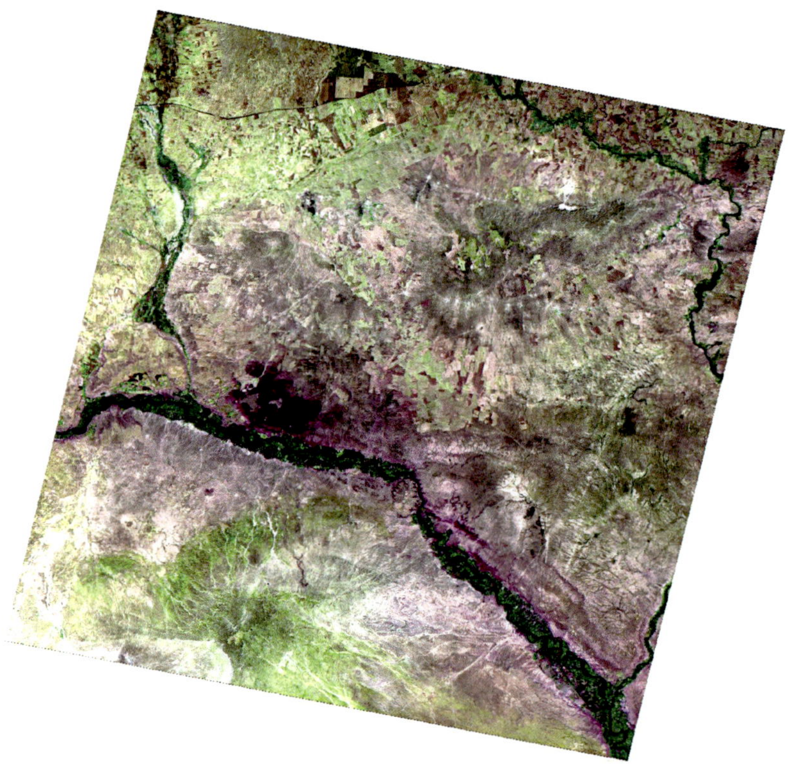

Fig. 3. Landsat-2 MSS-image 185/35 (WRS-1), date 27.06.1975, channel 574.

Fig. 4. Landsat-2 MSS-image 185/36 (WRS-1), date 27.06.1975, channel 574.

Fig. 5. Landsat-4 MSS-image 172/35 (WRS-2), date 12.07.1983, channel 574.

Fig. 6. Landsat-4 MSS-image 172/36 (WRS-2), date 12.07.1983, channel 574.

Fig. 7. Landsat-4 TM-image 172/35 (WRS-2), date 01.09.1990, channel 741.

Fig. 8. Landsat-4 TM-image 172/36 (WRS-2), date 01.09.1990, channel 741.

Fig. 9. Landsat-7 ETM-image 172/35 (WRS-2), date 11.08.2000, channel 741.

Fig. 10. Landsat-7 ETM-image 172/35 (WRS-2), date 11.08.2000, panchromatic channel.

Fig. 11. Landsat-7 ETM-image 172/36 (WRS-2), date 26.05.2001, channel 741.

Fig. 12. Landsat-7 ETM-image 172/36 (WRS-2), date 26.05.2001, panchromatic channel.

Fig. 13. Landsat-7 ETM-image 172/halfway 35 and 36 (WRS-2), date 29.01.1999, channel 741.

Fig. 14. Landsat-7 ETM-image 172/halfway 35 and 36 (WRS-2), date 29.01.1999, panchromatic channel.

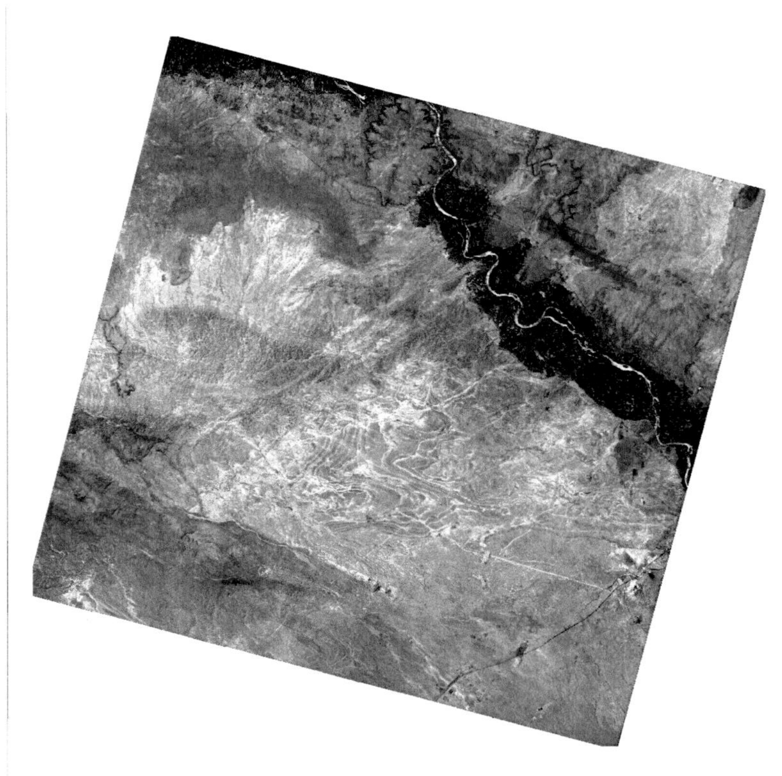

Fig. 15. SPOT-5 PAN-image, place 126-278, date 11.8.2006.

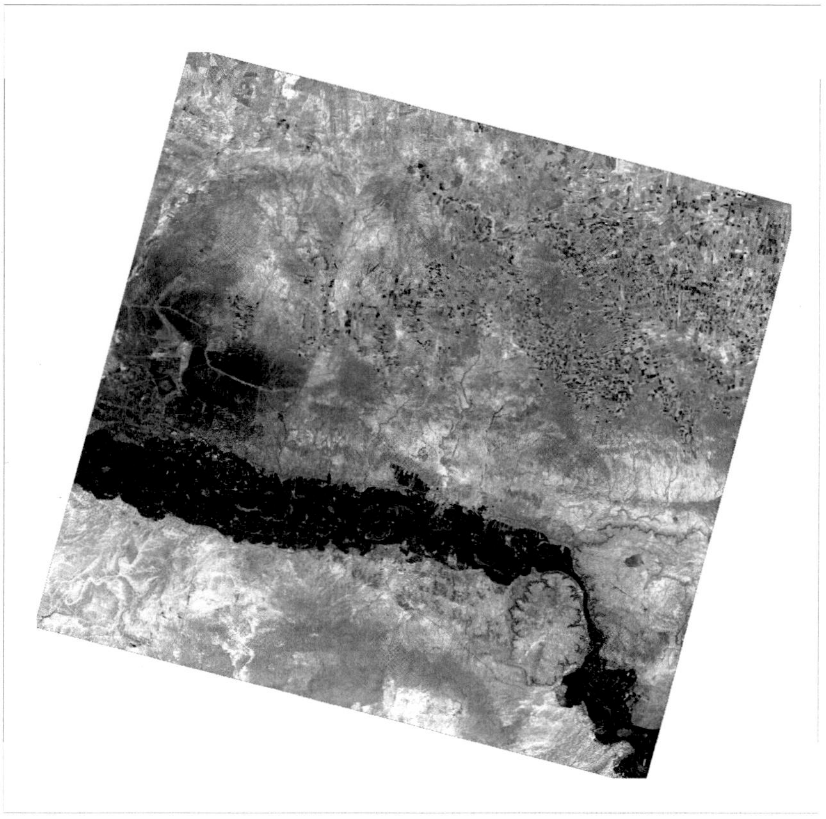

Fig. 16. SPOT-5 PAN- image, place 126-278 date 20.6.2006.

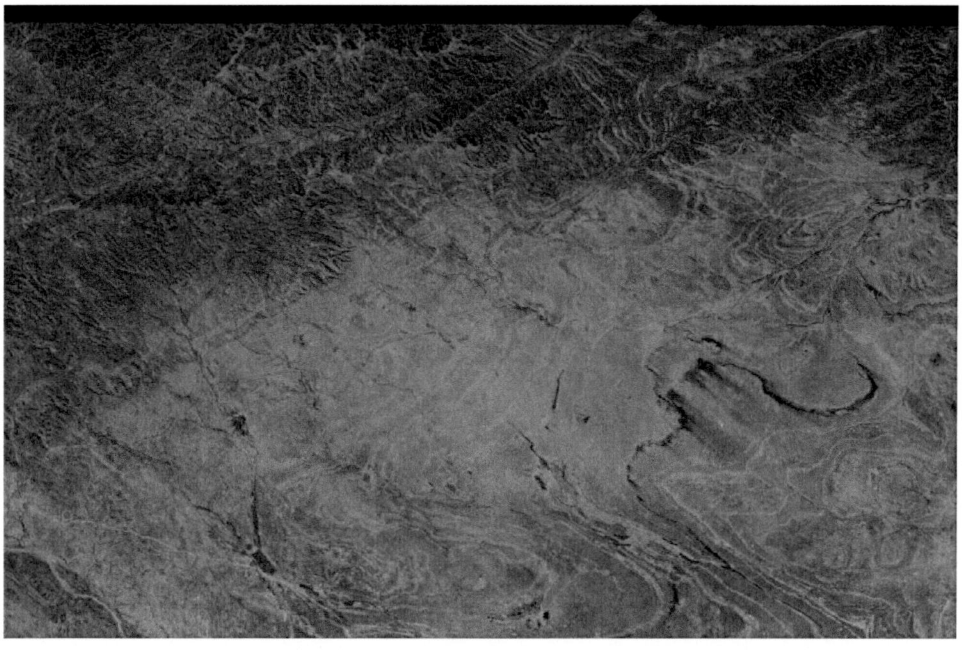

Fig. 17. Part No. 105 (ca. 10% from whole image) of CORONA photograph, date 28.06.1966.

Fig. 18. Part No. 205 (about 10% from whole image) of CORONA photograph, date 22.01.1967.

Fig. 19. QuickBird pan-sharpened multispectral image, date 15.07.2003. © Eurimage.

2.3. Digital Elevation Models (DEM)

The DEM (Digital Elevation Model) data was used to remove shadow areas from satellite image interpretation results, to carry out 3D- visualizations (still images and flight movies) of the ground, and to execute GIS-based site catchment analyses and viewshed analyses.

2.3.1 SRTM-DEM

The Shuttle Radar Topography Mission (SRTM) provides DEM at the resolution levels of 30 and 90 m covering the earth between latitudes 60N and 57S measured in February 2000. The DEM has been constructed by using synthetic aperture radar (SAR) interferometry, meaning that two radar images have been taken from slightly different positions, and the surface height has been determined by use of phase differences between images. The DEM was measured using two frequencies (C- and X-bands).

The acquired C-band DEM with the pixel size of 90 m largely covers the area of Jebel Bishri (Fig. 20.). WGS84 is used for both the horizontal and vertical datum. This means that ellipsoidal heights are provided. The absolute geolocational error within Eurasia is 8.8 m (90% errors in metres), with an absolute height error of 6.2 m and relative height error of 8.7 m. These are continent-wide averages determined through the use of kinematic GPS-transects.

Rodriguez *et al.* (2005) provide more detailed information about the error characteristics of the SRTM-DEM. The C-band SRTM-DEM was downloaded from the Global Land Cover Facility. X-band DEM with 30 m pixel size covered partly the area of Jebel Bishri. This and the ASTER-DEM data were used because of the better spatial resolution that they offer, and they were available earlier than the C-band SRTM-DEM. The drawback of X-band data is that the swathe width is 45 km, meaning that there are gaps between neighbouring orbits. The vertical accuracy should be better than 16 m absolute and 6 m relative, and the horizontal accuracy better than 20 m (Rabus *et al.* 2003). In Fig. 21. the coverage of a SRTM-DEM mosaic is shown as a red area beside the coverage of ASTER-DEMs on top of a Landsat panchromatic image. As mentioned previously, the X-band of the SRTM-DEM was provided by DLR (The German Aero Space Center) to SYGIS as a project member of the DLR project. Figs. 22. – 30. illustrate the individual X-band SRTM tiles.

2.3.2 ASTER-DEM

ASTER (Advanced Spaceborne Thermal Emission and Reflection Radiometer) is a multispectral imager which was launched onboard a Terra satellite in 1999. ASTER has 14 spectral bands from visible to thermal infrared region and their spatial resolution varies from 15 m to 90 m. There is also a single telescope looking backward in the near infrared region (channel 3B, 0.78 – 0.86 μm, spatial resolution 15 m) to give stereoscopic capacity for visibility in the along-track direction. Digital elevation model generation algorithm uses only instrument and satellite ephemeris data. The geodetic map projection is WGS84 with pixel spacing at 30 m. The horizontal and vertical accuracy should be better than 50 m and 15 m, respectively (ASTER, 2002). Two ASTER-DEMs (Figs. 21., 31. and 32.) were acquired through the NASA Earth Observing System Data Gateway (http://edcimswww.cr.usgs.gov/pub/imswelcome/).

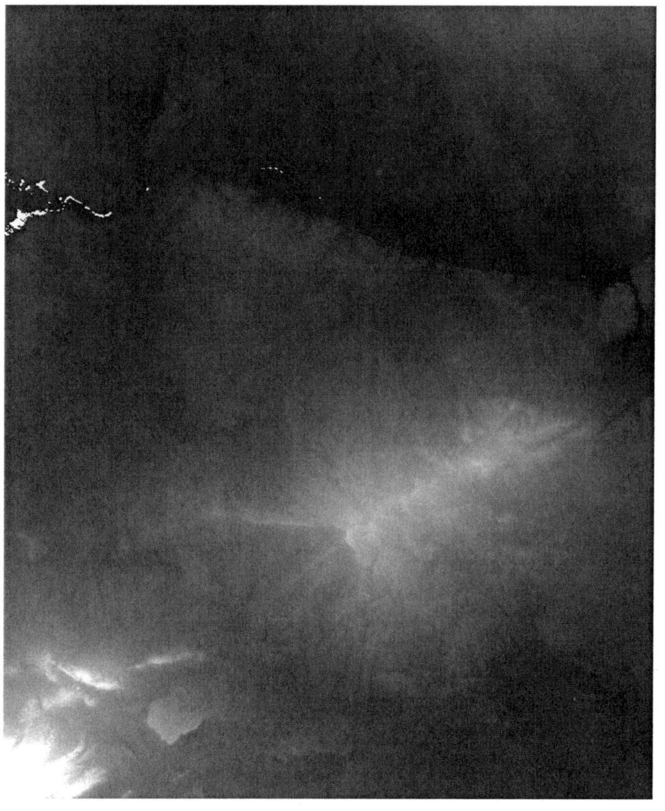

Fig. 20. The C-band SRTM-DEM of Jebel Bishri.

Fig. 21. The coverage of X-band SRTM-DEM (red area) and ASTER-DEMs (green area taken 24.9.2003 and blue area 16.10.2003) on top of Landsat ETM panchromatic image.

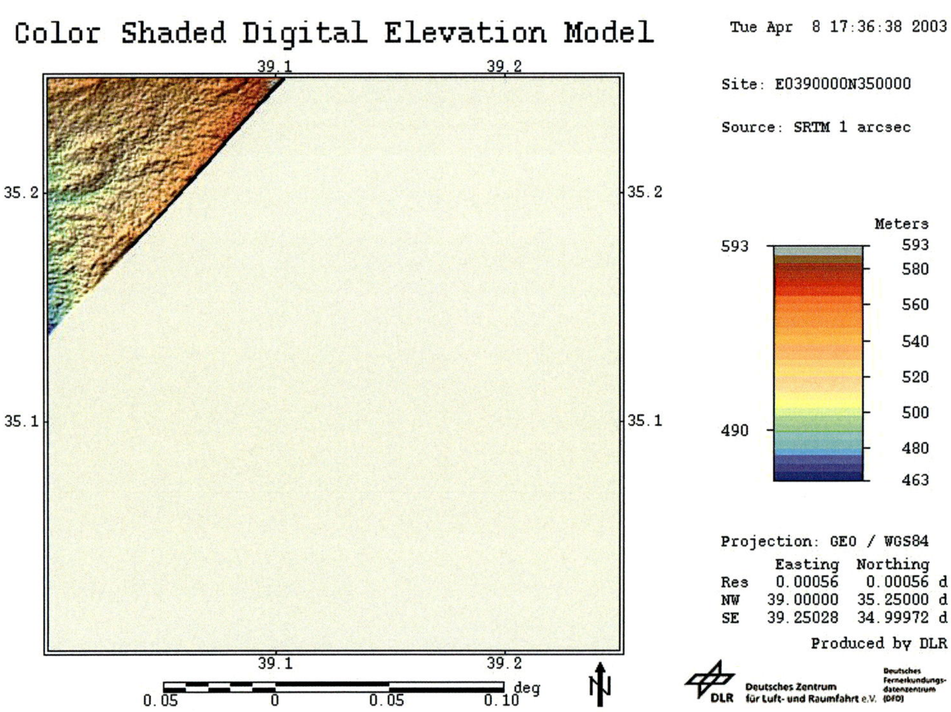

Fig. 22. The X-band SRTM-DEM tile.

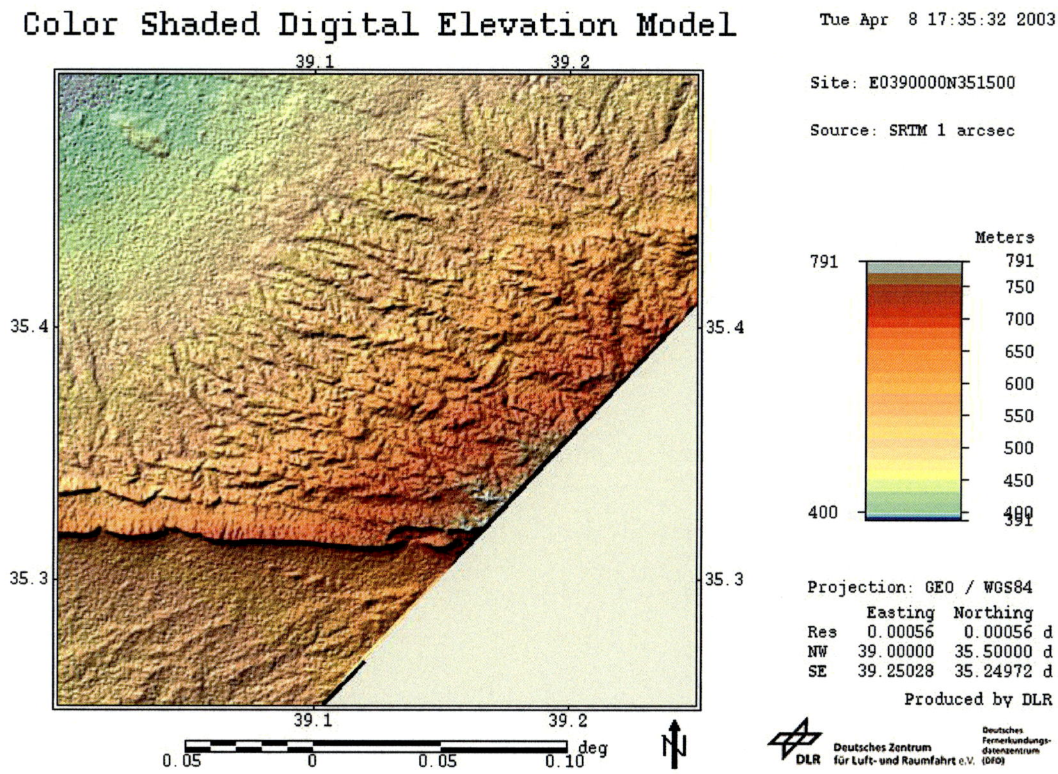

Fig. 23. The X-band SRTM-DEM tile.

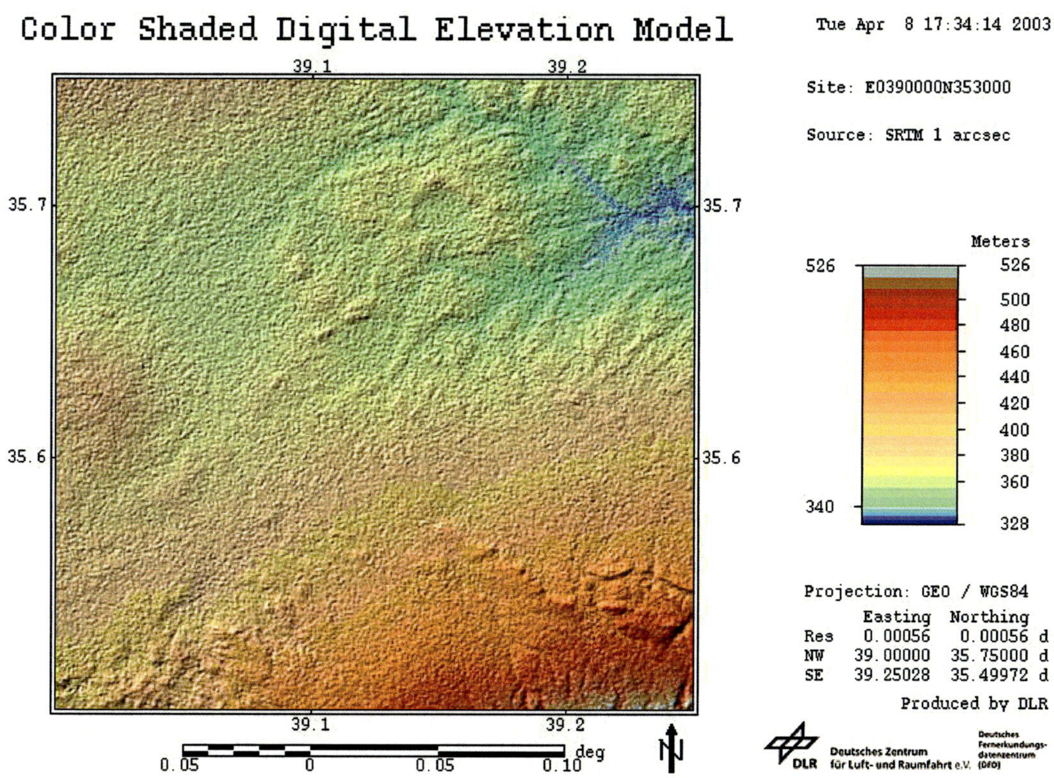

Fig. 24. The X-band SRTM-DEM tile.

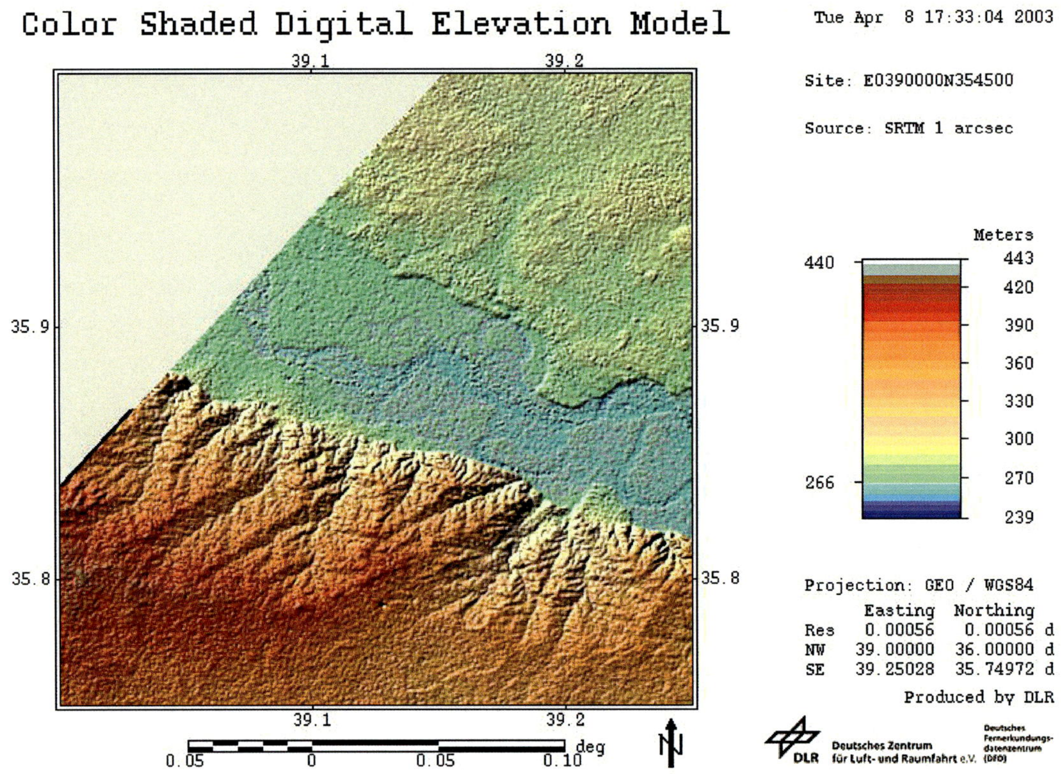

Fig. 25. The X-band SRTM-DEM tile.

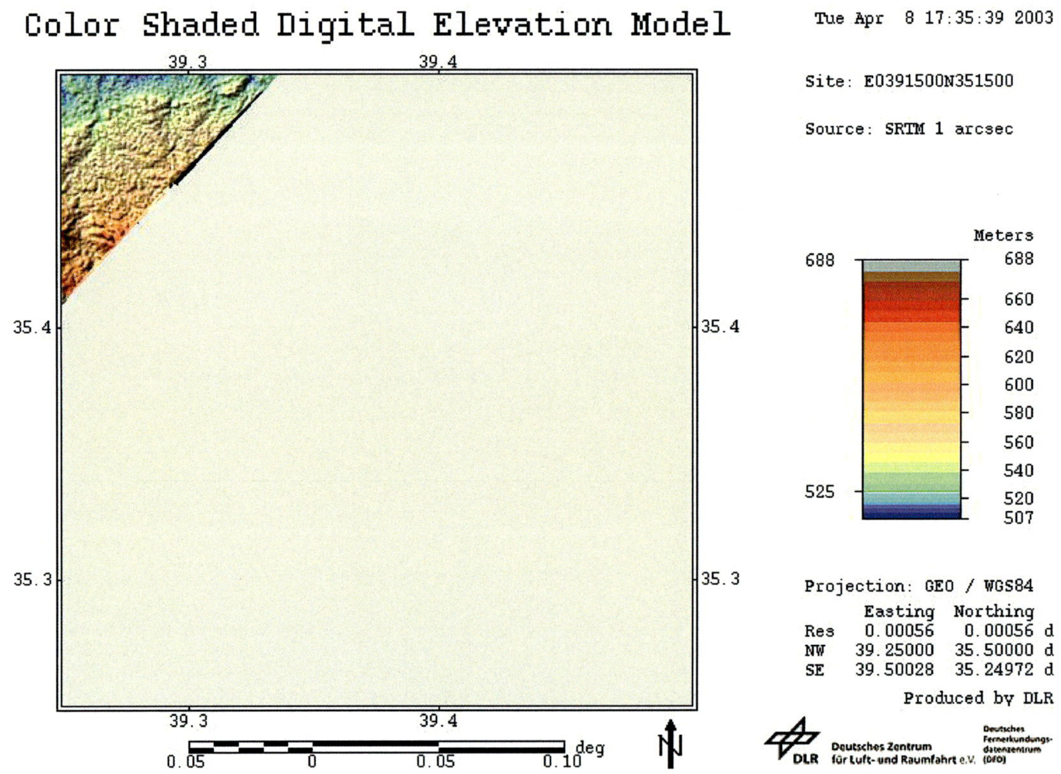

Fig. 26. The X-band SRTM-DEM tile.

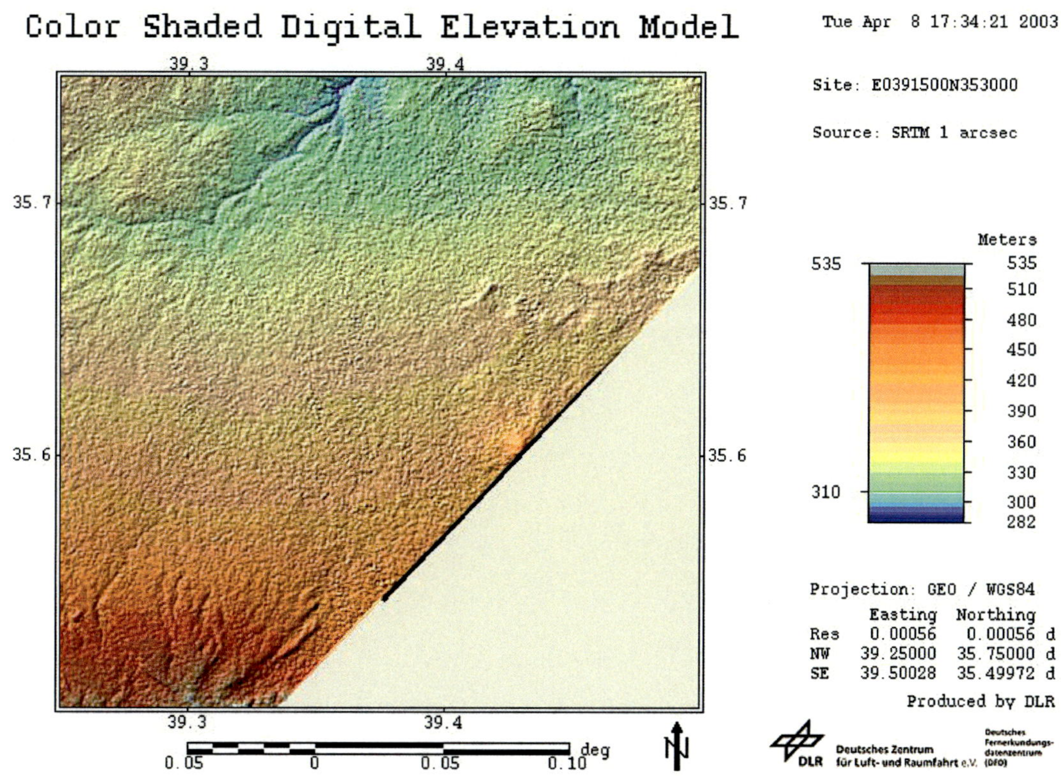

Fig. 27. The X-band SRTM-DEM tile.

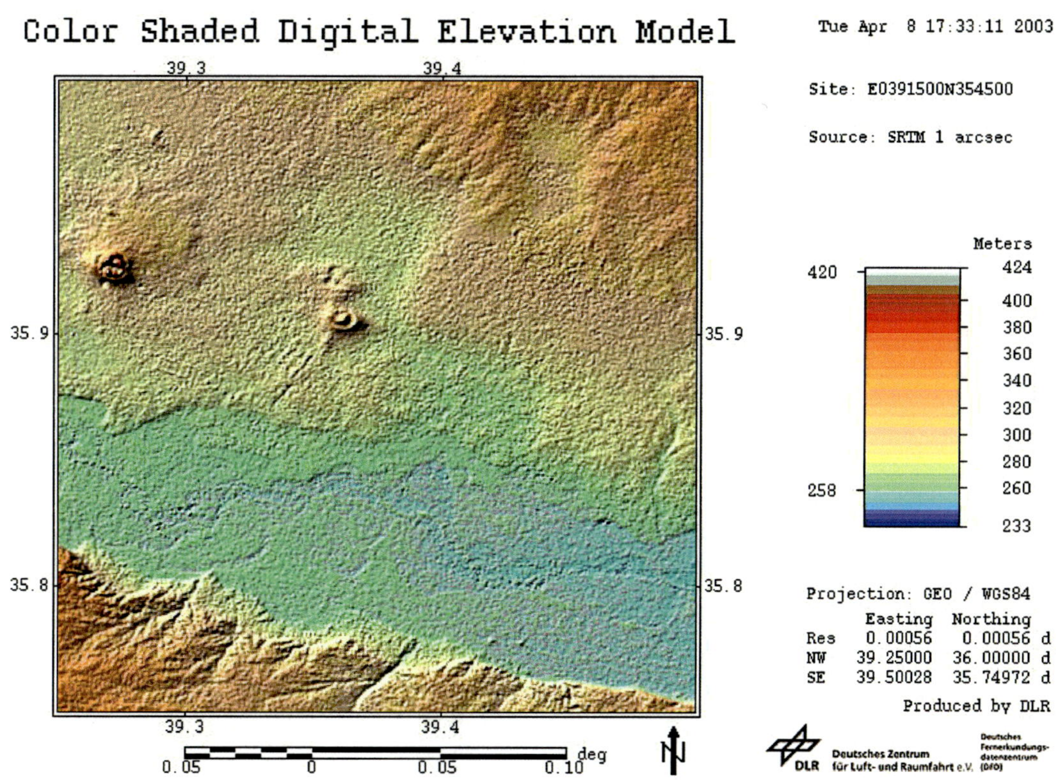

Fig. 28. The X-band SRTM-DEM tile.

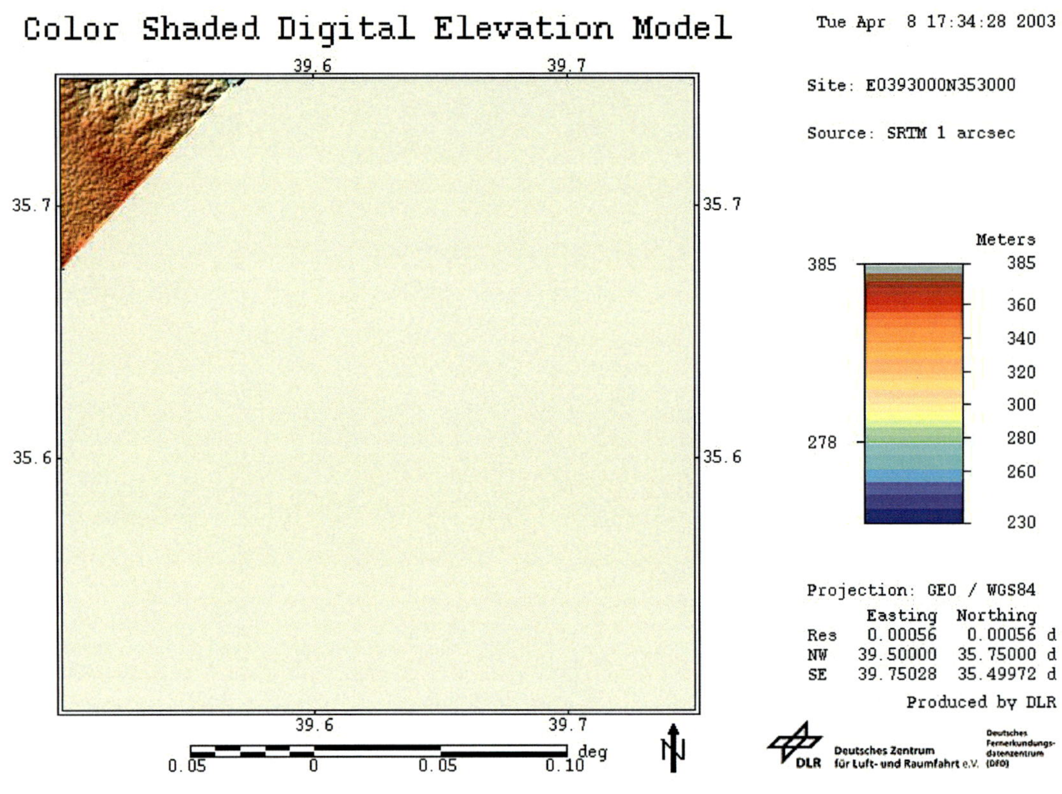

Fig. 29. The X-band SRTM-DEM tile.

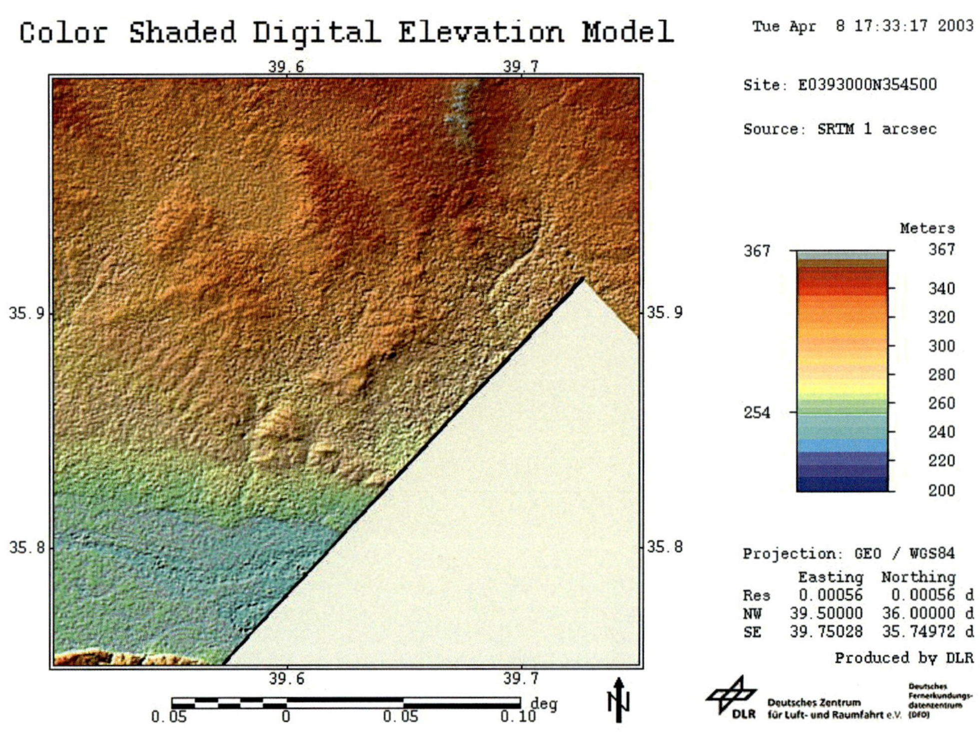

Fig. 30. The X-band SRTM-DEM tile.

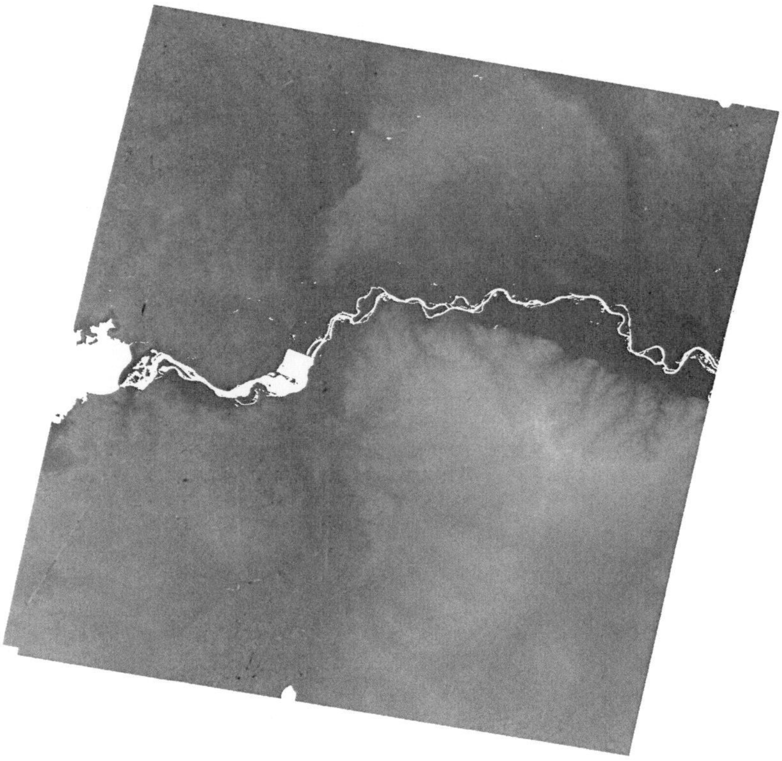

Fig. 31. ASTER-DEM, date 24.09.2003.

Fig. 32. ASTER-DEM, date 16.10.2003.

2.4. Pre-processing of the Data

2.4.1 Georeferencing of the Satellite Images

The georeferencing of the chosen Landsat-7 ETM image taken on 29.01.2000 was performed by using the panchromatic channel of the Landsat ETM image as a master image and georeferencing a Landsat ETM multispectral image, SRTM30 m data and ASTER-DEMs to the master image, using ground control points. The master image was georeferenced at Novosat Ltd using ground control points measured from UK military aviation maps (TPC G-4C, TPC G-4D, Scale: 1: 500, 000, UK 1998) and first degree polynomial model. Other images were georeferenced using 20–30 ground control points, first degree polynomial transformation, and nearest neighbour interpolation. The root mean square errors were less than one pixel. Later, the Landsat ETM image dating to 29.01.2000 was orthorectified using SRTM90m digital elevation model data and eight ground control points measured by using GPS.

The orthorectification of the GeoCover TM images was performed using ground control points, with as good a digital elevation model as possible, and then by processing larger mosaics each time. First, neighbouring images were tied together using common points. Then geodetic control points (usually 300 per mosaic) were located manually. Finally orthorectification was performed using triangulation process and digital elevation model data. Unfortunately the spatial resolution of digital elevation model data covering the study area is as low as 1 km. The orthorectification of the GeoCover MSS and ETM images was performed similarly in all cases, but ground control points were measured automatically by correlating the orthorectified TM-mosaic with a corresponding MSS or ETM image. The geopositional accuracy of mosaics was assessed using independent ground control points. The requirement was that Root-Mean-Square-Errors should be less than 50 m for the TM and ETM images and 100 m for MSS images. Failing that the data was reprocessed (Tucker *et al.* 2004).

Geometric correction was carried out for the chosen SPOT images using second degree polynomial model. Ground control points were acquired from the orthorectified Landsat-7 panchromatic image. Cubic convolution interpolation was used in rectification. The films of the CORONA photographs were digitized according to the required study areas. Smaller areas of these digitized CORONA photographs were georeferenced using the Landsat ETM panchromatic channel as a master image, in addition to first degree polynomial transformation with 10 – 15 ground control points for each area. Complete CORONA photographs were not referenced due to the difficult imaging geometry produced by the panoramic camera used.

2.4.2 Filtering of the Satellite Images

Gradient images were computed from the previously mentioned QuickBird image as well as the CORONA photographs in order to enhance linear features such as *wadis,* roads and paths, or other signs of human activity such as houses or ruins. Gradient images were computed using a Sobel gradient (Gonzales and Wintz 1987) with 3 x 3 filtering window for channels 1, 2 and 3. Figs. 33.-35. present examples of these gradient images.

Fig. 33. Part of a QuickBird image which has been filtered using Sobel-filtering.

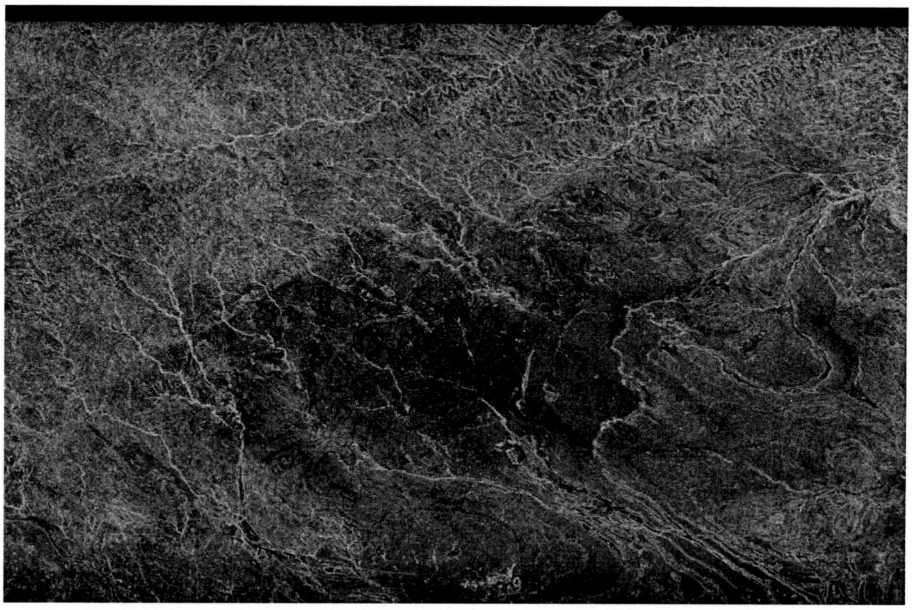

Fig. 34. Part of a Sobel-filtered CORONA photograph taken on 28.6.1966.

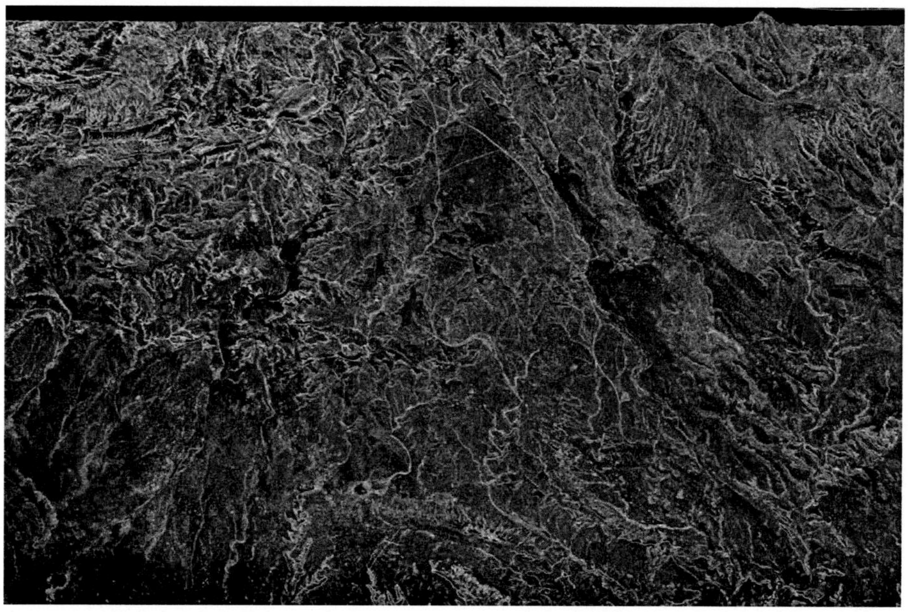

Fig. 35. Part of a Sobel-filtered CORONA photograph, date 22.01.1967.

2.4.3 Matching of the DEMs

As mentioned previously, SRTM30m-DEM data covered only part of the study area, and therefore it was decided to augment the DEM by ASTER-DEM data. The coverage of the DEMs can be seen in Fig. 21., where red represents the coverage of the SRTM30m-DEM, and green and blue represent the coverage of the ASTER-DEMs.

Because the heights of the SRTM30m-DEM are from the surface of ellipsoid WGS84 (Rabus *et al.* 2003), and the heights of the ASTER-DEM were relative, there was a need to study the heights of different DEMs more carefully.

Compared to the STRM30m-DEM, the heights of the ASTER-DEM 1 (the light green small triangle visible on the top in Fig. 21.) were 46.9 m lower and root-mean-square-error (RMSE) between the DEMs was 48.5 m. The heights of the ASTER-DEM 2 (the darker blue small triangle visible below in Fig. 21.) were 22.8 m lower than those of the SRTM-DEM, and RMSE was 27.0 m.

It was decided to adjust the ASTER-DEMs to the STRM30m-DEM. Fig. 36., upper row, illustrates the relationship of the pixels of the ASTER and SRTM30m-DEMs. After removing the average difference, (i.e., bias), between the DEMs, it was noticed that the slope of the linear regression equation between the SRTM30m and the ASTER-DEMs were not equal to 1, meaning that there was scaling factor in the ASTER-DEMs which needed to be removed (Fig. 36., middle row). Therefore the adjustment of the ASTER-DEMs was based on the regression equations between the SRTM30m and ASTER-DEMs. Correlations between the DEMs were high, the correlation coefficients being 0.9855 for the ASTER-DEM 1 and 0.9952 for the ASTER-DEM 2. After the adjustment (Fig. 36.), the RMSE between the SRTM-DEM and the ASTER-DEM 1 was 7.8 m; and for the ASTER-DEM 2 it was 10.6 m.

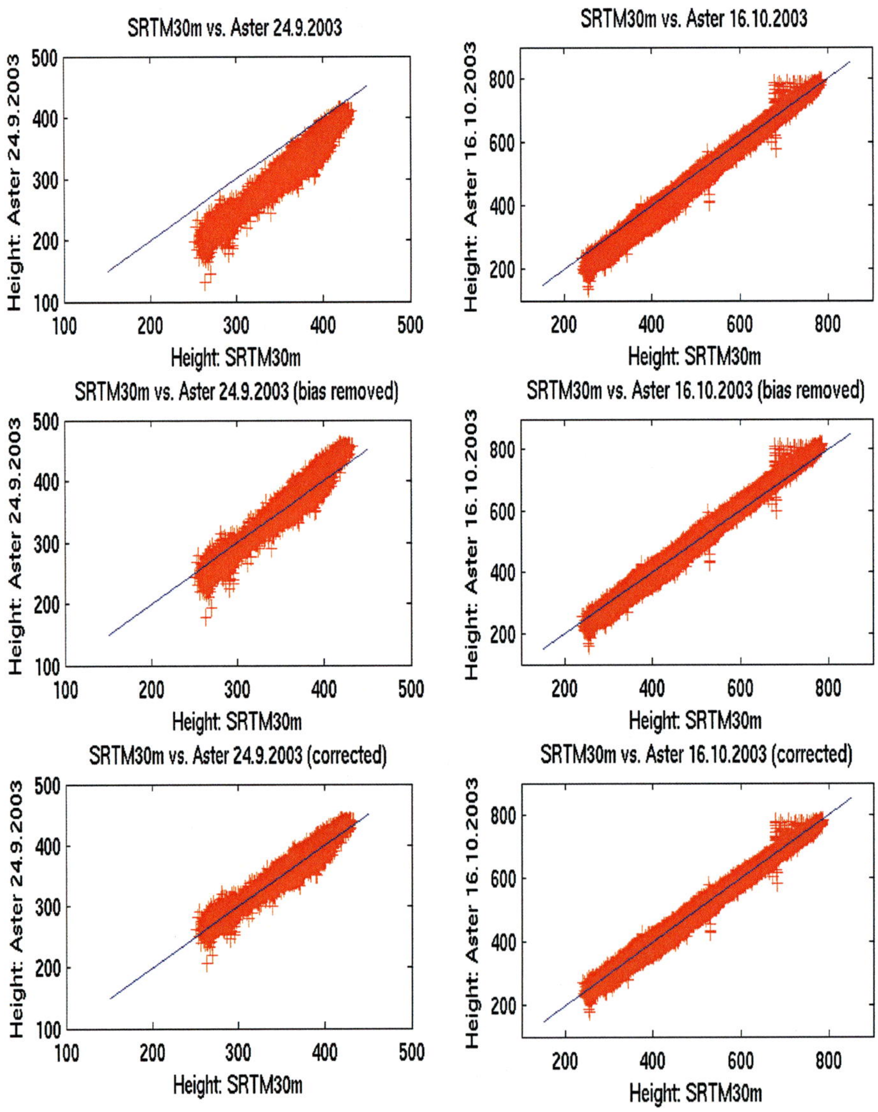

Fig. 36. This figure displays the relationship of pixels from SRTM30m and ASTER-DEMs. The upper row shows the uncorrected situation; the middle row shows bias removed; and the lower row displays the required correction, by use of linear regression.

2.5. The Application of the Data

Landsat satellite images have provided especially good data sources for our environmental observations. Cluster analysis offers the means for closer scrutiny of the environmental features, such as vegetation cover, or the movements of sand areas and the desert fringe. In our project the orthorectified panchromatic channel of the Landsat-7 ETM with the resolution of 15 m has also been applied for field mapping using the ERDAS MapSheets program. However, because of the relatively low spatial resolution of the Landsat images, it is practically impossible to detect sites or the structural remains of mobile groups, such as hunter-gatherers or pastoral nomads, who have been typically living in arid zones. Sedentary sites, such as tells, and monumental remains are more visible instead.

SPOT images, of which some bear higher spatial resolution (from 5 to 20 m) than the Landsat images, have enabled us to study tells, ancient mounds of villages and towns, as well as more recent village sites on the Euphrates and detect their layouts.

CORONA declassified satellite photographs from the 1960s providing spatial resolution even at 1.8 m have been of especial help in studying changes over decades, in delineating the borders of *wadis*, fields or ploughing areas; and in observing river channel changes as well as ancient road alignments which have disappeared during agricultural activities or new building projects in modern times. (See applications of the CORONA satellite photographs in archaeological prospecting in the Near East, e.g., Kennedy 1998; Ur 2003; Lönnqvist and Törmä 2003; Philip *et al.* 2005 and Mottram and Menere 2005).

Due to the high spatial resolution (at 0.6 m) of the QuickBird satellite images, the best opportunity for detection of individual sites was enabled. By this method several types of sites used by mobile groups were located. However, the nature of the sites should be checked on the ground, as images alone can lead to some misinterpretation. Still, even if several sites of mobile groups are detectable, open accumulated sites and rock shelters are generally not discernable with the QuickBird images either.

Initially our project was accepted to NASA's world monitoring program through the German Aerospace Centre (DLR) in 1999. As previously mentioned, the DLR project membership has provided Digital Elevation Model data (DEM for short) from the Shuttle Radar Topography Mission (SRTM) 2000. It has enabled our project to construct environmental 3D landscape models of the Euphrates valley and the northern parts of Jebel Bishri by fusing the DEMs with Landsat satellite image data. Further modelling has been carried out by ASTER-DEMs acquired from NASA. A cross-section of the Euphrates river *graben* was produced by using radar data with ASTER images. Site catchment analyses and viewshed analyses were also produced by using the radar data.

Bibliography

Aldenderfer, M. (1996) Introduction, in *Anthropology, Space, and Geographic Information Systems*, ed. by Aldenderfer, M. and Maschner, H.D.G., Spatial Information Systems, New York – Oxford: Oxford University Press, pp. 3-18.

Altmaier, A. and Kany, C. (2002) Digital surface model generation from CORONA satellite images, in *ISPRS Journal of Photogrammetry & Remote Sensing,* Vol. 56, pp. 221–235.

ASTER *User's Guide* (2002) Part III: DEM Product (L4A01) (Ver.1.0). December 2002. Japan: ERSDAC Earth Remote Sensing Data Analysis Center.

Banning, E.B. (2002) *Archaeological Survey, Manuals in Archaeological Method, Theory, and Technique.* New York: Kluwer Academic/Plenum Publishers.

DigitalGlobe (2004) = *DigitalGlobe Basic Imagery.* http://www.digitalglobe.com/

Gonzalez, R.C. & Wintz, P. (1987) *Digital image processing.* Reading, MA: Addison-Wesley.

Kennedy, D. (1998) Declassified satellite photographs and archaeology in the Middle East: case studies from Turkey, in *Antiquity*, Vol. 72, pp. 553-561.

Kramer, H.J. (1996) *Observation of the Earth and Its Environment: Survey of Missions and Sensors* 3rd enlarged ed. Berlin: Springer.

Lönnqvist, M. and Törmä, M. (2003) SYGIS – The Finnish Archaeological Project in Syria, in Proceedings of the XIXth International Symposium CIPA 2003 (The ICOMOS & ISPRS Committee for Documentation of Cultural Heritage): New Perspectives to Save Cultural Heritage, Antalya, Turkey, ed. by Altan, M.O., *The ISPRS International Archives of the Photogrammetry, Remote Sensing and Spatial Information Sciences,* Vol. XXXIV-5/C15, pp. 609-614.

Mottram, M. and Menere, D. (2005) Preliminary Surface Investigations in the Wadi Abu Qalqal Region, North Syria, in *Mediterranean Archaeology*, Vol. 18, 161-174.

Philip, G., Abdulkarim, M., Newsom, P., Beck, A., Bridgland, D., Bshesh, M., Shaw, A., Westaway, R. and Wilkinson, K. (2005) Settlement and Landscape Development in the Homs Region, Syria, Report on Work Undertaken 2001-2003, in *Levant*, Vol. 37, pp. 21-42.

Rabus, B., Eineder, M., Roth, A. and Bamler, R. (2003) The shuttle radar topography mission—a new class of digital elevation models acquired by spaceborne radar, in *ISPRS Journal of Photogrammetry & Remote Sensing* 57, pp. 241–262.

Rodriguez, E., Morris, C.S., Belz, J.E., Chapin, E.C., Martin, J.M., Daffer, W. and Hensley, S. (2005) *An assessment of the SRTM topographic products. Technical Report JPL D-31639.* Pasadena, California: Jet Propulsion Laboratory.

Scollar, I. (1990) *Archaeological Prospecting and Remote Sensing.* Topics in Remote Sensing 2. Cambridge University Press: Cambridge.

Tucker, C.J, Grant, D.M. and Dykstra, J.D. (2004) NASA's Global Orthorectified Landsat Data Set, in *Photogrammetric Engineering and Remote Sensing*, Vol. 70, pp. 313-322.

Ur, J. (2003) CORONA Satellite Photography and Ancient Road Networks: A Northern Mesopotamian Case Study, in *Antiquity*, Vol. 77, pp. 102-115.

3. THE COVERAGE IN MEDIUM- AND MICRO-SCALE: CLOSE-UPS TO THE GROUND AND BENEATH

Minna Lönnqvist

Fig. 1. An ancient jug protruding from Tell Tibne at the northern foot of Jebel Bishri. Photo: Kenneth Lönnqvist 2005.

3.1. Scales of Archaeological Studies

Archaeological surveying can be seen as art of making measurements of the positions of human-made remains or features left by humans of their past activities on the Earth's surface and presenting the data either graphically or numerically. In our case we have used geodetic approaches to remote sensing and plane surveying on the ground. Geodetic surveying takes into account the Earth's curvature and is applied to macro-scale approaches, whereas plane surveying is here applied to the ground studies of chosen plots in the area of Jebel Bishri (cf. Bannister *et al.* 1998: 1).

S. Kowalewski (1990) and L. Wandsnider (1998), like many others, have emphasized the benefits of the full-coverage survey in a regional scale. Modern surveys are also often aimed to more holistic landscape approaches. The aims go beyond the single site analyses and bring evidence of variations in site types and their dispersion in space. Our Finnish project SYGIS has been able to carry out a full-coverage remote sensing study with the Landsat and SPOT images over the region of Jebel Bishri. From the outset in 2000, while prospecting with the Landsat images in Finland, it looked as if Jebel Bishri would not yield much in the way of archaeological remains. Looking also at A. Poidebard's (1934) maps and aerial photographs, one may easily draw the conclusion that the mountain is largely devoid of past human activity. But that is an illusion that results from the coverage of the aerial studies and the scale of the utilized satellite data sources.

As mentioned earlier in the previous chapter, the CORONA satellite photographs have given rise to the possibility of zooming even into the scale of 1.8 m (in our case 2.7 m), and the QuickBird images in a chosen 8 km x 8 km plot in the central area of the mountain reveals sites with even 0.6 m spatial resolution. Although smaller and flimsier structures can be traced with these high resolution satellite images, bits and pieces of artefacts can only be traced, recorded and documented at the ground level by surveying. On the ground one can change the perspective to remains of human activities in a *medium-scale* on a place, site or feature/structure level or in a *micro-scale* in sampling and collecting artefacts (Fig.

1.), even assessing *minute-scale* information with a microscope from a kernel of earth deriving from a piece of pottery or retouches of flint implements. Changing the perspective ranging from the satellite image scale to the microscopic world also alters our view on the past.

Our project has mainly worked on the *macro-* and *medium-* levels, and the *micro-scale* was applied in collecting small finds. The ground level surveying and capture of ground data are used for location, environmental, chronological and cultural information. It has been emphasized that a combination of different sensors, whichever method is chosen, is vital for detection of archaeological remains (Banning 2002: 40). So, the remote sensing in our survey is served by field observation and visual inspection with the naked eye and *vice versa.*

3.2. Choosing Areas for Ground Surveys

Choosing the ground survey areas according to satellite data sources has been beneficial, although relief patterns reflecting the topography had to be retrieved from topographical maps and remote sensed radar data. Apart from receiving orthorectified images from the panchromatic Landsat-7 ETM channel, 15 m spatial resolution of the images has provided a method for producing of field maps, in addition to being a tool for prospecting. The panchromatic channel of Landsat-7 ETM and SPOT images, have also been useful for closer study of the terrain and road networks.

A. Brown (1999: 46) emphasizes the validity of demarcating the study areas on the field from geomorphological grounds. J. Evans and T. O'Connor (1999: 96) recommend global or local grid based units, but if particular land or preservation type is targeted, then another kind of environment should also be included in study of the region. To assess the basic research problem concerning mobility and sedentism, geographical and environmental factors (cf. Vita-Finzi 1978: 29) serving as physiographic units were the most important parameters to us in choosing areas for systematic and intensive study on the ground from a scientific point of view. But the practical situations thrown up by the circumstances on the ground caused some changes to the original plans.

From the outset in 2000 we had an official permission that covered the whole area of Jebel Bishri, and with regard to the northwestern and western areas, that permission was unfortunately restricted from the year 2004 before we reached the intensive survey to the northwestern flanks. The Syro-Japanese expedition has since been working in this northwestern district of Jebel Bishri in 2007-2009.

Fig. 2. Screening with binoculars on the Euphrates side in February, 2003. Photo: Minna Lönnqvist.

During our general ground study of the region we travelled around the area of Jebel Bishri with cars with four-wheel drive applying "window-screening" and binocular surveying (Fig. 2.) to grasp the environmental nature of the region in all the major districts. This was also useful for gaining general knowledge of the broader picture, so that we could make preliminary selections from the survey areas and understand discoveries in the context of the whole region. Short visits to the central areas of the mountain were already carried out in 2000 and 2003.

We chose to study three different environmental zones on the ground varying from 1) desert-steppe-type, to 2) steppe, and 3) irrigated river valley with fluvial terraces (see Fig. 3.). In 2000-2003 our prospecting and surveys concentrated on *the western side of the mountain including the piedmont and central western areas*. The western area consists of chiefly desert-steppe type of environment these days. In 2003-2008 prospecting, surveying and GIS studies were carried out on the more steppic *central eastern areas, and on the northeastern edge at the Euphrates.*

The study of the available water sources concerns the past environment, and it has been important to understand what kind of circumstances have prevailed and surrounded the ancient groups of people who lived in the region. As we shall discuss these aspects in due course, there are also short-term changes that open up the possibility of various explanations as to why the local people have changed their life styles. Our purpose has been to see how the finds differ between the three environmental zones, e.g., are there differences between mobile and sedentary, or monumental and non-monumental districts? As shown the Landsat images are good for visual study and predictive modelling based on environmental features indicating water when searching for archaeological sites on the ground. For example, the green oases such as Taibe, El Kowm and Qdeir, as well as *wadis* with their specific signatures, are natural environments that reveal not only the presence of water, but also that human activity had taken place in these locations from the earliest periods.

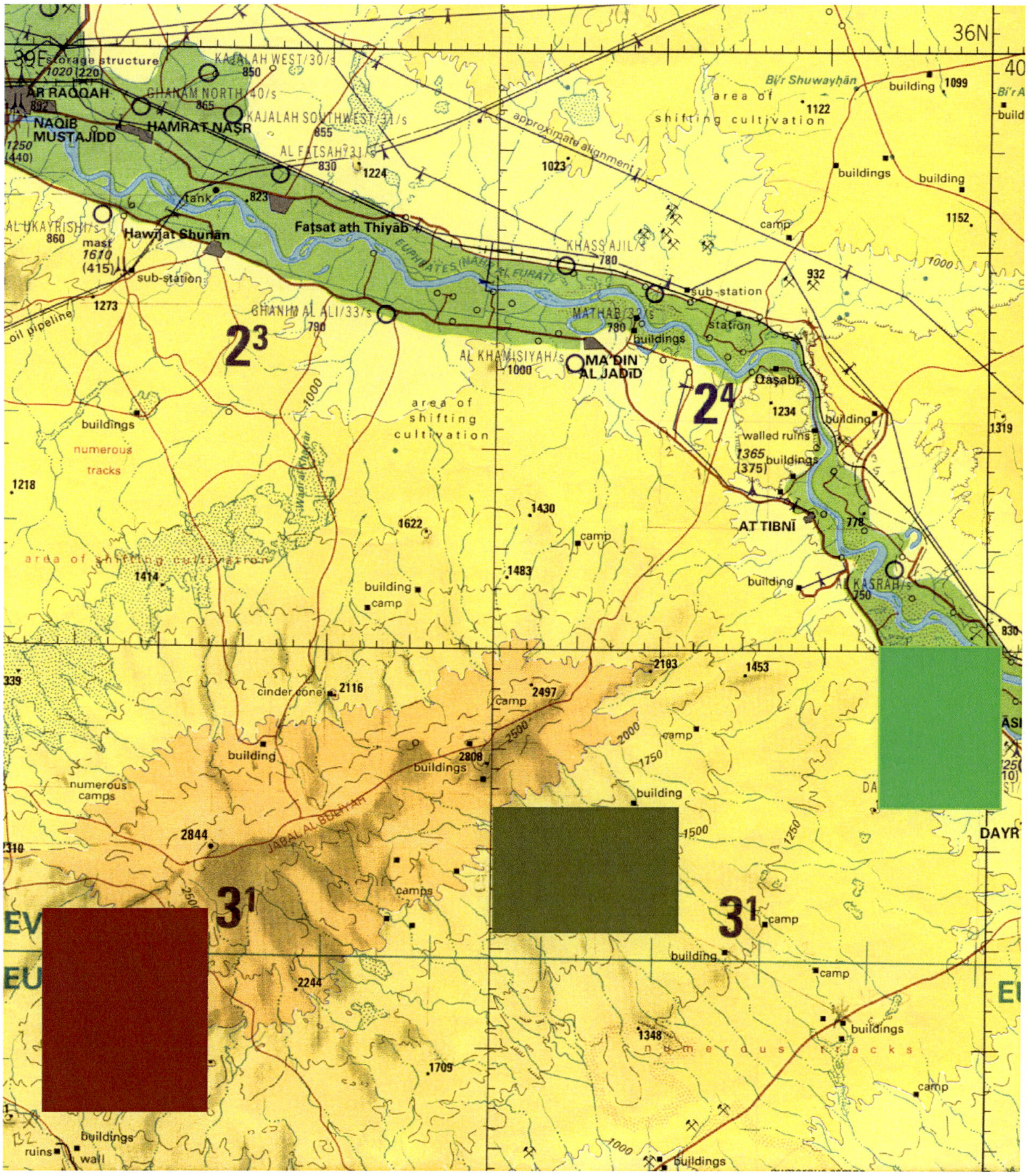

Fig. 3. Our main target areas in three different environmental zones during the ground surveys: brown indicating desert-steppe area in the districts of Qasr al-Hayr ash-Sharqi, Shanhas and Tar al-Sbai (2000-2003), olive green indicating steppe area including Nadra and Ash-Shujiri (2003-2008) and bright green indicating river valley between villages 'Ayyāsh and Mustaha (2003-2008). Basemap: © MOD Crown Copyright, 2010 published with the permission of the Controller of HMSO. See the names of studied areas and sites on Fig. 4.

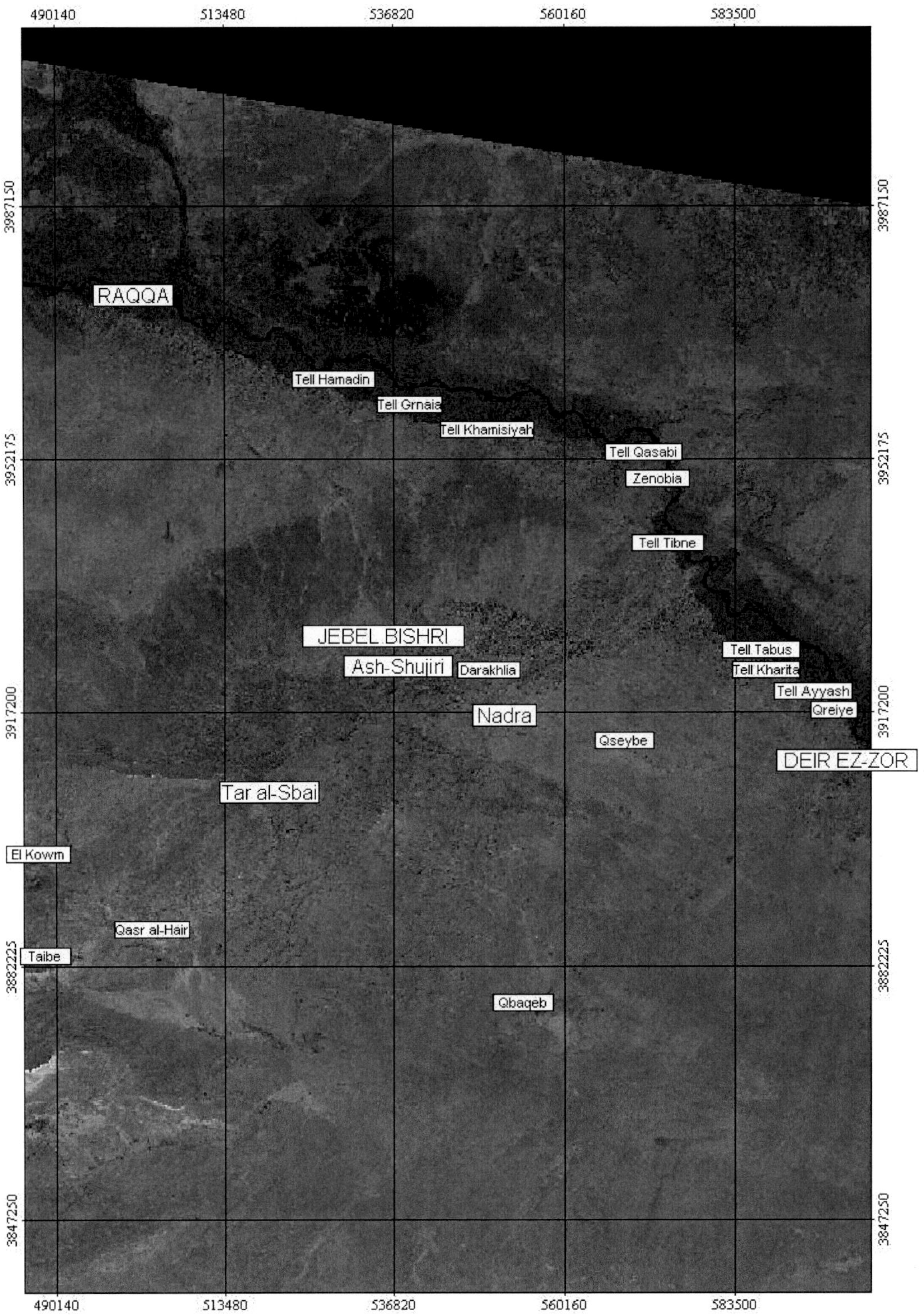

Fig. 4. The ground surveyed and visited areas or sites by the Finnish project SYGIS displayed and mapped on a Landsat satellite image in 2000-2010. Mapping by Minna Lönnqvist 2004.

3.3. Spatial Units of Study

Environmental or landscape approaches of the archaeological study in *macro-scale* required a larger-scale strategy also on the ground level. However, we have paid special attention to spaces and landscapes as contexts. "Space" appears as a more abstract and situational concept than "place" for which space provides a context (Tilley 1994: 15). "Landscape" is a spatial concept like "place" but bears an extra dimension of being an outdoor scenery compared to the concept of "place". Generally speaking in archaeology landscapes are seen as the result of interaction between environment and humans, their cultural meanings and values (Crumley and Marquardt 1990: 73-79), but some environmentalists would like to distance themselves from the human dimension in the concept, and therefore they prefer to restrict it to purely natural dimensions (see, e.g., Bottema 1999: 9). With the concept of "landscape" we here refer to physical dimensions of space in environment, which can denote purely natural dimensions of scenery or natural dimensions which have been modified by humans.

Our systematic *medium-scale* approach in the field concerned an intensive survey in chosen plots. The resolution of the field research included pedestrian style 15 metre interval study (see Fig. 6.) that covered specific transects from the chosen environmental zones (see Hole 1980: 34-35). The transects used for field walking were elongated rectangles (see, e.g., Figs. 5. and 7.), and the edges of these spatial units were therefore maximized, enabling us to trace the maximum number of sites. Although our approach has been regional and landscape orientated, a "site" as a traditional concept of an archaeological survey in *medium-scale*, has served here as a basic unit for field recording, documentation and mapping. The reason for this has been the traditional approach of cultural heritage management in which sites are basic units and as units can be more easily recognized for further protection and preservation.

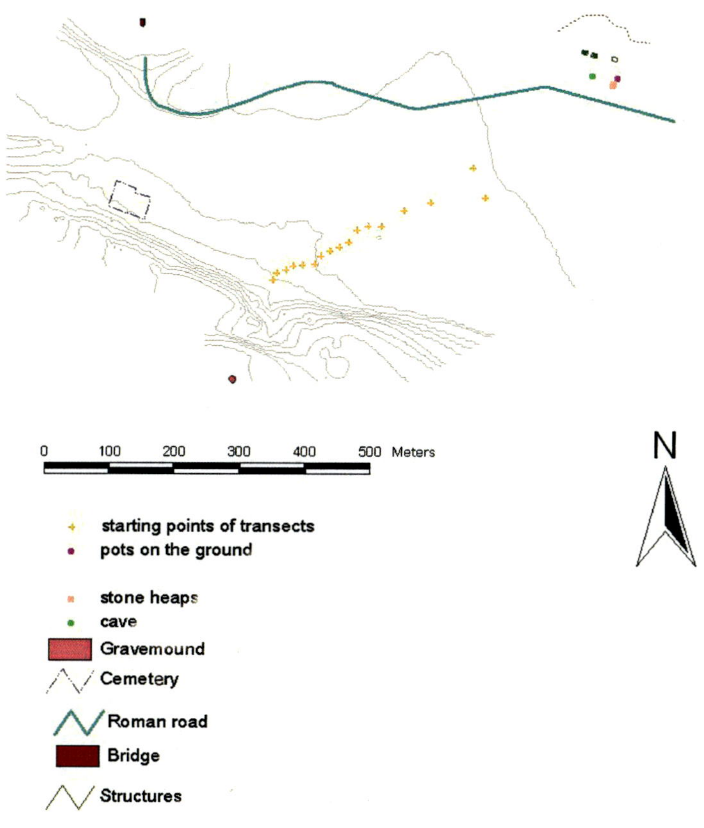

Fig. 5. Part of the intensively studied and mapped area on the piedmont terraces of the Euphrates side in the village of Kharita-Mustaha: sites H 1, H 13 and H 14. Transects were recorded with yellow crosses on the plateau H 1.

A site is sometimes loosely defined as "any past location/place which has been a locus of human activity" (cf. Banning 2002: 81-83), and for us the definition is not limited to the number of artefacts found in a square metre. At least three artefacts occurring in concentration were needed for labelling a location as a site. Therefore, our "sites" could include places which in some studies could just be considered artefact scatters. However, we did not have any parameters to differentiate scatters from sites during the fieldwork and mapping as this is often carried out in closer studies of the artefacts and on statistical grounds. Remote sensing emphasises structural remains, whereas open accumulated or open-air sites do not have any structures to be traced on the surface (Dunnell 1992). In addition, sometimes features/structures do not provide any artefacts on their surfaces from which to infer dating, but nevertheless we defined human-made features/structures as forming a site. (Cf. Binford 1992: 44-46). On the Euphrates side of Jebel Bishri geophysical examinations with Ground Penetrating Radar (GPR) were used for penetrating underneath the ground surface and to trace anomalies such as various structures.

Each period and site-type was approached as objectively as possible in order to provide basic and comparable accuracy. This objectivity not only identifies the intrinsic value given to each human activity in the past but also provides a necessary framework for gaining an overall picture of the cultural development in this region which has not been systematically surveyed before (Lönnqvist *et al.* 2006). Several surveys only record and document monuments and *tells* (ruined mounds of settlements) that are large in size, or merely search for Roman remains that are often regarded as priority targets of interest, when only the remains of civilizations are in focus. But in our view mobile sites such as those of the hunter-gatherers or pastoral nomads should not be regarded as inferior, and they should be given equal attention. The project has expressed a need for balancing the world heritage sites so that the remains of mobile societies would receive proper presentation; from the long perspective of the mobile past of people, in contrast to just the modern sedentary era (Lönnqvist and Törmä 2003). Small-scale societies especially provide perspectives on cultural development that was characteristic of steppic and desert region, from a long-term perspective. The distribution of sites may reveal hunting and ranging habits, territorialism and interaction.

3.4. Field Walking for Recording and Documentation on the Ground

It is important to note that the basic unit of recording, namely a "site" here, affects the survey strategy on the ground and consequently the site mapping. The sampling strategy included identifying all the sites falling in chosen transects divided in 15 m interval walking. Generally all the small finds were collected in the transects (Fig. 6.). However, the comprehensive systematic intra-site collecting of small finds at large sites such as at lithic workshops or tells were beyond our time-frame and capacity to study and store large amounts of finds. On the other hand, at such sites (Tell Tibne, Tell Tabūs and Tell Kharita) we walked all the surface areas through in an interval of 15 m at minimum to collect samples of artefacts. Our project did not have any permit for digging trial pits or carry out site excavations either in order to attain close stratigraphic evidence of the sites. Therefore no particular botanical, osteological or radiocarbon samples were collected at this stage.

Fig. 6. Ground surveying by field walking in desert environment on Jebel Bishri. Photo: Minna Lönnqvist 2006.

In the year 2000 our survey concentrated on the southwestern high ridges and the western piedmont areas around the Umayyad castle of Qasr al-Hayr ash-Sharqi (see Fig. 4.). This area was desert-steppe. We systematically field-walked the western plain between the Guesthouse of the Syrian General Directory of the Antiquities and Museums and Qasr al-Hayr, as well as the western part of the Bedouin village of Shanhas, 1.6 km northeast of Qasr. "Window screening" was carried out over the plain from Qasr al-Hayr to Tar al-Sbai (see Fig. 4.). Tar al-Sbai, a semi-circular escarpment on the western ridge of Jebel Bishri, was studied systematically in the area covering ca. 2.1 km x 100 m (Fig. 9.). There was a natural transect between a road following the direction of the edge and the line of the escarpment. Some sporadic checks were further made along the escarpment as well as visits to inner areas in order to understand the surroundings.

The area of Nadra which consists of steppe environment in the central eastern part of the mountain was our target in 2004 (see Figs. 3., 4. and 10.). Our expedition camp (UTM: 0537910, 3912307) situated a few hundred metres northeast of the Shula – Ash-Shujiri asphalt road ca. 696 m a.s.l. An area of ca. 20 km^2 was covered in the Nadra plain by field walking in addition to the surroundings of our desert camp area where ca. 10 km^2 were studied. Window screening and unsystematic visits were further carried out in the neighbourhood of the village of Ash-Shujiri. Prospecting with a QuickBird image was associated with the neighbourhood of Ash-Shujiri, and separate Roman sites such as a dam and a tower at Darakhlia as well as a Roman military camp were recorded in the vicinity of the village. Sites with Roman remains at Qseybe and Qebaqeb were separately targeted on the southeastern ridges and southern plains as well. (See Fig. 4.).

The intensive field walked area in 2003-2006 on the Middle Euphrates started first in 2003-2004 ca. 23 km northwest of Deir ez-Zor at the village of Kharita-Mustaha. The area covered extended over Tell Tabūs. In the years 2005-2006 our work on the Euphrates side of Jebel Bishri started ca. 12 km northwest from Deir ez-Zor and concentrated on the terraces between the villages of 'Ayyāsh, 'Ain Abu Jima and Kharita reaching the border of Kharita-Mustaha between Tell Kharita and Tell Tabūs. (See Figs. 4. and 11.). The area covering ca. 13 km in length and 500 m in width, a total of 6.5 km^2, was studied between 'Ayyāsh and Mustaha in 2003-2006 (see Fig. 11.). An ancient road and tombs surrounding the fort of Tabūs were studied deeper into the southern desert-steppe. Targeted short checks with a jeep were made on terraces along the way from Mustaha to Raqqa following the Euphrates valley. All the ancient tells, except Tell Ma'adan, on this Euphrates side of Jebel Bishri were mapped, and apart from Tell Tabūs, the tells of Tibne and Kharita (see Fig. 4.) were surface surveyed outside the intensively studied lots between 'Ayyāsh and Tabūs.

3.5. Data Capture on the Ground

After recognizing a site on the ground recording of the coordinates with GPS (Global Positioning System) was the main technique used to define the site in space for mapping; the coordinates identified having formed the basic parameters for working with GIS. UTM coordinates (37 s channel) are used in our survey in order to offer the global spatial identity for the site location. This is a numerical means to identify the position of an identified site and a way to further elaborate the data into graphics. In the following chapters, the sites for which coordinate information is not provided have not been studied by our survey teams. There are several reasons for this: we did not reach the areas during the permission which comprised the area; the areas were outside the permission of our studies or are under the study of other archaeological missions.

Documentation proceeded from recording the coordinates of the identified sites with computerised site documentation forms, EDM recording (Fig. 7.) where possible, measuring by hand, drawing by hand (Fig. 8.) and photographing each site both with digital and analogue cameras (BW film and colour slides). As additional documentation, traditional field books were used for notes.

Fig. 7. EDM recording at Tar al-Sbai. Photo: SYGIS 2000.

Fig. 8. Recording an ancient road with an EDM as well as by hand drawing in 2004. Photo from NW: Minna Lönnqvist 2004.

GPS way points were collected, for example, for indicating a road alignment on a satellite image. Coordinates were also specified as ground control-points for satellite image rectification purposes from various crossroads. In a range of places they were collected in order to define material signatures in satellite images.

The BW photographs are still necessary in addition to digital images as long as the future of storing digital data is suspect. Concentrations of small finds in open air and accumulated sites were recorded, after which the sampling of the small finds was carried out. As far as the structures were concerned, associated small finds were recorded and documented. Their specific locations in relation to structures were recorded in sample sites with EDM. Maps are used for visualizing spatial data, and they provide means to better understand spatial and topological relationships. Nowadays digital spatial data is largely handled in mapping (Kraak and Ormeling 1996: 1), and as explained in previous chapters, for us satellite images have provided a valuable basis. ArcView software has been used for producing maps from the captured EDM field data. Drawing by hand is a way to rapidly assess and record structures and features of a site. Hand-drawing reveals detail of the surfaces, as small features are not always preserved in digital recording and documentation.

The GIS regional perspective can be made up of transects, site distributions, sites and intra-site spatial analyses depending on the accuracy of specific data to be fed into databases. When concentrations of small finds are not available caution is required by not uncritically trusting in a few small surface finds as a definite means for dating the sites, because erosion through heavy sandstorms, winter rains with run-off mechanisms and other different post-depositional processes move the finds from their original locations – even in significant amounts. However, due to the lack of significant traffic or sedimentation in the area of the desert-steppe at the western edge, our experience is that the human impact has been very slight in moving small finds. This we observed as Palaeolithic workshops had largely remained undisturbed on the ground at some sites, throughout millennia.

In our survey the small finds have provided a *micro-scale* perspective mainly as a reference source for the site dating. In cases, where there was lack of surface finds, site-type comparisons in the Greater Southwest Arid Zone were one way to infer some tentative dating and functions of the site types. (Cf., e.g., Zarins 1992). Due to the primary need of the basic mapping of the site locations, and the large size of the area to be covered, intrasite analyses in *micro-scale* have not so far played any particular role in the study of our project, but the surface finds recorded with EDM *in situ* at sample sites opens the possibility of carrying out such analyses for particular purposes in the future.

Ground penetrating radar (GPR) was used for studying road structures and tracing buildings as well as tombs that are not necessarily visible on the ground surface (Fig. 12.). On the Euphrates side of Jebel Bishri concentrations of small finds served as indicators for possible buried sites on alluvial terraces. GPR was utilized to determine whether these concentrations held any value. Excavation or drillings are the only ways to experiment with and control the geophysical remote sensing results and recognise the nature of the anomalies under the ground, but unfortunately we did not hold a licence to carry out such studies.

Fig. 9. The intensively and systematically surveyed area (black bordered tilted rectangle) at Tar al-Sbai (in the year 2000). Mapping on a Landsat-7 ETM image. Minna Lönnqvist 2009.

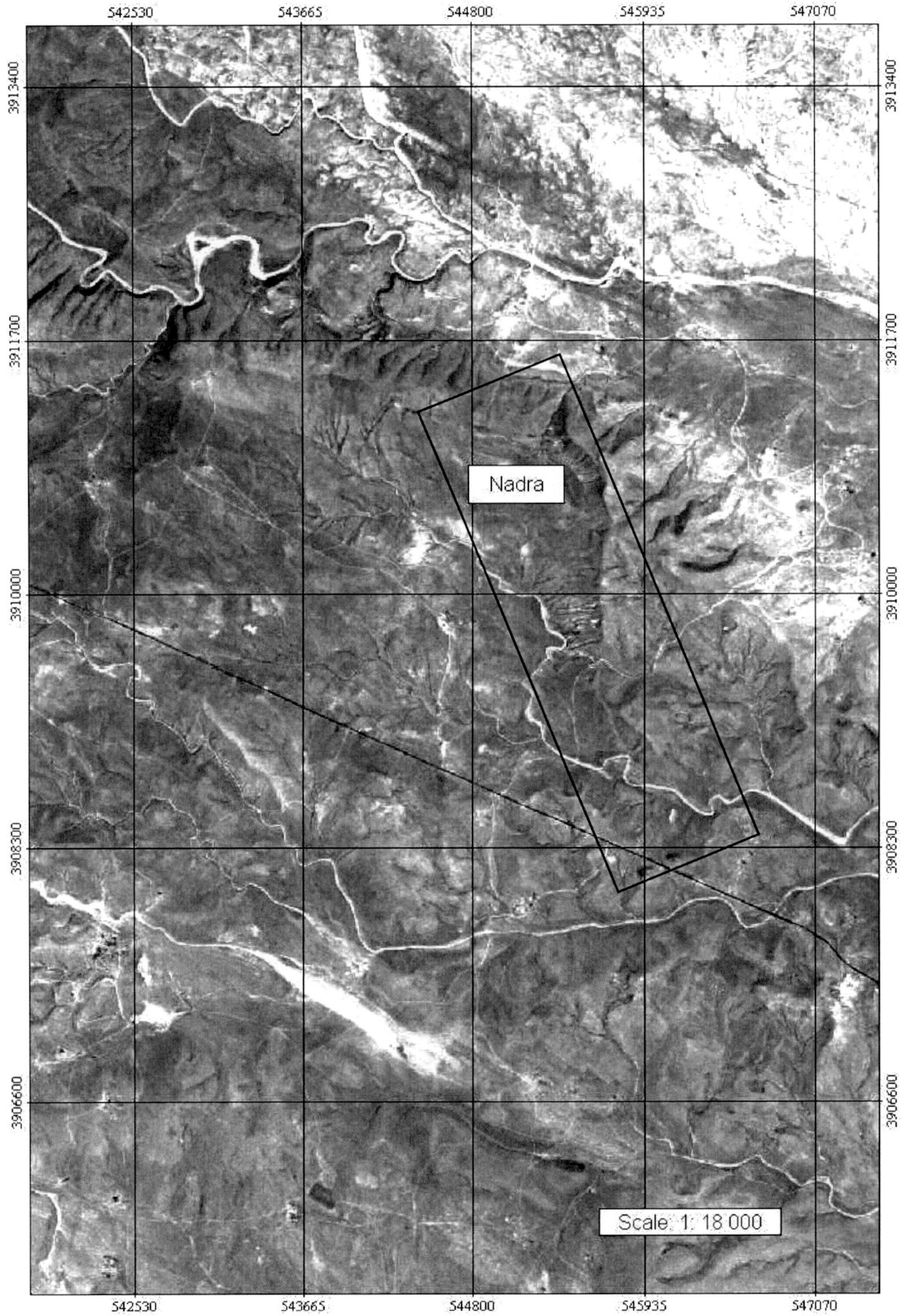

Fig. 10. The intensively and systematically surveyed area (black bordered tilted rectangle) at Nadra (in the year 2004) on the eastern central area of Jebel Bishri. Mapping on a SPOT image by Minna Lönnqvist 2009.

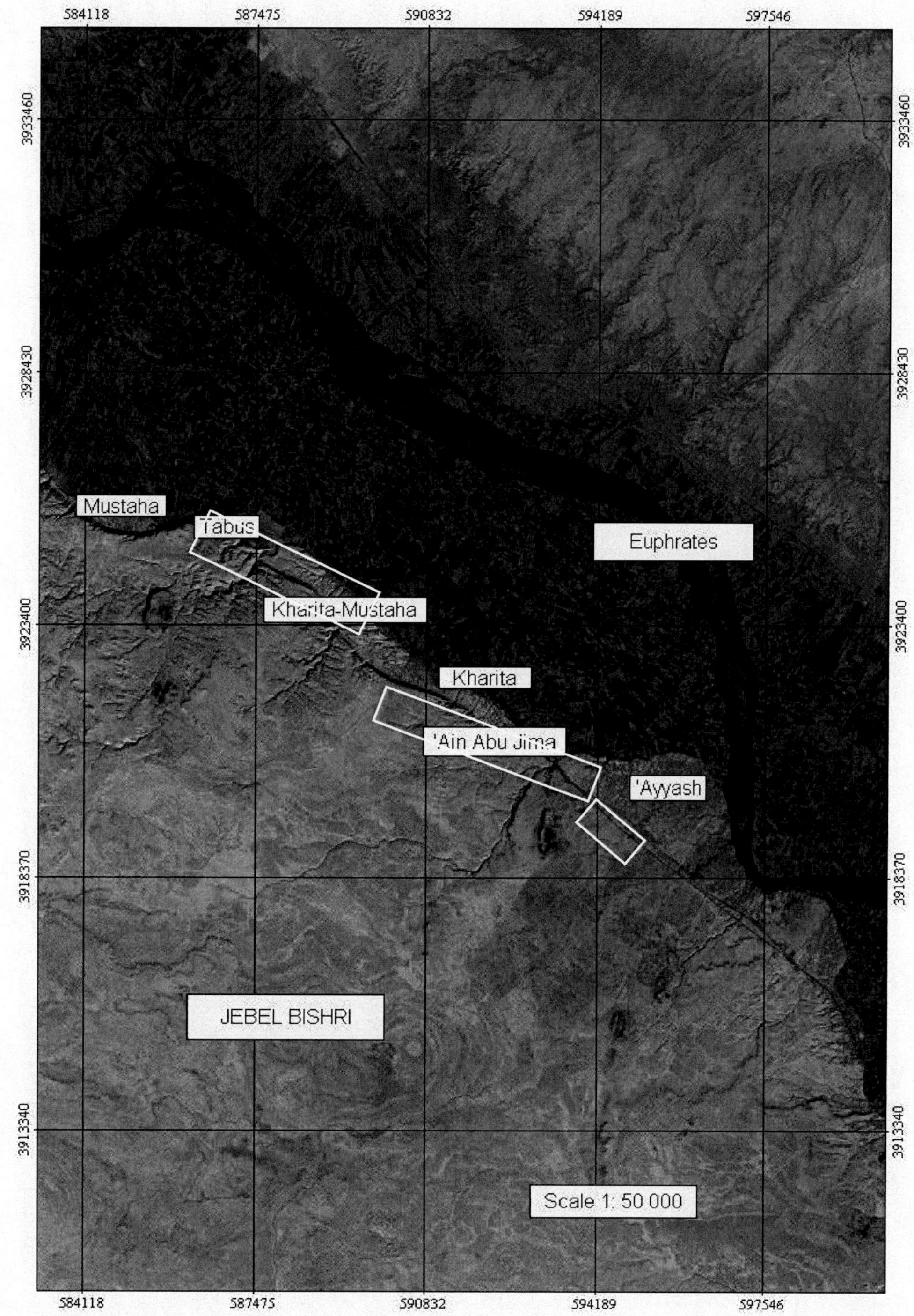

Fig. 11. Systematically surveyed areas (white bordered rectangles) between the villages of 'Ayyāsh and Mustaha (in the years 2003-2006) at the piedmont on the Euphrates side of Jebel Bishri as displayed on a Landsat-image. Mapping by Minna Lönnqvist 2005.

Fig. 12. Studying an ancient road with a GPR along the Euphrates. Photo: Gullög Nordquist 2004.

Fig. 13. Recording and documenting a Palaeolithic workshop above an abri at Tar al-Sbai. Photo: SYGIS 2000.

3.6. Artefacts: Collected Bits and Pieces from the Field

Small finds were collected into plastic bags in the traditional way, with site recording information written on them as well as on a cardboard label inside. Specific observations of the angles in which Palaeolithic finds were found were written on plastic bags and a 1 x 1 m concentration of debitage was recorded at Tar al-Sbai with an EDM and collected for further study (Fig. 13.). The finds were cleansed and put back into bags with tags (Fig. 14.). They all went through digital photographing, as well as documentation using BW film and colour slides (Fig. 15.) Finds were documented with their measurements. The most informative pieces were drawn, such as complete tools, pottery rims, handles and bases. The drawn pottery was recorded with the Munsell colour chart information. The directory of the Munsell numbers is given separately in this volume. The finds are now stored in the Palmyra Museum, and the documentation data will be stored in the archives of the University of Helsinki, Finland, in future, when the full research of all the finds has been completed.

Fig. 14. Packing washed pottery at the Guest House belonging to the General Directory of the Antiquities and Museums of Syria at the Temple of Bel in Palmyra. Photo: Minna Lönnqvist 2005.

Fig. 15. Photographing washed finds at the Museum of Deir ez-Zor. Photo: Minna Lönnqvist 2006.

3.7. Summarizing Remarks

To understand the methods used in our surveying and mapping helps to understand the collected data and its representative value. The macro-scale concerned the satellite image data that has been used both for prospecting, field surveying and mapping of the archaeological sites. Sites as basic recorded units are still used in the present study even if the approach is also landscape orientated. Coordinate information is emphasized both to benefit the General Directory of the Antiquities and Museums for its future protection and preservation work and for the benefit of the scientific identification of a site in space. This is an unambiguous way to provide the location and serves well in an area which is constantly under plundering, desertification and erosion. A site can disappear and without coordinates one may never find the original location. Also this may apply to those open accumulated sites of which all the finds were collected by our survey teams. At the sites where the amounts of small finds were very high only indicative finds were collected at the minimum of 15 m interval.

Bibliography

Bannister, A., Raymond, S. and Barker, R. (1998) *Surveying.* (Sixth edition). New York: John Wiley & Sons, Inc., Longman group edition, UK.

Banning, E.B. (2002) *Archaeological Survey.* Manuals in Archaeological Method, Theory, and Technique. New York: Kluwer Academic/ Plenum Publishers.

Binford, L.R. (1992) Seeing the Present and Interpreting the Past – and Keeping Things Straight, in *Space, Time and Archaeological Landscapes*, ed. by Rossignol, J. and Wandsnider, L., New York – London: Plenum Press, pp. 43-59.

Bottema, S. (1999) Landscape Archaeology and Reconstruction of the Mediterranean Environment based on Palynology, in *Environmental Reconstruction in Mediterranean Landscape Archaeology*, The Archaeology of Mediterranean Landscapes 2, ed. Leveau, P., Trément, F, Walsh, K. and Barker, G., Oxford: Oxbow Books, pp. 9-16.

Brown, A.G. (1999) Geomorphological Techniques in Mediterranean Landscape Archaeology, in *Environmental Reconstruction in Mediterranean Landscape Archaeology,* ed. Leveau, P., Trément, F., Walsh, K. and Barker, G., The Archaeology of Mediterranean Landscapes 2, Oxford: Oxbow Books, pp. 45-54.

Crumley, C.L. and Marquardt, W. H. (1990) Landscape: a unifying concept in regional analysis, in *Interpreting Space: GIS and Archaeology*, ed. by Allen, K.M.S., Green, S.W. and Zubrow, E.B.W., London – New York – Philadelphia: Taylor & Francis, pp. 73-79.

Dunnell, R.C. (1992) The Notion of Site, in *Space, Time and Archaeological Landscapes*, ed. by Rossignol, J. and Wandsnider, L., New York – London: Plenum Press, pp. 21-41.

Evans, J. and O'Connor, T. (1999) *Environmental Archaeology: Principles and Methods.* Midsomer Norton, Somerset: Sutton Publishing Ltd.

Hole, F. (1980) Archaeological Survey in Southwest Asia, in *Paléorient*, Vol. 6, pp. 21-44.

Kowalewski, S. A. (1990) Merits of Full-Coverage Survey: Examples from the Valley of Oaxaca, Mexico, in *The Archaeology of Regions: A Case for Full-Coverage Survey*, ed. by Fish, S.K. & Kowalewski, S.A. Washington D.C. – London: Smithsonian Institution Press, pp. 33-83.

Kraak, M.-J. and Ormeling, F. (1996) *Cartography: Visualization of Spatial Data.* Addison Wesley Longman Ltd.

Lönnqvist, M. and Törmä, M. (2003) SYGIS – The Finnish Archaeological Project in Syria, in Proceedings of the XIXth International Symposium *CIPA 2003* (The ICOMOS & ISPRS Committee for Documentation of Cultural Heritage) New Perspectives to Save Cultural Heritage, Antalya, Turkey, ed. Altan, M. Orhan, *The ISPRS International Archives of the Photogrammetry, Remote Sensing and Spatial Information Sciences,* Vol. XXXIV-5/C15, Istanbul, pp. 609-614.

Lönnqvist, M., Törmä, M., Lönnqvist, K., Whiting, M.S, Nuñez, M. and Okkonen, J. (2006) An Ethical Overview of Long-term Perspectivism, Experiences from an Archaeological Survey on Jebel Bishri in Syria, in *The e-volution of Information Communication Technology in Cultural Heritage, Where Hi-Tech Touches the Past: Risks and Challenges for the 21st Century,* Short papers from the join event CIPA/VAST/EG/EuroMed 2006, The 7th Symposium on Virtual Reality, Archaeology and Cultural Heritage VAST, 30 October – 4 November 2006, Nicosia, Cyprus, ed. by Ioannides, M. , Arnold, D., Niccolucci, F., Mania, K., Budapest: EPOCH Publication, pp. 108-113.

Poidebard, A. (1934) *La Trace de Rome dans le Désert de Syrie*: Le Limes de Trajan a la conquête arabe. Recherches aériennes (1925-1932). Texte. Atlas. Bibliothèque Archéologique et Historique. Tome XVIII. Paris: Paul Geuthner.

Tilley, C. (1994) *A Phenomenology of Landscape: Places, Paths and Monuments.* Oxford/Providence USA: BERG.

Wandsnider, L. (1998) Regional Scale Processes and Archaeological Landscape Unites, in *Unit Issues in Archaeology: Measuring Time, Space and Material,* ed. by Ramenofsky, A.F. and Steffen, A., Foundations of Archaeological Inquiry, Salt Lake City: The University of Utah Press, pp. 87-102.

Vita-Finzi, C. (1978) *Archaeological Sites in their Setting.* London: Thames & Hudson.

Zarins, J. (1992) Archaeological and Chronological Problems within the Great Southwest Asian Arid Zone, 8500-1850 B.C., in *Chronologies in Old World Archaeology,* Third Edition, 2 Vols., ed. by Ehrich, R.W., Chicago: The University of Chicago Press, Vol. 1, pp. 42-62.

4. MACRO-SCALE IN ENVIRONMENTAL AND LANDSCAPE STUDIES

Minna Lönnqvist, Markus Törmä, Kenneth Lönnqvist and Milton Nuñez

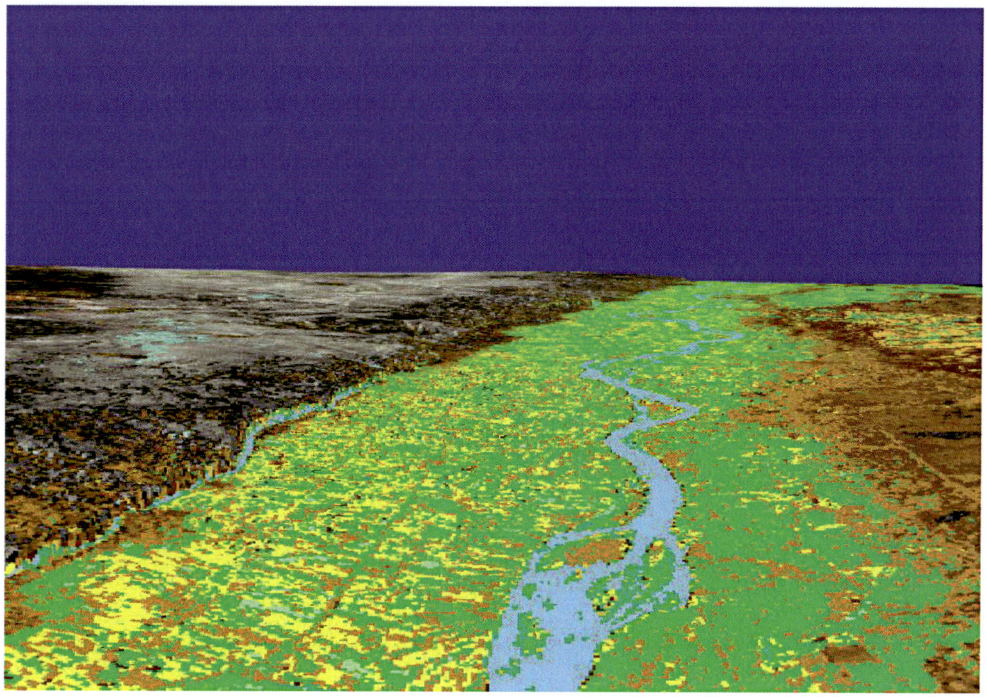

Fig. 1. 3D scene interpretation result, viewing the ridges of Jebel Bishri following the Euphrates river valley from the east to the west. The model constructed by Markus Törmä 2003.

4.1. Environmental Changes from the Long-Term Perspective

Geologically the formation of the Bishri block represents a northeast-oriented antiform in the Palmyride mountain belt the formation of which, at the northern end of the Arabian platform, dates from the Jurassic and Cretaceous periods (see Figs. 1.-3., Brew *et al.* 2003). Block mountains are produced by faulting and uplifting large blocks of rock (Webster 1994). Tectonic changes include uplift and subsidence (Rendell 2000), and the Bishri block is an uplift formation in the northern part of the Palmyride range (Brew *et al.* 2003). The block consists mainly of limestone including dolomites such as gypsum and marble beside sandstone, marl and basalt (Brew *et al.* 2003: 680, Table 1.). From the time of volcanic activity there are deposits of basalt in the central area of the mountain at Jebel Hemet Quebiba (Fig. 4.) as well as at the peninsulas of Halabiya (Fig. 5.) and 'Ayyāsh on the Euphrates side of the mountain.

Fig. 2. The block of Jebel Bishri seen from afar, from the side of the Syrian Desert. Photo: Minna Lönnqvist 2009.

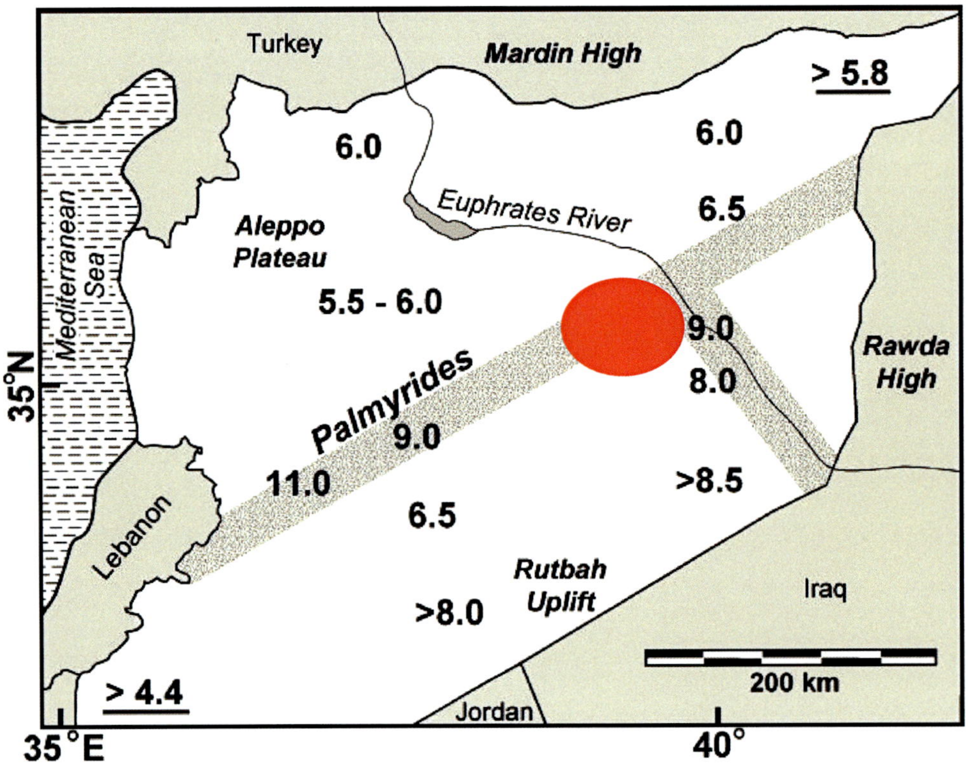

Fig. 3.1. The Palmyrides and the circled location of Jebel Bishri displayed in the northern part of the mountain belt on the map of Syria. Courtesy: M. Baranzangi and G. Brew.

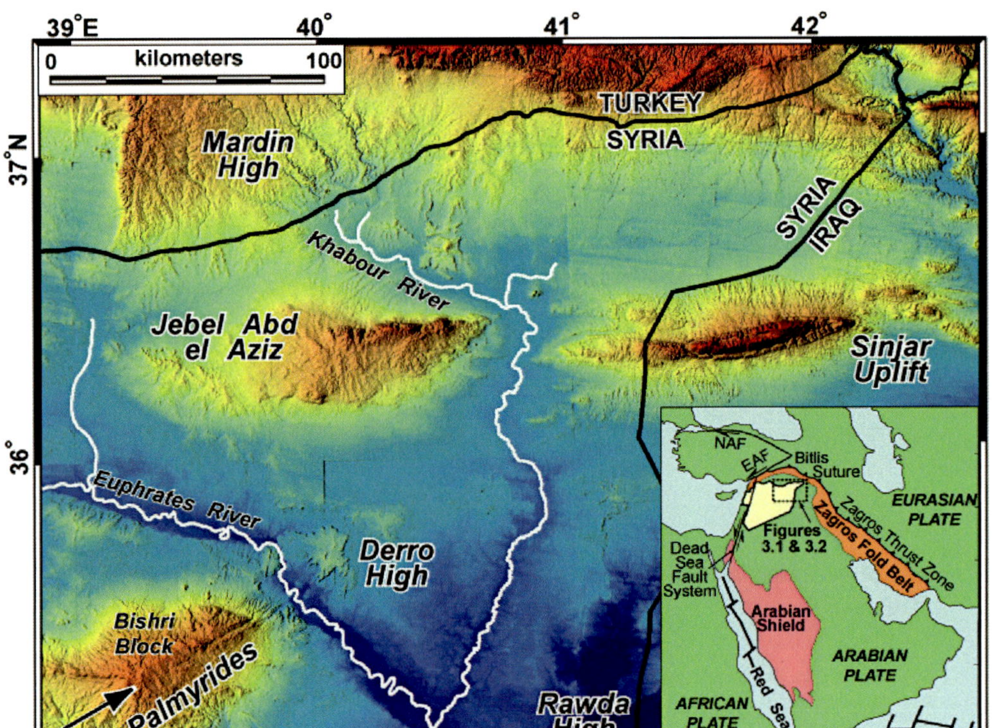

Fig. 3.2. Topographic map of the Bishri block and northeastern Syria. Courtesy: M. Baranzangi.

Fig. 4. Volcanic heights in the area of Jebel Hemet el-Quebiba in the central area of Jebel Bishri. Photo: Minna Lönnqvist 2008.

Fig. 5. 3D-scene mosaic of Landsat-4 MSS images 172/35 and 36, date 12.07.1983, channel 574 of the Euphrates river valley from the east to the west seen from the Jezira towards Jebel Bishri displaying the basaltic peninsula of Halabiya in the middle. Constructed by Markus Törmä 2006.

P. Beaumont (1993: 33) has classified areas that are typical of the arid zones: uplands, alluvial fan, alluvial plain and salt lakes or salt desert. Apart from the salt lakes these types of areas are largely compatible with the environment and landscape of the Jebel Bishri region today. Because of *sabkhas,* shallow salt plains, on the western piedmont and salt outcrops on the Euphrates side of the mountain, salt has been collected in the region since antiquity. Nowadays oil is drilled by the Syrian Petroleum Company on the mountain, and because of bitumen – and natural asphalt - asphalt is produced in the area. The fluvial terraces on the Euphrates side especially include marl, and beside different types of stone, bitumen and oil the natural resources which the region has consisted of ochre, iron ore and white sand. The marl has been used for pottery and white sand for manufacturing glass in Aleppo. (Musil 1928: 173).

It is clear that semi-arid and arid environments usually build up certain kinds of economies and cultural responses. Subsistence patterns develop with the environment. They are relatively enduring in areas with a continuous environmental history. Like farmers herders cause significant changes in the environment. The alteration of the environment can in extreme cause climatic changes (see Butzer 1982: 123-156). Monitoring changes from the past to the present in an environmental border zone like Jebel Bishri opens vistas for human response to the natural conditions in space and time. Human groups have found different strategies so that they can adapt to marginal regions. Archaeologists can monitor past change by reconstructing past environments and landscapes putting them into wider scale and perspective in relation to the present situation.

Nowadays the area of dry lands covers ca. 40 % of the surface of the globe, and currently depending on the definition, the deserts cover ca. 20-26 % of the Earth's surface (UNEP). Before the visible human impact on the Earth the coverage of steppes has been thought to have consisted of ca. 40 % while currently steppes cover approximately 16-24 % (UICN) of the Earth's surface. To gain a full understanding of ancient situations pertaining to the landscape of Jebel Bishri reconstruction of palaeoclimates and past environments is required. Climatic conditions largely define past ecosystems, and the evolution of landscapes is therefore dependant on climatic changes (Reitz *et al.* 1996: 11), which generally in early prehistory must have been chiefly caused by nature but in some points in the past may have already been caused by humans.

According to previous palaeoclimatological studies, the region of Jebel Bishri could have belonged to steppe and desert-steppe environment since the Terminal/Late Upper Palaeolithic (25,000 - 20,000 BP) and Epipalaeolithic periods (20,000 - 10,300 BP) (cf. van Zeist and Bottema 1982: 222-232, Figs. 14.10-14.13; see datings of the periods in Cauvin 2000; McLaren *et al.* 2007: 196-197). However, some recent studies have suggested that even if the previous research has showed steady continuity in this part of the climate of the Greater Southwest Asian Arid Zone during the Holocene, there have been periods of droughts and moist intervals. A "punctuated equilibrium" is seen as reflecting the present state of the Holocene climate that indicates various small changes (see, e.g., Algaze and Poumelle 2003). Scales of environmental change can be divided into long-term and short-term changes. Cooling or warming climates appear as long-term phenomena, but these phenomena are difficult to date accurately. (Dincauze 2000: 67-71). Archaeology provides a means to understand these long-term phenomena. Our study will elucidate conditions in the region of Jebel Bishri from the archaeological point of view by looking at the nature, continuity and change of the human impact in the environmental context. The types of the sites and their continuity and change, however, are not necessarily always dependent on environment but can be depending on social/ethnic choices and political situations. Short-term views on environmental changes are provided by our studies through satellite image data and employing ethnoarchaeological approaches on the ground (see ch. 15.).

4.2. The El Kowm Basin on the Western Piedmont

The oasis of El Kowm on the western piedmont of Jebel Bishri (see Figs. 6., 7.) comprises dozens of artesian springs. The springs are saturated with mineral salts and the water flowing in them has the temperature of + 30° C. The annual precipitation of the area is ca. 120 mm on average. The basin of the oasis is a 20-25 km wide ranging gap in the Palmyrides at the altitude of ca. 450 m a.s.l. bordered by low hills which rise to the north. (Besançon *et al.* 1981: 34; Le Tensorer *et al.* 2006; 2007: 624). This area has offered a passage for migrating animals, such as Persian gazelles (*Gazella subgutturosa*), and has attracted them with its perennial water sources throughout the past (Legge and Rowley-Conwy 1987: 79; Akkermans and Schwartz 2003: Fig. 2.13).

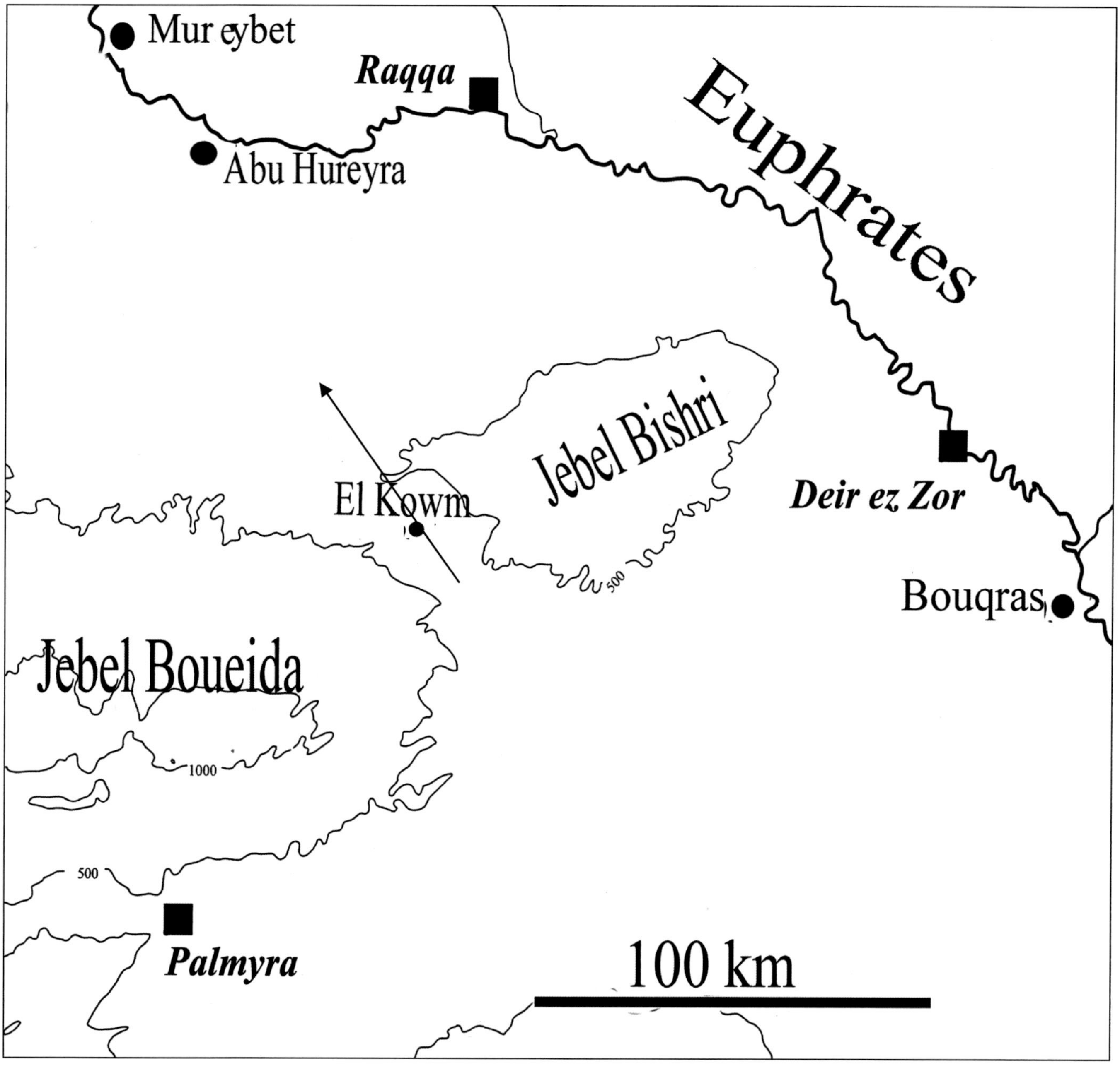

Fig. 6. The location of the prehistoric site of El Kowm in relation to the prehistoric sites of Abu Hureyra, Mureybet and Bouqras. The arrow is displaying the direction of the route of Persian gazelles through the Levantine corridor. (Basemap: the Syro-Japanese Tell el-Kerkh project; cf. Legge and Rowley-Conwy 1987: 79;mapping by Minna Lönnqvist 2010).

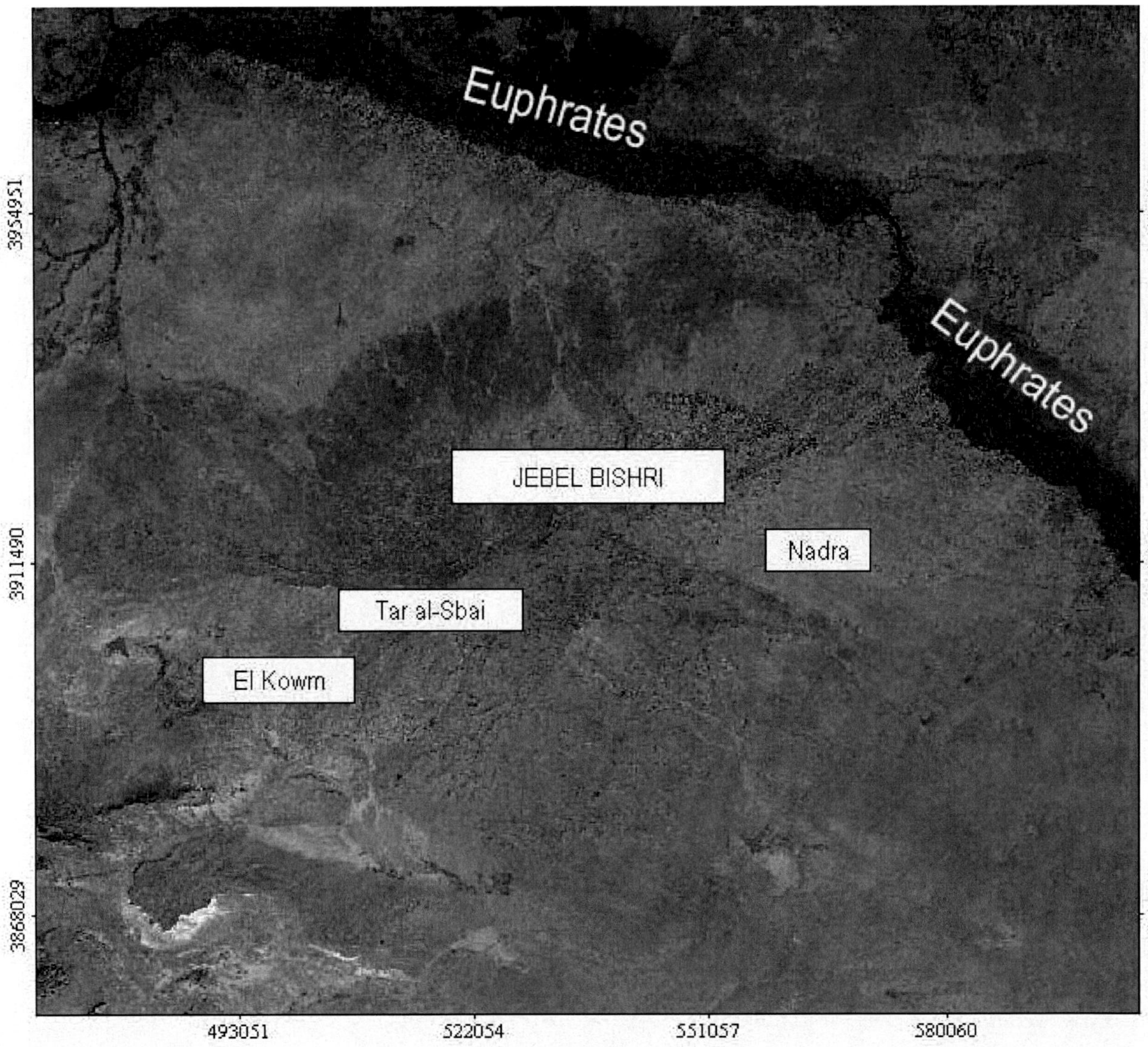

Fig. 7. A satellite map (scale 1: 100 000) displaying the oasis of El Kowm and the palaeolake of Nadra in the region of Jebel Bishri. Mapping by Minna Lönnqvist 2010.

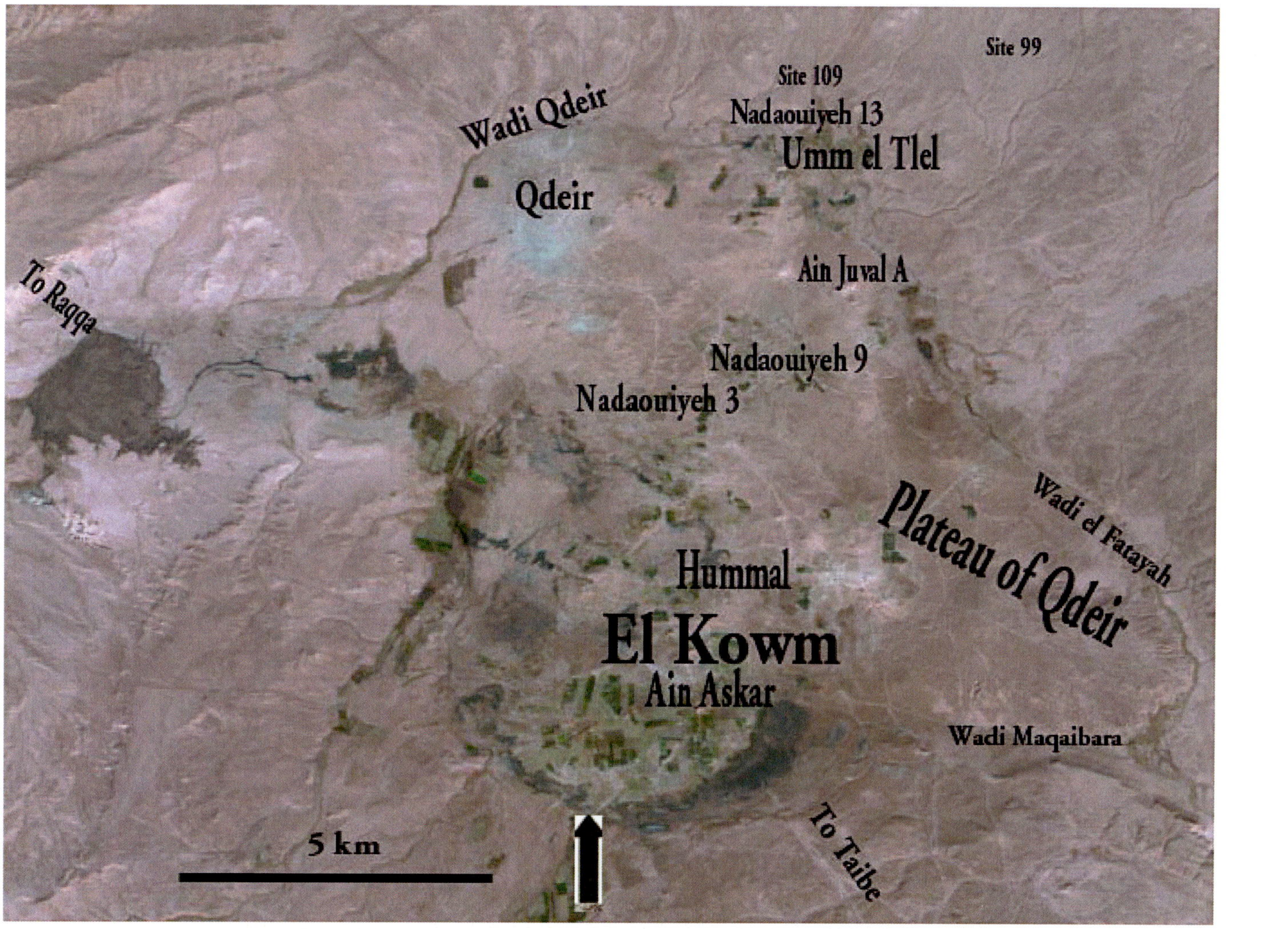

Fig. 8. Prehistoric site areas in the El Kowm basin identified by numerous archaeological missions before the Finnish studies on the western piedmont of Jebel Bishri. Based on Besançon et al. 1981.

Between Jebel Bishri and the oasis the flint strata of the Lower Eocene covers over 200 km². Jurassic formations, such as those to be found in the southern Bishri fault (Brew *et al.* 2003: 680), provide high quality flint, and therefore the areas in the NE-Palmyrides, including Jebel Bishri range, have offered an abundance of raw material for flint tool processing throughout the Stone Age. We can call this an ancient "Silicon Valley" – an ancient technological hub using flint. The best material that was used came from the northern part of the El Kowm basin. The basin of the *wadis* Faïdah and Fatayah are divided by the plateau of Qdeir which covers an area of ca. 3 km x 4 km and rises ca. 490 m a.s.l. A large area of salt plain, *sabkha,* that is situated in the northwestern side of the El Kowm basin covers an area of ca. 20 km². (See the sites in Fig. 8.; Besançon *et al.* 1981: 35; Cauvin 2000; Le Tensorer *et al.* 2006, 2007). The finds from the basin on the western plain provide *inter alia* the Oldowan-like pebble tools belonging to the Levantine Lower Palaeolithic period also present at Ubediyah in Israel (1,500 000-200,000 BP) and Achaeulian industries of the Lower Palaeolithic period (see Muhesen *et al.* 2006; Le Tensorer *et al.* 2007: 628-629; Besançon *et al.* 1981: Fig. 2.). However, Achaeulian industries coeval with the Levalloiso-Mousterian phase of the Middle Palaeolithic period have been also encountered at the El Kowm basin (Le Tensorer *et al.* 2007: 625).

The Palaeolithic sites of Hummal situated next to 'Ain Askar in the El Kowm basin (Fig. 8.) and the site of Umm el-Tlel next to the oasis of Qdeir offer evidence of steppe and desert fauna such as *Camelus dromedarius, Gazella, E. hemioechinus* and *Strutio, as well as Equids* (zebra, wild-ass, half-ass), *Bos primigenius*, *Dicerhorinus hemioechinus* and *Alcephaline* dating as far back as to the Mousterian levels. Recent finds include bones of a giant camel found in the Mousterian levels of El Kowm. (Griggo 1998; Cauvin 2000; Le Tensorer *et al.* 2006). A broken

Levallois point of flint has also been found embedded in the neck of a wild donkey at Umm el-Tlel testifying to the local hunting of this animal during the Middle Palaeolithic period (Boëda *et al.* 1999).

4.3. A Palaeolake Area at Nadra and Present Water Sources

In the area of Jebel Bishri the large *wadis* and Palaeolithic finds discovered along them apparently indicate that generally during the Pleistocene the landscape was composed of more reliable waterways. During our survey geomorphological signs of a palaeolake at the plain of Nadra in the central eastern area of the mountain were identified, echoing the past landscapes (Figs. 9., 10.). Nowadays Nadra forms an extensive valley with a large NE-SW-oriented seasonal *wadi* marking the alignment of a stream northwest from Jebel Nozairat al-Bishri (681 m a.s.l.). The edges of the lake area seem to follow the 600 m elevation contour line in the west, and the area seems to continue towards Al-Kharata plain in the southeast. This depression is surrounded by Palaeolithic tools witnessing to the existence of ancient shorelines. In our studies the edges of the Nadra plain offered small Palaeolithic rock shelter sites and accumulated open air sites (see more in the following chapter.). These sites in the neighbourhood of Nadra provide the evidence of a hunter-gatherer landscape during the Pleistocene period. That the sites lying below 600 m a.s.l. at Nadra did not seem to produce any Palaeolithic finds provides us a provisional *terminus* dating for the existence of the palaeolake.

Fig. 9. An interpretation of the Landsat-7 ETM satellite image, and the location of Nadra. Note the desertification line extending from the west. Interpretation by Markus Törmä, mapping by Minna Lönnqvist 2004.

Fig. 10. The area and edges of a palaeolake at Nadra. Photo: Kenneth Lönnqvist 2004.

T. Kusky and F. El-Baz (2000: 213-235) have studied the relationship of the uplift of the Syrian Arc Fold Belt that was under active tectonic evolution in the Late Cretaceous period. The drainage evolution and floodplain development in associations with the fluctuations of the Plio-Pleistocene period was scrutinized in the region. Temporary palaeolakes were recognized to have been formed by the Syrian Arc folds behind natural dams.

The most well-known palaeolake in the neighbourhood of the Palmyrides is located in the Palmyra basin offering Palaeolithic cave sites such as the Douara cave that has been studied *inter alia* by Y. Sakaguchi (see e.g. Fujii *et al.* http://www.um.utokyojp/publish_db/Bulletin/no29/no29003.html). However, the Middle Palaeolithic fauna from the Douara cave indicates that the environment already then belonged to the boundary between the Boreal Eurasiatic and Saharo-Sindian zone. This is supported by the fact that the Douara faunal assemblage was dominated by Saharo-Sindian species: *Camelus*, *Gazella*, *Hemioechinus*, *Psammomys*, *Meriones*, *Gerbilhis*, and *Allactaga*.

The results of the botanical evidence from the Middle Palaeolithic have provided evidence of a few species: *Celtis* and *Boraginaceae* nutlets. The Palmyra region is devoid of *Celtis* nowadays; but it is present in the moister areas of Anatolia. S. Payne has concluded that the evidence from the Middle Palaeolithic period (200,000-45,000 BP, see dating in McLaren *et al.* 2007) does not support the dating of the palaeolake of the Palmyra basin to the period in question, as the summers would have been long and dry, whereas winters were wet (Akazawa 1984). Furthermore, the Palmyra basin provides evidence of numerous climatic changes, and studies providing C-14 datings show that since the Isotope Stage 4 the basin has not been wet enough to form a palaeolake, although several short humid periods date from ca. 40,000/36,000 B.C., 27,000/24,000 B.C. and 18,000 B.C. (Besançon *et al.* 1997: 5-23). However, in the El Kowm basin evidence of abrupt climatic change in the form of torrential rains was gathered dating from 70,000 years back in time followed by cyclic evidence of arid and lacustrine phases from ca. 68,000 B.C. to 12,000 B.C. (Courty *et al.* 2006).

Further to the southwest of Palmyra the Damascus basin offers traces of another distinct ancient lake formation which is currently an oasis. The 'Azraq basin in Eastern Jordan forms a large area of limestone and basalt desert connected with the Syrian Hamada and Saudi Arabia, that also formed a palaeolake during the Pleistocene period. Its geomorphology dates from Cretaceous period and provides limestone, chalk and basalt, with stone industry of flint and other local archaeological material that has special interest for us, as it belongs today to the same Greater Southwest Asian Arid Zone as Jebel Bishri. Further south there is the palaeolake of the El Jafr Basin in southern Jordan with comparable features that include Lower and Middle Palaeolithic material at its ancient shore lines. In the Upper Palaeolithic period these areas in the Near East faced desiccation, which were followed by a humid face in the Late Pleistocene (see Garrard and Stanley-Price 1977: 109-112; Copeland 1988: 66-75). If the mountain area of Jebel Bishri was covered by terebinth-almond-woodland-type of steppe, collecting of nuts was also a possible activity, as well as harvesting of grasses and truffles.

Fig. 11. A landscape at Nadra with a lake and woodland steppe in the Palaeolithic era apparently looked like this landscape model reconstructing a lake in the area of the ancient Sahara. Source: FISA.

There is some evidence which might suggest that the traces of the palaeolake at Nadra as well as Palaeolithic activities in its area are contemporaneous with the Interglacial humid phase. For example, in the southern part of Jordan at Mudawarra, shell deposits have been obtained, and they indicate that a large lake existed in the area during the Oxygen Isotope Stage 5 (125,000-75,000 years ago), and this is roughly coeval with the evidence from El Kown. Additional samples have been acquired from the previously mentioned 'Azraq and El Jafr basins in Jordan and paleothemes from the Soreq cave in Israel. All these samples indicate that a humid phase existed during the Oxygen Isotope Stage 5. The lakes in the Nefud and the Rub al-Khali are apparently contemporary with this humid phase that took place during the Interglacial in the Pleistocene period, and which is also visible in the Sahara. Aridization started during the Isotope Stage 4 ca. 75,000-55,000 years ago (Sanlaville 2000: 15, see Fig. 7; Sanlaville 2002). It is thought that in the southern Levant a very dry period started in the late Middle Palaeolithic period during the Isotope Stage 3 ca. 58,000 B.C., and it continued to the earlier phase of the Upper Palaeolithic period. The climatic changes to more favourable warmer and wetter conditions could have started again ca. 30,000 B.C. in the Levant which roughly coincides with the Terminal Upper Palaeolithic and the beginning of the Epipalaeolithic era (Williams 2006: 327).

Near the southeastern Nadra plain a well called *Be'er el-Fajm* is known. We encountered large natural springs in this eastern plain (Figs. 12., 13.). Such springs also characterize the 'Azraq basin of Jordan which, as mentioned, had formed a palaeolake as well (Garrard and Stanley-Price 1977: 112). One water source at Nadra (UTM: 0551774, 3911802, ca. 530 m a.s.l.) is a completely natural fissure in gypsum ca. 6 m deep. Another site **I 28** (UTM: 0543776, 3913955, ca. 621 m a.s.l.) is originally a natural cavity hollowed out in gypsum bedrock by subterranean streams. It was 4.50 m in depth and 8 m in width with two dressed stone steps at the sides and collapsed stones at the bottom indicating a possible lid. Water harvesting sites with stone barrages were encountered at three points in *wadis* at Nadra. The first such site **I 14a** (UTM: 0546723, 3910264) had two stone barrages 13.5 m apart. The widths of the stone constructions were 2.7 m and 4 m. A single barrage **I 21b** (UTM 0545513, 3912930; 588 m a.s.l.) was 1.5-1.8 m wide and 1 m high, and **I 34** (UTM: 0546819, 3910057; 589 m a.s.l.) had two barrages: ca. 10 m in length, ca. 0.8-1 m in width and 1 m to 1.8 m in height standing 15 m apart.

Several water canals were constructed by the Romans on the desert-steppe plateau west of Jebel Bishri (see more details in ch. 11.). A. Musil (1928: 81-82) describes locations of several wells which he visited in the region of Jebel Bishri during his itinerary of 1908. Musil (1928: 172) mentions the watering station of ar-Rehub, also known as *Be'er Rehub*, situating 36 km southeast of Resafa-Sergiopolis on the northwestern foot of Jebel Bishri. In addition, we have traced on local maps a well called *Be'er Dabjan* in the northwestern piedmont, and close to it there is a *Be'er Ruman* (a Roman well?) at the northwestern part of the mountain. (Both of these wells were plotted on an Arabic engineering map of Syria, kindly provided by Khaled al-Assad, the Director emeritus of the Palmyra Museum, to our project.) *Be'er Ruman*, which we never visited on the ground, seems to refer to the Roman itinerary over Jebel Bishri which R. Mouterde and A. Poidebard (1945: 134) mention as a track leading from the Byzantine fortress-city Resafa-Sergiopolis to the fortress of Circesium on the Euphrates. They further mention an observation point connected with the watering site of Bîr Siqri or 'As-Šiqri in the central area of the mountain (Poidebard 1934; Mouterde and Poidebard 1945: 134). This site is apparently identical with the present-day pastoral village of Ash-Shujiri, which is largely abandoned now. Nearby we have recorded a large stepped Roman water harvesting site with a ruined tower. (See ch. 12. and 15.).

Fig. 12. Natural spring in gypseous rock at Nadra in the southeastern plains of Jebel Bishri.
Photo: Gullög Nordquist 2004.

Fig. 13. Natural spring in gypseous rock at Nadra.
Photo: Gullög Nordquist 2004.

According to Mouterde and Poidebard (1945: 134), wells are set in every 30 km distance along the transhumant pasturage of Jebel Bishri. Moreover recently the Syrian Petroleum Company has mapped a dozen of wells, all in the approximate radius of 20-25 km inside the Jebel Bishri region, a number of which do not, however, contain abandoned and suspended wells (Brew *et al.* 2003). (See the distribution of wells on a map Fig. 15.).

Fig. 14. A well at an abandoned Bedouin village (site I 1) in the area of Nadra. Photo: Christian Meyer 2004.

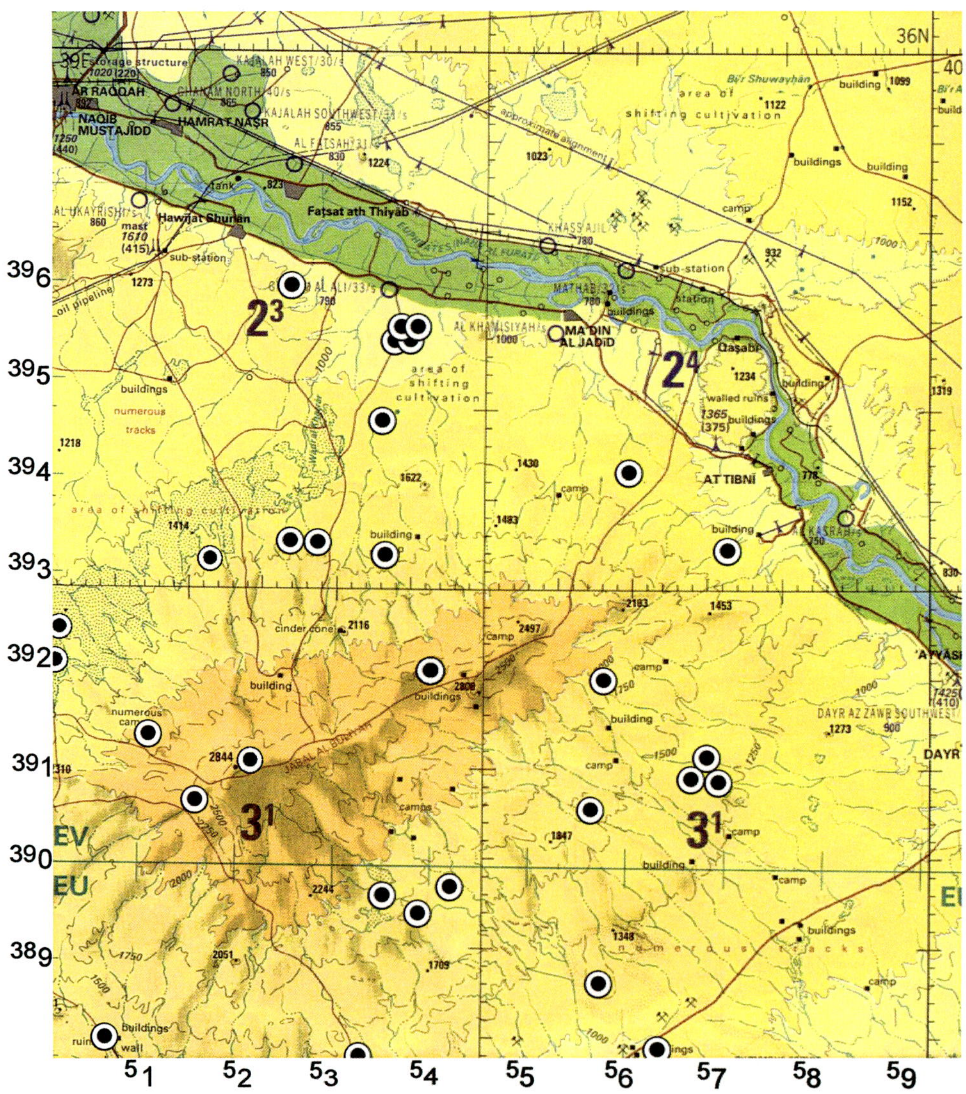

Fig. 15. Distribution of wells in the area of Jebel Bishri, including suspended wells. Based on the data collected by SYGIS and Brew et al. 2003. Basemap: © MOD Crown Copyright, 2010 published with the permission of the Controller of HMSO.

According to ACSAD, the gypseous bedrock is very difficult for digging wells, and the wells are strikingly few in number on the top areas of the mountain. From Nadra further to southeast through Al-Kharata an apparent Roman well was documented at Qseybe (see ch. 12.). The area of Qebaqeb on the southern piedmont is associated with a Roman military site with a water canal and the area is full of wells; nearly every household has its own well. (See ch. 11.).

Even if the sheep/goats prefer the distance of 10 km at minimum between watering points, in the cases when the local shrubs contain succulents, they can tolerate longer distances. It is known that sheep/goats can go for three days without water while donkeys can last up to four days and camels twelve days. (Heathcote 1983: 93-94).

4.4. Tracing and Reconstructing Late Quaternary Palaeoenvironments

Because no stratigraphical studies were allowed to be conducted in our research on Jebel Bishri, all specific environmental reconstructions of the past vegetation cover remain highly hypothetical and are retrieved from more general studies carried out in the neighbourhood. The evidence from the Ghab valley in northwestern Syria is based on the evidence of tree pollen and herbaceous pollen ratios of *Artemisia* and *Chenopodiaceae* and show considerable fluctuations between 50,000-12,000/11,000 BP. This is a period of alternating forest expansion and steppe environment interpreted as the result of changes from drier to moister intervals. At the beginning of the Holocene forest expanded in western Syria including species such as Quercus, Pistacia, Olea and Ostrya in the upland areas, and the increased humidity in 10,000-8000 BP, the period of the greatest forest expansion, ended in a decrease in humidity ca. 8000 BP (Bottema and van Zeist 1981). Datings of the analyses from the Mediterranean Sea cores are based on AMS (Accelerator Mass Spectrometry) that were compared with the pollen data. There are two distinct points: the highest level of herbaceous pollen such as *Chenopodiaceae* was reached in the Late Quaternary dated to 9000-8000 B.C. (Rossignol-Strick 1993: 135). In contrast, according to some studies, a sharp retreat of the woodlands in Syria has also been dated to have started at that time (see, e.g., Moore *et al.* 2000: 81), which in archaeological terms coincides with the transition from the Epipalaeolithic to the early Neolithic period. Palaeoenvironmental studies comprising oxygen-isotope comparisons and pollen data at Lake Gölhisar in Turkey from the Early Holocene also imply drier conditions from ca. 8600-6800 cal-B.C. High amounts of pistacia pollen can indeed be traced again to ca. 7000-5500 B.C. in the Eastern Mediterranean region, while isotopic fluctuations implying aridity and humidity oscillations belong to the period of ca. 6800-3100 cal-B.C. (Eastwood *et al.* 2006). On the other hand, B. Geyer and J. Besançon (1996: 8-10) have discussed the so-called Holocene Optimum in 11,000-6000 BP and its impact on the emergence of the settled life in the Euphrates valley. There was an apparent humid period that eased the settled life, at least during the Chalcolithic Halaf and Al-Ubaid periods.

Although the presented evidence indicates to the fluctuations of dry and moist periods, it needs to be emphasized that the information from the Mediterranean zone is not directly applicable to the desert-steppe region of Jebel Bishri in inner Syria as far as the continuous forest cover is concerned. Even if there was woodland steppe, the continuous forest did not reach the area during the Holocene any more. In any event, woodland steppe with scattered trees and grass could still have dominated the area of Jebel Bishri ca. 15,000-11,000 BP (Moore *et al.* 2000: 78-79), when Geometric Kebaran groups started to spread to more arid zones in 14,500-13/12,500 BP (Bar-Yosef 1987: 230; see the dating of the Geometric Kebaran in McLaren *et al.* 2007: 171-198) in the Epipalaeolithic times. The highland-lowland interaction clearly developed further due to these kinds of circumstances, if not already during the Palaeolithic period. The high areas provided surface water in the spring and early summer. (Henry 1989: 156, 169-170). The change from the long-term Geometric Kebaran camp life to more settled Natufian sites, including hamlets, has been explained by temperature changes by the final Pleistocene temperature affecting the diffusion of the wild cereals to higher areas. The Epipalaeolithic Natufian culture (13,000/12,500-10,300 BP) played a key role in the emergence of agriculture and sedentism (e.g., Bar-Yosef 1998; Cauvin 2000), but in regard to these processes the region of Jebel Bishri has traditionally been seen as a marginal area compared to the coastal Mediterranean Levant. The core area of the Natufian impact covers some hundred kilometres bordered by the Mediterranean Sea to the west and limited to the desert steppe to the east (Bar-Yosef 1998: Fig. 1.). However, the existence of the Epipalaeolithic groups on the mountain, and the possible initial agricultural endeavours in this desert-steppe area will be further assessed in the following chapter.

The transition from the Pleistocene to the Holocene period led to changes in the environment from the Glacial to Post-Glacial period and increase in the human exploitation of the environment. There was rising humidity at the beginning of the Holocene in the Eastern Mediterranean region (Avner *et al.* 1994: 285; cf. Rossignol-Strick 1993: 135-152), but its impact on the arid zone is suspect. The last 11,000 years has entailed exploitation of the Euphrates river fan with the emergence of agriculture (see Moore *et al.* 2000) followed by gradual inception of artificial irrigation techniques in these arid areas during the Neolithic period.

From the mid-Holocene onwards there is combined evidence that shows generally drier conditions, but from the Classical - Byzantine times there is evidence of rising humidity (Eastwood *et al.* 2006). Palaeoclimatological studies in the area of the Golan Heights have not provided very distinctive climate changes from 4500 B.C. up till present, although the pollen data shows higher anthropogenic impact of settlement life during the Chalcolithic/Early Bronze Age and during Hellenistic, Roman and Byzantine times as well as Crusader period (Neumann *et al.* 2007). Anyhow, collected proxy soil data from Northern Syria indicates that at the end of the Early Bronze Age and the beginning of the Middle Bronze Age there was an abrupt climatic change (Courty and Weiss 1997). This desiccation has been seen as a major cause of disturbances in the settlement life in the area at that time, and the Berlin 4ICAANE (Arz *et al.* 2004a and 2004b; Bar-Matthews and Ayalon 2004) brought substantial new evidence of this climate change from different parts of the Near East. On the other hand E. Peltenburg (2007: 247-248), for example, has called for more nuanced interpretation in which the development was not necessarily homogenous and acknowledges that local variations existed. For example, the evidence from Jerablus Tahtani on the Upper Euphrates testifies to the abandonment of the site caused by flooding instead of desertification in the end of the Early Bronze Age. G. Algaze and J. Poumelle (2003: 116) have paid attention to a similar situation taking place at Titrisis in Turkey ca. 2100 B.C. De-urbanisation of the area was followed by two centuries of wetter conditions. However, aridization and flooding in different regions are seen as typical features in the present climate change.

Fig. 16. A plot of acacia trees at Ash-Shujiri in central ridges of Jebel Bishri. The soil is reddish and ferrous in this area. Photo: Minna Lönnqvist 2006.

Fig. 17. Green fields of the Euphrates valley, where grazing of the village pastoralists takes place. Photo: Minna Lönnqvist, Nov. 2006.

4.5. 3D Modelling of the Present Landscape

In order to visually and spatially approach the environmental contrasts of the region of Jebel Bishri we constructed 3D landscape models displaying the mountain and the Euphrates valley. DEMs with satellite image data interpretation were merged using PCI Geomatica 9.1. Perspective scenes were saved and a fly-over animation was constructed. Figs. 1., 18. and 19. present the interpretation results concerning the merging of DEMs with Landsat-7 ETM image data. One is virtually able to follow the stream of the Euphrates up towards Raqqa, or approach from the Jezira the northern area of Jebel Bishri with the fly-over animation of the landscape models: (see a fly-over animation on a video clip: www.helsinki.fi/hum/arla/sygis).

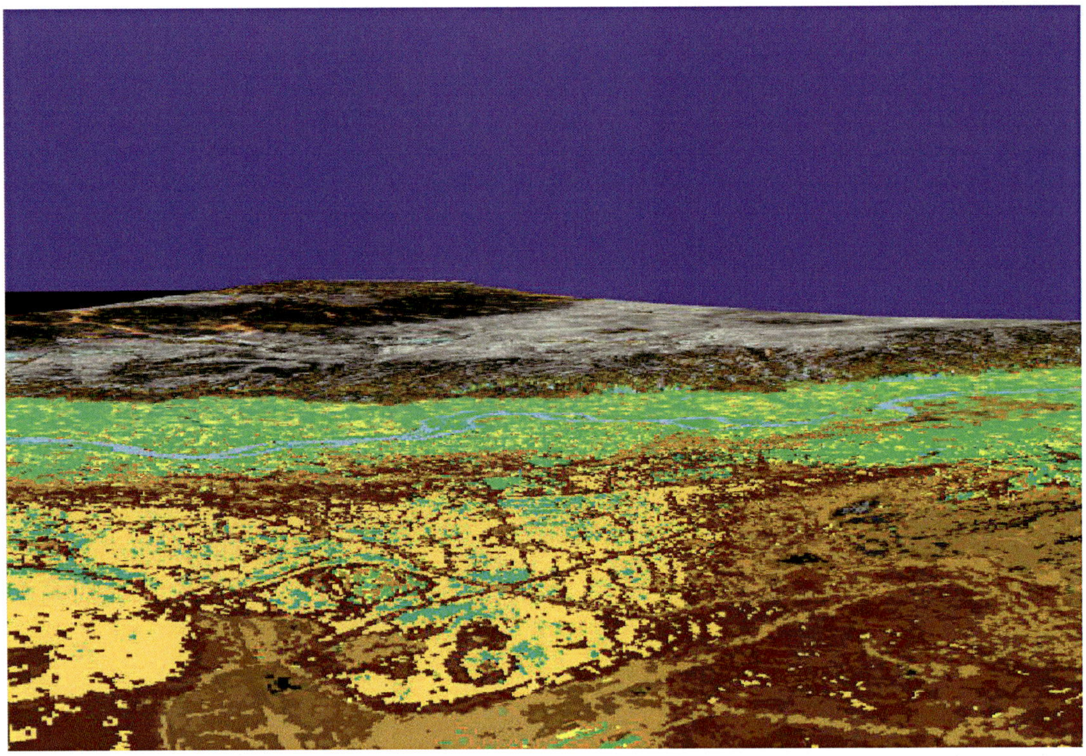

Fig. 18. This is 3D-scene of an interpretation result, viewing from the Jezira towards the Euphrates valley, flanked by Jebel Bishri in the background. Constructed by Markus Törmä 2003.

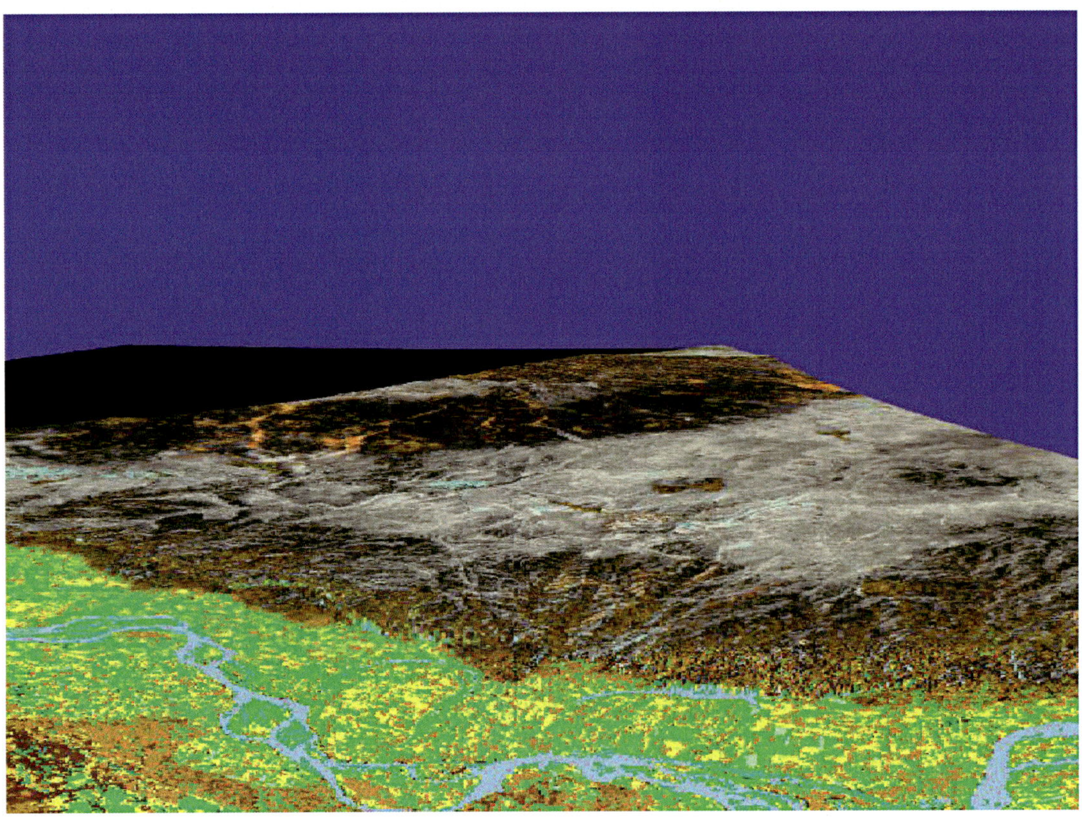

Fig. 19. This is a 3D-scene of an interpretation result, viewing Jebel Bishri from the north to the south. Note the tilting drainages descending from Jebel Bishri towards the Euphrates valley. Model constructed by Markus Törmä 2003.

4.6. The Present Vegetation Cover, Erosion and Desertification

H. Pabot's environmental model concerning Lebanon and Syria illustrates a hypothetical but plausible drastic retreat of the woodlands and vegetation cover (Thirgood 1981: 86). Nowadays continuous forests constitute only 3 % of Syria concentrating in the governorates of Lattakia and Tartous on the Mediterranean region (NAPC 2006: 25). Very isolated islands of terebinth-almond-woodland steppe were found in the neighbourhood of Jebel Bishri in the 1970s (Moore *et al.* 2000: 60), and our project only traced some scattered pockets such as those to be found at Ash-Shujiri in the central area of the mountain (see Fig. 16.). There are accounts from earlier travellers that the slopes of Jebel Bishri had trees (Grabar *et al.* 1978: 4), although there is no clarity concerning the density of their possible occurrence. There is neither record how much wood has been used as fuel and building by the Romans or Bedouins in the region. Some alterations in the tree populations can be compared to the present situation on the Euphrates side fringes in pilot C. Héraud's (1922) mappings from the early twentieth century. The maps show that poplar orchards were frequent in the past, but are not so dense today, although some poplars appear, and fruit trees are grown on the river plain. There are also earlier accounts of tamarisk woods in the neighbourhood. The plains are mainly cultivated with cotton, corn and barley today.

Already A. Poidebard took advantages of studying changes in vegetation cover by remote sensing. As mentioned, he paid special attention to using optimal seasons and daytimes in his aerial surveys in order to detect ancient remains in the region in the 1920s (see Brooks and Johannes 1990: 166-167). In our project we have used remote sensing in detecting environmental changes as contexts of possible archaeological situations. Images concerning the changes in the vegetation cover were computed using red channels of Landsat satellite images (channel 5 in MSS, 3 in TM and ETM); and their near-infrared channels (channel 6 in MSS, 4 in TM and ETM). The purpose of using red channel was to find changes concerning bare ground and mineral soil, especially sand, whereas the purpose of using near-infrared channel was to find changes related to the vegetation cover. Figs. 20. – 22. present change images between an MSS image from 27.6.1975, an MSS image from 12.7.1983, a TM image from 1.9.1990 and an ETM image from 29.1.1999. The difference between the Landsat ETM image red channel and the red channel of another Landsat image are displayed using red and blue primary colours in these image changes. The difference between the ETM image near-infrared channel and the other image near-infrared channel is shown using green as a primary colour. Therefore, green colour in a change image means that the amount of vegetation has increased between image acquisitions, and the reddish colour signifies that the amount of vegetation has decreased. The pixel size used was 60 m, when comparing the MSS and ETM images, and 30 m, when comparing the TM and ETM images. Most changes seem to happen at the river valley due to the different times of the growing season. Changes in the vegetation outside the river valley are small, but we can notice that the period between 1974 and 1999 (Fig. 20.) has meant decrease in the steppe vegetation, which is not to be understood simply due to a seasonal change. Between 1983 and 1999 (Fig. 21.) there has been a loss of steppic vegetation in the central area of the mountain, and the changes in the river valley seem to be due to crop harvestings. Greenness is especially visible in the western and northwestern piedmont areas between 1990 and 1999 (Fig. 22.), whereas northwestern slopes have lost steppe vegetation. The red colour in the river valley may be the result from crop harvesting again.

Fig. 20. A change image between an MSS image, date 27.6.1975, and an ETM image, date 29.1.1999.

Fig. 21. A change image between an MSS, date 12.7.1983 and an ETM image, date 29.1.1999.

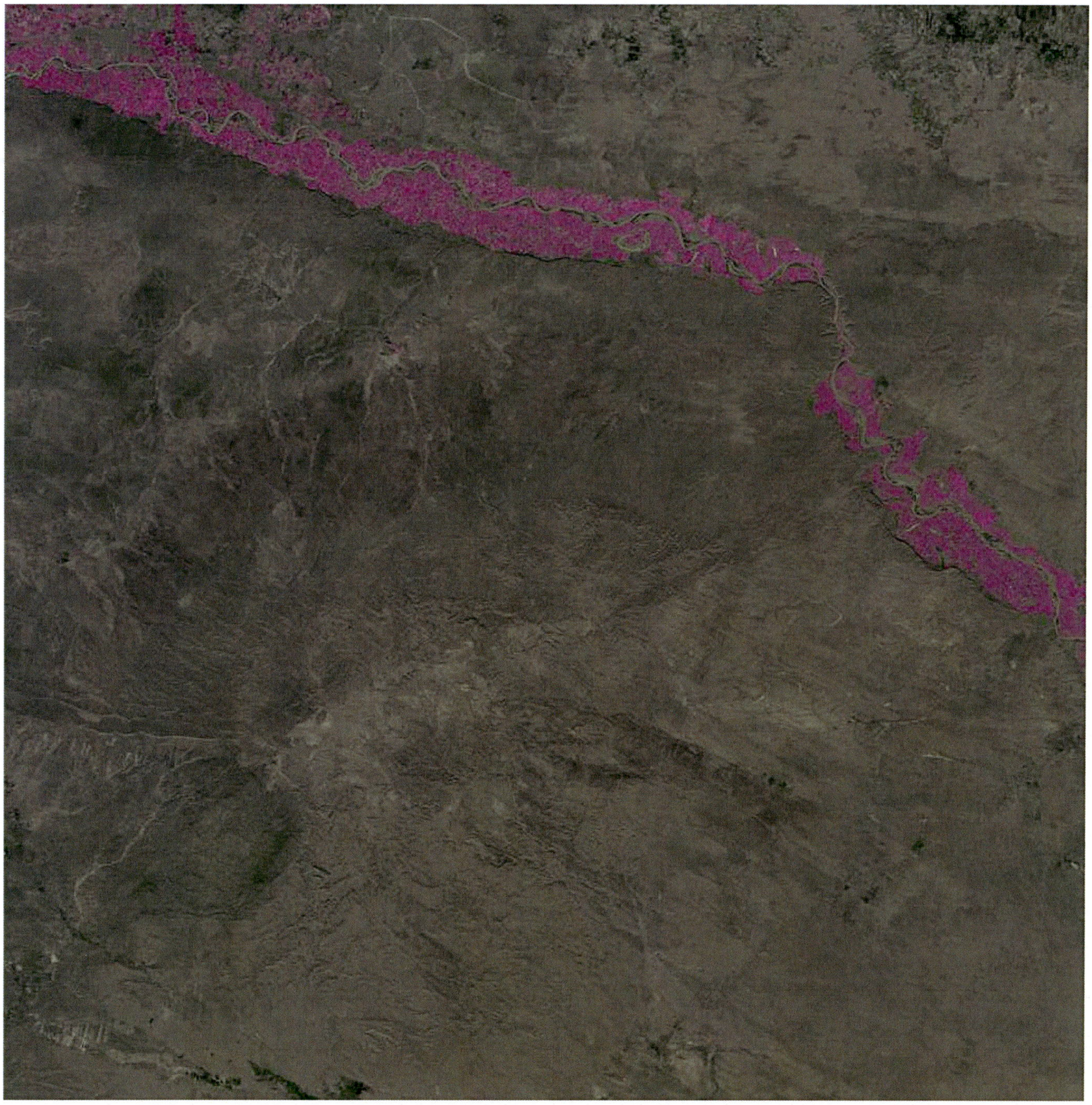

Fig. 22. A change image between a TM image, date 1.9.1990 and an ETM image, date 29.1.1999.

Degradation of the rangeland may be tied with changes in climate and human impact (UNCDD 1994). According to R.W. Dennell (1982: 51), grazing of sheep and goats may already have caused changes in the steppe vegetation cover in Syria and southwestern Iran, around 6000-4000 B.C. Ploughing with machinery and the easier application of the steppe for agricultural purposes have been regarded as factors that cause over-exploitation. But also drillings of wells for further grazing have deteriorated the situation. (Treacher 1993).

During the last 40 years degradation of the range land in general has been a particular problem in the Syrian steppe for which the Jebel Bishri desertification combating project set up by ACSAD with German co-operating parties is an example (see, e.g., Hoffman and Geerken 1997). Together with our own desertification analyses (Lönnqvist and Törmä 2004) these offer some case studies (see more in ch. 15). The differences in the distribution and movement of the sand veil can be clearly detected in the Landsat satellite images presented in ch. 2 of this volume. However, this area specified as Zone 5, like the drastically over-grazed steppe of Halabiya, still offered feather-grasses in the 1970s (Moore *et al.* 2000: 65).

The lack of vegetation cover increases erosion and therefore denudation. The heavy winds typical of the area move the sand and increase weathering in vulnerable soft stone materials such as lime stone, sand stone, gypsum and marble that are common in the region and can easily disintegrate. (See, e.g., NAPC 2006: 40). Wind erosion is typical of arid zones, and it is limited to dry, fine and loose sediments that are not protected by vegetation cover (Butzer 1964: 188). Morphological evidence goes beyond topographical evidence (Vita-Finzi 1978: 14) and has an important role in the study of landscape evolution. It is apparent that the summit areas of Jebel Bishri have been higher compared to the upland type of hills they are today as weathering has affected them. At Tar al-Sbai at the western edge of the mountain provides Palaeolithic rock shelters (*abris*), the roofs of which have collapsed due to erosion (see more in ch. 5.). Erosion also causes disturbance to archaeological finds and their locations; and has to be taken into account when interpreting depositional and post-depositional processes. (See, e.g., Wainwright 1992: 227-241; Banning 2002: 72-73). Artefact displacement is constant with sandstorms and water currents. The finds move along hill slopes that are aided by run-off mechanisms.

Because the long-term perspective belongs to the approaches of the Finnish project SYGIS to understand the environmental processes, attention was paid to the degradation of the range land and role of agricultural activities in the neighbourhood (see ch. 15.). Different satellite data sources from various years have provided possibilities to detect short-term changes in the desertification process of the mountain and river channel migration of the Euphrates. Comparisons have been made with old road alignments, changes in patterns of river channels, tent camp locations and shifting cultivation between the CORONA photographs and the Landsat images. They have provided indications to trace buried sites and old road networks. (Lönnqvist and Törmä 2004; 2006). The monitoring and comparison of environmental changes through the satellite data provide understanding of post-depositional processes, e.g., submerged layers, sand coverage or preservation of archaeological remains in the area. Monitoring these changes offers us the means to follow the development of the area, find new sites buried by environmental changes; and thus provide fresh choices for the planning of new strategies for the preservation of traditional economies and the cultures of the region (Lönnqvist and Törmä 2004).

ACSAD has been monitoring and seeking to encounter desertification in the area, by ploughings and tree plantation. In our studies ratio images CH7/CH1 were also computed from TM and ETM Landsat images, and their difference images were computed (Figs. 23. and 24.). Sandy soils should appear as bright areas in this channel ratio (cf. Hoffman and Geerken 1998). In the change image Fig. 24. the TM ratio image is shown using green primary colour and the ETM ratio image using red and blue primary colours. Therefore, areas which have been sandy in both images are shown as white, green areas correspond to increase of sandy areas in 1990 and reddish areas to increase of sandy areas in 1999. It should be noted that the seasons of images were different, so some changes are due to the seasonal differences in vegetation. Also, some detected changes can be due to the misregistration between images.

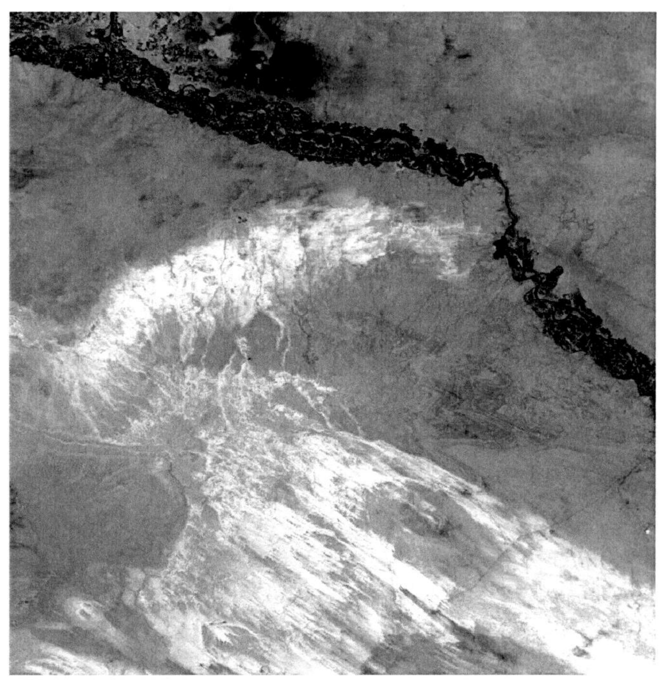

Fig. 23. Channel ratio 7/1 image computed from TM image, date 1.9.1990.

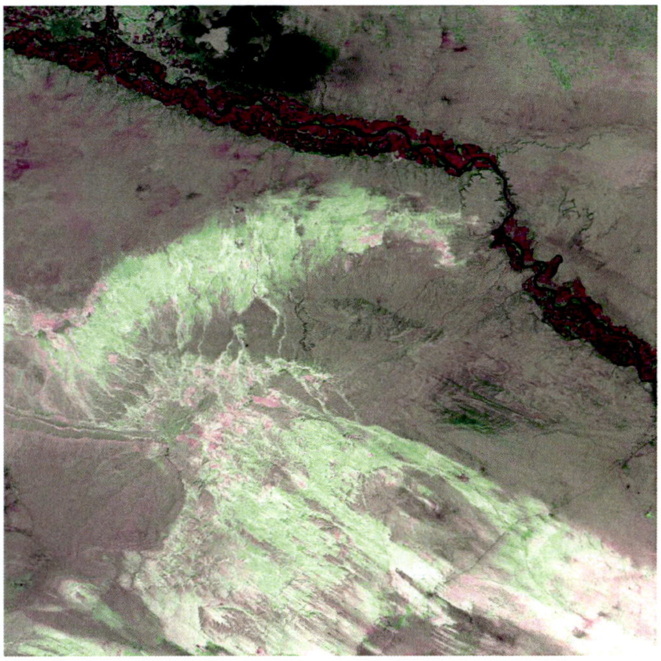

Fig. 24. A change image of channel ratio 7/1 computed from ETM image, date 29.1.1999 and TM image, date 1.9.1990.

However, the expansion of the sand cover has been most rapid during the past three decades monitored by ACSAD and SYGIS (Hoffman and Geerken 1998; Lönnqvist and Törmä 2004; 2006), and it had deteriorated the grazing grounds of the local nomads. Similar kinds of problems that are apparently not so rapid may have been faced by local people for thousands of years. It is known that when the vegetation of the steppe-type region is already scanty, the effects of grazing are clearly traceable in a short period. (Evans 1978; Evans and O'Connor 1999: 78). However, presently the rehabilitation programs of the steppe have had a positive impact for the grazing grounds of Jebel Bishri during the most recent years (ESCWA 2007; see ch. 15.).

While dealing with sustainable development the concept of sustainability denotes time scale that requires long-term thinking. Only archaeology can provide broader means for understanding the humans in their environment from a longer time perspective reaching from the prehistoric times to the present. Archaeology can offer background and models for planning sustainable development for future undertakings in fragile environments. The archaeological research of Jebel Bishri provides us with an indirect perspective on the environment from the cultural remains not only from the Pleistocene but also to the Holocene situations up till recent years.

4.7. Humans in Changing Environments of the Euphrates River Channel Area and its Terraces

Alluvial systems are special in their preservation mechanisms. Data concerning the changes in these systems is also important for archaeology. (Brown 1997: 2-3). The systems preserve information of climatic conditions and types of cultures. As previously mentioned, the northern and northeastern edges of Jebel Bishri follows the Euphrates river bank for over 100 km. The Euphrates is the most important and longest river in Syria stretching ca. 675 km in length through the country. Its water sources, however, are situated in Turkey and they cause high waters in the spring. Before the Tabqa dam was built in Syria in the 1970s the water flow of the river was ca. 840 m^3/s on average, the months with highest water flow being April and May. In spring when the snows melt in Anatolia that may cause a flow of up to 5000 m^3/s with the lowest readings taken in autumn (Wirth 1971: 109-110). The Euphrates *graben* was formed after the emergence of the Bishri block, and nowadays the river provides approximately 80 % of the water supply for Syria (ACSAD 2003).

The river valley forms a distinct spatial change in the landscape from the desert-steppe of the mountain. The river valley is also affected by degradation, and irrigation connected with pumping has caused ca. 17.6 % of the valley to be affected by salinization. Thus over 50 % of the valley is under salinization process that has resulted in thousands of hectares of arable land becoming useless. (ACSAD 2003; NAPC 2006: 40). Over half of the last century the irrigation technology has spread from the river valleys to the steppe, and it has changed the landscape of northeastern Syria in many ways (Hole and Zaitchik 2006).

The Euphrates *graben*, offers geomorphological evidence that dates the sedimentations to the Mesozoic, but mainly to the Cenozoic period. With ASTER DEM-tiles we constructed a cross-section of the Euphrates valley (Fig. 25.) at the village of Kharita ca. 22 km northwest of Deir ez-Zor, demonstrating the width of the river valley with terraces and the height of Jebel Bishri on the bank seen from the northwest. (Lönnqvist *et al.* 2007).

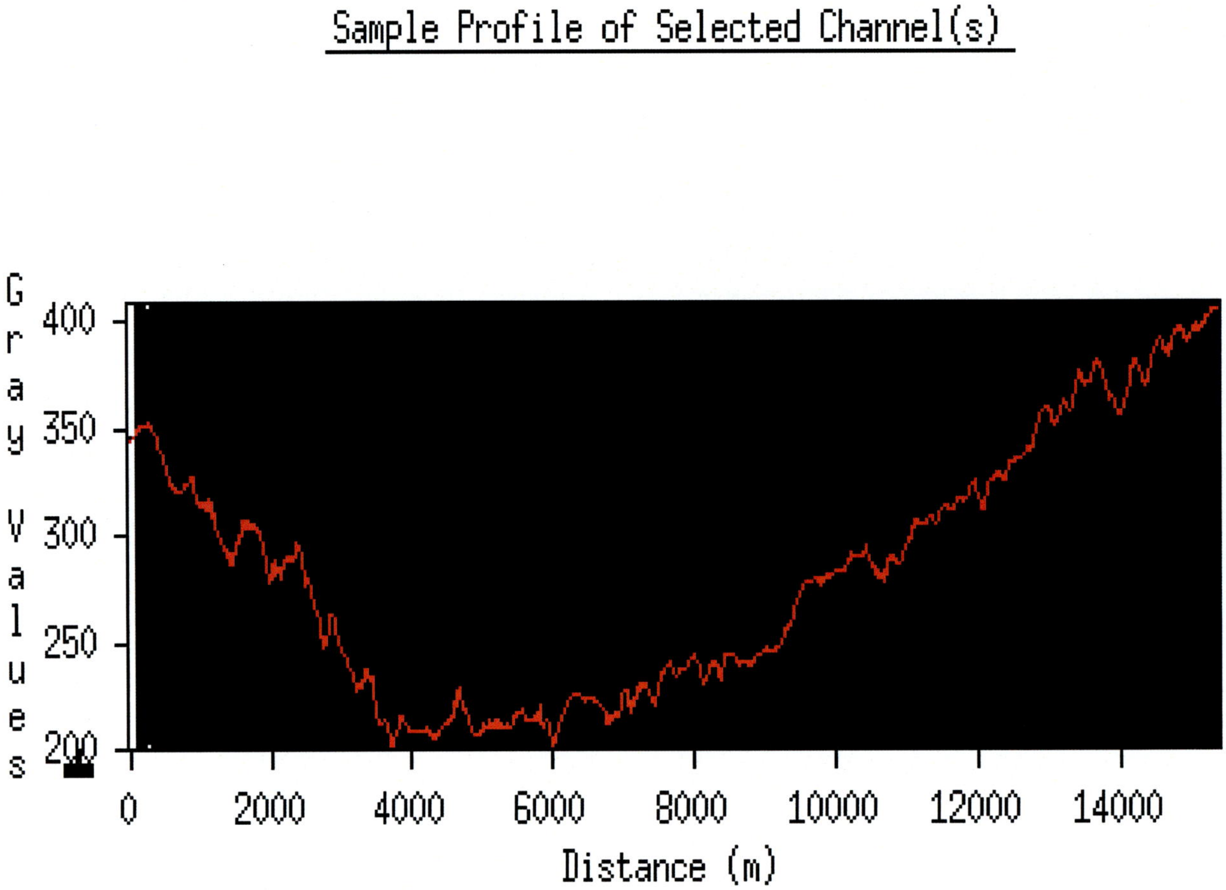

Fig. 25. A cross-section of the Euphrates valley seen from NW in the area of the village el-Kharita constructed by using ASTER-DEM data. Constructed by Markus Törmä and Jaakko Latikka 2007.

4.7.1 River Channel Changes

The Euphrates river channel changes are recognizable as part of the landscape evolution, and they affect the floodplain sedimentation process. These channel changes have altered the settlement life and agricultural endeavours throughout millennia. By comparing the CORONA satellite photographs from the 1960s with the Landsat images from the 1990s the changes of the river channel are clearly detectable during the past thirty years (Fig. 26.). The river valley is affected by degradation as well, and as mentioned, by salinization that causes problems for agriculture. The Tabqa dam, the construction of which was completed in the 1970s, has made changes to the river course and caused its flow to cease so that the rate and shape of the short-term changes between the 1960s and 1990s are not directly comparable with the ancient river channel changes caused by the natural water flow. (Lönnqvist *et al.* 2007).

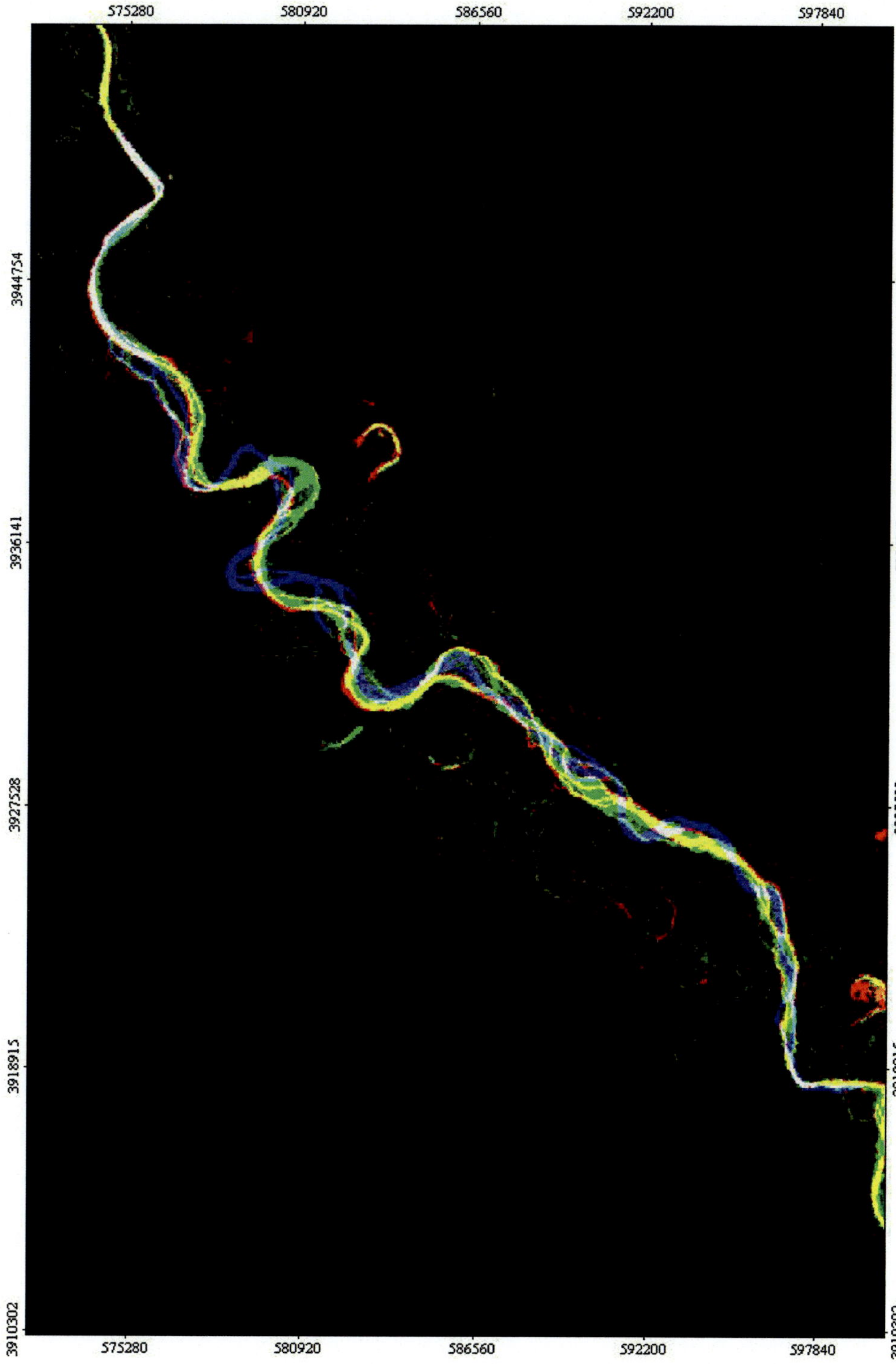

Fig. 26. The analysis of the river channel changes between the peninsula of Halabiya and the peninsula of 'Ayyāsh and in different decades covering a period of 35 years. CORONA satellite photographs from 1966 (blue colour), Landsat MSS 1983 (green colour) and ETM 2001 (red colour), over 25 years. The common channel alignments in 1983 and 2001 are expressed in yellow, in 1983 and 1966 in light green; and finally in 2001 and 1966 in purple. White colour shows the common alignment during all the years.

4.7.2 Terraces and Plateaus

Terrace formations also belong to floodplain evolution. Stream changes from aggradation to degradation cause incision into the old floodplain, forming a fluvial terrace. Terraces can be caused by climatic changes, changes in available sediments, changes in hydrological catchment, tectonic activity and base-level transformations. Floodplain "islands" or terraces can also submerge or disappear into silts when alluviation raises the floodplain. (Brown 1997: 31-39).

Fig. 27. A fluvial terrace, called a "plateau", including wadis connected to Jebel Bishri in the north. The green Euphrates valley can be seen in the background. Photo from S: Kenneth Lönnqvist 2003.

Fig. 28. Grazing on a higher plateau near Raqqa where Palaeolithic flints and Early Bronze pottery were identified on this Pleistocene terrace. Photo from SW: Kenneth Lönnqvist 2003.

In their prehistoric surveys concentrating on Palaeolithic finds on the Middle Euphrates J. Besançon and P. Sanlaville (1981) carried out geomorphological studies on the fluvial terraces between Raqqa and Deir ez-Zor. The Euphrates staircase of terraces between Birecik in Turkey and Abu Kemal in the Syrian-Iraqi border zone has been recently divided into six distinctive formations (Qf0-QfVI) (Sanlaville 2004: 128, e.g., Fig. 2.). The Middle Euphrates area following Jebel Bishri is associated with three of these major terraces QfI-QfIII: 1) the uppermost terrace is called a *plateau*, 2) a medium terrace is situated next towards the river channel and 3) a base terrace lies nearest to the river channel (see also Sanlaville 2004: 131, Fig. 7.). The *plateaus* (see Figs. 27. and 28.), formed during the Pleistocene, were widely used as grazing ground in the past just like today. In the Holocene era the Euphrates valley had gone through the latest geomorphological evolution: the accumulation of soils in the bottom of the alluvial area provided fertile land and possibilities for human occupation during the Neolithic period (Geyer and Besançon 1996: 8-9). These areas in the Middle Euphrates valley already bore distinctive names during the Amorite Age as testified in the Mari archives from the early second millennium B.C. (see Lafont 2000).

A British-Syrian team carried out geological investigations (Westaway *et al.* 2005: 31-33; Westaway *et al.* 2007) contemporaneously with our surveys on the right bank of the Euphrates in 2005-2007. The aim of these British-Syrian studies was to date the Middle Euphrates terrace deposits. On the basalt peninsulas of ʻAyyāsh and Halabiya protruding from Jebel Bishri the team has redated deposits with different radiometric methods: 1) argon-argon (Ar-Ar), 2) uranium-series (U-series) dating and 3) optically-stimulated luminescence (OSL). The datings redefine the chronology of the Lower and Middle Palaeolithic artefacts discovered in the area. For recordings the dates for the banks of the Euphrates at Halabiya the team obtained ~ 2800-2100 ka equivalent to QfIII. For the village terrace of ʻAyyāsh the obtained dating for QfI is ~425-415 ka, being far older than was previously suggested stratigraphically, namely ~140-70 ka. The new dating provides the *terminus post quem* for the Lower Palaeolithic finds associated with ʻAyyāsh and partly for the area of ʻAin Abu Jimā.

Ancient structures that were indicated by concentrations of pottery were traced with Ground Penetrating Radar (GPR) by our survey teams on the plateau of **H 1.** It was expected that some of the structures had submerged and disappeared in the accumulated sediment. These post-depositional processes drastically differ from the evidence of surface sites in the desert-steppe areas where sedimentation was meagre.

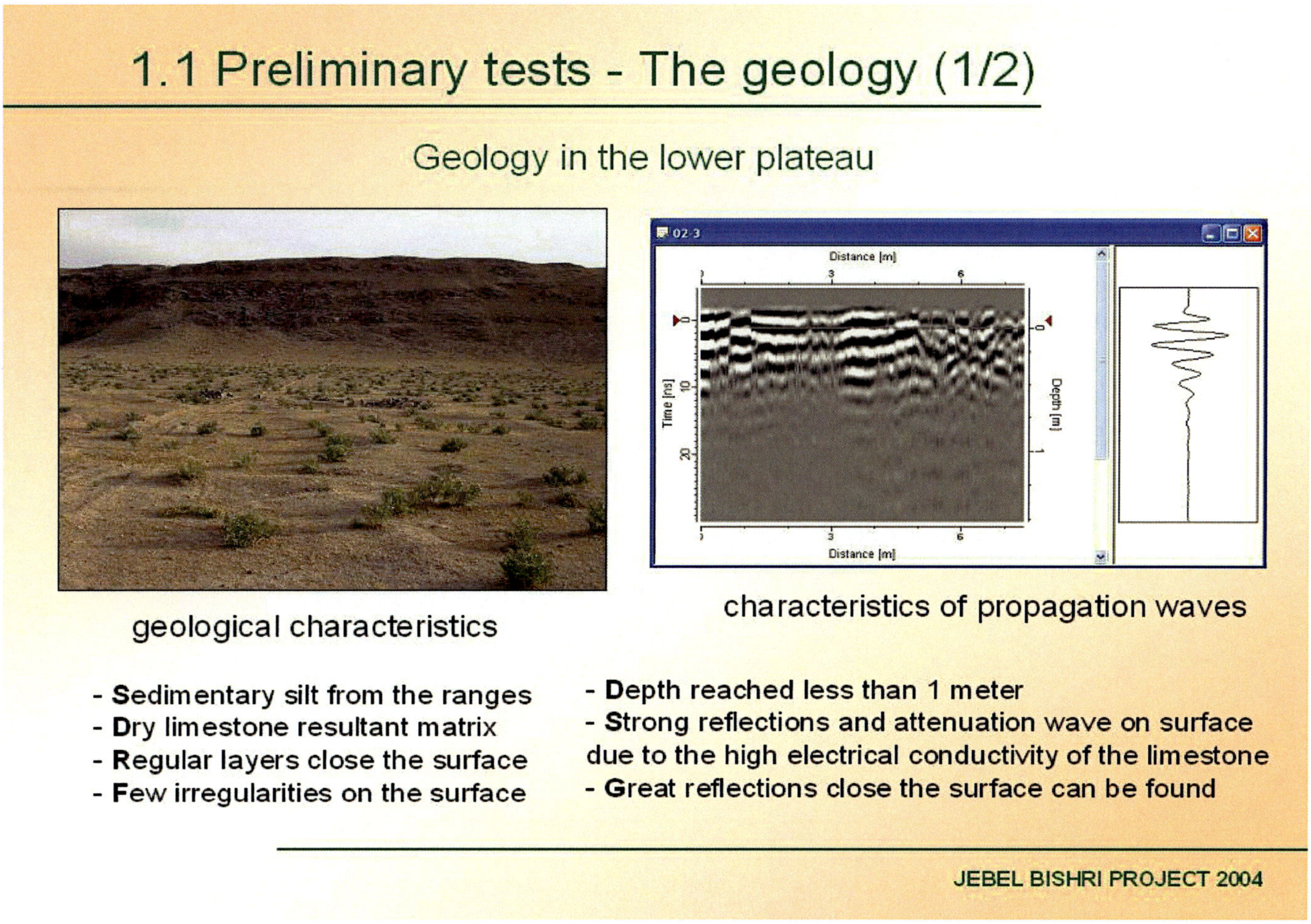

Fig. 29. Analyses of the plateau H 1 soils with ground penetrating radar by Josep Pedres Rodez.

4.8. Conclusions

Jebel Bishri clearly provides a spatial boundary between the Euphrates river and the Syrian Desert. This table mountain called the "Bishri block" is a northeastern plunging formation in the Palmyrides and has been connected to the formation of this mountain belt and the cultures surrounding it. Nowadays it belongs to the Greater Southwest Asian Arid Zone, and permanent rain-fed cultivation between the 100 mm to 200 mm isohyets is practically impossible, when the precipitation reaches only 200 mm once in a dozen years. The spatial boundary applies to environmental contrasts that are well visible in the landscape changing from a desert-steppe through steppe to alluvial and irrigated agricultural areas as well as from plains to mountains and valleys. The Euphrates, with its changing river course and different terrace formations, was also clearly a convenient zone for fishing from the earliest times to the present. The river fans became used for agriculture already in the Late Epipalaeolithic and Early Neolithic period, and breeding grounds for domesticated animals along the Neolithic period. Here and there in the area of Jebel Bishri, including the southern, western and eastern piedmonts, are oases that are fed by springs and wells.

The contrasts have naturally affected the economy of the local people for millennia. During the Pleistocene the piedmonts, especially to the west, seem to have been providing small rivers, some apparently were seasonal. This is an area of gazelle movement, and the oasis of El Kowm clearly attracted animals as well as early human hunters. It seems that during the Pleistocene there were moister and lacustrine conditions which are also represented in the geological formation of Nadra, the site of a palaeolake in the central eastern area. The palaeolake could have existed in the Middle Palaeolithic period, and it probably provided some fishing opportunities. However, at the beginning of the Holocene the continuous forest did not reach Jebel Bishri.

The Holocene may only have comprised some changes in the environment from woodland steppe to the present desert-steppe on the mountain of Jebel Bishri. Archaeological evidence in the coming chapters will provide some additional but tentative and indirect evidence of the circumstances in different periods through the oscillation of subsistence economies and cultural remains. The data indirectly illustrates the possibilities for hunting, gathering, pastoralism and agriculture in the neighbourhood during various periods.

Bibliography

ACSAD (2003) *Inventory Study and Regional Database on Sustainable Vegetation Cover Management in West Asia (TN 2).* The Arab Center for Studies of Arid Zones and Dry Lands. Damascus: ACSAD.

Akazawa, T. (1984) The Ecology of the Middle Paleolithic Occupation at Douara Cave, Syria, in *Paleolithic Site of Douara Cave and Paleogeography of Palmyra Basin in Syria, Part IV: 1984 Excavations*, ed. by Akazawa, T. and Sakaguchi, Y., The University Museum, The University of Tokyo, Bulletin No. 29, Tokyo.

Akkermans, P.M.M.G. and Schwartz, G.M. (2003) *The Archaeology of Syria: From Complex Hunter-Gatherers to Early Urban Societies (ca. 16,000-300 BC)*. Cambridge World Archaeology. Cambridge: Cambridge University Press.

Algaze, G. and Poumelle, J. (2003) Climatic Change, Environmental Change, and Social Change at Early Bronze Age Titriş Höyük: Can Correlation and Causation, in *From Villages to Towns, Studies Presented to Ufuk Ensin*, ed. by Özdoğan, M., Hauptmann, H., Başgelen, Istanbul: Arkeoloji ve Sanat Publications, pp. 103-128.

Arz, H.W., Lamy, F. and Paetzold J. (2004a) Multidecadal to Centennial-scale Climate Variability in the Holocene as Recorded in Northern Red Sea Sediments, in *American Geophysical Union, Fall Meeting 2004, abstract* #U42A-01.

Arz, H., Pätzold, J. and Müller, P. (2004b) Multi-decadal to Centennial Climate Variability in the Mid-Holocene and the 4.2 ka Event as Recorded in Shaban Deep Brine Sediments, Northern Red Sea, in *4ICAANE, 4th International Congress on the Archaeology of the Ancient Near East, 29 March- 3 April 2004, Abstracts*, p. 83.

Avner, U., Carmi, I. and Segal, A. (1994) Neolithic to Bronze Age Settlement of the Negev and Sinai in Light of Radiocarbon Dating: A View from the Southern Negev, in *Late Quaternary Chronology and Paleoclimates of the Eastern Mediterranean*, ed. by Bar-Yosef, O. and Kra, R.S., Ann Arbor, Michigan: Radiocarbon, pp. 265-300.

Banning, E.B. (2002) *Archaeological Survey.* Manuals in Archaeological Method, Theory, and Technique. New York: Kluwer Academic/ Plenum Publishers.

Bar-Matthews, M. and Ayalon, A. (2004) Evidence from Speleothem Isotopic Records for Abrupt Climatic Changes During the Last 12,000 Years and their Impact on Human Settlements in the Eastern Mediterranean Region, in *4ICAANE, 4th International Congress on the Archaeology of the Ancient Near East, 29 March - 3 April 2004, Abstracts*, p. 83.

Bar-Yosef, O. (1987) Late Pleistocene Adaptations in the Levant, in *The Pleistocene Old World, Regional Perspectives, Interdisciplinary Contributions to Archaeology*, ed. by Soffer, O., New York – London: Plenum Press, pp. 210-236.

Bar-Yosef, O. (1998) The Natufian Culture in the Levant: Threshold to the Origins of Agriculture, in *Evolutionary Anthropology*, Vol. 6/5, pp. 159-177.

Beaumont, P. (1993) *Drylands: environmental management and development.* Routledge: London.

Besançon, J., Copeland, L., Hours, F., Muhesen, S. and Sanlaville, P. (1981) Le Paléolithique d'El Kowm, Rapport préliminaire, in *Paléorient,* Vol. 7/1, pp. 33-55.

Besançon, J. and Sanlaville, P. (1981) Aperçu Géomorphologique sur la Vallée de l' Euphrate Syrien, in *Paléorient,* Vol. 7/2, pp. 5-18.

Besançon, J. and Sanlaville, P (1991) Une Oasis dans la Steppe Aride Syrienne: La Cuvette d'El Kowm au Quaternaire, in *Cahiers de l'Euphrate* 5-6, pp. 11-32.

Besançon, J., Delgiovine, A., Fontugne, M., Lalou, C., Sanlaville, P. et Vaudour, J. (1997) Mise en Evidence et Datation de Phases Humides du Pléistocene Supèrieur dans la Région de Palmyre (Syrie), in *Paléorient*, Vol. 23, pp. 5-24.

Boëda E., Geneste J.M., Griggo C., Mercier N., Muhesen S., Reyss J.L., Taha A. and Valladas H. (1999) A Levallois point embedded in the vertebra of a wild ass: hafting projectiles and Mousterian hunting weapons, in *Antiquity*, Vol. 73, pp. 394-402.

Bottema, S. and Van Zeist, W. (1981) Palynological Evidence for the Climatic History of the Near East, 50, 000 – 6,000 BP, in *Préhistoire du Levant: Chronologie et organisation de l'Espace depuis les Origines jusqu'au VIe millénaire,* Colloques Internationaux du Centre National de la Recherche Scientifique No. 598., ed. by Cauvin, J. and Sanlaville, P., Paris: Editions du CNRS, pp. 111-132.

Brew, G. , Best, J., Barazangi, M. and Sawaf, T. (2003) Tectonic Evolution of the NE Palmyride mountain Belt, Syria: the Bishri Crustal Block, in *Journal of the Geological Society,* Vol. 160/5, pp. 677-685.

Brooks, R.R. and Johannes, D. (1990) *Phytoarchaeology.* Historical, Ethno- & Economic Botany Series. Vol. 3. T. R. Dudley general ed. Portland, Oregon: Dioscorides Press.

Brown, A.G. (1997) *Alluvial Geoarchaeology: Floodplain archaeology and environmental change.* Cambridge Manuals in Archaeology. Cambridge: Cambridge University Press.

Butzer, K. (1964) *Environment and Archaeology: An Introduction to Pleistocene Geography.* London: Methuen & Co.

Butzer, K. (1982) *Archaeology as Human Ecology:* Method and Theory for a Contextual Approach. Cambridge: Cambridge University Press.

Cauvin, J. (2000) *The Birth of the Gods and the Origins of Agriculture.* Trans. Watkins, T. Cambridge: Cambridge University Press.

Copeland, L. (2004) The Paleolithic of the Euphrates Valley in Syria, in *From the River to the Sea, The Palaeolithic and the Neolithic on the Euphrates and the Northern Levant,* Studies in honour of L. Copeland, ed. by Aurenche, O., La Mière, M. and Sanlaville, P., BAR International Series, No. 1263, Oxford: Archaeopress, pp. 19-61.

Courty, M.-A., Boëda, E., Emery-Barbier, A., Griggo, C., Mercier, N., Théry-Parisot, I. and Valladas, H. (2006) The Relative Timing of Human Occupation and Palaeoenvironmental Changes in the High-resolution Sequence of Umm El Tlel: A Palaeolithic site in the Syrian desert, in *Geophysical Research Abstracts,* Vol. 8, 01815.

Courty, M.-A. and Weiss, H. (1997) The Scenario of Environmental Degradation in the Tell Leilan Region, NE Syria, during the Late Third Millennium Abrupt Climate Change, in *Third Millennium B.C. Climate Change and Old World Collapse,* ed. by Nüzhet Dalfes, H., Kukla, G. and Weiss, H., NATO ASI (Advanced Science Institutes) Series, Berlin –New York: Springer, pp. 107-147.

Dennell, R.W. (1982) Archaeology and the Study of Desertification, in *Desertification and Development: Dryland Ecology in Social Perspective,* ed. by Spooner, B. and Mann, H.S., London: Academic Press, pp. 43-60.

Dincauze, D.F. (2000) *Environmental archaeology: Principles and Practice.* Cambridge: Cambridge University Press.

Eastwood, W.J., Leng, M.J., Roberts, N. and Davis, B. (2006) Holocene climate change in the eastern Mediterranean region: a comparison of stable isotope and pollen data from Lake Gölhisar, southwest Turkey, in *Journal of Quaternary Science,* Vol. 22, pp. 327-341.

ESCWA (2007) = *Economic and Social Commission for Western Asia. Land Degradation Assessment and Preventation: Selected Case Studies from the Escwa Region.* New York: United Nations.

Evans, J.G. (1978) *An Introduction to Environmental Archaeology.* London: Elek Books Ltd.

Evans, J. and O'Connor, T. (1999) *Environmental Archaeology: Principles and Methods.* Midsomer Norton, Somerset: Sutton Publishing.

Garrard, A.N. and Stanley-Price, N.P. (1977) A Survey of Prehistoric Sites in the Azraq Basin, Eastern Jordan, in *Paléorient,* Vol. 3, 1975-1977, pp. 109-123.

Geyer, B. and Besançon, J. (1996) Environnement et occupation du sol dans la vallée de l'Euphrate Syrien durant le Néolithique et le Chalcolithique, in *Paléorient,* Vol. 22/2, pp. 5-15.

Grabar, O., Holod, H., Knustad, J. and Trousdale, W., *et al.* (1978) *City in the Desert: Qasr al-Hayr East.* An Account of the Excavations Carried out at Qasr al-Hayr East on Behalf of the Kelsey Museum of Archaeology at the University of Michigan, with the Help of Harvard University and the Oriental Institute of Chicago. Vols. I and II. Harvard Middle Eastern Monographs 23-24. Cambridge, Mass: Harvard University Press.

Griggo, C. (1998) La Faune Moustérienne du Site d'Umm el Tlel (Syrie): Étude Préliminaire, Fouilles 1991-1993, in *Cahiers de l'Euphrate,* 8, pp. 11-25.

Heathcote, R.L. (1983) *The Arid Lands: Their Use and Abuse*. Themes in resource management. London and New York: Longman.

Henry, D.O. (1989) *From Foraging to Agriculture: The Levant at the End of the Ice Age*. Philadelphia: University of Pennsylvania Press.

Héraud, C. (original 1922, reprint 1988) *Une Mission de Reconnaissance de l'Euphrate en 1922*. Première Partie: Les Cartes. Damas: Institut Français de Damas.

Hoffman, K. and Geerken, R. (1997) Degredations- und Desertifikationsprozesse in den Bishri-Bergen Syriens, in *Spektrum der Wissenschaft*, No. 2, pp, 71-73.

Hoffman, K. and Geerken, R. (1998) Multisensor analysis of desertification processes in the Bishri Mountains (Syria), in *Future Trends in Remote Sensing*, ed. by Gundmandsen, P., Rotterdam: Balkema, pp. 73 - 76.

Hole, F. and Zaitchik, B.F. (2006) Policies, plans, practice, and prospects: irrigation in northeastern Syria, in *Land Degradation and Development*, Vol. 18/2, pp. 133-152.

King, F.B. and King, J.E. (1996) Interdisciplinary Approaches to Environmental Reconstruction, An Example from the Ozark Highland, in *Case Studies in Environmental Archaeology*, ed. by Reitz, E.J., Newsom, L.A., Scudder, S.J., Interdisciplinary Contributions to Archaeology, New York: Plenum Press, pp. 71-85.

Kusky, T. and El-Baz, F. (2000) Neotectonics and fluvial geomorphology of the Northern Sinai Peninsula, in *Journal of African Earth Sciences*, Vol. 31, pp. 213-235.

Lafont, B. (2000) Irrigation Agriculture in Mari, in *Rainfall and Agriculture in Northern Mesopotamia*, (MOS Studies 3), Proceedings of the Third MOS Symposium (Leiden 1999), ed. by Jas. R.M., Publications de l'Institut historique-arhéologique néerlandais de Stamboul, Vol. LXXXXVIIII, sous la direction Mellink, M.J., de Roos, J., Roodenberg, J.J. et Veenhof, K.R. Istanbul: Nederlands Historisch-Archaeologisch Instituut te Istanbul, pp. 129-145.

Le Tensorer, J.-M., Muhesen, S. and Schmid, P. (2006) *Research on the Paleolithic of the El Kowm Area*. Abstract.

Le Tensorer, J.-M, Jagher, R., Rentzel, P., Hauck, T., Ismail-Meyer, K., Pümpin, C. and Wojtczak, D. (2007) Long-term Site Formation Processes at the Natural Spring Nadaouiyeh and Hummal in the El Kowm Oasis, Central Syria, in *Geoarchaeology*, Vol. 22, pp. 621-639.

Legge, A.J. and Rowley-Conwy, P.A. (1987) Gazelle Killing in Stone Age Syria, in *Scientific American*, Vol. 257, pp. 76-83.

Lönnqvist, M. and Törmä, M. (2004) Different Implications of a Spatial Boundary, Jebel Bishri between the Desert and the Sown in Syria, in *The ISPRS XXth Congress Proceedings*, ed. by Altan, M.O., *The ISPRS International Archives of the Photogrammetry, Remote Sensing and Spatial Information Sciences*, XXXVB, Istanbul, pp. 897-902.

Lönnqvist, M. and Törmä, M. (2006) Observing Changes and their Effects in Nomadic Environment, Remote-sensing and Archaeology for Sustainable Development on Jebel Bishri in Syria, in *Environmental Geoinformatics and Modelling, Proceedings of ICEM '05: International Conference on Environmental Management (28-30 Oct. 2005*) IV, ed by Reddy, M.A., Hyderabad.

Lönnqvist, M., Törmä, M., Lönnqvist, K., Nuñez, M., Okkonen J. and Latikka, J. (2007) The Euphrates Channel Changes and Archaeology along Jebel Bishri in Syria, in *Proceedings of the XXI Symposium, CIPA 2007, Athens, Anticipating the Future of the Cultural past*, Vol. 1, ed. Georgopoulos, A., the International Archives of Photogrammetry, Remote Sensing and Spatial Information Sciences, Vol. XXXVI-5/C53, pp. 465-470.

McLaren, S., Reynolds, T., Gilbertson, D., Grattan, D., Hunt, C., el-Rishi, H., Barker, G. and Duller, G. (2007) Pleistocene environments and human settlement, in *Archaeology and Desertification: The Wadi Faynan Landscape Survey, Southern Jordan*, by Barker, G., Gilbertson, D.D. and Mattingly, D.J., Wadi Faynan Series 2, Levant Supplementary Series 4, Oxford: Oxbow Books, pp. 177-198.

Moore, A.M.T. (1991) Abu Hureyra 1 and the Antecedents of Agriculture on the Middle Euphrates, in *The Natufian Culture in the Levant*, ed. by Bar-Yosef, O. and Valla, F.R., International Monographs in Prehistory, Archaeological Series 1, Ann Arbor, Michigan: International Monographs in Prehistory, pp. 277-294.

Moore, A.M.T., Hillman, G.C. and Legge, A.J. (2000) *Village on the Euphrates: From Foraging to Farming at Abu Hureyra*. Oxford: Oxford University Press.

Mouterde, R. and Poidebard, A. (1945) *Le Limes de Chalcis: Organisation de la steppe en Haute Syrie Romaine.*

Bibliothèque archéologique et historique. Tome XXXVIII. Documents aériens et épigraphiques. Texte and Atlas. Paris: Paul Geuthner.

Muhesen, S., Le Tensorer, J.-M. and Abdul Rahman, A. (2006) Recent Palaeolithic Discoveries from el-Kowm (Hummal) Syria, in *Abstracts of the 5th International Congress on the Archaeology of the Near East, 3-8 April 2006*, p. 49.

Musil, A. (1928) *Palmyrena: A Topographical Itinerary*. American Geographical Society. Oriental Explorations and Studies No 4. New York: American Geographical Society.

NAPC (2006) = Ministry of Agriculture and Agrarian Reform, NAPC, National Agricultural Policy Center, Working Paper No. 2, Agricultural Research and Extension in Syria: Present Situation and Policy Recommendations with the Support of Project GCP/SYR/006/ITA – Phase II.

Neumann, F., Schölzel, C., Litt, T., Hense, A. and Stein M. (2007) Holocene vegetation and climate history of the northern Golan heights (Near East), in *Vegetation History and Archeobotany*, Vol 16, pp. 329-346.

Peltenburg, E. (2007) Diverse Settlement Pattern Changes in the Middle Euphrates Valley in the Later Third Millennium B.C.: the Contribution of Jerablus Tahtani, in *Sociétés humaines et changement climatique à la fin du troisième millénaire: une crise a-t-elle eu lieu en Haute Mésopotamie?* Actes du Colloque de Lyon, 5-8 décembre 2005, ed. by Kuzucuoğlu, C. and Marro, C., Varia Anatolica XIX, Institut Français d'Etudes Anatoliennes - Georges Dumezil, Paris: De Boccard, pp. 247-266.

Perrot, J. (1993) Remarques Introductives, in *Paléorient*, Vol. 19/1, pp. 9-21.

Poidebard, A. (1934) *La Trace de Rome dans le Désert de Syrie*: Le Limes de Trajan a la conquête arabe. Recherches aériennes (1925-1932). Texte. Atlas. Bibliothèque Archéologique et Historique. Tome XVIII. Paris: Geuthner.

Reitz, E.J., Scudder, S.J., and Scarry C.M. (1996) *Case Studies in Environmental Archaeology*. Interdisciplinary Contributions to Archaeology. New York: Springer.

Rendell, H. (2000) Tectonic frameworks, in *Arid Zone Geomorphology: Process, Form and Change in Drylands*, ed. by Thomas, D.S.G., Second edition, Chichester: Wile & Sons, pp. 13-22.

Rossignol-Strick, M. (1993) Late Quaternary Climate in the Eastern Mediterranean, in *Paléorient*, Vol. 19/1, pp. 135-152.

Sanlaville, P. (2000) *Le Moyen-Orient arabe. Le milieu et l'homme*. Paris: Armand Colin.

Sanlaville, P. (2002) Changements climatiques et évolution des plaines alluviales dans le sud du Levant durant les stades isotopiques 5 (125-75 Ka) et 4 (75-55 Ka), in *Paléorient*, Vol. 28, pp. 15-26.

Sanlaville, P. (2004) Les Terasses Pléistocènes de la Vallée de l'Euphrate en Syrie et dans l'Extrême du Sud de la Turquie, in *From the River to the Sea, The Palaeolithic and the Neolithic on the Euphrates and in the Northern Levant, Studies in honour of Lorraine Copeland*, ed. by Aurence, O., Le Mière, M. and Sanlaville, P., Maison de L'Orient et de la Méditerrannée Jean Polloux, BAR International Series 1263, Oxford: Archaeopress, pp. 115-127.

Thirgood, J.V. (1981) *Man and the Mediterranean Forest: A History of Resource Depletion*. New York: Academic Press.

Treacher, T.T. (1993) Policy Issues in Livestock Production in Arid Regions and the Management of Extensive Grazing Lands, in *Strategies for Sustainable Animal Agriculture in Developing Countries, Proceedings of the Expert Consultation on Strategies for Sustainable Animal Agriculture in Developing Countries,* Rome, Italy, 10-14 December 1990, ed. By Treacher, T.T. and Mack, S., Rome: FAO (Animal Production and Health Division), pp. 99-161.

UICN = The World Conservation Union, North Africa Program, Education and Biodiversity Conservation Project. Paper 34E. *Steppes*.

UNCCD 1994 = United Nations Convention to combat desertification in countries experiencing serious drought and/or desertification, particularly in Africa. A/AC.241/27. Paris.

UNEP = United Nations Environment Program.

Wainwright, J. (1992) Assessing the Impact of Erosion on Semi-arid Archaeological Sites, in *Past and Present Soil Erosion, Archaeological and Geographical Perspectives*, ed. by Bell, M. and Boardman, J., Oxford: Oxbow Monograph 22, pp. 227-241.

Van Zeist, W. and Bottema, S. (1982) Vegetational History of the Eastern Mediterranean and the Near East during the Last 20.000 Years, in *Palaeoclimates,*

Palaeoenvironments and Human Communities in the Eastern Mediterranean Region in Later Prehistory, Part ii, ed. by Bintliff, J.L. and van Zeist, W., Oxford: BAR International Series, No. 133 (ii), pp. 277-323.

Webster (1994) = *Webster's New World Dictionary*. UK: Simon & Schuster.

Westaway, R., Bridgland, D., Challis, K. and White, M. (2005) A Dating Framework for the Lower and Middle Palaeolithic in Levantine Rivers, in *CBRL 2005, Newsletter of the Council for the British Research in the Levant,* pp. 31-33.

Westaway, R., Bridgland, D., Daoud, M., Radwan, Y., Abu Romieh, M., Yassminh, R. and Al Ashkar, A. (2007) Investigations of Euphrates terraces and related archaeology, NE Syria, 2006-2007, in *CBRL 2007, Newsletter of the Council for the British Research in the Levant 2*, pp. 82-89.

Williams, J.K. (2006) The Levantine Aurignacian: a closer look, in *Towards a Definition of the Aurignacian, Proceedings of a Symposium Held in Lisbon, Portugal, June 25-30, 2002,* ed. by Bar-Yosef, O. and Zilhão, J., Trabalhos de Arqueologia 45, Lisboa: Instituto Portugués de Arqueologia, pp. 317-352.

Wirth, E. (1971) *Syrien: Eine Geographische Landeskunde.* Wissenschaftliche Länderkunden. Band 4/5. Darmstad: Wissenschaftliche Buchgesellschaft.

Vita-Finzi, C. (1978) *Archaeological Sites in their Setting.* London: Thames & Hudson.

5. A PERSPECTIVE ON SPATIO-TEMPORAL MOBILITY OF HUNTERS AND GATHERERS

Minna Lönnqvist, Milton Nuñez, Kenneth Lönnqvist, Markus Törmä with Contributions by Margot Stout Whiting, Jari Okkonen and Michael Herles

Fig. 1. Tar al-Sbai at the western edge of Jebel Bishri. Photo from W: Jean-Pierre Matar 2000.

5.1. Palaeolithic Sites on the Western Piedmont

The western piedmont of Jebel Bishri offers countless Palaeolithic sites with tools chiefly representing Achaeulian and Mousterian traditions of the Lower Palaeolithic (ca. 450,000-200,000 BP) and Middle Palaeolithic (200,000-45,000 BP) periods (e.g., Besançon *et al.* 1981; see the datings in McLaren *et al.* 2007) comparable to the finds from the neighbouring El Kowm basin. At the sites such as Hummal, Qdeir and Umm el-Tlel in the basin the Upper Palaeolithic period is also testified (Le Tensorer *et al.* 2006; Le Tensorer *et al.* 2007: 628; Ploux and Soriano 2003). In the previous chapter we already referred to the high quality flint that is occurring in natural deposits in the region conveniently at disposal for the prehistoric inhabitants. We have dubbed the western piedmont area as an ancient "Silicon Valley" with its distributions littered with stone artefacts as evidence of an ancient technological "hub". The beginning of the Lower Palaeolithic in the region has extended according to the latest pebble tool finds discovered in recent years, probably to 1 million years back in time (Le Tensorer *et al.* 2006). Therefore the depth of cultural past is felt surrounding the explorers on the plains beneath Jebel Bishri.

Fig. 2. A view from Tar al-Sbai towards the western piedmont. Photo: Jean-Pierre Matar 2000.

During our survey of the year 2000 on the western piedmont a Palaeolithic site **A 1** was identified between

Qasr al-Hayr ash-Sharqi and the Syrian Department of Antiquities Guest House, a few hundred metres north of the Qasr. Some find samples from Palaeolithic sites ca. 5.5 km northeast of Qasr al-Hayr ash-Sharqi in the area of Wadi Suq were also recorded. They are the sites **A 8** – **A 11** and **B 2**. In the area of the modern Bedouin village of Shanhas at the courtyard complex **A 6a** (see more in ch. 15) ca. 1.6 km northeast of the Qasr we also detected past Palaeolithic activities on the limestone outcrops of this piedmont area.

Fig. 3. The locations of El Kowm, Qasr al-Hayr and Tar al-Sbai dealt with in the present chapter. Mapping: Minna Lönnqvist on Landsat-7 ETM image. Eurimage ©.

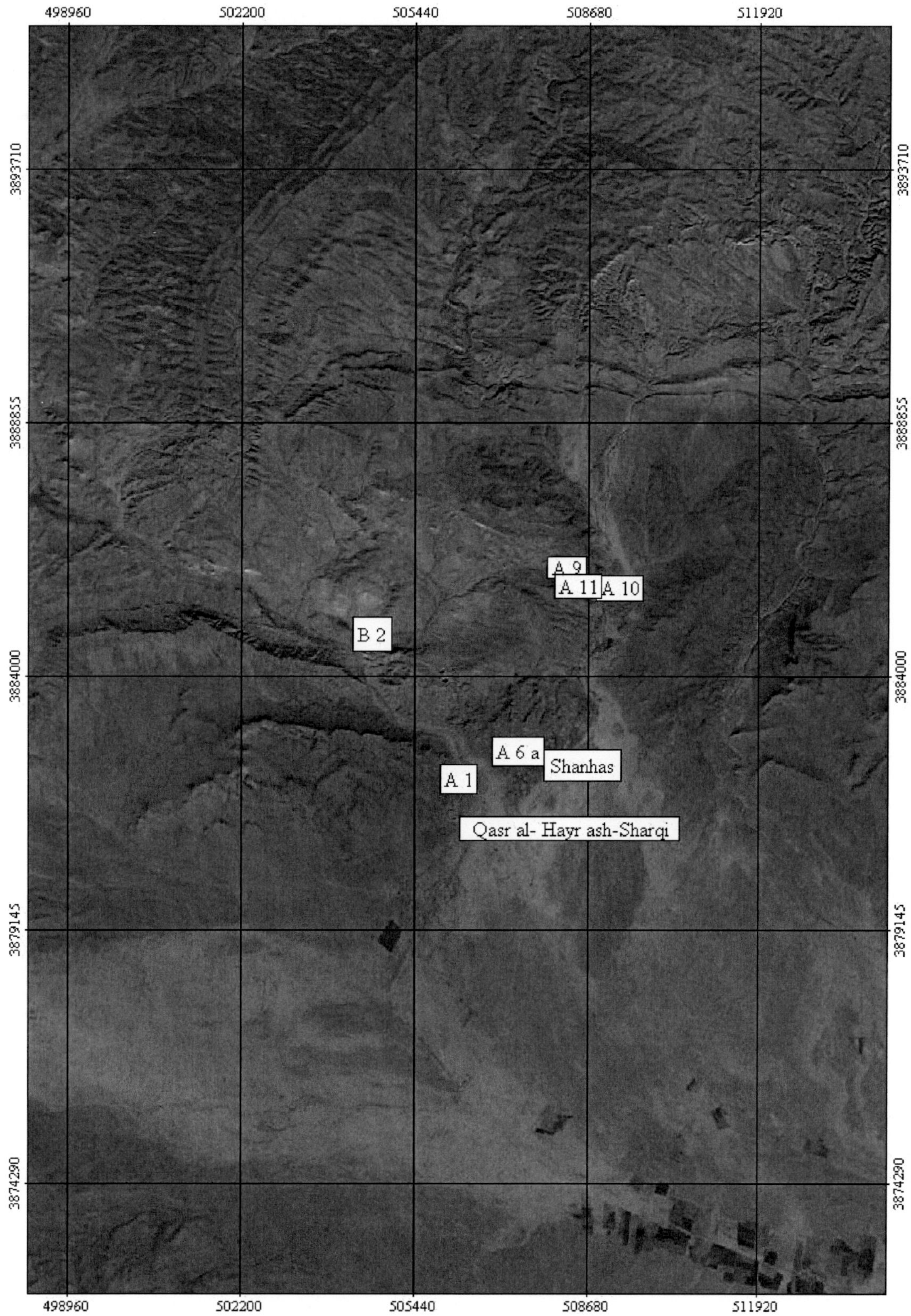

Fig. 4. Palaeolithic sites around Qasr al-Hayr ash-Sharqi on the western piedmont of Jebel Bishri. Scale: ca. 1:50,000. Mapping by Minna Lönnqvist 2009.

5.1.1 Prehistoric Sites on the Western Piedmont

Site A 1
Open Air Site

UTM: 0505934, 3881500, ca. 390 m a.s.l.
Size: 119 m x 27 m

Site **A 1** is an open air site situated on a very low hill with a *wadi* running on its eastern side. It produced tools representing the Middle Palaeolithic period. The local variations of the Early Middle Palaeolithic period are known as the Hummalian and Yabrudian traditions typical of the El Kowm basin area (cf. Besançon *et al.* 1981; Le Tensorer *et al.* 2006). In the tools from site A 1 there are one biface, 11 scrapers, one borer-scraper, one burin, one retouched side-scraper, one denticulated tool as well as four cores and eight flakes. Several tools seem to have been influenced by the Yabrudian tradition.

Fig. 5. Worked flint finds from site A 1. Photo: Milton Nuñez 2004.

Fig. 6. Worked flint finds from site A 1. Photo: Milton Nuñez 2004.

Fig. 7. Worked flint finds from site A 1. Photo: Milton Nuñez 2004.

Fig. 8. Worked flint finds from site A 1. Photo: Milton Nuñez 2004.

Fig. 9. Worked flint finds from site A 1. Photo: Milton Nuñez 2004.

Fig. 10. Worked flint finds from site A 1. Photo: Milton Nuñez 2004.

Fig. 11. A flint core with a borer/pick edge from site A 1. Photo: Milton Nuñez 2004.

Fig. 12. Worked flints from site A 1. A core, scrapers, a cleaver and a Levallois-type point (lower row, to the far right). Photo: Milton Nuñez 2004.

Site A 9
Open Air Site

UTM: 0508606, 3986349
Size: undefined

Site **A 9** produced Middle Palaeolithic tools and debitage: one retouched blade, a side-scraper and flakes including a flake-scraper.

Site A 10
Open Air Site

UTM: 0509253, 3886067
Size: undefined

Site **A 10** produced retouched Levallois flakes and a core featuring a notch from the Middle Palaeolithic period.

Site A 11
Open Air Site

UTM: 0508554, 3886464
Size: undefined

Site **A 11** produced five scrapers, one being a flake-scraper, and two side-scrapers of which one might be a Levallois point. Some small Neolithic implements were also included in the finds at the site.

Fig. 13. Middle Palaeolithic finds from site A 11 representing a Levallois-related technique and tools transitional from the Middle to the Upper Palaeolithic period. Photo: Milton Nuñez 2004.

Site B 2
Open Air Site

UTM: 0504815, 3884501
Size: undefined

Site **B 2** provides evidence of activities mainly from the Upper Palaeolithic and possibly Epipalaeolithic/ Neolithic Age. Most of the finds were scrapers. In addition there is one borer/burin among the finds (see Fig. 14.). B 2 is also the site of a Roman/Byzantine tomb that will be described in ch. 11.

Fig. 14. Worked flint finds from site B 2 dating from the Upper Palaeolithic period as well as small ones from the Epipalaeolithic/ Neolithic Age. Photo: Milton Nuñez 2004.

Site A 6b
Open Air Site

UTM: 0507533, 3882331
Size: undefined

The tools discovered at a courtyard complex at the Bedouin village of Shanhas represented Levallois-related technique and signs of Yabrudian tradition from the Middle Palaeolithic period. There were at least four scrapers among the finds, one borer, and two to three cores representing detached portions of which one might be a broken biface. (See some finds in Fig. 15.)

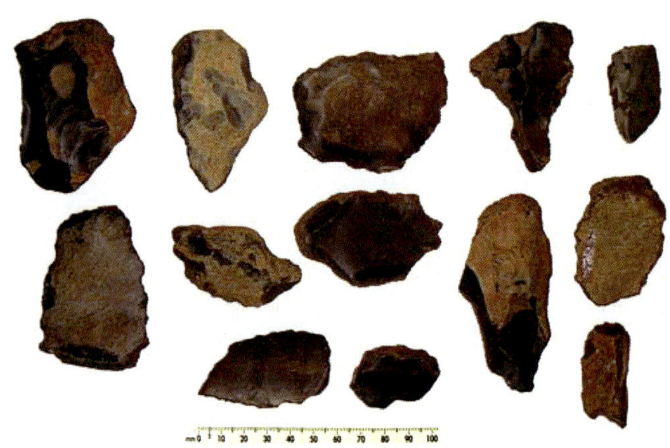

Fig. 15. Worked flint finds from the Middle Palaeolithic site A 6a at the village of Shanhas. Photo: Milton Nuñez 2004.

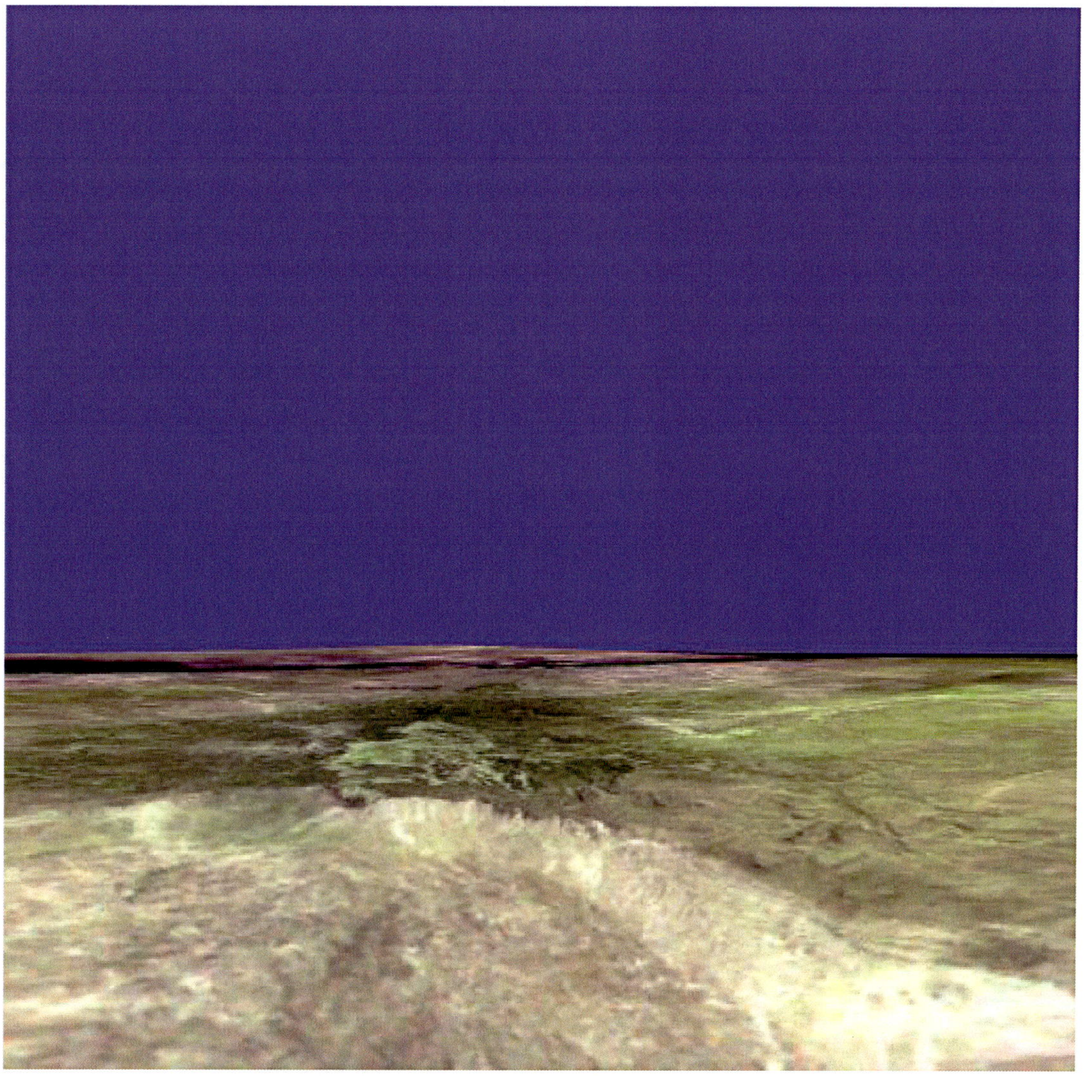

Fig. 16. Computerized 3D landscape scene of Tar al-Sbai from the southwest, the mosaic of Landsat-4 MSS images 172/35 and 36, date 12.07.1983, channel 574 from south to north. Note the visibility to the Euphrates valley behind. Constructed by Markus Törmä 2006.

5.2. Tar al-Sbai: a Semi-circular Escarpment with Sites of Mobile Groups

As already mentioned, Tar al-Sbai (the Circle of the Sba'a tribe), also known as Jabal Shabah, forms a semicircular escarpment at the western edge of Jebel Bishri (Figs. 1., 16.). The name of the area indicates to the border zone between the Sba'a and Fid'ān Bedouin tribes populating the region of Jebel Bishri, the latter chiefly governing the western part of the mountain today. The large ca. 40 km long escarpment includes ca. 4 km long arc of porous limestone that is identified as Tar al-Sbai. The area generally lies between the elevation contour lines of the 700 and 800 m a.s.l.

Our expedition sent beforehand a small group by a jeep to carry out "window screening" in the region of Tar al-Sbai on the 7th of July, 2000. The visit resulted in identifying the nature of the sites and recording of the first prehistoric finds at E 2, E 3 and E 4 (see Figs. 18., 19. and 20.), although the last two sites were not necessary sites in the real sense but represent stray finds. A group executing the systematic field walking followed the preliminary visit of the small group.

Tar al-Sbai is like a thematic archaeological park consist of ancient remains left by mobile groups: it is exemplified by intersecting sites of hunter-gatherers and pastoral nomads along the mountain edge (Fig. 17.). Actually the escarpment has produced the densest area of sites and structures associated both with hunter-gatherers and pastoral nomads on Jebel Bishri. The sites at Tar al-Sbai are offering visibility and different perspectives and vistas over the large desert plateau beneath. The fact that the mountain edge has served as a popular ground among hunters for millennia is testified not only by rock shelters but also by numerous crescent-shaped hunting blinds constructed of piled stones in the area. Beside the hunters' sites the numerous pastoral nomadic sites mainly consist of burial grounds formed by cairns/*tumuli* beside stone enclosures and megalithic stone structures of different kinds (Lönnqvist *et al.* 2006), many of them apparently connected with ancestral rituals (see ch. 8.). In this chapter we only deal with the sites that can be distinguished as belonging to hunters and gatherers.

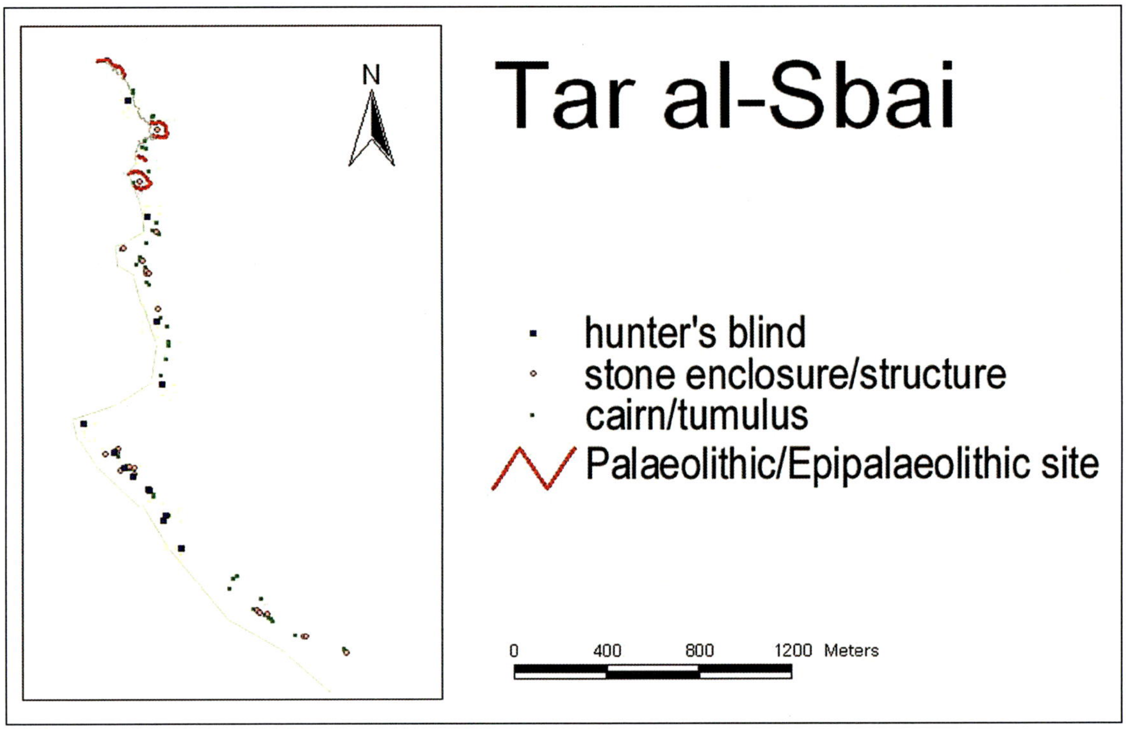

Fig. 17. Distribution of the types of archaeological sites and structures at Tar al-Sbai. The order of the red Palaeolithic/Epipalaeolithic sites is from up to downwards: C 1, B 5, C 5 and C 7. Mapping by Jari Okkonen with the EDM data captured by Kirsi Lorentz, Donald Lillqvist and Mervi Saario 2000.

5.2.1 Prehistoric Sites near Tar al-Sbai

Site E 2
Open Air Site

UTM: 0517946, 3907031
Size: undefined

Site **E 2**, which was identified during the unsystematic preliminary visit, produced evidence of blade and bladelet industry and microlithic lunates from the Epipalaeolithic period. Among the five blades there were two side scrapers. There were three retouched bladelets and two different kinds of scrapers related to the Pre-Pottery Neolithic B period. The finds thus present Epipalaeolithic and Neolithic activities at the site.

Site E 3
Stray find?

UTM: 0520138, 3903162
Size: undefined

A Lower Palaeolithic cleaver was discovered at E 3.

Site E 4
Stray find?

UTM: 0517686, 3901743
Size: undefined

A Levallois core was discovered at E 4.

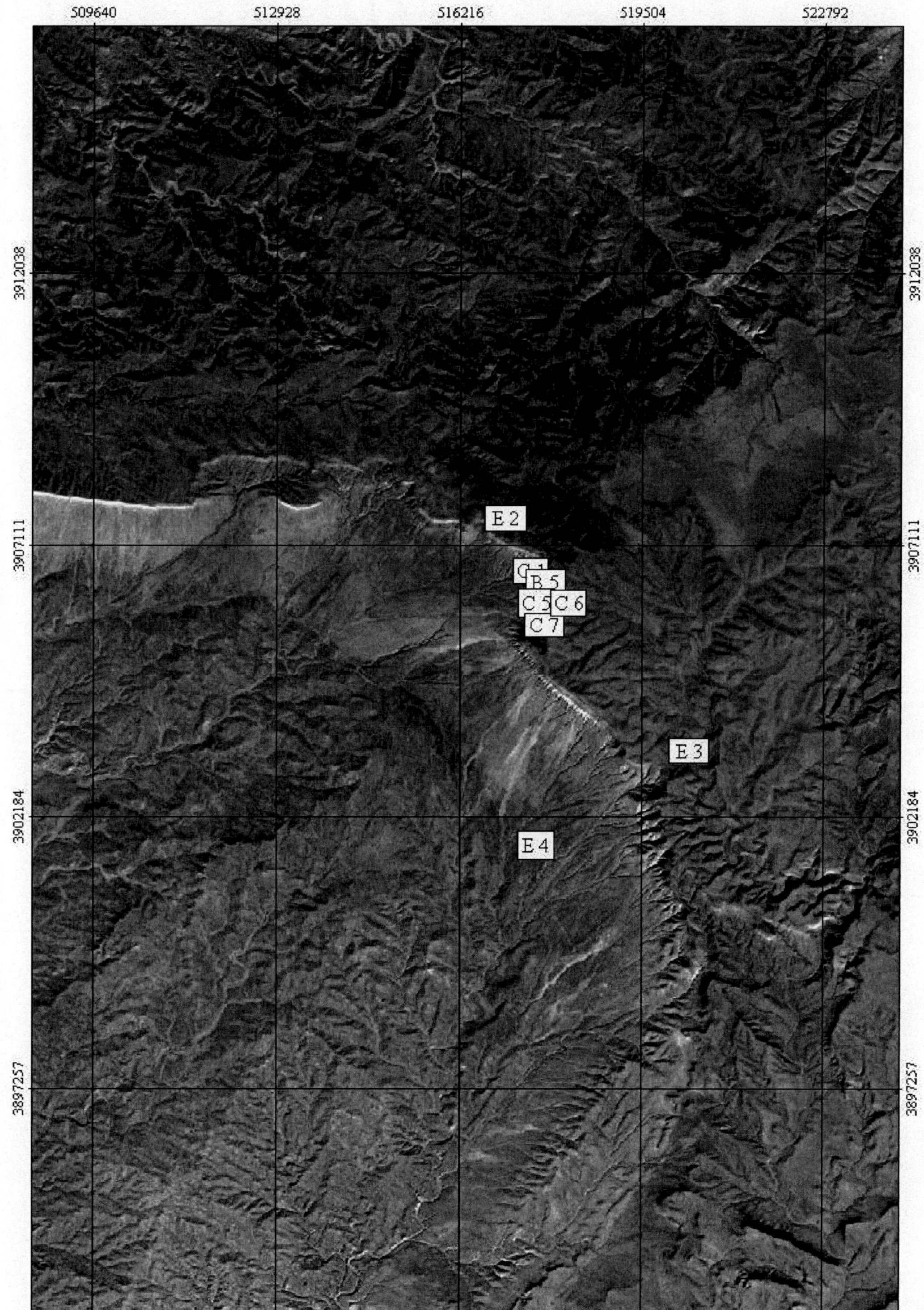

Fig. 18. Palaeolithic and Epipalaeolithic sites or stray-finds at Tar al-Sbai. Scale ca. 1:50,000. Mapping by Minna Lönnqvist 2009.

Fig. 19. Worked flint finds from site E 2. Photo: Photo: Milton Nuñez 2004.

Fig. 20. A core from E 4. Photo: Milton Nuñez 2004.

Fig. 21. Surveying the area of Tar al- Sbai above rock shelters. Photo: Margot Stout Whiting 2000.

5.2.2 Rock Shelters (Abris) and Flint Workshops at Tar al-Sbai

The whole expedition team following the preliminary visit to Tar al-Sbai consequently systematically studied a narrow area which covered ca. 100 m x 2100 m at the edge (see Fig. 18.). The captured UTM coordinates during this systematic survey extending from the northwest to the southeast of the first recorded site at Tar al-Sbai start with site **C 1** (UTM: 0517906 and 3907065) and end at the last site **A 34** (UTM: 0518052, 3904927). **D 4** (UTM: 0520209, 3903263) was an additional site observed at the edge outside this area providing more recent Bedouin tombs (see ch. 8.). Several *abris* were identified at the escarpment of Tar al-Sbai already during the preliminary visit. The area of Palaeolithic and Epipalaeolithic sites extended over the stretch of ca. 1 km in length (see Figs. 17. and 18.) These *abris* have originally been shallow rock shelters existing side by side and have especially been suitable for human occupation. The type of the rock shelters was originally associated with the French Upper Palaeolithic culture. Occupational debris usually accumulates on the floor of the shelter, but in our case the erosion which had caused the collapse of the roofs had made the floors vulnerable to the mechanics of water, sand and winds, so that practically most of the material had swept away into the valley beneath. (See, e.g., Evans 1978: 101-102).

Fig. 22. The area of rock shelters at Tar al-Sbai. Photo from NW: Jean-Pierre Matar 2000.

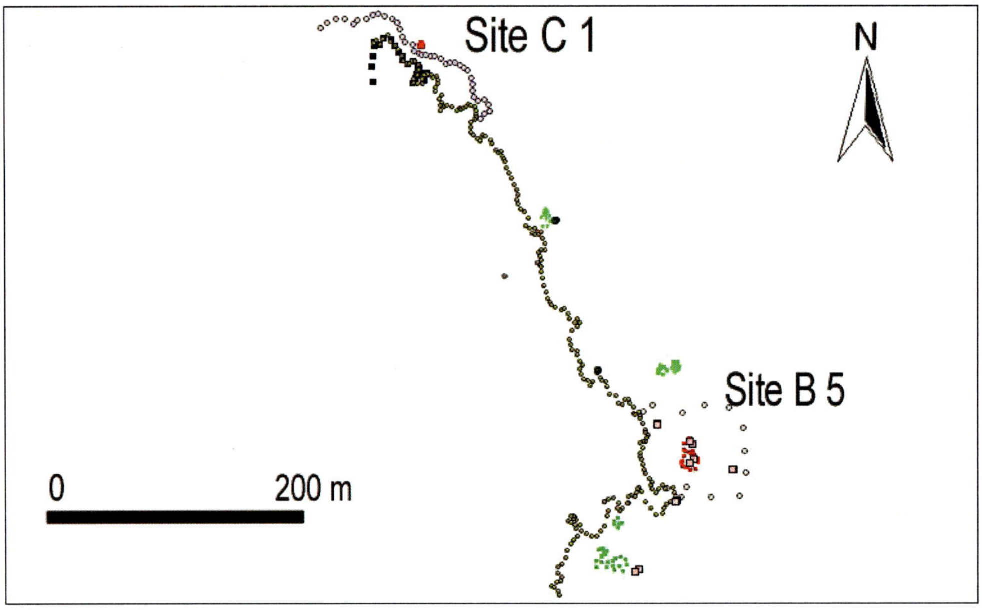

Fig. 23. The locations of site C 1 and site B 5 at the edge of Tar al-Sbai.
Grey dots indicate the extensions of the Palaeolithic/Epipalaeolithic sites, green dots indicate cairns/tumuli and red ones stone enclosures.
EDM mapping by Kirsi Lorentz, Donald Lillqvist, Mervi Saario and Jari Okkonen 2000-2004.

Fig. 24. The abri site C 1, a lorry used by our expedition serves in the background as a scale.
Photo from S: Margot Stout Whiting. 2000.

Site C 1
Rock Shelter

UTM: 0517906, 3907065
Size: 175 m x 18 m

Site **C 1** is a Palaeolithic/Epipalaeolithic collapsed *abri* (Figs. 23. and 24.) with an apparent workshop on it. The site is an elongated one following the mountain edge; it produced flint tools and debitage in quantity. Clactonian industry was identified in one tool, indicating to some Lower Palaeolithic influence. The finds also included six flakes of which one was bifacial and one represented Levallois-technique from the Middle Palaeolithic period and had probably been used as a denticulated tool. The finds also included several blades (seven recorded pieces) of which three were categorized as knives (both sides used) and one as a burin. There were three borers, five scrapers, bladelets and one microlithic lunate as well as two cores with handles. The industry indicated that there had been Upper Palaeolithic, Epipalaeolithic (Kebaran and possibly Natufian influence) and Neolithic activities at the sites. Collected debitage consisted of 29 pieces. (See samples of the finds in Figs. 25.-28.).

Fig. 25. Worked flint finds including Kebaran perforators from site C 1. Photo: Milton Nuñez 2004.

Fig. 26. Worked flint finds from site C 1 including two borers (two items in the lower row to the right). Photo: Milton Nuñez 2004.

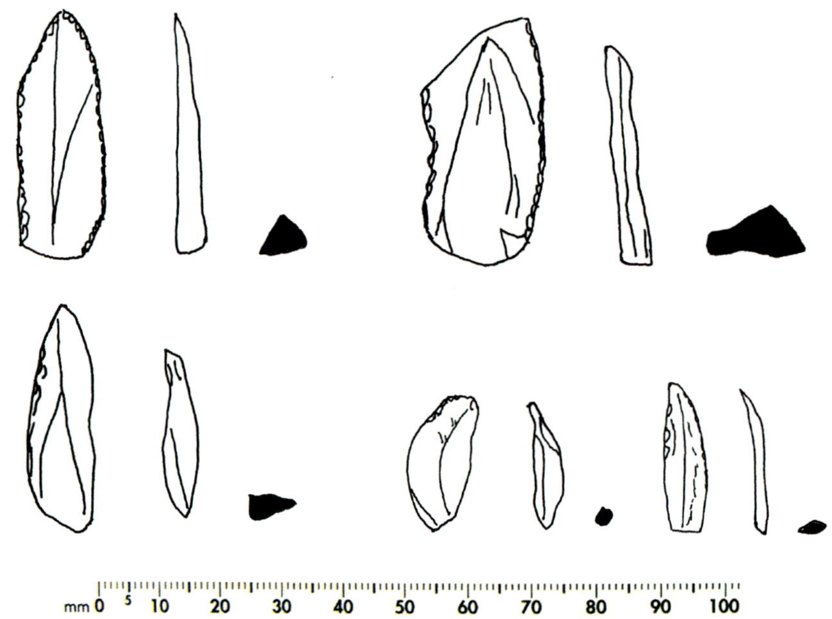

Fig. 27. Worked flint finds from site C 1.

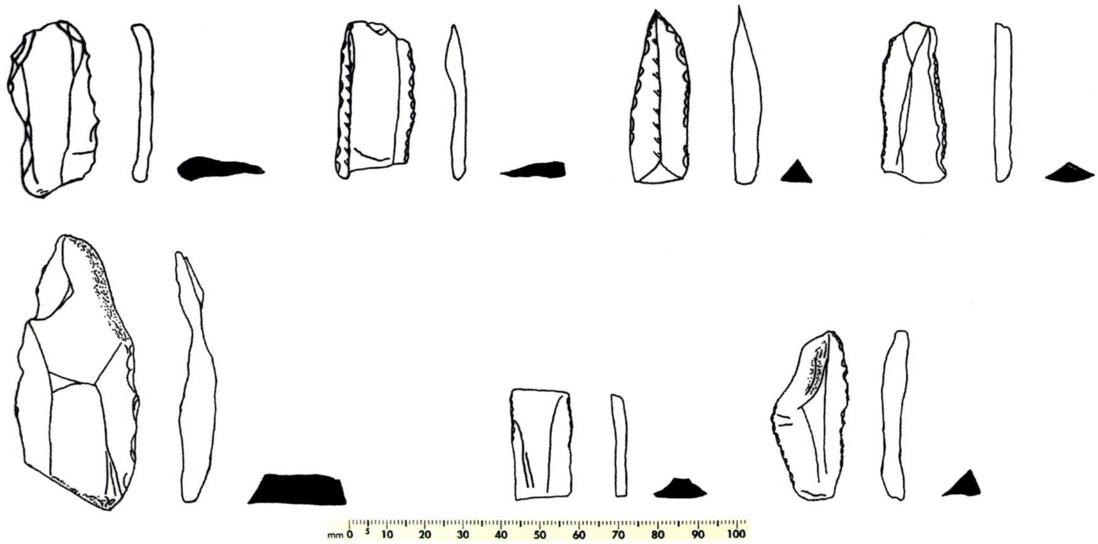

Fig. 28. Worked flint finds from site C 1.

Fig. 29. Collapsed rock shelter site B 5 of the abri type. Photo from W: Margot Stout Whiting 2000.

Site B 5
Rock Shelter

UTM: 0517950, 3906502
Size: 46.7 m x 13.5 m

Site **B 5** is also an *abri* site (see Figs. 20., 23. and 29.) with a collapsed roof in line with an apparent workshop above it. Altogether 14 blades and blade tools, including three end-scrapers and one side-scraper, were collected at the site, besides three other types of side-scrapers. One bladelet was identified as a backed knife, another as a knife with two edges. There was one burin which seemed to have served as a "perforator". One Canaanean blade was also included in the implements. The finds were dated to the Epipalaeolithic to the Neolithic period and Early Bronze Age. Microlithic finds included Kebaran and Natufian implements such as one small and one large lunate. Three bipolar cores were also identified among the finds. (See samples of finds in Fig. 30.).

Fig. 30. Worked flint finds from the abri site B 5. Photo: Milton Nuñez 2004.

Site C 5
Rock Shelter

UTM: 0518067, 3906770
Size 10 m x 29 m

Site **C 5** is another Palaeolithic/Epipalaeolithic *abri* (see Fig. 20.). Above, on a hill, there has been a workshop, the material of which had run down with watern currents along a small *wadi* to the *abri* site. The site has produced seven blades, seven scrapers of which four are side-scrapers, some from the Upper Palaeolithic period. In addition, seven Epipalaeolithic lunates and one trapezoid indicate the Geometric Kebaran impact at the site. Two tetrahedral cores were also identified among the finds. (See samples of finds in Figs. 31. and 32.).

Fig. 31. Worked flint finds from site C 5. Photo Milton Nuñez 2004.

Fig. 32. Worked flint finds from site C 5. Photo: Milton Nuñez 2004.

Fig. 33. Rock shelter C 7 beneath the workshop C 7. Photo from W: Minna Lönnqvist 2000.

Site C 7
Rock Shelter and Workshop

UTM: 0518077, 3906718
Size: 9.40 m x 4.20 m

Site **C 7** is also an *abri* (Fig. 33.) above which there is a flint workshop. The concentration of finds appeared in the centre of the workshop area covering 3 m x 2-5 m. We collected 174 tools, cores and flakes. There were 34 cores of which several are prismatic, 31 scrapers or burins and 19 microliths, including lunates like at the previously mentioned sites. In addition, a few flakes were collected. The quantity and variety of tools is very extensive. Some of them were unfortunately heavily crusted. The site has produced 20 blades and blade tools, mostly side-scrapers, six borers/awls of various kinds. The workshop area provides a mixture of worked flint finds covering the time-span from the Upper Palaeolithic and Epipalaeolithic (the Geometric Kebaran) to the Pre-Pottery Neolithic period. (See Figs. 34.-36.)

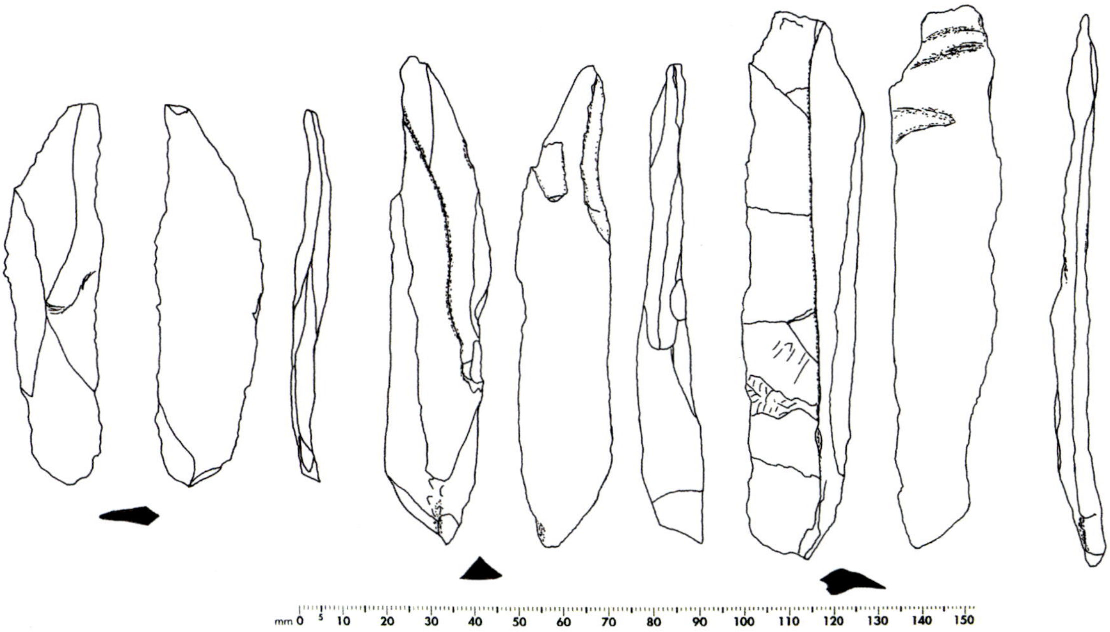

Fig. 34. Worked flint finds from site C 7. Drawn by Kenneth Lönnqvist 2004.

Fig. 35. Worked flint finds from site C 7. Photo: Milton Nuñez 2004.

Fig. 36. Worked flint finds from site C 7. Photo: Milton Nuñez 2004.

5.2.3 Evidence of Possible Contacts

The Palaeolithic sites at Tar al-Sbai seem mainly to belong to the Upper Palaeolithic period, but - as mentioned - some random evidence of the Lower and Middle Palaeolithic activities was identified in the area as well. Epipalaeolithic and Neolithic impact was also visible. However, some tools may have moved from upper areas to the mountain edge by natural forces. The connection of Tar al-Sbai with the vast ancient "Silicon Valley" and the oases beneath on the western piedmont, which have been providing numerous prehistoric sites, has to be taken into account.

The evidence points to the existence of prehistoric hunters, who were looking for game such as gazelles, ibexes and hares on Jebel Bishri. As previously indicated, Upper Palaeolithic sites are present in the El Kowm basin at Hummal and Qdeir Plateau as well as Umm el-Tlel on the western piedmont beneath Tar al-Sbai (see the locations of these prehistoric sites in the previous chapter; Le Tensorer *et al.* 2006; Le Tensorer *et al.* 2007: 628 and Ploux and Soriano 2003). The late Upper Palaeolithic starts the transition to microlithic tools. The Levantine Upper Palaeolithic period is especially exemplified in the Ahmarian and Aurignacian industries (Belfer-Cohen and Goring-Morris 2002: 57-68; Gilead 1991 and see also the Levantine Aurignacian in Williams 2006) including some local variations, like those present in the finds of Umm el-Tlel (see Ploux and Soriano 2003: 5-34). The assemblages of the finds from Tar al-Sbai are also comparable to the late Upper Palaeolithic and Epipalaeolithic finds at Palmyra basin (cf. Fujimoto 1979). However, the so-called Ksar Akil points (cf. e.g., Ohnuma *et al.* 1988), the typical implements of the Upper Palaeolithic period, are void in the material of Tar al-Sbai. In any event, the escarpment may represent an area for seasonal hunting grounds used by the inhabitants of Palaeolithic base camps at El Kowm and Umm el-Tlel nearer to perennial water sources which continued through the Epipalaeolithic and Neolithic periods. This would therefore indicate that the first steps towards the highland-lowland interaction, "*transhumance*" in D.O. Henry's line (1989: 167-170)[1], took place in

[1] *Transhumance* is a term primarily applied to pastoral societies and their seasonal movements from base villages in lowlands

habitat tracking in the region. The interaction may have already started among Palaeolithic hunter-gatherers and was continued by Epipalaeolithic foragers and later by pastoralists continuing up until today in the arid areas of the Levant.

Was the choice of locating sites at Tar al-Sbai only a convenient approach to prey, or a matter of taste for the impressive landscape? The Upper Palaeolithic meant transition in human thinking as far as art and religion were concerned (see, e.g. Mithen 1996: 171-210; Lewis-Williams 2002). Locating the sites at the mountain edge with a good visibility to the plateau beneath must have had some significance for the hunter-gatherers of the period, as we have not encountered comparable concentrations of the Upper Palaeolithic/Epipalaeolithic traditions in the inner areas of Jebel Bishri or on the Euphrates bank. The development of mind and totemistic thinking (Lévi-Strauss 1966) may have had some impact in the choice of the sites with imposing landscape. However, from the hunter's point of view it needs to be taken into account, as earlier mentioned, that the seasonal route of Persian gazelles is situated below in the Levantine corridor including the El Kowm basin (see, e.g., Akkermans and Schwartz 2003: 37). The route must have been known to hunters throughout the past.

Some bladelets and microburin technology of Tar al-Sbai resemble the Mushabian complex related to the arid zone in Jordan. It was related to and contemporaneous with the Geometric Kebaran but lasted longer and seems to follow more the line of the arid zone. (See Bar-Yosef 1987: 230; Henry 1989: 125-149). This also represents adaptation to drier conditions in the arid zone (Henry and Shen 1995). Madamaghan industry is a local feature of the Mushabian in southern Jordan. However, it needs to be taken into account that microburins can be produced as debitage of Upper Palaeolithic bladelets (Bordeur *apud* Henry 1974: 390). Microlithic points, some comparable to the Mushabian industry related to the Geometric Kebaran phase, were encountered at Tar al-Sbai. The so-called La Mouillah points of the Madamaghan industry seem to have also spread as far as to the 'Azraq basin, although the points are typical of the Madamaghan sites in southern Jordan (Henry 1974: 389; Henry 1995). Otherwise, points or lance heads were not very visible in the finds of Tar al-Sbai, but instead more visible concentrations of scrapers, knives and burins were found. This indicates that the sites at Tar al-Sbai were mainly used for processing carcases, hides, bones and antler. One may therefore venture to ask whether the sites at Tar al-Sbai were basically used as base sites for processing hunted animals and for foraging in the Palaeolithic/Epipalaeolithic times. The lunates indicate that reaping of some plants may have taken place in the area as well, although some have recently suggested that lunates were used as points in hunting. I. Gilead (1991) suggests that the Upper Palaeolithic people were organized in small bands of mobile foragers, who were involved in acquiring ungulate meat and plant foods.

5.2.4 From Camping to Huts and Houses at the Mountain Edge

As previously mentioned, the transition from the Upper Palaeolithic to the Epipalaeolithic had taken place ca. 22,000 – 20,000 BP. The evidence of this transitional period indicates that the first major steps towards sedentarization were taken after the long continuum of the Palaeolithic mobility at Tar al-Sbai. The change from the Epipalaeolithic Kebaran to the Natufian phase represents a major shift from a mobile hunter-gatherer life style to more sedentary way of life in the Levant. (See, e.g, passim Bar-Yosef and Valla ed. 1991). But as the oases of El Kowm, Qdeir and Umm el-Tlel testify, the Palaeolithic way of life already included some degree of semi-sedentary camping. A mobile group can change to a more sedentary lifestyle from one reason or another.

The chief subsistence economies of the people of the Epipalaeolithic era were still hunting and gathering. Different seeds, grasses and nuts were collected. At Tar al-Sbai lunates, trapezoids and microburin technique are present (cf. Henry 1974: 389-398). Our finds now extend the boundary of the Levantine Epipalaeolithic cultures further to the northeast on the fringes of desert-steppe regions. The Geometric Kebaran influence seems to have extended from its central areas of the Levantine woodlands to the arid zone during a moister climatic interlude. (Bar-Yosef 1987: 230). Geometric Kebarians were foragers who had adapted to semi-desert environments (Henry 1989: 171) like that of Jebel Bishri. They lived in bands estimated to have included 15-17 people on average (Henry 1989: 173). This phase is also attested at the oasis of El Kowm on the western piedmont (Cauvin and Coqueugniot 1988: 270-282). Epipalaeolithic finds were also recently found at Hummal in the El Kowm basin (Le Tensorer *et al.* 2007: 628). The evidence from El Kowm and Jebel Bishri seems further to add to our knowledge of the early highland-lowland interaction during the Epipalaeolithic period.

---

to grazing lands in highlands, but, for example, D.O. Henry (1989) uses the term to illustrate seasonal movement of hunter-gatherers over different environments/habitats already in the Middle Palaeolithic times (Henry 1995: 373) .

The Natufian culture, the phase of which is related to that of the Kebaran, emerged sometime ca. 13/12,500-12,000 B.C. (Cauvin 2000), and it also adapted to semi-arid conditions in its latest phase in the Negev and Sinai. Natufian related sites are found at El Kowm as well, and the process of sedentarization and shift to agriculture is especially featured in the earliest occupation phases of Abu Hureyra (Moore *et al.* 2000) up the Euphrates some 100 km northwest of the central areas of Jebel Bishri (see more also in Anfinset 2008). As previously mentioned, Abu Hureyra on the Euphrates is a site of special importance for the beginning of agriculture (see Cauvin 1991b and Moore *et al.* 2000). In any event, the microlithic lunates typical of the Natufian culture are more elongated in their shape at Abu Hureyra than in the Mediterranean forest belt, and this also applies to the lunates of Tar al-Sbai.

Jebel Bishri belongs to the same arc in the arid zone as Sinai, the Negev and eastern Jordan but has not been previously systematically studied to provide evidence of the Epipalaeolithic impact and spatial mobility of the groups in the area. The animals hunted in the Epipalaeolithic Levant include gazelles, ibex and hare. The plants exploited include nuts, pulses, wild cereals and tubers (Bar-Yosef 1987: 230-231). In recent studies it has been noted that in the Central Highlands of Jordan the period of 18,00-9000 B.C. was a period of alternating moist and dry intervals, the duration of which was ca. 2000-3000 years (Barker *et al.* 2007). This period coincides with the Epipalaeolithic Kebaran and Natufian cultures and seems to be coeval with the spread of these cultures to the areas that had been more arid before.

Site C 6
Huts

UTM: 0518087, 3906742
Sizes: a) 8.70 m x 10.85 m, b) 4.55 m x 7 m, c) 1.30 m x 1.40 m

Huts and houses with clear entrances and thresholds were documented at Tar al-Sbai. One hut/house compound site **C 6** situated next to the Palaeolithic *abri* site **C 5** that had produced Epipalaeolithic artefacts. Site C 6 comprises three curvilinear hut formations (a.-c.), including one clear entrance (1.4 m in width) with a threshold opening to the south. These hut bases also had an attached stone-constructed crescentic hunting blind (see more on the hunting blinds in paragraph 5.2.5 below) on one wall, which may be a later addition using the stones from the hut but which could be contemporary as well. Inside a passing track there was a square stone construction and on the other side a stone wall of three parallel stone courses. Whether these structures belong to the hut compound is not clear. The finds in association with C 6 consisted of a small flint point, a blade, a scraper and a perforator. This site seems to have been in use during the Pre-Pottery Neolithic period, but if associated with the site C 5 it may have already been built in transition from the Epipalaeolithic to the Neolithic period as well, although such huts are very rare during the former period.

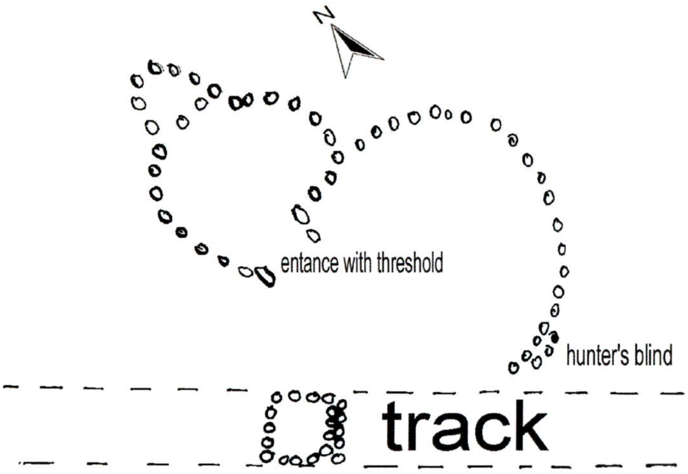

Fig. 37. A schematic plan of huts and other stone structures at site C 6. Not in scale. Drawing by Minna Lönnqvist 2000.

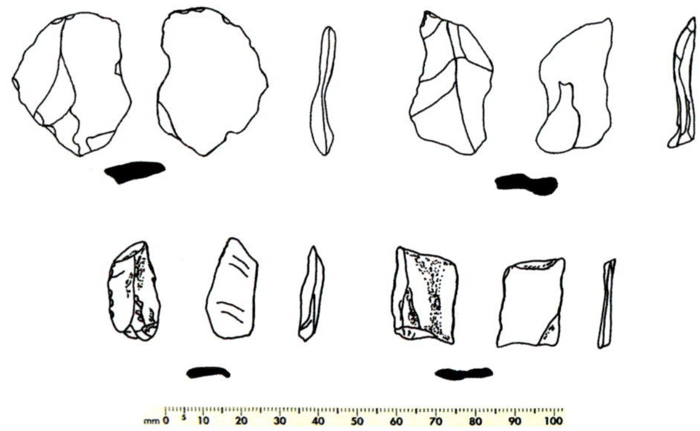

Fig. 38. Finds from site C 6.

5.2.5 Hunting Blinds

Small hunting blinds were found in continuous series along the mountain edge at Tar al-Sbai (see the distribution map above in Fig. 17.). These crescent shaped fences are built of natural limestones (varying 20-50 cm in diameter) piled with dry masonry technique in as a barrier ca. 0.7 m -0.8 m in height on average. The width the construction is generally 2 m, and the depth is ca. 1.7 m. The shelter was built for a hunter to hide behind the fence. Interestingly, traces of paths of the hunted animals were regularly visible in front of the blinds on the slope.

Fig. 39. A hunting blind site C 2.
Photo from N: Margot Stout Whiting 2000.

Altogether 12 hunting blinds at the altitude of ca. 702-710 m a.s.l. were recorded and documented at the edge: **C 2** (UTM 0517971, 3906886; see Fig. 39.), **A 15** (UTM: 0518015, 3905919)**, A 21** (UTM: 0518020, 3905640; see Fig. 40.), **A 25** (0517827, 3905445; 704 m a.s.l.; see Fig. 41.) two structures at **A 28** (UTM: 0517939, 3905199; see Fig. 42.), **A 30** (UTM: 0517951, 3905178), two structures at **A 33** (UTM: 0518002, 3905073) and **A 34** (UTM: 0518052, 3904927; ca. 709 m a.s.l.) and **B 6** (UTM: 0518016, 3906375; see Fig. 43.). As previously mentioned, one was attached to the hut bases at **C 6**. No finds were associated with any of the blinds but the hut bases at **C 6** produced worked flints. One hunting blind, **C 2**, seemed to be associated with a hunter's rock shelter (UTM: 0517997, 3906876). Small stone circles that looked like hearths were associated with the blinds at the sites **A 29** and **A 33,** but no surface finds were discovered in them either. It is likely that the blinds have been used to hunt gazelle and ibex. Although the contextual evidence of the small finds would associate the nearest sites with the Epipalaeolithic and Neolithic Age, it is more plausible that these kinds of structures have been used in quite recent times by the local Bedouins (see, e.g., Madoun 1996). However, small points of the Epipalaeolithic period could have been used in this kind of close range hunting as well.

As far as hunting methods are concerned beside hunting blinds there are kites, large V-shaped animal traps built of stone walls of several dozen metres and ending in an enclosed pit area in the neighbourhood of Jebel Bishri. These kinds of traps were already used in the gazelle-killing in the earliest settlements of Abu Hureyra (Legge and Rowley-Conwy 2000: 439) but have also been quite recently used by Bedouins as well. Kites are referred to in A. Poidebard's (1934; see Fig. 44.) studies of the Roman eastern *limes* zone, which concentrated on ancient tells, monuments and roads in Syria. One of the kites appears in the passage associated with Taibe and El Kowm on the western piedmont of Jebel Bishri. The construction can be associated with the afore-mentioned special corridor (ch. 4.) between two mountain ridges which the gazelle flocks entered during their seasonal migration route in the piedmont of Jebel Bishri. The wanderings of the gazelles took place between Arabia and the Euphrates (Legge and Rowley-Conwy 2000: 439). El Khabra at Qdeir in the El Kowm basin has provided evidence of a PPNB hunting post with beautiful flint points (Cauvin, J. 1991). From the Chalcolithic period there is a remarkable deposit of gazelle bones recovered at El Kowm 2 - Caracol (Vila 1991).

Fig. 40. A hunting blind site A 21.
Photo from NW: Juhana Saukkonen 2000.

Fig. 41. A hunting blind site A 25.
Photo from NE: Juha Pakkala 2000.

Fig. 42. A hunting blind site A 28. Photo from N: Juhana Saukkonen 2000.

Fig. 43. A hunting blind site B 6.
Photo from E: Margot Stout Whiting 2000.

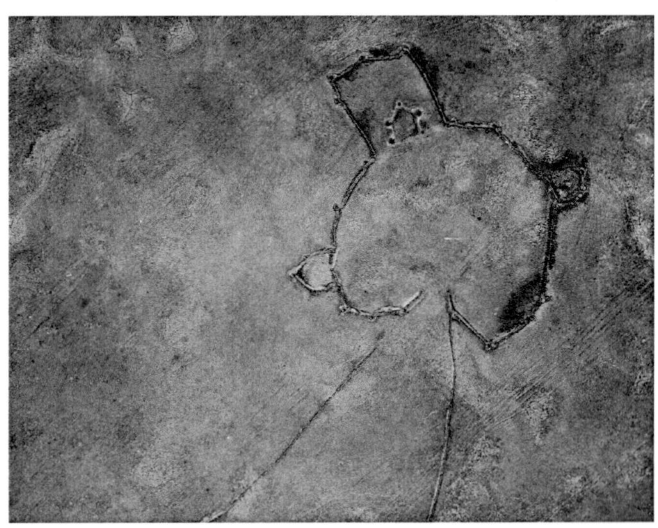

Fig. 44. An aerial photograph of a kite near Damascus.
Source: Poidebard 1934.

5.3. Palaeolake Shores at Nadra

The palaeolake area of Nadra (see ch. 4, Fig. 45.) nowadays forms a high plateau ca. 8 km to the southeast of the largely abandoned Bedouin village of Ash-Shujiri on Jebel Bishri and ca. 3 km northwest of Jebel Nozairat al-Bishri. The area can be reached driving ca. 30 km along the main asphalt road from Deir ez-Zor towards Palmyra. Then one needs to turn to the northwest at the village of Shula, which leads to Jebel Bishri, and continue ca. 30 km along the asphalt road. The high plateau of Nadra is surrounded by hills and pierced by *wadis*. In the valley our survey concentrated on the western edge of the plateau and the field walked area covered ca. 10 km x 3 km in the neighbourhood, reaching as close as to 4 km to the village of Ash-Shujiri located in the summit area.

The Palaeolithic sites, which we discovered and which situated at the edge of the palaeolake at Nadra and above the 600 m contour line, are sites **I 13, I 27, I 19** and **I 37**. As mentioned in the previous chapter, they provide possibilities for a tentative *terminus* dating of the ancient shore line of the lake as the hunter-gatherer sites inside the palaeolake are post-Palaeolithic, the majority being pastoral sites with structures to be described in the following chapter. The surveys in eastern Jordan, such as those in the 'Azraq basin, the area of another palaeolake, provide comparable archaeological finds from the Lower Palaeolithic to Chalcolithic times (see, e.g., Garrard and Stanley-Price 1975-1977; Garrard *et al.* 1994). These sites are also somewhat reminiscent in their environment to Jebel Bishri.

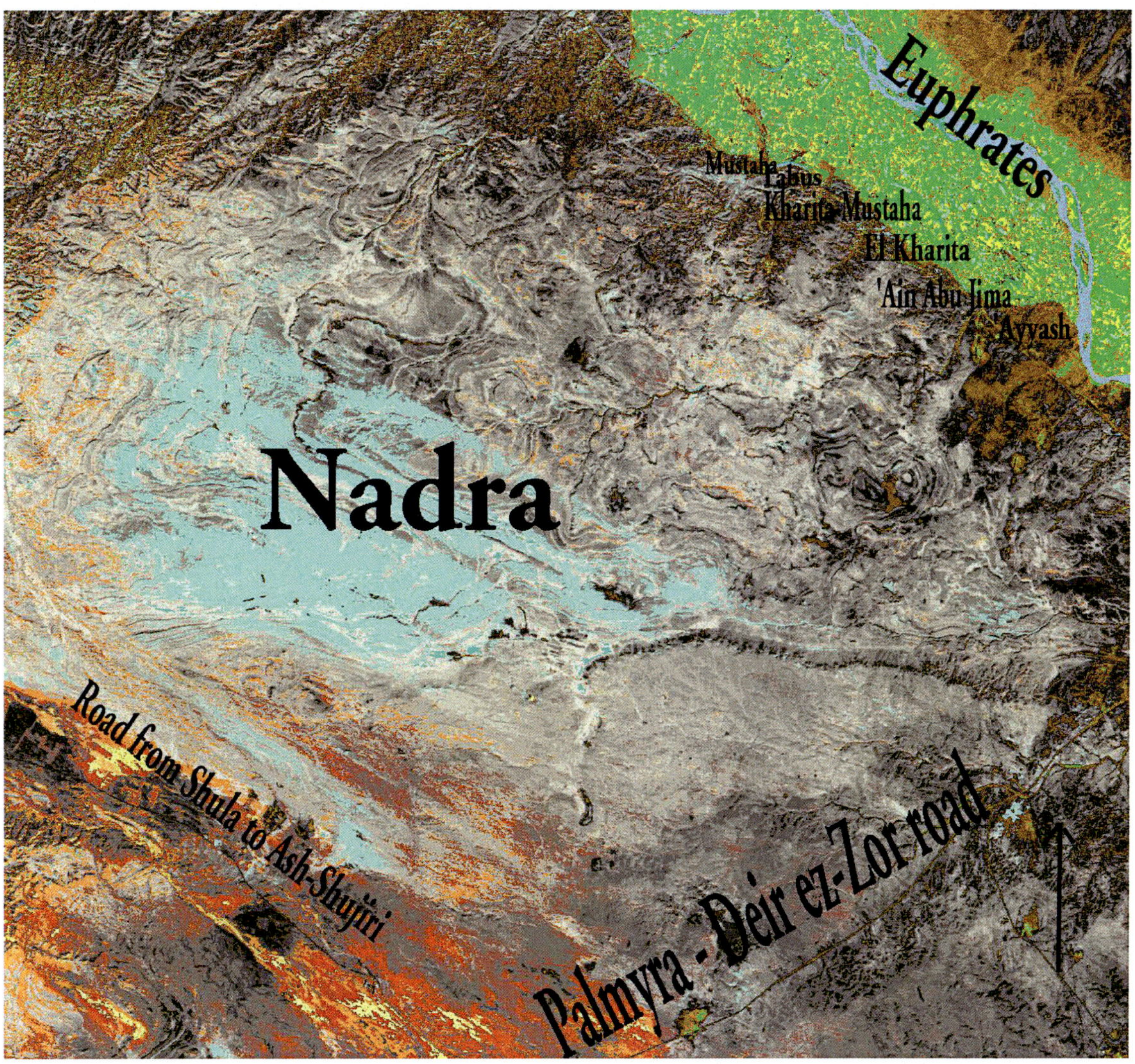

Fig. 45. The area of the palaeolake at Nadra visible in a clustered Landsat-7 ETM image taken in 2000. The clustering result by Markus Törmä.

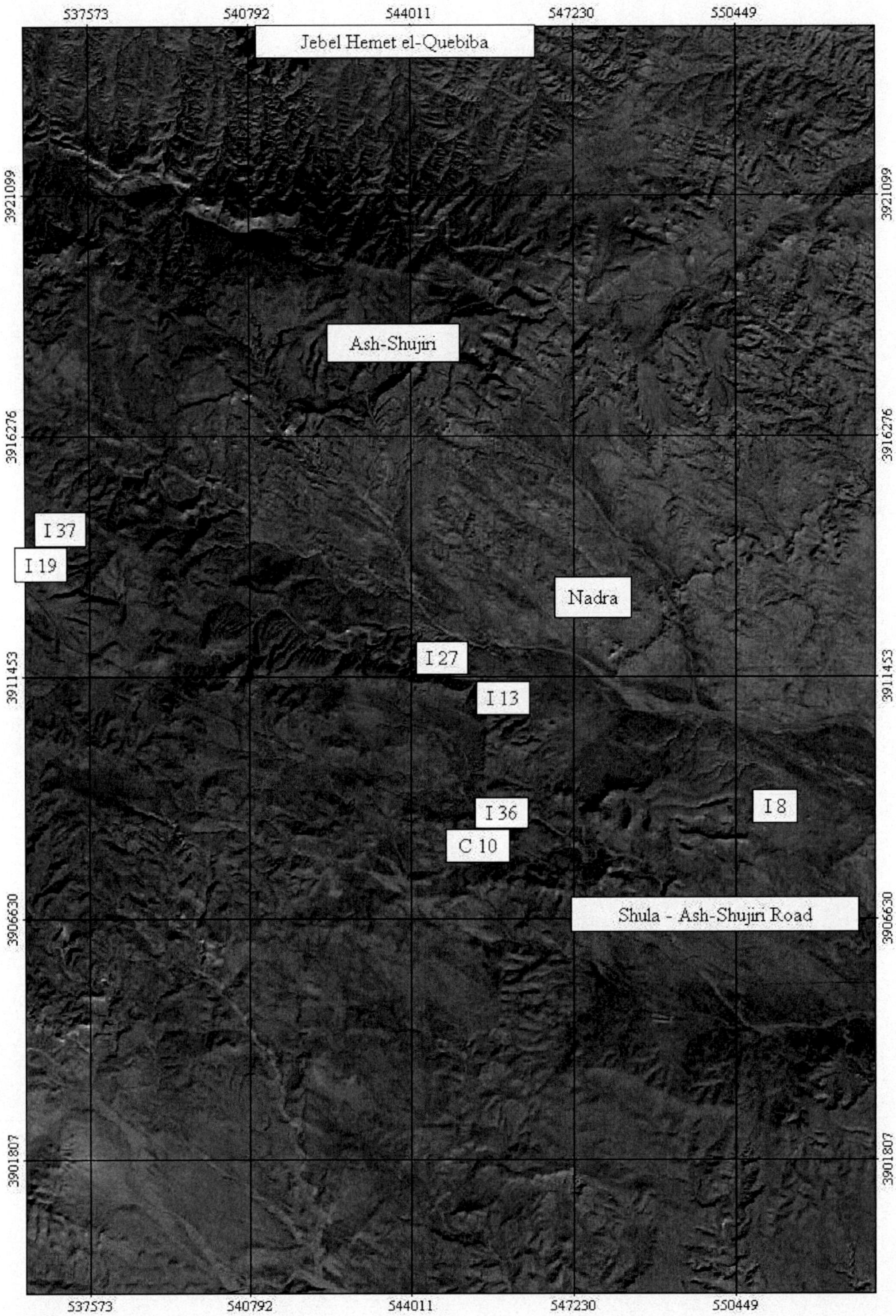

Fig. 46. Sites which produced Palaeolithic finds around Nadra. Scale: 1: 50,000. Mapping by Minna Lönnqvist 2009.

Site I 13
Open air Site

UTM: 0546557, 3910187; 627 m a.s.l.
Size: undefined

Site **I 13** (Fig. 46.) is an open air Palaeolithic site situated on a ridge south of an abandoned and ruined Bedouin village **I 1** (see more in ch. 15.). The Palaeolithic site incorporates a later cairn/*tumulus* tomb. An Achaeulian handaxe (see Figs. 47. and 48.), either representing the Lower or the Middle Palaeolithic tradition, was found. In the latter era comparable axes are also known. There were also cores, a fossilized bone and a point - reminiscent to the Upper Palaeolithic Ksar Akil type point (cf. Ohnuma *et al.* 1988) - scrapers and a possible Geometric microlith.

Fig. 47. Worked flint finds including three cores, one handaxe, one broken (Ksar Akil type?) point, scrapers and a Geometric microlith from site I 13. Photo: Kenneth Lönnqvist 2004.

Fig. 48. A small Achaeulian handaxe from site I 13 from the district of Nadra. Because of its small size the axe may also belong to a later Achaeulian tradition in the Mousterian phase. Photo: Kenneth Lönnqvist 2004.

Site I 27
Open Air Site

UTM: 0544260, 3913041; 602 m a.s.l
Size: ca. 75 m x 60 m

Site **I 27** (Fig. 46.) is an open accumulated lithic site located on a plateau above the confluence of two wadis. The site forms like a "cape" with a *wadi* surrounding it. Lithics include quartzite and flint, and there is a quartzite manufacturing area at the eastern edge. The worked stone implements date from the Palaeolithic, Epipalaeolithic and Neolithic periods.

Site I 19
Open Air Site

UTM: 0537957, 3912474; 713 m a.s.l.
Size: 100 m x 100 m

Site **I 19** (Fig. 46.) is an open accumulated site which situated on the southern slope of a small hill ca. 150-200 m west of the 2004 expedition camp (see ch. 3) at the edge of the palaeolake. The site produced a number of flint tools dating from the Palaeolithic and Epipalaeolithic periods, and some seem to date from the Neolithic period as well (see samples of finds in Figs. 49., 50.).

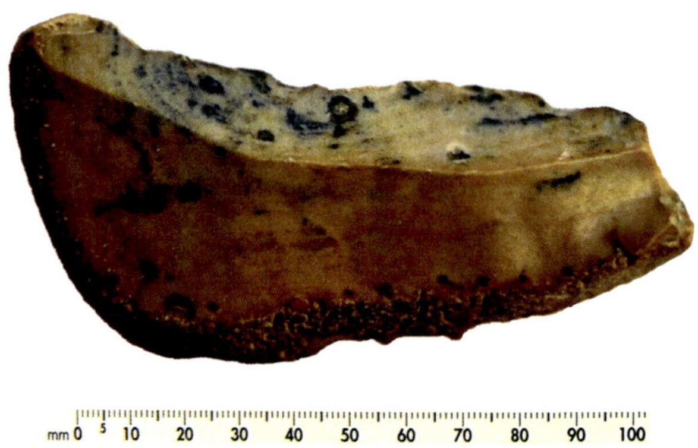

Fig. 49 A side-scraper from site I 19. Photo: Kenneth Lönnqvist 2004.

Fig. 50. Worked flint, such as a core, an awl and scrapers from site I 19. Photo: Kenneth Lönnqvist 2004.

Site I 37
Rock Shelter

UTM: 0537782, 3913311; ca. 687 m a.s.l.
Size: 2 m x 3-4 m

Site **I 37** (Fig. 46.) is a rock shelter located ca. 1 km northeast of the 2004 expedition camp (see the location in ch. 3.). the location falls outside our systematically surveyed area in Nadra. The site is situated on a northern slope of a mountain ridge along a *wadi*. It produced Palaeolithic and Epipalaeolithic as well as Neolithic/Chalcolithic and Bronze Age flint implements including a tabular scraper and a Canaanean blade (see samples in Fig. 51.).

Fig. 51. Worked flint finds from site I 37. Photo: Kenneth Lönnqvist 2004.

5.4. The Oasis of Qseybe

Site K 6
Open Air Site

UTM: 0568664, 3911075; 392 m a.s.l.
Size: undefined

To the southeast of the palaeolake there is an oasis called Qseybe, where the site **K 6** (see ch. 3., Fig. 4.) produced finds from the Roman period (see ch. 12 in the Al-Kharata plain). However, the site had an earlier prehistoric occupation as well. This area was outside our systematically surveyed districts and was thus studied as a separate entity.

A Hummalian point (cf. Le Tensorer *et al.* 2007: Fig. 4.; Copeland 1985:189) was collected from the surface. This could indicate first signs of highland-lowland interaction already in the Middle Palaeolithic period. The Hummalian point represents evidence of close contacts between El Kowm and Qseybe in the period. There also were signs of Neolithic and Chalcolithic activities exemplified by perforators and tabular scrapers. (See finds in Fig. 52.).

Fig. 52. Lithic industry from the Qseybe site K 6, a weapon in the upper row to the far left is a Hummalian point. Photo: Kenneth Lönnqvist 2006.

5.5. Hunter-Gatherer Sites and Stone Tools on the Euphrates Side of Jebel Bishri

Floodplains like the terraces of the Euphrates were typically used by early humans, especially by the Lower Palaeolithic hunter-gatherers (see Evans 1978: 90). Our intensive survey on the Euphrates side of Jebel Bishri extended from the village of ʻAyyāsh, ca. 12 km northwest of Deir ez-Zor continuing through the villages of ʻAin Abu Jimā, El-Kharita and Kharita - Mustaha to Tell Tabūs. As far as we have studied the region of Jebel Bishri on the ground, the Lower Palaeolithic sites – apart from the El Kowm basin on the western piedmont - are concentrating in the Euphrates river valley. The tools, which J. Besançon and P. Sanlaville (1981) had earlier recovered in their survey between Sabkha and ʻAin Abu Jimā, were of the Achaeulian-type. L. Copeland (2004) has also recently published her findings from the surveys of 1978 and 1979 on the Pleistocene terraces beneath Jebel Bishri. In her survey Copeland (2004: 27, 36-45) discovered Late Achaeulian artefacts from the area of ʻAin Abu Jimā formation like S. Muhesen (1993) had earlier done. The ʻAin Abu Jimā (QfII) formation is a very important and voluminous terrace continuing as far as to the Sajour (Sanlaville 2004: 118). As we discussed in the previous chapter, the British and the Syrians (Westaway *et al.* 2005 and 2007) have produced new radiometric datings in the area of ʻAyyāsh extending the dating of the terrace QfI to ~425-415 ka. This gives a much older date to the deposit than was earlier suggested stratigraphically, namely ~140-70 ka. The new dating provides the *terminus post quem* for the Lower Palaeolithic finds associated with ʻAyyāsh and partly for the area of ʻAin Abu Jimā. (See map in Fig. 54.).

Apart from one possible collapsed rock shelter site above a Pleistocene terrace H 1 near Tell Tabūs, all the prehistoric sites that we found between the village of ʻAyyāsh and Tell Tabūs appeared to be open accumulated sites. Some of the sites, however, were actually situating on the ridges of Jebel Bishri and not on the actual terraces of the river valley. There were Lower and Middle Palaeolithic artefacts, cores and debitage as well as some possible Epipalaeolithic and Neolithic implements. An Achaeulian-type of pick (Figs. 55. and 56.) from the Lower Palaeolithic period and a biface (Fig. 57.) were discovered at the ʻAyyāsh - ʻAin Abu Jimā area as stray finds. As far as we know, this pick is the first

Middle Achaeulian artefact at the 'Ayyāsh – 'Ain Abu Jimā area as the oldest artefacts earlier recovered by L. Copeland (2004) and S. Muhesen (1993) in the area date from the Late Achaeulian period. However, Copeland (2004: 38) notes that some cores from 'Ain Abu Jimā were thought to be from the Middle Achaeulian period. Previously comparable Middle Achaeulian artefacts are known from Hamadine and Mad'an situating between Raqqa and Halabiya at the foot of Jebel Bishri (Copeland 2004: 33-38), whereas Late Achaeulian artefacts are known all the way from Raqqa to Deir ez-Zor (Besançon and Sanlaville 1981; Copeland 2004). Comparable picks have been also found during a geomorphological prospection in the Tartous area in Syria (cf. Besançon *et al.* 1994: 11-12, Fig. 6). Copeland (2004: 53) has further observed that Middle Achaeulian artefacts have been found in various ecological zones in the Levant, but most seem to be near ancient beaches or river banks. This is in accordance with our finds that were discovered on the terraces of the Euphrates river. As previously indicated, apart from the Middle Euphrates area, the nearest sites belonging to the Middle and Late Achaeulian tradition are in the El Kowm basin on the western piedmont of Jebel Bishri where they are numerous. They are known at Nadaoiyeh Aïn Askar and Hummal. At Nadaoiyeh in the El Kowm basin the Achaeulian culture can be followed over a period covering ca. 400,000 years. (Le Tensorer *et al.* 2006; Le Tensorer *et al.* 2007: 625).

Fig. 53. Site J 13 on a terrace next to the village of 'Ayyāsh before 'Ain Abu Jimā. The village can be seen in the background. Photo from W: Michael Herles 2005.

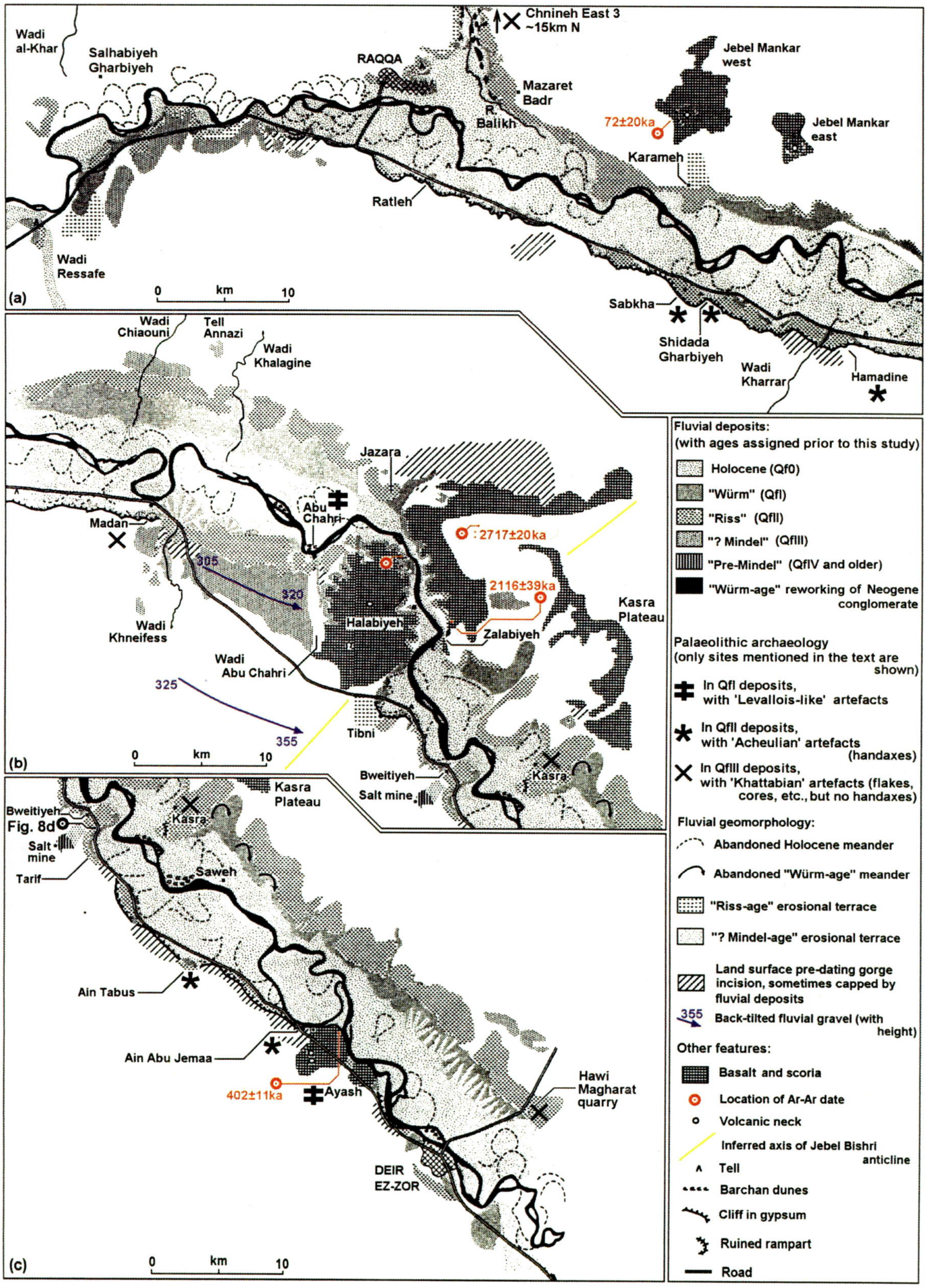

Fig. 54. The Euphrates terraces between Raqqa and Deir ez-Zor. Note that our findings add new evidence of the Middle Achaeulian tradition at the 'Ayyāsh - Ain Abu Jimā sector, Levallois tradition at Tabūs and mappings of tells. Courtesy: Westaway et al. 2007 based on the map by Besançon and Sanlaville 1981.

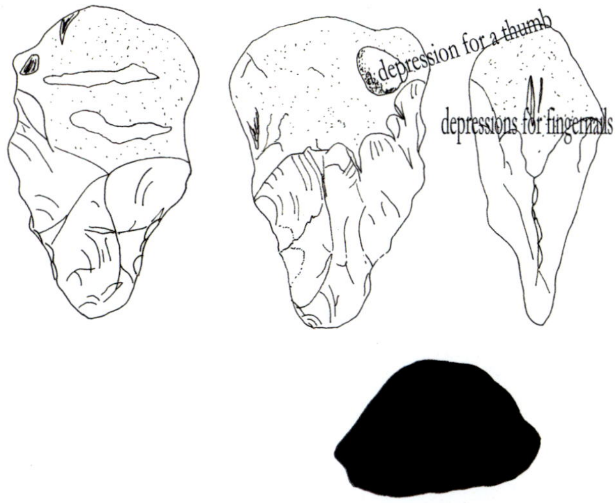

Fig. 55. A Lower Palaeolithic pick (representing the Middle Achaeulian tradition) discovered as a stray find (UTM: 0592683, 3921228) at the 'Ayyāsh - 'Ain Abu Jimā area. Scale 1:2. Drawing by Minna Lönnqvist 2006.

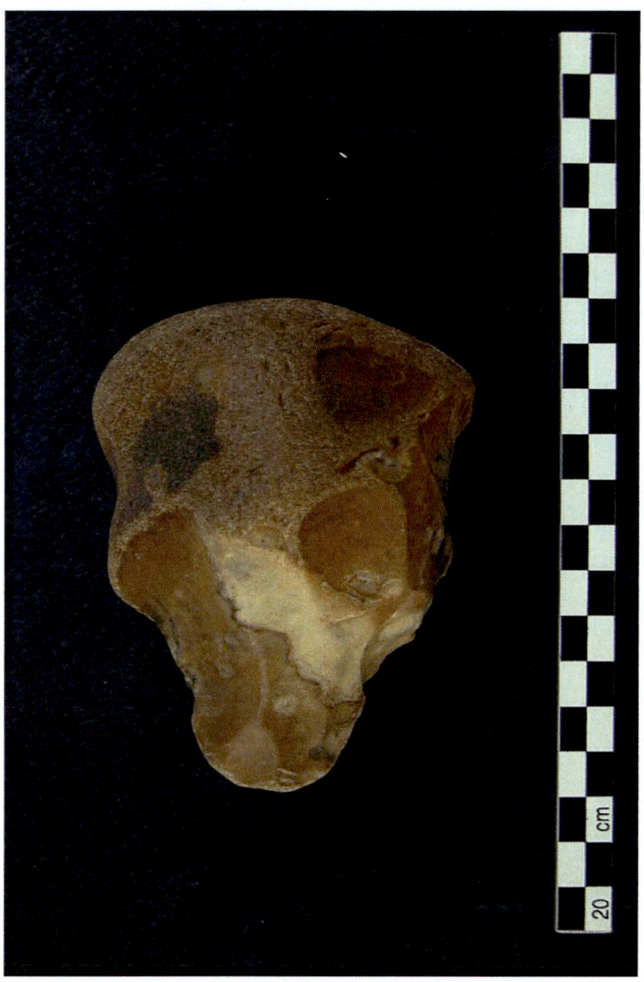

Fig. 56. A photograph of the Middle Achaeulian pick found as a stray find at 'Ain Abu Jima – 'Ayyāsh area and presented in the drawing above. Photo: Michael Herles 2005.

Fig. 57. A biface as a stray find from 'Ayyāsh. Photo: Michael Herles 2005.

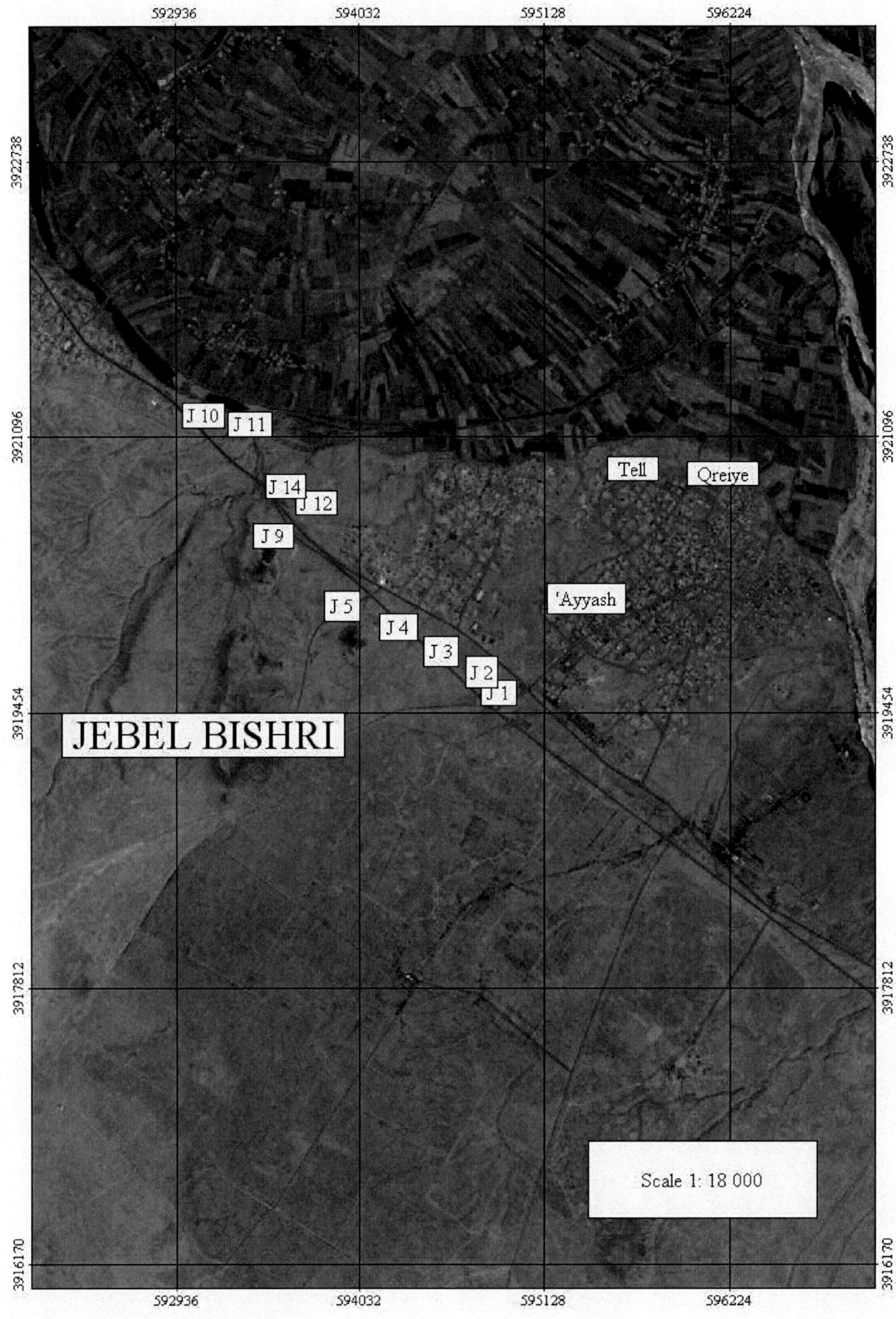

Fig. 58. Palaeolithic sites in the vicinity of the village of ʿAyyāsh. Mapping on a SPOT image by Minna Lönnqvist 2010.

Fig. 59. Site J 4 in the area of 'Ayyāsh. Photo from SSW: Michael Herles 2005.

Sites J 1 - J 5, J 9
Open Air Sites

UTM: 0595032, 3919265; UTM: 0594 872, 3919389;
UTM: 0594649, 3919584; UTM: 0594526, 3919679;
UTM: 0593827, 3920245; UTM: 0594113, 3920045;
239 - 247 m a.s.l.
Sizes: undefined

Sites from **J 1** to **J 5** (see sites on the map in Fig. 58. and especially the landscape of J 4 in Fig. 59.) as well as site **J 9** are situated near 'Ayyāsh in the alignment area of the Deir ez-Zor – Aleppo road so that the first four sites fell in between two separate lanes of the road, and the sites J 5 and J 9 were on the southern side of the road (see Fig. 58.). The finds in the vicinity of 'Ayyāsh constituted of flint concentrations of undefined sizes in the bedrock area covered by silty sands. Sites J 4 and J 5 produced both Achaeulian and Levallois-types of flint objects and debitage from the Lower and Middle Palaeolithic periods (see Figs. 60. and 61.). Site J 9 produced Palaeolithic artefacts as well (see Fig. 62.).

Fig. 60. Palaeolithic finds from site J 4 at 'Ayyāsh. Photo: Michael Herles 2005.

Fig. 61. Palaeolithic worked flint finds from site J 5 in the vicinity of 'Ayyāsh. Photo: Michael Herles 2005.

Fig. 62. Worked flint finds from site J 9 in the vicinity of 'Ayyāsh.
Photo: Michael Herles 2005.

Fig. 63. Site J 11 on a terrace producing Middle Palaeolithic finds between the villages of 'Ayyāsh and 'Ain Abu Jimā. Photo from W: Michael Herles 2005.

Sites J 10 and J 11
Open Air Sites

0593109, 3921225 and UTM: 0593185, 3921140;
226-237 m a.s.l
Sizes: J 10: 50 m x 80 m and J 11: undefined

Sites **J 10** and **J 11** (Figs. 58. and 63.) were situated on a gypseous ridge protruding to the river valley just in the border zone of the villages 'Ayyāsh and 'Ain Abu Jimā on the western side of a S-N aligned *wadi*. These sites produced plenty of worked flints. The size of the site J 10 was large, covering an area of ca. 50 m x 80 m. There were tools resembling Levallois-technique from the Middle Palaeolithic period at the sites.

Sites J 12 and J 14
Open Air Sites

UTM: 0593991, 3920688; UTM: 0593793, 3920920;
244 m a.s.l.
Sizes: undefined

Sites **J 12** and **J 14** were on the eastern side of the *wadi* closer to 'Ayyāsh. These sites including Bronze Age and Islamic cairn tombs also produced Palaeolithic finds like a large chopper tool and two Levallois-types of flake objects, the latter ones discovered at J 14. (See Figs. 64. and 65.).

Fig. 64. Palaeolithic finds from Site J 12 at 'Ayyāsh. Photo: Michael Herles 2005.

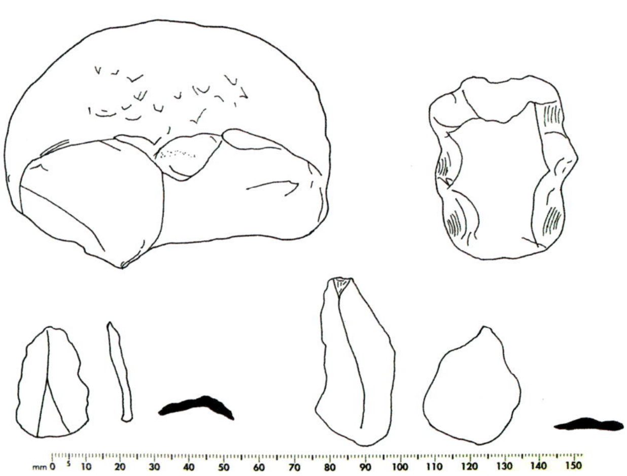

Fig. 65. Palaeolithic finds including two Levallois-type flint objects below from site J 14.

A ridge area belonging to the piedmonts of Jebel Bishri was covered by clayish silt between the villages 'Ain Abu Jimā and Kharita – Mustaha. Although there were Lower Palaeolithic finds in the 'Ayyāsh – 'Ain Abu Jimā area, in our survey the Levallois-type industries of the Middle Palaeolithic were, however, most pronounced among the Stone Age finds in the stretch between 'Ain Abu Jimā and Tell Tabūs. These prehistoric sites from **L 2** to **L 13** with a few pieces of apparent Bronze Age pottery were identified near 'Ain Abu-Jimā and El-Kharita. (Fig. 66.). Nearly all of the finds were of poor quality due to the ongoing weathering process in this desert type of environment.

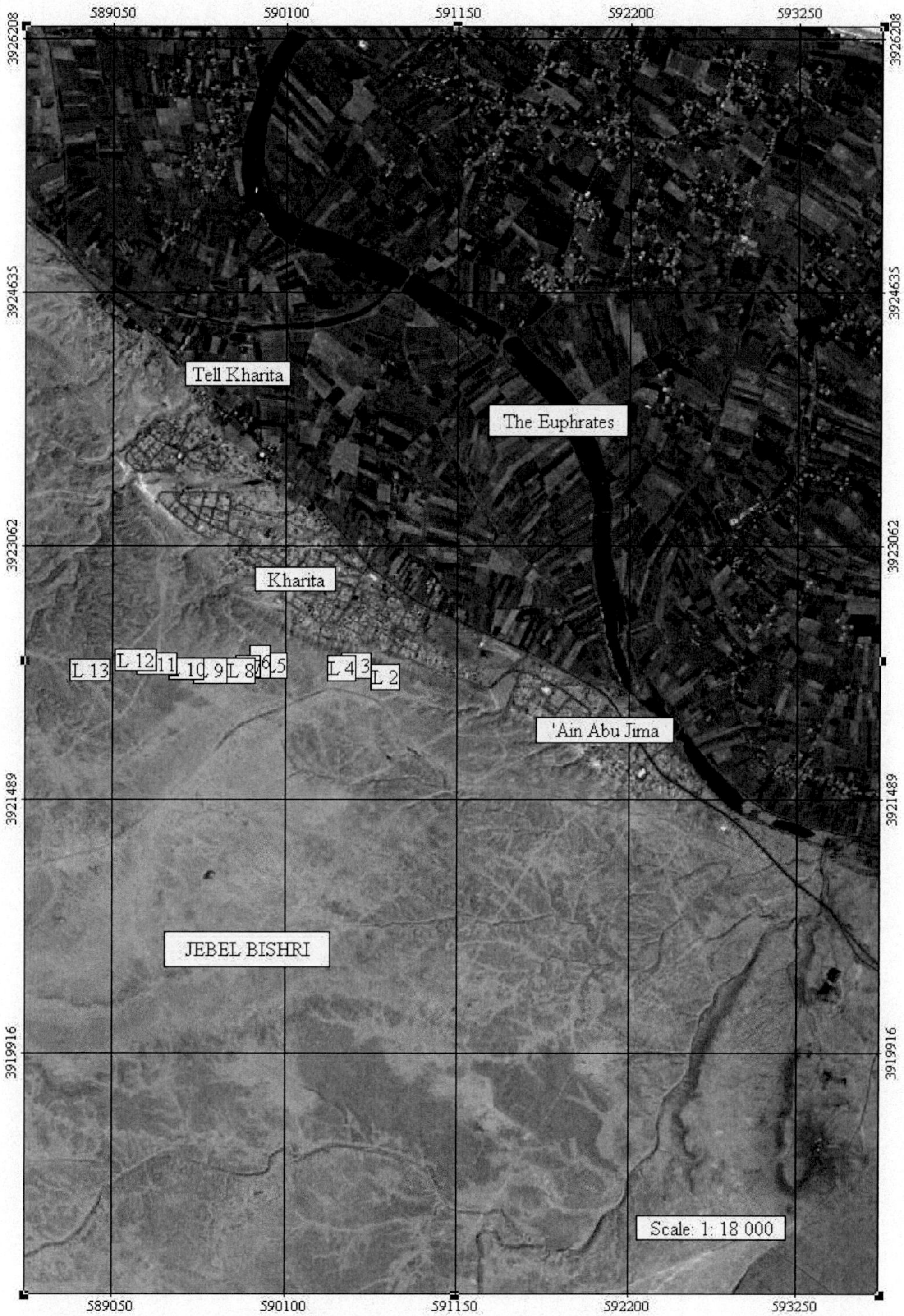

Fig. 66. Palaeolithic sites near the villages of 'Ain Abu-Jimā and El Kharita. Mapping on a SPOT image by Minna Lönnqvist 2010.

Sites L 2- L 4
Open-air Sites

UTM: 0590597, 3922084; UTM: 0590486, 3922129
UTM: 0590480, 3922232; 288 m- 294 m a.s.l.
Sizes: ca. 30/50 m x 30 m

Sites **L 2**, **L 3** and **L 4** were all situated between the villages of 'Ain Abu Jimā and El-Kharita on the gypseous ridge of Jebel Bishri overlooking the Euphrates valley (see Fig. 66.). They were located between a road leading from 'Ain Abu Jimā to Jebel Bishri and the Euphrates river valley. The sites produced Middle Palaeolithic flint tools and debitage. The site L 3 was one of the few ones on the Euphrates side on the stretch from 'Ayyāsh to Mustaha (see Figs. 67. and 68.) that apart from Palaeolithic finds produced indications of Epipalaeolithic Natufian activities. That could provide additional evidence that the Epipalaeolithic hunter-gatherers were active as far as the edge of the mountain on the Euphrates bank. K. Kohlmeyer (1984: 104) mentions a possible Natufian site in the confluence of the Euphrates and the Balikh rivers. In addition, Abu Hureyra (Moore *et al.* 2000) and Mureybet (Cauvin, M.-C. 1991) further up the Euphrates over 100 km to the northwest are the sites which, however, are the most famous ones for the Epipalaeolithic finds in the Middle Euphrates area.

Fig. 67. Worked flint finds from site L 3. Photo: Kenneth Lönnqvist 2006.

Fig. 68. Worked flint finds from site L 3. Photo: Kenneth Lönnqvist 2006.

Sites L 5 - L 9
Open Air Sites

UTM: 0589867, 3922154; UTM: 0589806, 3922076;
UTM: 0589678, 3921809; UTM: 0589482, 3921729;
UTM: 0589678, 3921809;
287-296 m a.s.l.
Sizes: L 5: 50 m x 100 m; L 6: 20 m x 30 m; L 7: 30 m x 50 m; L 8: 30 m x 50 m; L 9: 50 m x 80 m

Site **L 5** to **L 9** (Fig. 66.) were on a gypseous ridge area covered by Middle Palaeolithic flints. At site **L 9** there might also be a graveyard represented by bulbous formations on the ground covering an area of 80 m x 50 m.

Site L 10
Open Air Site

UTM: 0589343, 3921728; 250 m a.s.l.
Size: ca. 20 m x 30 m

Sites L 11 - L 13
Open Air Sites

UTM: 0589220, 3921826; UTM: 0588721, 3921601;
UTM: 0588505, 3921850;
281 m a.s.l.
Sizes: ca. 30 m x 50 m

All the sites from **L 4** to **L 11** (Fig. 66.) providing Palaeolithic flints were in a flat area consisting of silty sands, whereas sites **L 12** and **L 13** had several sandy hills and a NW-SE oriented *wadi* on the northern side of the road leading from 'Ain Abu Jimā to Jebel Bishri. To the north there were possible tombs on hills, but the area was very weathered to say anything definite about the nature of the tombs. Site L 12 produced both heavily weathered pottery pieces and Palaeolithic flints.

Fig. 69. A stitched panoramic view from a ridge of Jebel Bishri over the plateau H 1, a Pleistocene terrace, beneath Tabūs on the Euphrates side of Jebel Bishri. Photo from S: Kenneth Lönnqvist 2004.

Site H 1
Plateau

UTM: NW 0587782, 3924814; NE 0588385, 3924630; SE 0588017, 3924508: SW 0587641, 3924709; 241-252 m a.s.l.
Size: 300 m x 50 m

A Pleistocene terrace **H 1**, a plateau (Fig. 69.), which we have briefly indicated in previous chapters, is situated ca. 23 km west of Deir ez-Zor just beneath Tell Tabūs (see the location in ch. 3, Fig. 4.). Tell Tabūs is crowned by the Late Roman – Byzantine fort on the southern side of the Deir- ez-Zor – Aleppo -road. The area is inside the village of El-Kharita - Mustaha between the villages of El-Kharita and Mustaha. It is limited to the mountain to the south, the Euphrates valley to the north and two *wadis* to the east and west. The extensions of the plateau area are ca. 300 m x 150 m. As mentioned previously (ch. 3), we divided the plateau into eighteen transects and field walked it through from east to west. The surface finds included worked flints (Fig. 70.), pieces of pottery from different periods and other small finds (see ch. 10.). Among the flints there were some Palaeolithic finds of the Levallois-type industry, such as cores, tools and debitage. Some worked flints seemed also to date from the Neolithic period. We observed that up on the western slope of the mountain there were possible sites of rock shelters that had disappeared due to erosion, and the recovered Palaeolithic tools had consequently been washed by water currents down to the plateau. The area must be in the neighbourhood of 'Ain Tabous (see Fig. 54.), which L. Copeland (2004: 39) had studied in the 1970s, although we did not identify any spring related to the toponym "'Ain" in the area. Tell Tabūs itself is situated 1 km up and west of the plateau. Copeland designated the finds from the site of 'Ain Tabous as Proto-Levallois cores or debitage. Further to the northwest J. Besançon and P. Sanlaville (1981: 10) recovered Achaeulian tools at Tibne like Copeland had done (2004: 27). We also identified some flint tools at Tell Tibne, which however seem to date from the Chalcolithic period.

Fig. 70. Worked flint finds from plateau H 1. Photo: Kenneth Lönnqvist 2006.

We shortly studied some terraces near Raqqa which provided evidence of tools representing Levallois-technique. The Mousterian industry was present on the Pleistocene terraces studied by the Syro-Japanese mission between Raqqa and the northern ridge of Jebel Bishri. The sites include Barayt Tell Hammam and, 2 km east of it there is a rock shelter site called Metba'a (Al-Maqdissi and Ohnuma 2008: 151).

5.6. Towards Sedentarization on the Euphrates Side of Jebel Bishri

We did not discover any Upper Palaeolithic sites on the Euphrates side of the mountain, but as mentioned, some sites produced possible evidence of Epipalaeolithic activities and gradual sedentarization. Neolithic evidence was very meagre between 'Ayyāsh and Tell Tabūs as well, although the steppic area of Jebel Bishri (see also more in the following chapter) and El Kowm have been influenced by Neolithic activities. Only two ovoid lime plaster bowls marked as site **H 25** (UTM: 0588564, 3924794; 229 m a.s.l.) (see Fig. 71.; both 38 cm in diameter; the preserved height of one was 38 cm and the other was 15 cm) and apparent Neolithic implements in the neighbourhood of the Pleistocene terrace H 1 east of Tabūs indicate to some Neolithic impact in our research area on the Euphrates, apparently dating from the PPNB phase. On the other hand, along the Middle Euphrates there are other Neolithic sites recorded and even excavated by other missions. For example, Bouqras, ca. 60 km southeast and beyond Deir ez-Zor, is a major prehistoric site on the right bank of the Euphrates dating from the PPNB period. It represents a hunters' and shepherds' society, so still at that stage hunting was an important subsistence economy beside the evolving domestication process (Akkermans *et al.* 1983). Another site 80 km northwest is on the right bank of the Euphrates towards Raqqa discovered by K. Kohlmeyer (1984: 104-105) in his survey. The site was labelled No. 38. It produced finds dating from the Pre-Pottery Neolithic and Pottery Neolithic periods. The Syro-Japanese mission recently studying archaeological sites near Raqqa has found some Neolithic evidence on the Euphrates side as well. Further up the Euphrates Abu Hureyra has especially provided important evidence from the Pre-Pottery Neolithic period (Moore *et al.* 2000). Apart from the Pre-Pottery Neolithic sites on Jebel Bishri itself (see ch. 6.), on the desert side El Kowm basin also offers sites with Pre-Pottery Neolithic evidence (e.g. Cauvin 2000), which will be further discussed in the following chapter.

Fig. 71. Two lime plaster, or the so-called white ware, bowls discovered at plateau H 1. Photo: Kenneth Lönnqvist 2000.

The sites which have especially produced comparable lime plaster bowls, the so-called *vaiselles blanches* or white ware, dating from the PPNB, are El Kowm 1, El Kowm 2–Caracol and Abu Hureyra as well as Bouqras. All the sites are inside the radius of ca. 100 km from our site H 1. Interestingly, lime plaster was used for producing house floors, vessels and funerary attributes at Abu Hureyra 2 (Moore *et al.* 2000: 263-264, 267, 201-202), for floors and vessels at El Kowm (Dornemann 1986: 57-58) and Bouqras (de Conteson and Van Liere 1966). All of the sites were sedentary at the time, but also providing additional evidence of pastoral development. The PPNB period still falls in the time before the introduction of pottery but the production process of lime plaster vessels included similar feature to the pottery production process and specialisation, which are also often connected with pronounced sedentism. The whole process of producing lime plaster needs some specialisation and a more or less sedentary way of life. It is worth emphasizing that because of its gypsum the area of Jebel Bishri is a convenient source of the raw material for lime and gypsum based production.

5.7. Conclusions and Discussion

Jebel Bishri offers a perspective on mobile habitation and the development of the human minds through the Palaeolithic period: a million to hundreds and tens of thousands of years old open accumulated sites are showing that our historical perspective is only a glimpse of the long succession of mobile habitation in the area. The Palaeolithic era covers ca. 98-99 % of the archaeological time-scale of the one million years in the region. It is therefore quite understandable that the majority of the sites in the region of Jebel Bishri, taking into account the piedmonts, consists of countless Palaeolithic sites. However, the recording and documenting of all these open air sites, and the amount of flint artefacts, debitage and chert in an area covering ca. 1 million hectares, are naturally beyond the scope of any individual intensive surface survey. We have presented only some indicative finds from our empirical field work in small plots (see Fig. 72.) in different parts of the mountain for the purpose of site mapping and understanding the activities in different environmental sectors. Jebel Bishri is associated with the so-called Levantine corridor that participated in peopling the areas of Europe by modern humans.

As far as we have studied the region on the ground, the Palaeolithic sites are concentrating on the vicinity of the oasis of El Kowm and its basin on the western piedmont as well as on the terraces facing the Euphrates. Although other missions have traced even earlier activities in the area of El Kowm, the Euphrates area produced the oldest evidence of our survey: a Middle Achaeulian artefact from the Lower Palaeolithic period in the 'Ayyāsh – 'Ain Abu Jima area. An Achaeulian handaxe, the "Swiss army knife" of the Lower Palaeolithic, which in our case, however, may belong to the Achaeulian tradition of the Mousterian phase (cf. the reference to the Achaeulian industries in the Levalloiso-Mousterian levels at El Kowm in Le Tensorer *et al.* 2007: 625) was discovered in the central area of the mountain. This axe was found with a fossilized bone at the edge of the Nadra plain in connection with the shoreline of the palaeolake. Some vague examples of Lower Palaeolithic activities were also traced at Tar al-Sbai at the western edge of the mountain.

The Middle Palaeolithic, beside the Levallois-technique, appeared in the local variations such as the Hummalian and Yabrudian traditions on the western piedmont comparable to the sites in the El Kowm basin in general. A few possible Levallois-related points on the piedmont witness to hunting activities in the small finds of the large plateau during the Middle Palaeolithic Age. The Middle Palaeolithic evidence is further present in the district of Nadra at the edges of the palaeolake, in the oasis site of Qseybe and on the Pleistocene terraces of the Euphrates river valley. The Hummalian point found at Qseybe on Jebel Bishri is a proof (based on the local technique of Hummal) of the highland-lowland interaction between the El Kowm basin and Jebel Bishri already in the Middle Palaeolithic Age. Middle Palaeolithic evidence is also most pronounced in the prehistoric finds on the Euphrates side of Jebel Bishri. It is puzzling, why Upper Palaeolithic is invisible in the finds of the Euphrates side so far. Does this indicate that only the corridor through the El Kowm basin was used during the Upper Palaeolithic period as a point of access to the north? Based on present knowledge the indications of the Upper Palaeolithic activities seem, apart from some finds from the Nadra district, to be restricted to Tar al-Sbai on the mountain proper. Tar al-Sbai with the western piedmont could have retained close contacts with the western oases, possibly such sites as Hummal, the sites in the Qdeir plateau and Umm el-Tlel. Around Jebel Bishri there seems to be no limit to the raw material.

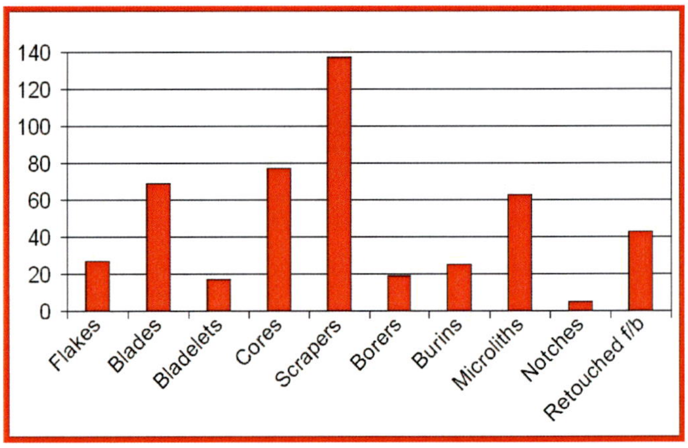

Table 1. An overall statistics of the flint industry at the studied sites on the western piedmont and Tar al-Sbai.

The stone industry of the sites on the western piedmonts and Tar al-Sbai mostly involved different kinds of scrapers (see Table 1.) that illustrate the nature of hunters' activities. At Tar al-Sbai a few points represent actual hunting itself in the Upper Palaeolithic, Epipalaeolithic and Neolithic periods. However, the existence of numerous hunting blinds attests that hunting could have been a viable activity on the slopes. On the other hand, the evidence supports the assumption that the prehistoric *abris* at Tar al-Sbai were chiefly used for butchering and processing carcasses as well as working with animal bone and antler. It is apparent that hunting took place in a kilometre's or two's radius, after which the animals were processed at the rock shelter sites. Some foraging had apparently been taking place at the sites during the Epipalaeolithic and Neolithic periods, too.

Our interpretation also involves observing the dynamics in the finds reflecting the changes in the hunter-gatherers' world in general (Binford 1983: 23). The Epipalaeolithic period of the Kebaran and Natufian phases is clearly visible in the finds in the neighbourhood of Jebel Bishri especially at Tar al-Sbai (see Fig. 72.), and this involves a more pronounced transition into semi-sedentism in the Natufian era. Apart from Tar al-Sbai, Nadra also provides sites which are transitional from the Epipalaeolithic to the Neolithic period. Those sites will be dealt with in the following ch. 6. The act of sedentarization can become seasonal, as among semi-sedentary hunter-gatherers and pastoral nomads (Kent 1989). The use of microliths means transition in stone industry and in ways of thinking. M. Neeley (2002: 48-50) describes this change of technology in the Levant as pertaining to forager behaviour with resource procurement and mobility strategy that is becoming more site-focused. Although the change to the microliths may denote alterations in the hunting methods and processing methods of carcasses, in addition to the more varied ways of collecting wild flora of different types, the light weight and easy processing are closely connected with mobility.

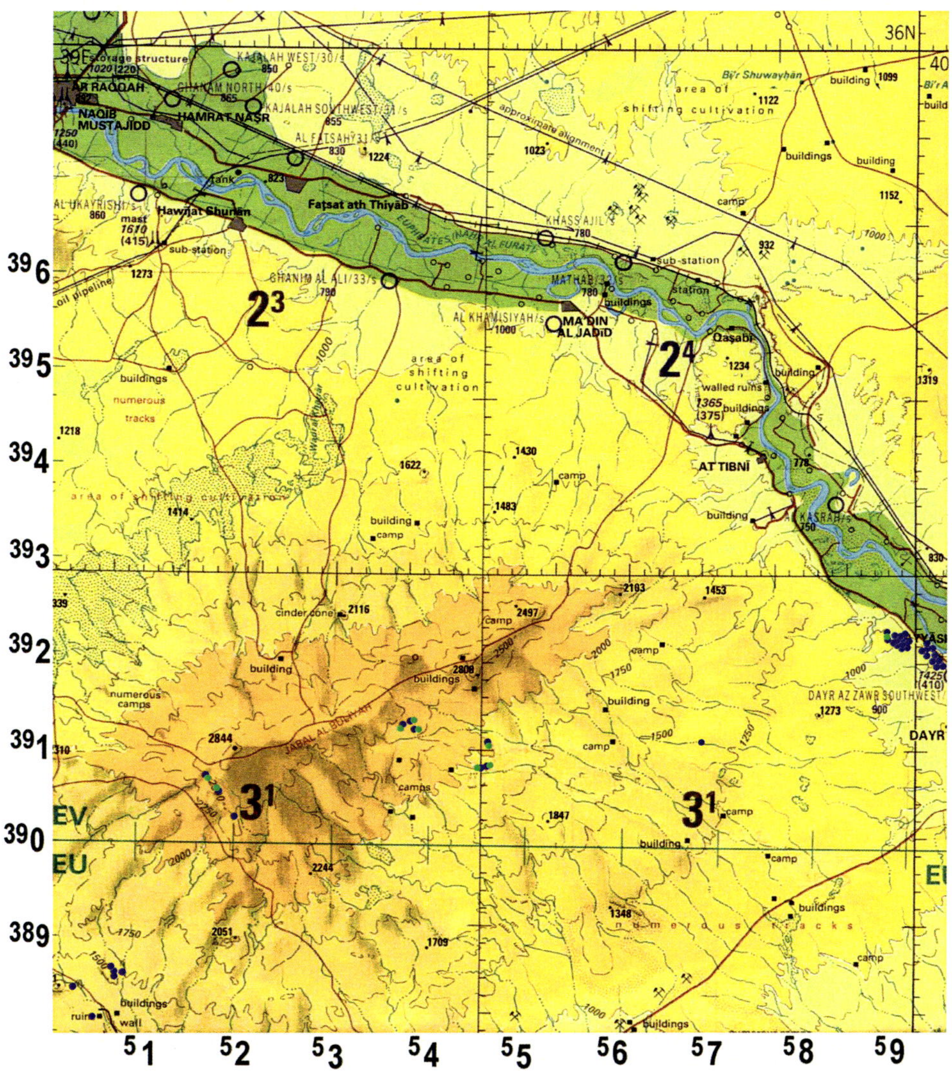

Fig. 72. The distribution of the Palaeolithic (blue) and Epipalaeolithic (green) finds in our survey districts on Jebel Bishri are presented here. Heights are in feet. Mapping: Kenneth Lönnqvist 2009. Basemap: © UK MOD Crown Copyright, 2010 published with the permission of the Controller of HMSO.

S. Kent (1991) has especially studied the effects of mobility on site size and structure among hunter-gatherers. Ecological factors are known to affect the site structure and mobility among those groups. At any rate, the environmental factors influence not only movement and sedentarization but also the size and structure of the sites reflecting group organisation. S. Kent's studies indicate that anticipated mobility had a significant impact on site structure and inventory compared to actual mobility, subsistence strategies, seasonal occupation and ethnicity. With many others, she points out that increased sedentism increases complexity. Groups of hunter-gatherers may consist of bands, tribes, clans or even chiefdoms based on kinship ties (Sahlins 1968). The complexities of their living areas are determined by the size of the group, the sedentariness and the social structure. However, the hunter-gatherers of the Palaeolithic and Epipalaeolithic times have been thought to mainly have formed bands compared to more complex hunter-gatherers of our contemporary natives. It is clear that the more sedentary the sites become, the more their sizes also tend to increase, and therefore the degree of sedentism not only affects the site size but also the permanence of the structures (Kent 1991: 33, 39).

Arid and unpredictable environmental conditions are shown to be linked with social behaviour in the form of relatively low population densities, high mobility, portable technology and fluidity in social organisation (Gould *et al.* 1972 *apud* Bird and Bliege Bird 2005: 86-87). In societies like those of the hunter-gatherers and nomads, the high degree of mobility causes abandonments and reoccupations of specific locations due to seasonal changes and the availability of resources. The material inventories, like tools and accommodation, are designed to serve mobility (cf. Schiffer 1987: 93). A number of studies have argued that there is a contrast between residential mobility and sedentism in diversity of assemblage, i.e., the number of different tool types found in an assemblage (Veth 2005: 106). The diversity has also been categorized, reflecting the fact that base camps have greater diversity than maintenance camps and extracted camps (Taylor and Smith 1978, Goodyear *et al.* 1983 *apud* Savage 1990: 343). In this scheme the sites in the El Kowm basin clearly appear as base camps, whereas the sites at Tar al-Sbai could have served as a maintenance area from which extracted camps spread deeper into Jebel Bishri.

Bibliography

Akkermans, P.A., Boema, J.A.K., Clason, A. T., Hill, S. G., Lohof, E., Meiklejohn, C., Le Miére, M., Molgot, G.M.F., Roodenberg, J.J., Waterbolk-van Rooyen, W., and van Zeist, W. (1983) Bouqras Revisited: Preliminary Report on a Project in Eastern Syria, in *Proceedings of the Prehistoric Society* , Vol. 49, pp. 335-372.

Akkermans, P.M.M.G. and Schwartz, G.M. (2003) *The Archaeology of Syria: From Complex Hunter-Gatherers to Early Urban Societies (ca. 16,000-300 B.C.)*. Cambridge World Archaeology. Cambridge: Cambridge University Press.

Al-Maqdissi, M. and Ohnuma, K. (2008) Preliminary Reports of the Syria-Japan Archaeological Joint Research in the Region of Ar-Raqqa, Syria, 2007, in *Al-Rāfidān*, Vol. XXIX, pp. 117-196.

Anfinset, N. (2008) Towards a Specialised Food Production – A Look at Jebel Bishri in *Jebel Bishri in Context, Introduction to the Archaeological Studies and the Neighbourhood of Jebel Bishri in Central Syria,* Proceedings of a Nordic Research Training Seminar in Syria, May 2004, ed. Lönnqvist, M., British Archaeological Reports, International Series 1817, Oxford: Archaeopress, pp. 1-13.

Barker, G., Gilbertson, D.D. and Mattingly, D.J. (2007) *Archaeology and Desertification: The Wadi Faynan Landscape Survey, Southern Jordan.* Wadi Faynan Series 2. Levant Supplementary Series 4. Oxford: Oxbow Books.

Bar-Yosef, O. (1987) Late Pleistocene Adaptations in the Levant, in *The Pleistocene Old World: Regional Perspectives, Interdisciplinary Contributions to Archaeology*, ed. by Soffer, O., New York – London: Plenum Press, pp. 210-236.

Bar-Yosef, O. and Valla, F.R. (ed.) (1991) *The Natufian Culture in the Levant.* International Monographs in Prehistory. Archaeological Series 1. Ann Arbor, Michigan: International Monographs in Prehistory.

Belfer-Cohen, A. and Goring-Morris, N. (2002) Why Microliths? Microlithization in the Levant, in *Thinking Small: Global Perspectives on Microlithization*, ed. by Elston, R.G. and Kuhn, S.L., Archeological Papers of the American Anthropological Society, Vol. 12, pp. 57-68.

Besançon, J., Copeland, L., Hours, F., Muhesen, S. and Sanlaville, P. (1981) Le Paléolithique d'El Kowm, Rapport préliminaire, in *Paléorient,* Vol. 7/2, pp. 33-55.

Besançon, J., Copeland, L., Muhesen, S. and Sanlaville, P. (1994) Prospection géomorphologique et préhistorique dans la region de Tartous (Syrie), in *Paléorient*, Vol. 20, pp. 5-19.

Besançon, J. and Sanlaville, P. (1981) Aperçu Géomorphologique sur la Vallée de l' Euphrate Syrien, in *Paléorient,* Vol. 7/2, pp. 5-18.

Binford, L. (1983) *In Pursuit of the Past: Decoding the Archaeological Record.* Berkeley and Los Angeles. California: University of California Press.

Bird, D.W. and Bliege Bird, R. (2005) Evolutionary and Ecological Understandings of the Economics of Desert Societies: Comparing the Great Basin USA and the Australian Deserts, in *Desert Peoples: Archaeological Perspectives*, ed. by Veth, P., Smith, M. and Hiscock, P. Oxford: Blackwell Publishing, pp. 81-99.

Cauvin, J. (1991) El Khabra: un poste de chasse néolithique du PPNB final de faciès Qdeir (Oasis d'El Kowm, Syrie), in *Cahiers de L'Euphrate* 5-6, pp. 47-53.

Cauvin, J. (2000) *The Birth of the Gods and the Origins of Agriculture.* Trans. Watkins, T. Cambridge: Cambridge University Press.

Cauvin, M.-C. (1991) Du Natoufien au Levant Nord? Jayroud et Mureybet (Syrie), in *The Natufian Culture in the Levant*, ed. by Bar-Yosef, O. and Valla, F.R., International Monographs in Prehistory, Archaeological Series 1, Ann Arbor, Michigan, pp. 295-314.

Cauvin, M.-C. and Coqueugniot, E. (1988) L'Oasis d'El Kowm et le Kébarien Géometrique, in *Paléorient,* Vol. 14/2, pp. 270-282.

Copeland, L. (1985) The Pointed Tools of Hummal Ia, in *Cahiers de l'Euphrate* 4, pp. 177-189.

Copeland, L. (2004) The Paleolithic of the Euphrates Valley in Syria, in *From the River to the Sea, The Palaeolithic and the Neolithic on the Euphrates and the Northern Levant,* Studies in honour of Lorraine Copeland, ed. by Aurenche, O., La Mière, M. and Sanlaville, P., Maison de L'Orient et de la Méditerranée Jean Pouilloux, BAR International Series, No. 1263, Oxford: Archaeopress, pp. 19-61.

De Contenson, H. and van Liere, W.J. (1966) Premier sondage à Bouqras en 1965, Rapport préliminaire, in *Les Annales Archéologiques Arabes Syriennes*, Vol. 16, pp. 181-192.

Dornemann, R.H. (1986) *A Neolithic Village at Tell el Kowm in the Syrian Desert.* Studies in Ancient Oriental Civilization No. 43. Chicago: The Oriental Institute of the University of Chicago.

Evans, J.G. (1978) *An Introduction to Environmental Archaeology.* London: Elek Books Ltd.

Fujimoto, T. (1979) Upper Paleolithic and Epi-Paleolithic assemblages in the Palmyra Basin, in *Paleolithic Site of the Douara Cave and Paleogeography of Palmyra Basin in Syria, Part II: Prehistoric Occurrences and Chronology in Palmyra Basin*, ed. by Hanihara, K. and Akazawa, T., Tokyo: University of Tokyo Press, pp. 77-130.

Garrard, A.N. and Stanley-Price, N.P. (1975-1977) A Survey of Prehistoric Sites in the Azraq Basin, Eastern Jordan, in *Paléorient*, Vol. 3, pp. 109-123.

Garrard, A., Baird, D., Colledge, S., Martin, L. and Wright, K. (1994) Prehistoric Environment and Settlement in the Azraq Basin: an Interim Report on the 1987 and 1988 Excavation Seasons, in *Levant,* Vol. XXVI, pp. 73-109.

Gilead, I. (1991) The Upper Paleolithic period in the Levant, in *Journal of World Prehistory*, Vol. 5, pp. 105-154.

Gould *et al.* 1972 *apud* Bird, D.W. and Bliege Bird, R. (2005) Evolutionary and Ecological Understandings of the Economics of Desert Societies: Comparing the Great Basin USA and the Australian Deserts, in *Desert Peoples: Archaeological Perspectives*, ed. by Veth, P., Smith, M. and Hiscock, P., Oxford: Blackwell Publishing, pp. 81-99.

Henry, D.O. (1974) The Utilization of the Microburin Technique in the Levant, in *Paléorient*, Vol. 2/2, pp. 389-398.

Henry, D.O. (1989) *From Foraging to Agriculture: The Levant at the End of the Ice Age.* Philadelphia: University of Pennsylvania Press.

Henry, D.O. (1995) *Prehistoric Cultural Ecology and Evolution: Insights from Southern Jordan.* Interdisciplinary Contributions to Archaeology. New York: Plenum Press.

Henry, D.O. and Shen, C. (1995) The Madamaghan Sites, in *Prehistoric Cultural Ecology and Evolution: Insights from Southern Jordan*, ed. by Henry, D.O., Interdisciplinary Contributions to Archaeology, New York: Plenum Press, pp. 295-318.

Kent, S. (1989) Cross-cultural perceptions of farmers as hunters and the value of meat, in *Farmers as Hunters*, ed. by Kent, S., New Directions in Archaeology, Cambridge: Cambridge University Press, pp. 1-17.

Kent, S. (1991) The Relationship between Mobility Strategies and Site Structure, in *The Interpretation of Archaeological Spatial Patterning,* Interdisciplinary Contributions to Archaeology, ed. by Kroll, E.M., Price, T.D., Springer: New York – London, pp. 33-60.

Kohlmeyer, K. (1984) Euphrat-Survey, Sie mit Mitteln der Gerda Henkel Stiftung durchgeführte archäologische Geländebegehung im Syrischen Euphrattal, in *Mitteilungen der Deutschen Orient-Gesellschaft,* Vol. 116, pp. 95-11.

Le Tensorer, J.-M., Muhesen, S. and Schmid, P. (2006) *Research on the Paleolithic of the El Kowm Area (Syria).* Abstract.

Le Tensorer, J.-M, Jagher, R., Rentzel, P., Hauck, T., Ismail-Meyer, K., Pümpin, C. and Wojtczak, D. (2007) Long-term Site Formation Processes at the Natural Spring Nadaouiyeh and Hummal in the El Kowm Oasis, Central Syria, in *Geoarchaeology,* Vol. 22, pp. 621-639.

Legge, A.J. and Rowley-Conwy, P.A. (2000) The Exploitation of Animals, in *Village on the Euphrates: From Foraging to Farming at Abu Hureyra,* by Moore, A.M.T., Hillman, G.C. and Legge, A.J., pp. 423-471.

Lévi-Strauss, C. (1966) *The Savage Mind.* The Nature of Human Society Series. Chicago: The University of Chicago Press.

Lewis-Williams, D. (2002) *The Mind in the Cave.* London: Thames & Hudson.

Lönnqvist, M. and Törmä, M. (2003) SYGIS – The Finnish Archaeological Project in Syria, in Proceedings of the XIXth International Symposium *CIPA 2003* (The ICOMOS & ISPRS Committee for Documentation of Cultural Heritage) New Perspectives to Save Cultural Heritage, Antalya, Turkey, ed. by Altan, M.O., *The ISPRS International Archives of the Photogrammetry, Remote Sensing and Spatial Information Sciences,* Vol. XXXIV-5/C15, Istanbul, pp. 609-614.

Lönnqvist, M. with the contributions by Törmä, M., Nuñez, M., Lönnqvist, K. Stout Whiting, M., Okkonen, J., Riihiaho, H. and Nissinen, M. (2006) Archaeological Surveys of Jebel Bishri, The Preliminary Report of the Finnish Mission to Syria 2000-2004, in *KASKAL,* Vol. 3, pp. 205-242.

Madoun, M.A. (1996) *The Silk Road and Palmyra, Battering Cultures.* Damascus: Al Namir.

McLaren, S., Reynolds, T., Gilbertson, D., Grattan, D., Hunt, C., el-Rishi, H., Barker, G. and Duller, G. (2007) Pleistocene environments and human settlement, in *Archaeology and Desertification: The Wadi Faynan Landscape Survey, Southern Jordan,* by Barker, G., Gilbertson, D.D. and Mattingly, D.J., Wadi Faynan Series 2, Levant Supplementary Series 4, Oxford: Oxbow Books, pp. 177-198.

Mithen, S. (1996*) The Prehistory of the Mind: A Search for the Origins of Art, Religion and Science.* London: Thames & Hudson.

Moore, A.M.T., Hillman, G.C. and Legge, A.J. (2000) *Village on the Euphrates: From Foraging to Farming at Abu Hureyra,* Oxford: Oxford University Press.

Muhesen, S. (1993) Chapitre II: L'Acheuléen récent évolué de l'Oronte, in *Le Paléolithique de la vallée moyen de l'Oronte* (Syrie), BAR International Series, No. 587, Oxford: Archaeopress.

Neeley, M. P. (2002) Going Microlithic: A Levantine Perspective on the Adoption of Microlithic Technologies, in *Archeological Papers of the American Anthropological Society,* Vo. 12, No. 1, pp. 45-55.

Ohnuma, K. with contributions by C.A. Bergman and M.H. Newcomer (1988) *Ksar Akil, Lebanon: A Technological Study of the Earlier Upper Palaeolithic* Levels *of Ksar Akil.* Vol. III: Levels XXV-XIV. BAR International Series. No 426. Oxford: BAR.

Ploux, S. and Soriano, S. (2003) Umm el Tlel, Une Séquence du Paléolithique Supérieur en Syrie Centrale, Industri des Lithiques et Chronologie Culturelle, in *Paléorient,* Vol. 29/2, pp. 5-34.

Poidebard, A. (1934) *La Trace de Rome dans le Désert de Syrie,* Le Limes de Trajan a la conquète arabe, recherches aériennes (1925-1932), Texte, Atlas, BAH T. XVIII. Paris: Geuthner.

Sahlins, M.D. (1968) *Tribesmen.* Foundations of Modern Anthropology Series. Englewood Cliffs, New Jersey: Prentice-Hall, Inc.

Sanlaville, P. (2004) Les Terasses Pléistocènes de la Vallée de l'Euphrate en Syrie et dans l'Extrême Sud de la Turquie, in *From the River to the Sea, The Palaeolithic and the Neolithic on the Euphrates and in the Northern Levant, Studies in honour of Lorraine Copeland,* ed. by Aurence, O., Le Mière, M. and Sanlaville, P., Maison de L'Orient et de la

Méditerranée Jean Pouilloux, BAR International Series, No. 1263, Oxford: Archaeopress, pp. 115-127.

Savage, S.H. (1990) Modelling the Late Archaic Social Landscape, in *Interpreting Space: GIS and Archaeology*, ed. Allen, M.S., Green, S.W., Zubrow, E.B.W. London – New York – Philadelphia: Taylor & Francis, pp. 330-355.

Schiffer, M. (1987) *Formation Processes of the Archaeological Record.* Albuquerque: University of New Mexico Press.

Smith, A.B. (2005) Desert Solitude: The Evolution of Ideologies Among Pastoralists and Hunter-Gatherers in Arid North Africa, in *Desert Peoples: Archaeological Perspectives*, ed. by Veth, P., Smith, M. and Hiscock, P., Malden, Oxford and Carlton: Blackwell Publishing, pp. 261-275.

Westaway, R., Bridgland, D., Challis, K. and White, M. (2005) A Dating Framework for the Lower and Middle Palaeolithic in Levantine Rivers, in *CBRL 2005, Newsletter of the Council for the British Research in the Levant,* pp. 31-33.

Westaway, R., Bridgland, D., Daoud, M., Radwan, Y., Abu Romieh, M., Yassminh, R. and Al-Ashkar, A. (2007) Investigations of Euphrates terraces and related archaeology, NE Syria, 2006-2007, in *CBRL 2005, Newsletter of the Council for the British Research in the Levant 2*, pp. 82-89.

Veth, P. (2005) Cycles of Aridity and Human Mobility: Risk Minimization Among Late Pleistocene Foragers of the Western Desert, Australia, in *Desert Peoples: Archaeological Perspectives*, ed. by Veth, P., Smith, M. and Hiscock, P. Oxford: Blackwell Publishing, pp. 100-115.

Vila, E. (1991) Note sur un dépôt de gazelles dans une fosse Uruk d'El Kowm 2 – Caracol (Syrie), in *Cahiers de l'Euphrate* 5-6, pp 55-60.

Williams, J.K. (2006) The Levantine Aurignacian: a closer look, in *Towards a Definition of the Aurignacian, Proceedings of a Symposium Held in Lisbon, Portugal, June 25-30, 2002*, ed. by Bar-Yosef, O. and Zilhão, J., pp. 317-351.

6. VISTAS ON ANCIENT PASTORAL NOMADISM IN THE CENTRAL AREAS OF THE MOUNTAIN

Minna Lönnqvist, Kenneth Lönnqvist, Markus Törmä
with Contributions by Margot Stout Whiting and Milton Nuñez

Fig. 1. Pastoral corrals at Nadra on Jebel Bishri. Photo: Christian Meyer 2004.

6.1. Transition to Pastoral Nomadism

Mobile societies live in ecological time, which is cyclic. Eco-time consists of periods following the yearly movement of the sun. Hunters and fishers are dependent on seasonal availability of subsistence like pastoral nomads for the availability of grazing areas. People can move from a camp to a hamlet or a village according to the season and the availability of food. (Aveni 2002: 148-152). In the societies of hunters and pastoralists the symbiosis of humans with animals is central. Hunter-gatherers like pastoral nomads depend on animals in their subsistence economy.

Goats and sheep are the main livestock reared in the region of Jebel Bishri today, although grazing of dromedaries is frequent. As previously mentioned, the Bedouin tribes occupying the area today are all originally camel nomads (see more in ch. 14). Donkeys and dogs are kept as aid in the rearing of livestock. Goats/sheep were apparently domesticated in the late Epipalaeolithic or Pre-Pottery Neolithic A (PPNA) period in the Near East. The earliest ninth millennium B.C. dating of the first domesticated caprovides from the Zagros area of Iraqi Kurdistan (Perkins 1964: 1565-1566) have, however, been challenged in recent years (see, e.g., Zeider 1999: 12). In any event, pastoral nomadism based on goat and sheep rearing became a livelihood in the Levantine region in the Pre-Pottery Neolithic B (PPNB) period (see, e.g., Rosen 1988; Zarins 1989a; Bar-Yosef – Khazanov eds. 1992; Peters *et al.* 1999).

The transition to pastoral nomadism in the Pre-Pottery Neolithic B (PPNB) period is especially attested in the neighbourhood of Jebel Bishri. Recent studies in Wadi Rum in Jordan have also provided additional evidence of pastoralism in the early Neolithic periods (Albert and Henry 2004). In the neighbourhood of Jebel Bishri the closest sites providing the actual evidence of pastoral nomadism from the PPNB period are situated in the oases of El Kowm (Cauvin 1981, 1990; Zarins 1989a; Stordeur *et al.* 1991) and Qdeir-1 on the western piedmont (Stordeur 1993; Stordeur and Taha 1996), and they are among the earliest ones in the Near East. Stratigraphy from El Kowm 2–Caracol attests to cyclic occupational past typical of nomadic cultures: there are layers witnessing destructions, abandonments and reoccupations (Stordeur *et al.* 1991: 37-39). These are processes comparable with the modern history of the Bedouin cultures in the neighbourhood (see ch. 14. and 15.) The transition to agriculture and animal domestication at Abu Hureyra up the Euphrates is attested from the Epipalaeolithic to the PPNB Neolithic layers, the first domesticated cereals dating from ca. 9000 B.C. (Moore *et al.* 2000). Another previously mentioned PPNB site in the neighbourhood is Bouqras down the Euphrates (Akkermans *et al.* 1983), and sites in the Palmyra basin provide PPNB evidence as well (Fujii *et al.* 1987). Jebel Bishri connected with the Palmyrides is therefore located between three notable PPNB sites, namely El Kowm, Abu Hureyra and Bouqras. A. Moore (1982: 16) has demonstrated how PPNB sites in the region are more numerous in steppic areas and seem to be related to pastoral nomadism in the region.

In the case of the El Kown basin the question concerning the development of pastoral nomadism contemporaneously with irrigation agriculture is especially intriguing. This is the nature of the symbiosis of the small livestock rearing with the agricultural life, village bases on the Euphrates and the surrounding oases today. Evidence of artificial water channels from the site of El Kowm 2 - Caracol (Stordeur 2000: 46) and provisionally water cisterns from Jericho in the Jordan valley (Kenyon 1981: 7)[1] indicate to first steps in the introduction of drainage and irrigation technology in an oasis. Both sites were linked with desert environment and have produced evidence of the development of early agriculture and animal husbandry during the Pre-Pottery Neolithic period. The evidence of drainage technology at El Kowm has been dated to ca. 7000-6500 B.C. and is therefore apparently earlier than the channel irrigation system recorded in the Samarran site of Choga Mami in Iraq (cf., e.g., Oates and Oates 1976:64).

S.H. Lees and D.G. Bates (1974) earlier offered a model of the transition from mixed rain-fed agriculture and pastoralism to irrigation agriculture and nomadic pastoralism. The authors of the model saw the symbiosis of the irrigation farming and livestock economy as the advent of the specialized nomadic pastoralism which is a more advanced form of pastoralism than the initial stage. The specialized pastoral nomadism, which is based on the secondary products, such as milk, hair and skin, was earlier thought by A. Sherratt (1981; 1983) to have developed in the Chalcolithic and Early Bronze Ages. However, nowadays the secondary exploitation of animals has been agreed to go back to the Neolithic times. Evidence to support this has especially come from Turkey (Grigson 1995; Kansa *et al.* 2009).

J. Cauvin (2000: 196-197) rejects the idea that irrigation agriculture served as a platform on which pastoral nomadism itself was initially developed. Also, in our view, one needs to explore more deeply the transition to pastoralism in arid areas - such as the region of Jebel Bishri in particular - to find out whether the transition in the more desert-steppe areas took different directions and represented an even earlier alternative form of mobility than village herding in oases and river valleys. A.J. Legge and P. Rowley-Conwy (1987) note that in the Abu Hureyra district, gazelle killing and extinction of the species in the desert-steppe areas stimulated the transition to pastoral nomadism in connection with the local hunter-gatherers. This would pose the question whether pastoral nomadism in arid areas developed directly from the hunter-gatherer phase and semi-sedentarism along *wadis* (cf. Henry 1995: 373). Beside Jordan, some evidence of this kind of transition can be sought from Judea, the Negev and Sinai (Rosen 1984; Horwitz 2003). Also the question of the nature of the *transhumant* type of pastoralism at this stage needs to be studied more closely and, as will be discussed, the earliest pastoralism in the area did not necessarily have to represent the *transhumant* type.

As far as the beginning of camel nomadism in the region is concerned, I. Köhler-Rollefson (1993), a specialist in camels in archaeozoology, has argued for the early domestication of camels. She holds that the domestication of dromedaries in the Near East took already place in the Early Bronze Age, while even earlier evidence can be sought in Central Asia. Camel caravans may therefore have been traversing the Arabian and Syrian Deserts in the neighbourhood of Jebel Bishri as early as in the Bronze Age, as the camel provides the only means for carrying long-distance burdens over wide desert areas. However, camel nomadism is an avenue to a fuller type of nomadism compared to the rearing of small livestock that is often more tied with agricultural areas.

6.2. Tracing Pastoral Sites and Structures

J. Zarins (1992: 49-51) has stated that, although the initial phase of pastoral nomadism dates from the Pre-Pottery Neolithic B (PPNB) and the Pottery Neolithic periods, the actual climax of the pastoral culture in the area of the Greater Southwest Asian Arid Zone took place in the Chalcolithic and Early Bronze Ages. In the period of this climax, cairns and stone circles of different

[1] However, O. Bar-Yosef doubts K.M. Kenyon's suggestion that some of the structures surrounding the PPNA stone tower would be water tanks. (See Bar-Yosef 1981: 161).

kinds appear as the major components of the nomadic technocomplex. This is also the case of the more steppic pastoral landscape of the Jebel Bishri proper. Other structural features which Zarins includes in the technocomplex are kites, the large animal traps referred to in the previous chapter, and cultic platforms as well as pillars. The latter cultic remains we have not, however, been encountered in the area of Jebel Bishri.

Attaining visibility of pastoral sites is a challenge for an archaeologist (see, e.g., Chang 2006: 193-195). Nomads use a lot of perishable materials which do not leave signs in the archaeological record. But the so-called site fixtures can be found and documented (Cribb 1991: 69). Because of their high spatial resolution aerial photographs - in our case only a couple of those provided by A. Poidebard – or satellite images such as QuickBird are data sources that could be used in prospecting flimsy pastoral sites, such as cairns/*tumuli* or stone enclosures, by remote sensing. IKONOS images have been successfully used by other projects to study cairns (Bradbury 2009). Radar images have been utilized with some success in other studies as well. Beside the systematically surveyed areas on the ground we carried out some prospecting of corral sites with a QuickBird satellite image (see ch. 7).

Specifically at Nadra (see Fig. 2, see more of the location and landscape in ch. 4.) in the central eastern area of Jebel Bishri and at Tar al-Sbai (see ch. 8.) at the western edge of the mountain many sites of pastoral nomads exemplified by cairns/*tumuli* and stone enclosures were recorded in our systematic ground surveys. The occurrence of cairns/*tumuli* is not exceptional in these areas, as the whole space of the mountain proper is evenly scattered by these tombs on higher ridges. Stone enclosures including corrals (Fig. 1.), i.e. animal pens (cf. Haiman 1992), even if frequent in the central areas, are not as numerous as cairns/*tumuli* covering the whole region.

The cairns/*tumuli* are constructed of limestone/gypsum cobbles varying from 20 cm to 30 cm in diameter; and the size of the tombs varies from ca. 3 m to 20 m in diameter, being generally 0.5 m to 1 m in height. Often these tombs are mixed structures of stones and earth, and this is why we use the designation "cairn/*tumulus*" instead of "cairn" or "*tumulus*".

The existence of stone enclosures in the neighbourhood of Jebel Bishri (Dussaud 1929; Buccellati and Kelly-Buccellati 1967; Zarins 1992) has been known for decades, but exact location information has not been published before our project (Lönnqvist and Törmä 2003; Lönnqvist *et al.* 2006). While most of these stone enclosures can be identified as corrals, some may be hut bases of camps, and others are parts of megalithic complexes with stone rings and rows. The sizes of the enclosures vary from ca. 3 m to 31 m in diameter (see more on the stone enclosures in ch. 7. and 8.). Stone-built corrals are common nomadic structures in the Near East, but some in Iran and Afghanistan have been made of mud (Cribb 1991: 96). In his aerial images Yann Arthus Bertrand also offers a vivid example of corrals of the present-day Peul nomads and their huts at Timbuktu in Mali (see: http://www.yannarthusbertrand2.org/index.php?option=com_datsogallery&Itemid=27&func=detail&catid=63&id=1513&p=1&l=1152). The layouts of these corrals and huts find parallels with the ancient sites of Jebel Bishri.

The reasons for the number, size and function of enclosures in a pastoral compound have been lucidly explained, for example, by M. Haiman (2002) in his ethnoarchaeological studies of pastoral sites in the Negev. According to Haiman's observations, the sheep were allocated to different enclosures according to age and their function in the pastoralists' subsistence system. The enclosures varied from 5 m to 31 m in diameter. An example site with six enclosures was divided so that one enclosure was for 30 lambs, one for 150 ewes that had not given birth yet, one was a sleeping area, one was allocated to 80 ewes already having lambs, one was for 40 up to 3 month old lambs that were not yet pastured and one for 16 sheep that were not pastured yet. Similar functions have also been observed at ancient sites by F. Eddy and F. Wendorf *et al.* (1999) during their surveys in Sinai.

Fig. 2. Sites I 1- I 38 and C 10 at Nadra and its neighbourhood.
Mapping by Kenneth and Minna Lönnqvist.

6.3. Transitional Sites from Mobile to Semi-Sedentary Hunter-Gatherers and Pastoral Nomads in the Central Area of the Mountain

Site C 10 including site I 36 Stone Enclosures/Corrals

UTM coordinates: 0545940, 3908253; ca. 590 m a.s.l.
Size of the site C 10 (including I 36): 80 m x 100 m

Sizes of the enclosures at C 10: from east to west as follows a) 12.1 m x 14.7 m, b) 13.1 m x 16.3 m, c) 10.5 x 16.7, d) 11.2 m x 8.8 m, e) 13.4 m x 17.2 m. (See a sketched plan in Fig. 63.)

The site **C 10** (see the location on map Fig. 2.) consists of at least five curvilinear stone enclosures and a mortar as a site fixture. The site was discovered during the spring of 2000 in the central area of the mountain. It is situated on the southern side of the Shula - Ash-Shujiri asphalt road that leads over the central part of Jebel Bishri in a NW-SE-direction. The site can be reached by driving ca. 30 km from Shula towards Ash-Shujiri. The stone enclosures can be found behind the first protruding rock formation in an otherwise undulating plain area. A few hundred metres to the south there is a large seasonal *wadi*, which has provided water for the site.

This site C 10 is associated with site **I 36** (UTM 0546150, 3908053, 595 m a.s.l.), which was discovered in 2004 on the other, northern side of the asphalt road (Fig. 2.). I 36 incorporates only one visible stone enclosure (5 m in diam; see a sketch in Fig. 66.) whereas others may have been destroyed by the road. The site C 10, including I 36, has produced Middle Palaeolithic, Epipalaeolithic and Neolithic tools. Therefore its nature from hunter-gatherer activities to semi-sedentarism and pastoral nomadism makes it clearly transitional. The site lies on the ancient shores of the palaeolake of Nadra discussed in the previous chapters. Natufian type microliths such as a lunate and some points were discovered at the site. We also collected cores, scrapers, awls, points, knives and perforators dating from the Neolithic period. A piece of obsidian may be associated with the Neolithic activities at the site. In addition, among the finds there was pottery from the Early Bronze Age and a fragment of a glass bracelet from the Byzantine/Islamic period as well. The surrounding cliff faces had Arabic inscriptions. (See more in ch. 14.). The site C 10/I 36 is important in understanding the shift from mobile hunter-gatherer culture to semi-sedentarism and pastoralism. The stone circles, possible corrals, are visible structures signifying the site C 10/ I 36. The date to which these structures in specific belong is difficult to assess without an excavation.

Fig. 3. Stone circles of the multi-period site C 10 and a seasonal wadi to the south. Photo from N above: Minna Lönnqvist 2000.

Fig. 4. A stone circle of the site C 10 still visible in the year 2008, while some of the circles had either been covered by sands or their stones had been scattered around. Photo from N above: Minna Lönnqvist 2008.

Fig. 5. A mortar found between stone enclosures of C 10. Photo from S: Minna Lönnqvist 2000.

Fig. 6. Palaeolithic worked flint finds from C 10. Photo: Kenneth Lönnqvist 2006.

Fig. 7. Flint implements from site C 10 including PPN evidence. Photo: Kenneth Lönnqvist 2004.

Fig. 8. Flint implements from site C 10, including PPNB arrow heads, scrapers, awls/borers and a piece of obsidian in the lower row to the far right. Photo: Kenneth Lönnqvist 2006.

Fig. 9. Worked flint finds including PPN perforators from site I 36 associated with site C 10. Photo: Kenneth Lönnqvist 2004.

Fig. 10. Pottery from site C 10. Photo: Kenneth Lönnqvist 2008.

Fig. 11. A general view over the site I 16 on the western bank of a large wadi at Nadra. Photo from N: Gullög Nordquist 2004.

Site I 16
Settlement

UTM: 0545733, 3912088; 573 m a.s.l.
Size: ca. 100 m x 100 m.
No visible structures

In the area of Nadra there is a multi-period semi-sedentary settlement site **I 16** that was already occupied in the prehistoric times (Fig. 11.). It is situated on the western bank of a large *wadi* (Fig. 2.), the bank forming a small hill. Ca. 200 m to the west rises a large ring-*tumulus* **I 17** overlooking the site. A mill-stone (size 25 cm x 31 cm, 6.7 cm in thickness) was discovered in the initial inspection at site I 16 (coordinates for the millstone UTM: 0545676, 3912170), but it unfortunately disappeared while documentation proceeded (possibly local Bedouins took it after having seen us working at the site).

Altogether 142 flint objects or flakes were collected at the site and 58 pottery sherds. The microlithic tools recovered from the site seem to include material that is transitional from the Epipalaeolithic[2] to the Pre-Pottery Neolithic (PPN) period. There are points reminiscent of the so-called Harifian type (Fig. 12.), which is connected with desert Natufians in the transition to the Neolithic period in the Negev, Sinai and the Irano-Turanian vegetation zone like Jebel Bishri. The animals hunted at Abu Salem and Ramat Harif in Sinai during the Epipalaeolithic included dorcas gazelles, wild sheep, onager and aurochs. Evidence of smaller animals consisted of dog/wolf, turtle, tortoise, gecko and ostrich eggs. The

[2] In the preliminary report Lönnqvist 2006 *et al.* there unfortunately is a mistake concerning the finds from site I 16, starting them already from the Palaeolithic Age, although they actually start from the Epipalaeolithic Age.

palaeobotanical remains at Har Harif included pistachio and olive, but there was obvious potential for cereals, lentils and legumes, as well as various bulbs, tubers, roots, truffles and grasses. The Harifians formed bands of two to three families, which were semi-sedentary in their mode of living (see Goring–Morris 1991).

At **I 16** Neolithic (Pre-Pottery Neolithic) blade and bladelet industry seems to have been active. Types include tools comparable to the El Kowm steps VIII and IX (see Dornemann 1986: 55, Pls. 31-33; cf. also Pirie 2007). Fauna discovered from the transitional Natufian - PPNA cave occupation of Iraq ed-Dubb in the Jordanian Highlands has included evidence of an emphasis on gazelle hunting, but the hunting strategies from the Natufian to the PPN seem to have changed somewhat (Edwards and Martin 2007). Here at our site we have evidence of gradual sedentarization in transition from the Epipalaeolithic Natufian to the PPNA/B periods. Borers and perforators of different kinds are a typical feature at several PPN sites like here at I 16 (Figs. 13. and 14.). It has been suggested that they were used for reed processing (Jenkins and Rosen 2007: 435). At our site I 16 along the large *wadi,* in the area of a previous palaeolake, reed could have been available in quantity. A number of scrapers such as tabular scrapers including fan-scrapers among the flint finds obviously date from the Chalcolithic and Bronze Ages (Fig. 15.), and they seem to indicate pastoral economy at the area in those periods. It has been noted that tabular scrapers are used for butchering animals, but that they are also especially suitable for shearing sheep (Henry 1995: 372). This would re-enforce evidence of the existence of specialised pastoral nomadism in the region. The finds from the site also include Chalcolithic, Early Bronze Age, Middle Bronze Age (Figs. 16.-18.) and Roman pottery, which clearly point to a multi-period use of the site. There is a piece of painted Al-Ubaid/Halaf pottery as well as pieces of a hole-mouth jar. Pastoralism is not only indicated in the flint finds but also in the types of the hole-mouth jars.

Fig. 12. Microlithic finds including a blade scraper, microburins and points from site I 16. Photo: Kenneth Lönnqvist 2008.

Fig. 13. Finds representing perforators, blade and bladelet industry from site I 16. Photo: Kenneth Lönnqvist 2004.

Fig. 14. Various flint objects from site I 16.
Photo: Kenneth Lönnqvist 2004.

Fig. 15. Scrapers from site I 16. Photo: Kenneth Lönnqvist 2004.

Fig. 16. Piece of painted Al-Ubaid pottery.

Fig. 17. Pieces of a hole-mouth jar. Photo: Kenneth Lönnqvist 2004.

Fig. 18. a.-c. Pottery pieces with comb-waved decoration from site I 16.
Photo: Kenneth Lönnqvist 2004.

Fig. 19. Measuring and studying a ring-tumulus I 17 at Nadra. Photo: Gullög Nordquist 2004.

Site I 17
A Megalithic Ring-tumulus

UTM : 0545933, 3912115; 593 m a.s.l.
Size of the ring-*tumulus*: 18.5 m x 19.5 m, max. h. 1.0 m
Constructed of field stones of ca. 0.8 m in diameter.

A large cairn/*tumulus* of the megalithic ring-*tumulu*s type (Figs. 19. and 20.) was found on top of a hill (Fig. 2.), some hundred meters northwest of the settlement I 16. There is a double labyrinth-type passage around the cairn, which makes the site an imposing ring-*tumulus*. Associated flint artefacts were found around the grave. The structure most likely dates from the Chalcolithic/Bronze Age, according to the associated finds and comparable structures in the Near East. B. Rothenberg and J. Glass (1992: 142) describe as follows the comparable structures that they discovered in Sinai: stone enclosures with a cist in the middle covered by a heap of stones and associated with the so-called Elatian lithics from the Chalcolithic Age. In the Golan Heights, cf. Rujm el-Hiri, a large ring-*tumulus* was also found. It provided Late Bronze Age secondary burials, but the earliest finds in the neighbourhood date from the Chalcolithic period (see Kochavi 1991: 182-183; Mizrachi and Zohar 1993: 1286-1287). After our studies at site I 17, a Syro-Japanese team has excavated some comparable ring-*tumuli* on the northwestern flank of Jebel Bishri (Fujii and Adachi 2010: 66-69). The tombs of their findings are dated to the Early and Middle Bronze Ages according to the types of the associated finds and C-14 datings. In our studies the nearby settlement I 16, which seems to be associated with the grave I 17 (Fig. 20.) that is clearly overlooking the settlement, and the settlement was occupied during the Chalcolithic and Bronze Age.

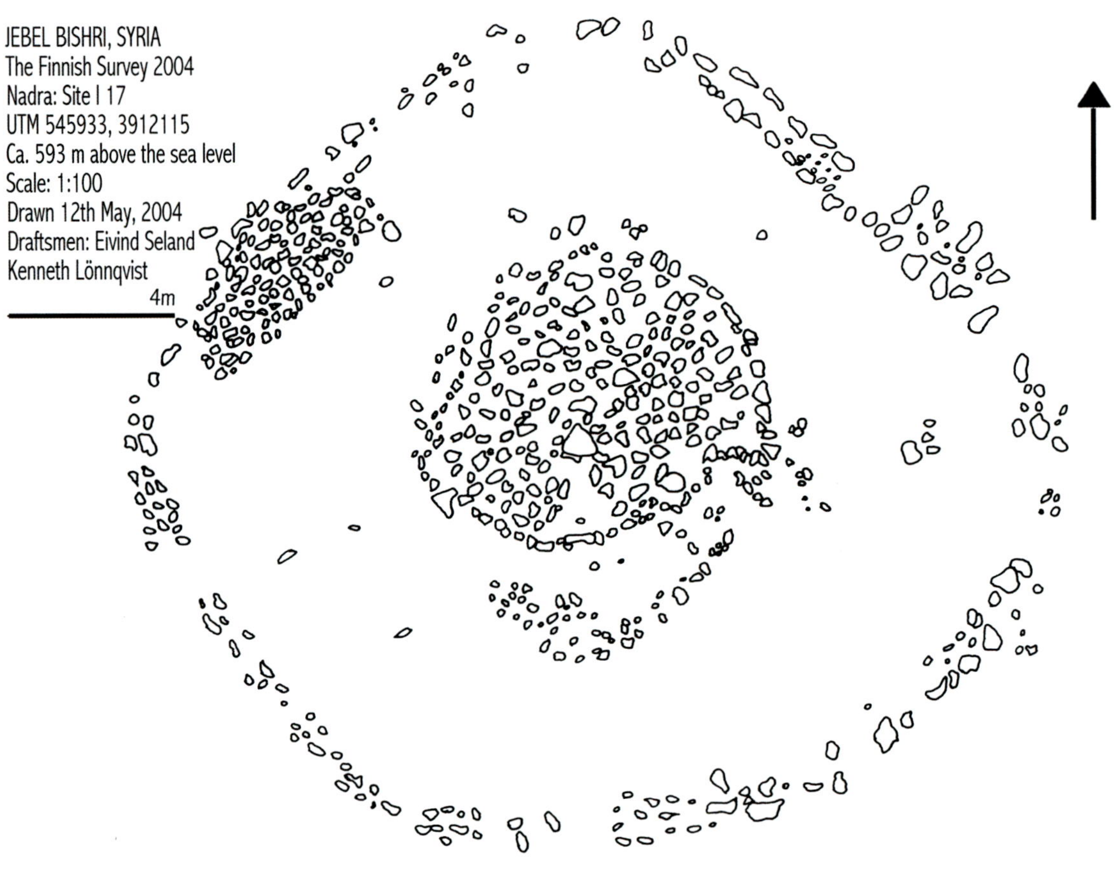

Fig. 20. The plan of the ring-tumulus at the site I 17.

Fig. 21. A blade and scrapers from site I 17. Photo: Kenneth Lönnqvist 2004.

Site I 8
Open Air Site

UTM: 0551925, 3910802; ca. 525 m a.s.l.
Size of the site: ca. 100 m x 100 m

This open accumulated site is also situated inside the palaeolake area of Nadra. It was found on the northwestern bank of the east-west-oriented *wadi* (Fig. 2.) which produced a number of flint artefacts and pottery sherds.

Flint finds include cores, a fan-scraper (Fig. 22.) and sickle blades. The pottery is reddish, handmade with a loop handle possibly dating from the Neolithic period. The finds indicate dating generally to the Epipalaeolithic, Neolithic and Chalcolithic periods like the occupation at site I 16. Roman pottery (see ch. 12.) and a piece of glass bracelet from the Byzantine/Islamic period with different colours and inlaid features were also found at the site (see ch. 14).

Fig. 22. Flint finds including cores, scrapers and a geometric microlith from site I 8. Photo: Kenneth Lönnqvist 2004

6.4. Cairns/*Tumuli* and Stone Enclosures/Corrals in the Pastoral Landscape of Nadra and Its Neighbourhood

Site I 2
Cairns/Tumuli (9 tombs)

UTM: 0545921, 3909111; ca. 616 m a.s.l.
Size of the hill site: circum sphere 500 m.

Sizes of the cairns/*tumuli*: a) max. 7.8 m in diam., h. 0.3 m; b) 4.2 m x 5.5 m, height 0.1-0.2 m; c) 3.0 m x 3.2 m, h. 0.1 m – 0.2 m; d) 4.2 m x 3.8 m, h. 0.1 m; e) 4.6 m x 5.0 m, h. 0.5 m; f) 3.4 m x 4.7 m, flat and heavily eroded; g) 3.3 m x 2.8 m, h. 0.2 m; h) 3.0 m x 3.1 m, h. 0.2 m; i) 7.0 m x 5.2 m, h. 0.8 m.

A burial site I 2 (a-i) with nine cairns/*tumuli* (Figs. 23.-25.) is situated on a high ridge following the 600 m elevation contour line by the northward mountain pass from the Shula - Ash-Shujiri road (Fig. 2.). Some of the tomb structures are badly scattered and their stones have been reused. No associated finds were discovered at the site.

Fig. 23. Tomb I 2e. Photo from SSE: Kenneth Lönnqvist 2004.

Fig. 24. Tomb I 2c.
Photo from NW: Kenneth Lönnqvist 2004.

Fig. 25. Tomb I 2i.
Photo from E: Kenneth Lönnqvist 2004.

Fig. 26. Site I 3. Photo from E: Kenneth Lönnqvist 2004.

Site I 3
Cairn/Tumulus

UTM: 0546817, 3909626; ca. 596 m a.s.l.
Size: 9 m in diam., h. 0.7 m

A partly destroyed cairn/*tumulus* (Fig. 26.) is located on a hillock south of a crossroads of two roughly east-west-aligned desert tracks under a hill. On the western side there is a *wadi* (Fig. 2.). No associated finds were encountered at this site either.

Site I 4
Cairn/Tumulus

UTM: 0546948, 3909304; ca. 608 m a.s.l.
Size: 8 m x 7.1 m, h. 1 m

This cairn/*tumulus* (Fig. 27.) in a good condition is situated on a limestone ridge on the western side of a desert track (Fig. 2.). No associated artefacts were encountered at the site either.

Fig. 27. Site I 4. Photo from E: Kenneth Lönnqvist 2004.

Site I 5
Cairns/Tumuli with
Ring-tumulus (4 tombs)

UTM: 0546966, 3909161; ca. 596 m a.s.l.
Size of the site: ca. 500 m x 50-70 m

Sizes of cairns/*tumuli:* a) 7.0 m x 6.3 m, h. ca. 0.5 m; b) 4.6 m x 3.5 m, h. ca 0.5 m; c) 8.2 m x 8.0 m, h. 0.3 m; d) 8.4 m x 5.5 m, h. 0.5 m

Four roundish cairns/*tumuli* (a-d: Figs. 28.-31.) are situated on a limestone ridge west of a desert track. There is a small depression to the north. (See Fig. 2.) One of the cairns/*tumuli* is destroyed by the robbing of stones for later Islamic graves found nearby. The largest cairn/*tumulus* in the south has a labyrinth type of circle around it forming a ring-*tumulus*. No associated artefacts were found at this site either.

Fig. 28. Tomb I 5a.
Photo from SE: Kenneth Lönnqvist 2004.

Fig. 29. Tomb I 5b.
Photo from SE: Kenneth Lönnqvist 2004.

Fig. 30. Tomb I 5c with a path finder, rujm marker, on top.
Photo from E: Kenneth Lönnqvist 2004.

Fig. 31. Tomb I 5d. Photo from S: Kenneth Lönnqvist 2004.

Site I 6
Cairn/Tumulus

UTM: 0547122, 3909010; ca. 609 m a.s.l.
Size of the cairn/*tumulus*: 6.6-7 m in diam., h. 0.5 m.

This cairn or *tumulus* is situated on a hill top east of a desert track and site I 5 (Fig. 2.). No associated artefacts were found at the site.

Fig. 32. Tomb I 6.
Photo from NW: Kenneth Lönnqvist 2004.

Site I 7
Cairns/Tumuli (2 tombs), Rujm Marker, Bottom of House and Stone Structure

UTM: a) cairn 551944, 3910910, b) house/tent base 0551964, 3910907 and c) cairn and rectangular grave 0552037, 3910796, d) 0552092, 3910598 stone structure of undefined nature; ca. 497-530 m a.s.l.

Sizes of the cairns: a) 5 m and c) 1.2 m in diam.; b) house 8.0 m x 3.6 m; d) semi-circular stone structure 3 m in diam.

This site consists of two cairns (a, c) and a rectangular stone foundation of a house (b) (Fig. 33.) that will also be discussed in ch. 12. dealing with the Roman remains of Nadra. The other cairn/*tumulus* had a path finder or the so-called *rujm* marker (Fig. 34.) on top. There is also a semi-circular stone structure (d) which might be a grave. No finds were identified inside the house, but outside to the east Late Roman pottery with some unidentifiable small reddish pieces of pottery and a notched flint were found (Fig. 35.).

Fig. 33. Foundations of a house (b) and a rujm marker at site I 7. Photo from E: Kenneth Lönnqvist 2004.

Fig. 34. A path finder, rujm marker, at site I 7. Photo from E: Kenneth Lönnqvist 2004.

Fig. 35. A notched flint artefact from site I 7. Photo: Kenneth Lönnqvist 2004.

Site I 9
Stone Enclosures (2 structures)

UTM: 0551776, 3910907; ca. 501 m a.s.l.
Size of the site: 16.5 m x 6.4 m

Sizes of the enclosures: a) 7.4 m x 6.4 m and b) 9.1 m x 6.4 m; constructed of large field stones of ca. 0.5 diam. (See a sketched plan in Fig. 63.)

This is a stone enclosure site in a small depression (Fig. 2.) with two adjoined semicircular structures or enclosures (Fig. 36.). No associated finds were encountered.

Fig. 36. Site I 9. Photo from SE: Kenneth Lönnqvist 2004.

Site I 10
Cairns/Tumuli (2 tombs) and Stone Enclosures (3 structures)

UTM: 0551639, 3911226 - 0551653, 3911253;
ca. 525 -532 m a.s.l.
Size of the site: 30 m x 20 m

Sizes of the structures: a) cairn/*tumulus* 5.1 m in diam., h. 0.2 m, b) cairn/*tumulus* 5.5 m in diam, h. 0.85 m; four stone enclosures: c) half-circle 4.5 x 6.5 m; d) open rectangular fence structure 15.5 x 8.5 m; e) half-circle ca. 13.6 m across, thickness of the walls being 0.8 m; f) an undefined additional stone enclosure 6 m in diam. (See a sketched plan in Fig. 63.)

There are two cairns/*tumuli* (Fig. 37.) and four stone enclosures (Figs. 38.-41.) on a limestone ridge overlooking the surrounding plain and a *wadi* (see Fig. 2.). Finds consist of Chalcolithic and Early Bronze Age pottery sherds (Fig. 42.) and a flint blade.

Fig. 37. A looted tomb (a) at site I 10.
Photo from S: Kenneth Lönnqvist 2004.

Fig. 40. Stone enclosure (b) of site I 10.
Photo from SW: Kenneth Lönnqvist 2004.

Fig. 38. Stone enclosure (c) of site I 10.
Photo from S: Kenneth Lönnqvist 2004.

Fig. 41. Worked flint from site I 10.
Photo: Kenneth Lönnqvist 2004.

Fig. 39. Weathered pottery pieces recovered from site I 10.
Photo: Kenneth Lönnqvist 2004.

Fig. 42. Pottery pieces recovered from site I 10.
Photo: Kenneth Lönnqvist 2004.

Site I 11
Cairns/Tumuli (4 tombs)
and Stone Enclosures (5 structures)

UTM: 0551461, 3912985 - 0551521, 3912895; ca. 523-557 m a.s.l. Size of the site: ca. 70 m x 150 m.

Sizes of the structures: cairns (a-d): a) 9.5 m x 8.0 m, h. 1 m b) 1.8 m x 2.5 m, h. 0.3 m, c) 7.7 m x 6.2 m, d) 4.1 m x 3.5 m , h. 0.7 m.; stone enclosures (e-i): e) 6.4 m x 6.0 m, f) 13.0 m x 12.2 m, g) 14.9 m x 18.2 m, h) 7 m x 8.6 m and i) 12 m x 11 m. (See the sketched plan of the enclosures in Fig. 63.)

This site includes four cairn/*tumulus* burials (Fig. 43.) which had been partly destroyed and five stone enclosures or corrals (Fig. 44.). The burials are on the top of a hill. One cairn/*tumulus* also has a path finder, the so-called *rujm* marker, on top (Fig. 43.). No associated finds were encountered at the cairns/*tumuli*. There is a series of connected rectangular enclosures, five in number situating from NW to SE (e-i). Associated finds consist of flints: one blade and one flake.

Fig. 43. Cairn (a) with a rujm marker on top at site I 11. Photo from W: Kenneth Lönnqvist 2004.

Fig. 44. A stitched panorama over the stone enclosures at site I 11. Photo from E: Kenneth Lönnqvist 2004.

Site I 12
Cairn/Tumulus, Stone Enclosures (2 structures)
and Stone Wall

UTM: 0552185, 3912482; ca. 543 m a.s.l.
Size of the site: ca. 40 m x 20 m

Size of the structures: a) cairn/*tumulus*: 11.2 m in diam., h. 1.0 m; enclosure (b-c): 12.7 m as well as 16.6 m in diam. and d) 4 m long wall. (See a sketched plan in Fig. 64.)

This site is situated on a hill top (Fig. 2.), where structures are oriented roughly from NW to SE with two stone enclosures or corrals associated with a cairn/*tumulus* (Fig. 45.) in the middle with an associated N-S oriented wall. The grave had been plundered. A flint blade dating from the Chalcolithic or Early Bronze Age was associated with the site.

Fig. 45. Cairn at site I 12.
Photo from N: Kenneth Lönnqvist 2004.

Site I 13
Ring-tumulus

UTM: 0546557, 3910187; 627 m a.s.l.
Size of the ring-*tumulus*: max. 6.2 m in diam., h. 0.3 m.

This is a cairn/*tumulus* with an inner and outer stone circle, making it the so-called ring-*tumulus*. It is situated on a ridge south of an abandoned and ruined Bedouin village **I 1** (see more in ch. 15; see the location in Fig. 2.). (At the site Palaeolithic flint tools, a handaxe of the Achaeulian type, cores, a fossilized bone and a Geometric Kebaran flint implement were found as described in the previous chapter).

Site I 15a-b
Stone Enclosure/Corral with Water Harvesting Site

UTM: 0545414, 3911581-545504, 3911420;
ca. 600-602 m a.s.l.

Sizes of the structures: a) enclosure: 31.5 x 28 m and b) the water harvesting site: 6 m x 4 m. (See a sketch plan of the enclosure in Fig 64.)

The site is situated on northeastern slopes of the E-W oriented hilltop along a *wadi* (Fig. 2.). This is an oval shaped stone enclosure. Associated finds consist of flints from the Chalcolithic Age and pottery from the Chalcolithic, Bronze Age and Roman period. A piece of greenish and decorated glass bracelet from the Byzantine/Islamic period was also found (see ch. 14) at the site. A water harvesting site with a barrage (constructed of limestone cobbles, 50 cm x 50 cm x 10 cm in size) is included to the site.

Site I 18
Open Air Site

UTM: 0545479, 3911838; ca. 570 m a.s.l.
Size: 600-700 m x 200-300 m.

This is an open accumulated site on the bank of a *wadi*. Finds consisted of some unidentifiable pieces of pottery and flint.

Site I 20
Stone Enclosures (4 structures)

UTM: 0544927, 3912428; ca. 590 m a.s.l.
Size of the site: ca. 30 m x 50 m with structures.

Sizes of enclosures (a-d): a) 12 m x 15 m; b) 7 m x 7 m; c) ca. 10 m x 5 m and 11 m x 10.3. The stones of the structures are generally 0.3 m – 0.4 m up to 0.7 m x 0.8 m in diam. (See a sketched plan in Fig. 64.)

Rectangular stone enclosures or house bases (Fig. 46.) of different sizes bordered by stone lines are situated on the northeast side of a large *wadi* sloping to the southwest (Fig. 2.). The finds consist of worked flint, pottery pieces and copper slag. In the flints there are *inter alia* Epipalaeolithic tools, which are obviously from an earlier phase of activities at the site. The pottery dates from different periods (Fig. 47.), e.g., from the Bronze Age.

Fig. 46. Site I 20. Photo from E: Kenneth Lönnqvist 2004.

Fig. 47. Pottery from site I 20. Photo: Kenneth Lönnqvist 2004.

Fig. 48. Cairn/tumulus at site I 22b. Photo from NW: Kenneth Lönnqvist 2004.

Fig. 49. A cairn/tumulus at site I 22c. Photo from NW: Kenneth Lönnqvist 2004.

Fig. 50. Worked flint finds from I 22 f. Photo: Kenneth Lönnqvist 2004.

Site I 21
Cairn/Tumulus and Stone Enclosures/Corrals (2 structures) with Water Barrage

UTM: 0545502, 3912971- 0545511, 3913001; 582-595 m a.s.l. Size of the site: ca. 100 m x 700 m.

Sizes of the structures: a) cairn/*tumulus*: 3.6 m in diam.; enclosures: b) 12.5 m x 15.0 m; c) 27.8 m x 13 m x 16 m; d) 6.0 m x 9.0 m; e) barrage: 23 m x 1.5-1.8 m, h. ca. 1 m, f) amorphous stone structures. (See sketched plan of the enclosures in Fig. 64.).

A cairn/*tumulus* is situated in between a small enclosure and two larger ones. The larger ones consist of a horseshoe–shaped and a crescentic stone enclosure situated on the slope of a *wadi*-stream (Fig. 2.). The stone enclosure I 21b adjoins I 21a to the north. The water barrage is situated to the west of the enclosures and is constructed over a *wadi*. Associated pottery dates from the Roman period (see more in ch. 12.).

Site I 22
Cairns/Tumuli (5 tombs) and Stone Ring

UTM: 0540078, 3911932; 674-712 m a.s.l. Size of the site: 100 m x 700 m

Sizes of the structures: cairns/*tumuli* (a - e): a) 7 m in diam., h. 0.5 m; b) 11 m in diam., h. 1.25 m; c) 9 m in diam., h. 0.9 m, plundered; d) 5 m in diam., h. 0.5 m; e) 6-7 m in diam., h. ca. 1 m; g) stone ring: 2 m in diam.

A northern high ridge ca. 2 km northeast of the 2004 expedition camp (see ch. 3.) had several cairns (Fig. 48. and 49.) above a *wadi* (Fig. 2.). Some of the cairns had been plundered and looked like they had cists inside. West of one (b) of the cairns there is a stone ring. One cairn/*tumulus* (d) serves as a good look-out point. I 22 f) produced some worked flints (Fig. 50.).

Site I 23
Stone Enclosure

UTM: 0537819, 3912408; 707 m a.s.l.
Size of the enclosure: max. 17 m in diam.; constructed of field stones 0.35 m x 0.40 m in size.

This stone enclosure was situated ca. 150 m northwest of the 2004 expedition camp (see ch. 3.) visible on the northern side of the Shula - Ash-Shujiri -road (Fig. 2.). The enclosure is 200 m west up a small slope behind the base tent. No artefacts were encountered at the site.

Fig. 51. Site I 23. Photo from W: Kenneth Lönnqvist 2004.

Fig. 52. Cairn/tumulus I 24 a.
Photo from S: Kenneth Lönnqvist 2004

Site I 24
Cairns /Tumuli (2 tombs) and Stone Enclosure

UTM: 0538241, 3911973; 705 m a.s.l.
Sizes of the structures: cairns/*tumuli*: a) 6 m in diam. and b) 6.5 m in diam; c) enclosure: 4 m x 5 m in size, constructed of stones 0.2 m x 0.5 m – 1.0 m in size.

This site is situated 350 m east of the 2004 expedition camp on a ridge (Fig. 2.). Two cairns (a-b) were found together with a stone enclosure (see a sketched plan in Fig. 65.) No associated finds were discovered at the site.

Fig. 53. Cairn/tumulus I 24 b.
Photo from SE: Kenneth Lönnqvist 2004.

Site I 25
Disturbed Graveyard

UTM: 0543949, 3913372; 607 m a.s.l.

This is a disturbed graveyard with circular cairns and a rectangular structure (1.5 m x 1.0 m in size) in a small sheltered valley. Some worked flints were discovered in association with the graves.

Site I 26
Ring-tumulus

UTM: 0543994, 3913239; ca. 598 m a.s.l.
Size of the ring-*tumulus*: max. 8.5 m in diam.

This is a disturbed cairn/*tumulus* situated on a high point between two *wadis* (Fig. 2.). The cairn has a clear outer ring constructed of large encrusted stones mixed with earth. Worked flints, including a Chalcolithic burin, were discovered in association with the structure (Fig. 54.).

Fig. 54. Worked flint from site I 26.

Photo: Kenneth Lönnqvist 2004.

Sites I 29/30
Stone Enclosures (3 structures)

UTM: 0543775, 3913502; ca. 612 m a.s.l.
Sizes of the enclosures (a-c): a) 18 m x 14 m;
b) 14 m x 19 m; c) 15.5 m x 8.5 m.
(See a sketch plan in Fig. 65.).

Sites I 29/30 (Fig. 2.) consist of curvilinear stone enclosures (see Fig. 55.) which are so close to each other that they can be considered to have formed a single site. The stone enclosure (I 29) has a clear 1.5 m wide entrance and a 1.0 to 1.5 m thick wall opposite to the entrance, like at the larger enclosure I 30. It seems that the functions of the walls in front of the entrances of the enclosures were to serve as porches to guide animals into the enclosures. No surface finds were encountered at the site.

Fig. 55. Site I 29/30. Photo from W: Kenneth Lönnqvist 2004.

Site I 31
Ring-tumulus,
Islamic Grave and Rujm Marker

UTM: 0543686, 3913508; ca. 625 m a.s.l.
Size of the ring-*tumulus*: 6.5 m in diam., outer circle 13.5 m in diam., ca. 0.7 m h; grave: 2 m x 0.5 m.

This site (see the location in Fig. 2.) is a large fairly badly preserved cairn/*tumulus* of the so-called ring-*tumulus* type, with an outer circle of 13.5 m surrounding the cairn. To the southeast there is an Islamic grave and a landmark as a path finder, the so-called *rujm* marker. No surface artefacts were encountered at the site.

Site I 32a-d
Stone Enclosures (3 structures)

UTM: a)-c) 0543561, 3913420, d) 0543562, 3913446; 621-625 m a.s.l. Sizes of enclosures: a) 7 m in diam., entrance in w: 0.7 m; b) 9.5 m x 14 m, entrances 4.5 m w. and 0.7 m w.; c) 3.0 m in diam., entrance 2.5 m w.; d) 17.5 m x 14 m, entrance 1.15 m w.
(See a sketch plan in Fig. 65.)

This site is situated on a slope to the south of a *wadi* (Fig. 2.). The site consists of three curvilinear stone enclosures or corrals (a-c) each having an entrance towards the east (see Fig. 56.) and a square stone enclosure/house base (d) having an entrance to the south. There is an entrance of 4.5 m in width between b and c. The square stone enclosure is ca. 50 m south of the curvilinear enclosures situating on a slope towards the south with an opening in the south. This structure is rather well preserved with two rows of stone walls still standing to the height of 0.7 m. There is a stone slab inside the square structure. One flint blade and a pottery sherd were discovered at the site.

Fig. 56. One of the stone enclosures at Site I 32.
Photo from E: Kenneth Lönnqvist 2004.

Site I 33
Cairns/tumuli (2 tombs) and Stone Enclosures
(4 structures)

UTM: 0538333, 3912899; ca. 616 - 687 m a.s.l.
Size of the site: 150 m x 20 m.

Sizes of cairns/*tumuli* (a-b): 2.2 m in diam. and 1.8 m in diam; (c-e) enclosures: 17.5 m in diam., 15.5 m in diam., partly destroyed by a *wadi;* l. N-S: 13.8 m w. E-W: 19.8 m (undressed stone blocks 0.3 m - 1.0 m in diam.) (See a sketch plan in Fig. 66.).

This is a multi-period site including large enclosures of ovaloid/circular shape situated along the northern edge of a *wadi* surrounded by several rock shelters without surface finds. It is ca. 800 m north of the 2004 expedition camp (Fig. 2.). There are two minor cairns/*tumuli* associated with the enclosures. One enclosure has an opening to the south. The site with enclosures has produced plenty of flint artefacts and some pieces of pottery. The flints in the area date from the Epipalaeolithic, Neolithic and Chalcolithic Ages. There is, *inter alia*, a fan-scraper and Canaanean blades indicating to Chalcolithic/Early Bronze Age activities. The pottery pieces are from various periods dating from the Chalcolithic to the Bronze Age, and the site also included Roman/Byzantine and glazed Islamic pottery. It is evident that some flints can be associated with the mentioned rock shelters in the area but had moved from their original locations.

Fig. 57. A stone enclosure at site I 33. Photo from S: Kenneth Lönnqvist 2004.

Fig. 58. Surface finds including microliths from the Epipalaeolithic period, PPNB blades, a perforator, scrapers, a tabular scraper, and Canaanean blades from site I 33. Photo: Kenneth Lönnqvist 2004.

Site I 35
Stone Enclosures (4 structures)

UTM: 0543344, 3913613; 616 m a.s.l.
Size of the site: 150 m x 20 m

Sizes of the enclosures (a-d): a) max. 18.5 m in diam.; b) 13.3 m x 10.5 m; c) 19.5 m x 12.5 m.; d) 12 m x 8 m; (stone blocks varying from 0.3 m x 0.5 m to 0.5 m x 1 m in size) (See a sketch plan in Fig. 66.)

A series of curvilinear stone enclosures is situated side by side aligned in the N-S direction in a small depression east of a desert track (Fig. 2.). Surface finds include pottery and flints. Pottery pieces seem to date from the Chalcolithic and the Bronze Age, and pieces of moulded pottery from the Roman/Byzantine era were also found (see more in ch. 12.).

Fig. 59. Site I 35. Photo from W: Kenneth Lönnqvist 2004.

Fig. 60. Pottery from site I 35. Photo: Kenneth Lönnqvist 2004.

Fig. 61. A scraper and a microburin from site I 35.
Photo: Kenneth Lönnqvist 2004.

Site I 38
Stone enclosure

UTM: 0537085, 3913155: ca. 700 m a.s.l.
Size of the enclosure: 9 m x 9 m
(See a sketched plan in Fig. 66.)

This U-shaped stone enclosure is situated ca. 850 m north of the expedition camp of 2004. It was found near a *wadi* below a cavern. The enclosure is opening towards a rock shelter. The *wadi* is parallel to site I 33. No finds were discovered at the site.

6.5. Types of Cairns/*Tumuli* and Stone Enclosures

Cairns/*tumuli* were located on hillocks often following the longitudinal alignment of hills or higher ridges on Jebel Bishri. They divide the space of the desert-steppe and steppe into fixed points like nodes in addition to simple stone heaps, *rujm* markers (see more in ch. 14), serving the desert travellers as path finders or signposts. In some cases cairns/*tumuli* were topped by *rujm* markers. As far as the surface appearances are concerned, some cairns/*tumuli* have surrounding stone courses or rings composing of the so-called ring-*tumuli* like the megalithic structure at site I 17 (Fig. 20.). Some have surrounding stone walls like casings (Fig. 62.) Because some cairns/*tumuli* had been looted, we were able to observe that their inner structures comprised cists.

Fig. 62. A cairn with a surrounding stone lining.
Photo: Margot Stout Whiting 2004.

In stone enclosures we encountered circle type enclosures of which some were clearly corrals, and square structures which may be either animal pens or house bottoms (see Figs. 63.-66.). We here present schematic drawings of the shapes of the stone enclosures from the district of Nadra. In the following two chapters, ch. 7. and 8., more examples of stone enclosures are presented from other parts of Jebel Bishri. It needs to be noted that at Nadra and its surrounding areas most of the stone enclosures are curvilinear reflecting the mobility of the sites. Five sites out of the 16 illustrated ones include rectangular structures.

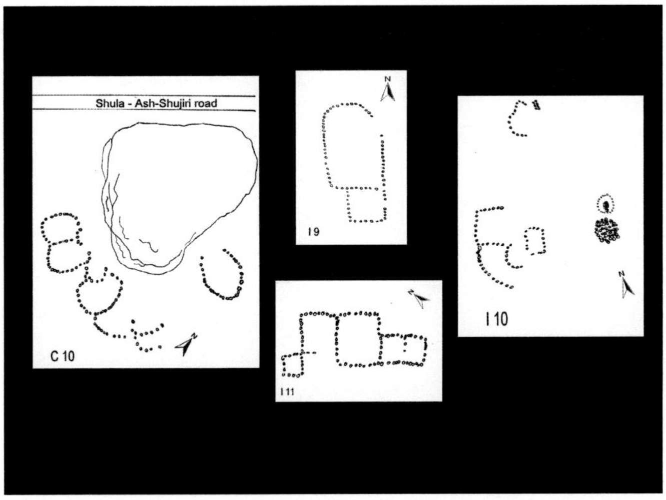

Fig. 63. Sketch plans of compound architecture at sites C 10, I 9, I 10 and I 11. Not into scale.

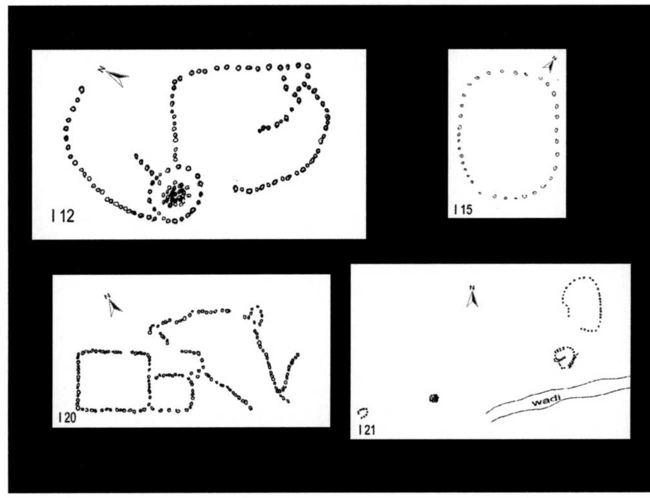

Fig. 64. Sketch plans of the compound architecture at sites I 12, I 15, I 20 and I 21. Not into scale.

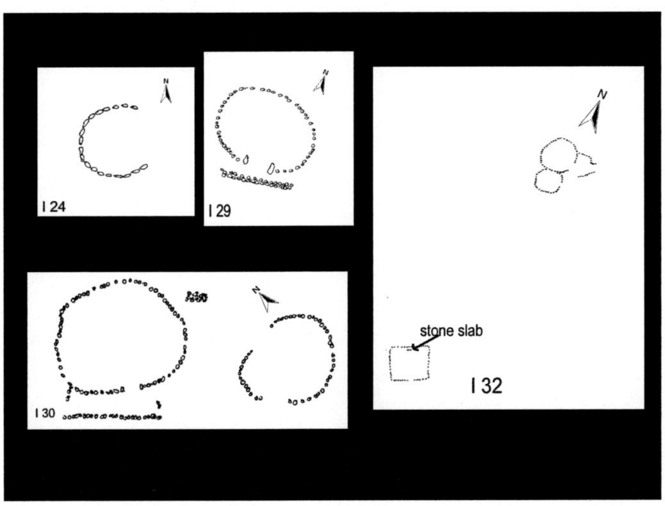

Fig. 65. Sketch plans of compound architecture at sites I 24, I 29, I 30 and I 32. Not into scale.

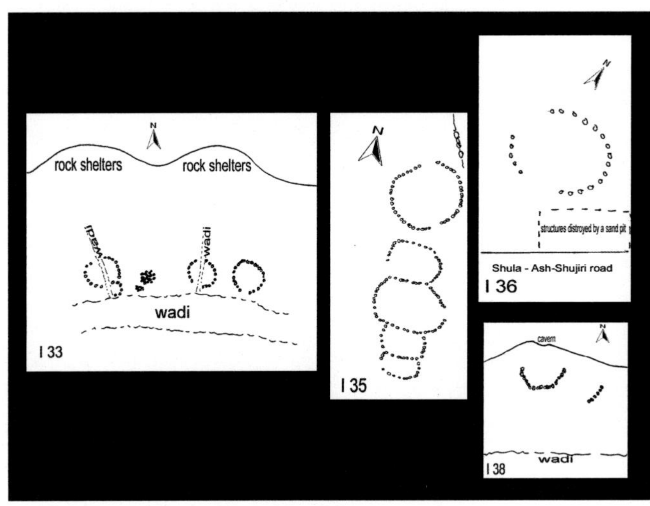

Fig. 66. Sketch plans of compound architecture at sites I 33, I 35, I 37, I 38 and I 39. Not into scale.

In nomadic cultures roundup fences such as corrals are quite common and they are sometimes attached to subsidiary corrals to receive animals separated from the herd (see Ingold 1980: 117, Fig. 18.). As mentioned earlier, ethnoarchaeological studies, e.g., by M. Haiman (2002) indicate that there were different functional uses for each enclosure unit.

6.6. Conclusions and Discussion

According to our observations, when using remote sensing in prospecting, structures of pastoral people exemplified in cairns/*tumuli* and stone enclosures of different kinds, are most difficult to detect. Being flimsy, these types of structures, like the abandoned Bedouin tent bases need data sources with the highest possible spatial resolution. As already mentioned, based on our remote sensing studies, "window screenings" and pedestrian ground surveys, we can state that on Jebel Bishri cairns/*tumuli* are the most common structural remains that we have identified and mapped. The second commonest structure type which we have encountered consists of stone enclosures such as pastoral corrals, i.e., animal pens, and megalithic rings of different kinds (see more in ch. 7. and 8.).

As mentioned previously, our research area for the ground survey at Nadra covered ca. 22 km^2 and another 10 km^2 in its neighbourhood. Most of the sites are associated with large seasonal water courses, i.e., *wadis*. Altogether 39 sites were recorded in the area during the years 2000 and 2004. According to our final studies, of the 39 sites ca. 32 sites can be classified as pastoral sites (all these pastoral sites, except a later Bedouin site I 1 which will be discussed in ch. 15., have been presented here in this chapter) with or without structures. Of the 32 pastoral most comprised structures, and only a few were open accumulated sites. Comparing the distribution of the pastoral sites to the studied area there were one to two sites in every km^2 on average. In Fig. 67. below a map presents the distribution of the pastoral sites in the district of Nadra. The majority of the pastoral sites in the area were already noted in our preliminary report (cf. Lönnqvist *et al.* 2006), although here the final classification of the finds has added a few more examples of pastoral sites.

Apart from the site C 10/I 36 in the neighbourhood of Nadra, a transition from the forager stage to pastoral nomadism can be detected at the sites such as I 16, in which Epipalaeolithic Natufian phase changes to the PPNA and PPNB phases. It seems that shifts to hypothetically moister environmental conditions during the so-called Holocene Optimum (see ch. 4.) had a minor impact if none in the region of Jebel Bishri as no actual larger scale sedentarization is attested during the Neolithic period – I 16 is a small settlement near a seasonal water course. The sites C 10/I 36 and I 16 may have served as seasonal hunting and pastoral bases in the highland area for the contemporary Epipalaeolithic and Neolithic sites of El Kowm, but there is still a possibility of independent pastoral development from the hunter-gatherer phase in higher altitudes. Indications of the continuity from the Epipalaeolithic through the PPNA period to the PPNB would enforce this possibility as there is a hiatus in the stratigraphic sequence of El Kowm during the PPNA.

The sites such as C 10/I36 and I 16 could indeed hold some keys, if excavated, to the transition from hunter-gatherer economy to pastoral nomadism, another economy based on mobility in the desert-steppe areas. The early pastoral period in Central Sahara offers some analogous possibilities of the shift from the pronounced hunter-gatherer time to pastoral mobility, indicating that herders might have lived aside hunter-gatherers (cf. Di Lernia 1999 *apud* Smith 2005: 262-263). Also hunting has probably served as a subsidiary economy of the past pastoral nomads in the area of Jebel Bishri - like we observed hunting among the Bedouins of the region today. The site I 16 is also topographically associated with the ring-*tumulus* site I 17 and continues to offer evidence of the occupation during the Chalcolithic and Bronze Age. The site seems to represent the development in pastoral nomadism, similar groups continuing to occupy the site from the Neolithic to the Chalcolithic and the Early/Middle Bronze Age. Tabular scrapers and hole-mouth jars indicate that the role of pastoral economy was pronounced at the site during the Chalcolithic and Bronze Age periods.

Even if there is previous occupation by hunters and foragers in the neighbourhood, the pastoral occupation is present in the majority of the sites at Nadra. One reason can be that most of the area belonged to a palaeolake and therefore was under water during the Palaeolithic era. The area of Nadra is still characterized by a large *wadi* associated with smaller subsidiary ones. Our studies on grazing potential and stone enclosure/corral sites at Ash-Shujiri above Nadra in the central area of Jebel Bishri (see ch. 7.) have enhanced our view of these two steppic areas as being best grazing grounds on the mountain differing from western desert-steppe areas such as Tar al-Sbai. The area studied by remote sensing in the neighbourhood of Ash-Shujiri covered 64 km^2.

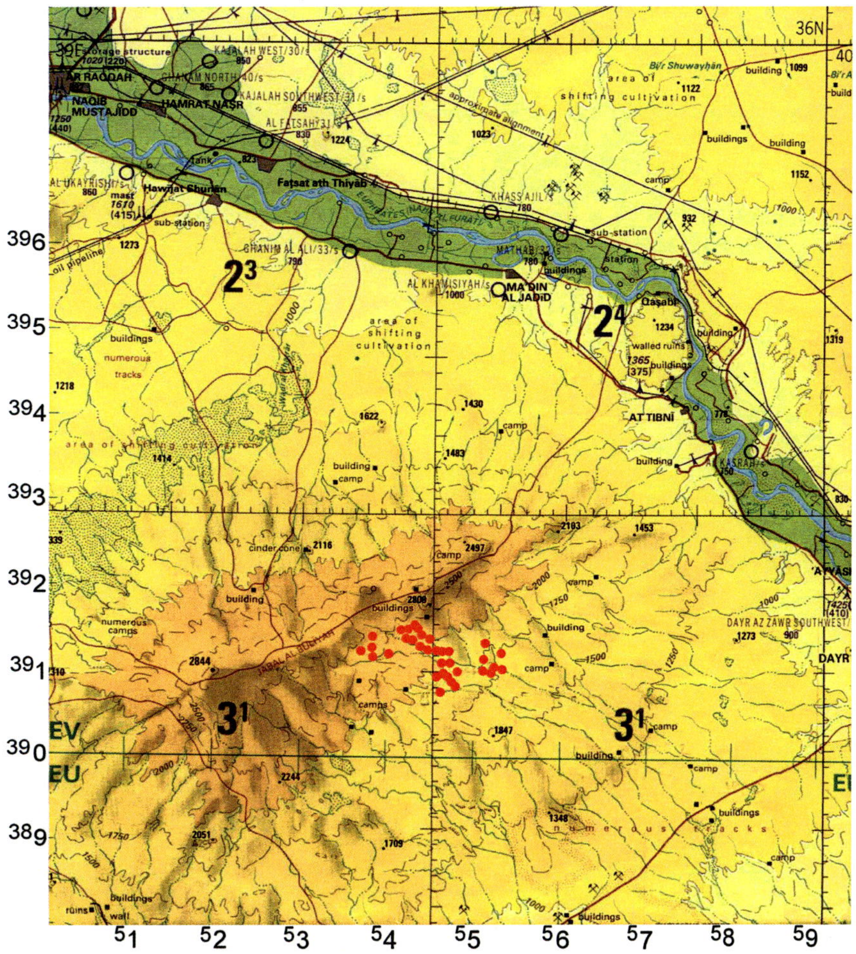

Fig. 67. The distribution of pastoral nomadic sites in the area of Nadra on Jebel Bishri dating from the Pre-Pottery Neolithic to the Bronze Age. Mapping by Kenneth Lönnqvist. Basemap © UK MOD Crown Copyright, 2010 published with the permission of the Controller of HMSO.

Of the 32 pastoral sites identified on the ground at Nadra and its surroundings 18 sites included graves, generally being cairns/*tumuli*, and 16 sites comprised stone enclosures of different kinds. Interestingly, in these numbers there are five to seven sites which included both cairns/*tumuli* and stone enclosures. At least from the time of the cairn/*tumulus* burials, sacred meanings can be associated with the choice of higher ground offering impressive views over Jebel Bishri. The choice of higher locations for cairn burials is typical in many parts of the world. As mentioned, the tombs also served as path finders, beside the specific built up stone heaps like *rujm* markers as separate constructions or on top of cairns/*tumuli*, for travellers in the desert.

The cairn/*tumulus* sites and stone enclosures on the mountain are mostly associated with chipped stone industry, often represented by scrapers and blade tools of different types. However, pottery is not totally absent, but it is scanty, mainly consisting of pieces of heavily weathered Chalcolithic, Bronze Age and Late Roman-Byzantine pottery. The earliest datings for cairns/*tumuli* in Near Eastern archaeology connect them with the Chalcolithic period (ca. 5000-3300 B.C.) (see, e.g., Zarins 1992: 50) becoming more common in the Early Bronze Age (3300-2000 B.C.) but the tomb type has been in use among the Bedouins up till present. The earliest pastoral corrals date from the PPNC/Chalcolithic and Early Bronze Age, but our studies indicate that an earlier and transitional use of these structures may have been possible in the PPNB period at the sites such as C 10/I 36. Both categories of structures, namely cairns/*tumuli* and corrals, belong to the pastoral technocomplex, as defined by J. Zarins (1992). Curvilinear forms are preferred in these types of structures, reflecting the high degree of mobility of the cultures.

Outside Jebel Bishri parallels for the circular enclosures can be sought for the Greater Southwest Asian Arid Zone throughout in Jordan, the Negev, Sinai and Israel. The Southern Ghors and 'Arabah archaeological project has especially found comparable circular enclosures in the survey areas in southern Jordan. For example, Site 129, which is situated between the Wadis of Umruq and Feifa consists of such circles, one being ca. 12 m and the other 8 m in diameter. But south of Wadi al-Ghuweib there are enclosures reaching even 25 m in diameter. (MacDonald *et al.* 1992: 56-60).

At the Chalcolithic sites in the Judayid Basin in southern Jordan there are structures which are comparable to those found on Jebel Bishri. These sites in Jordan are situated on the terraces and slopes of Jebel el-Jill and Jebel Queisa ca. 1200-1000 m a.s.l. The diameters of the enclosures at Jebel el-Jill vary from 3 m to 10 m, including a semi-subterranean pit house. Like the stone enclosure site C 10 on Jebel Bishri, these sites have structures of bimodal distribution and grinding slabs. At site J 14 on Jebel el-Jill the Chalcolithic layers consisted of drift sand overlying an apparent Epipalaeolithic horizon. Furthermore, at J 24 on Jebel Queisa the Chalcolithic layers overlie an early Neolithic unit. Therefore, both sites are of the long-term camp type in their nature (see Henry 1995: 345-374), resembling the situation at the curvilinear sites of Jebel Bishri some of which also seem to have a long-term character. The people of the comparable Chalcolithic sites at Wadi Hisma in southern Jordan based their subsistence economies on sheep/goat pastoralism with the addition of hunting and gathering resembling the Timnian sites in the Negev (Henry 1995: 358).

In the Negev and Sinai regions B. Rothenberg's identifications of the Elatian (4500-3500 B.C.) and Timnian (3500-2500 B.C.) phases, comprising different kinds of stone enclosures, are classic (see, e.g., Rothenberg and Weyer 1979; Rothenberg and Glass 1992). Numerous circular structures dating from the Chalcolithic and Bronze Age have also been studied by E. Anati (1986) in the area of Har Karkom in the Negev. Most of the circular enclosures and compounds identified during F. Eddy's and F. Wendorf's survey in Central Sinai were dated to the Chalcolithic Age. Some of them had still been in use during the Bronze Age. In association with the enclosures there were cairns and *tumuli* alike. Also kites and cultic platforms were encountered. The compound enclosures in Sinai have been interpreted as representing pens for sheep and goats to be kept safe during the nights against wild animals. Sheep and goats in those sites were mainly used for secondary products rather than for meat. (Eddy and Wendorf *et al.* 1999: 135). Numerous similar structures have been found in Arabia as well (Zarins 1989b: 136-139).

The life of ephemeral camping apparently created a need for the mobile people to have permanent resting places for their ancestors. Ancestors have a central meaning in patriarchal lineages and tribalism, and the ancestor cult is one of the earliest cults in humanity's past (Durkheim 1912). Tribalism as a social structure may be traced behind cairns/*tumuli*. Ring-*tumuli* appear as a special type of tomb structure in cairn/*tumulus* fields which we have recorded. This refers to the existence of social organisation based on status (cf. Sahlins 1968), and the apparent existence of a chiefdom system. The head of a family or the chief of a tribe could have been treated differently by giving special signs of status to a tomb in a graveyard. (Lönnqvist *et al.* 2010). In the Mari cuneiform texts from the second millennium B.C. there are special words used for cairns/*tumuli*, and several chiefs of the pastoralists are mentioned (see ch. 9.).

While cairns/*tumuli* appear as markers of the ritual worlds associated with the ancestors in the landscape, the corrals are equally pronounced structures for living domesticated goats/sheep (Lönnqvist *et al.* 2010). Interestingly the area of Nadra can be characterized as a steppe space featured by structures dedicated to ancestors and animals. S. Kent (1989: 15) has noted that in several primitive societies domesticated animals are considered to be comparable to domestic artefacts. In contrast, humans are placed in the same category as wild animals and plants. However, the domesticated animals as "objects" are socially, ritually, politically and economically more valuable than wild animals – they are nomads' property.

The Chalcolithic, Early Bronze Age and the Roman period appear to have been environmentally favourable periods in the steppic and high altitude areas in the Near East (see, e.g., Rothenberg and Glass 1992; 141; Geyer and Besançon 1996: 8-10; Neumann *et al.* 2007). Nomadic compounds form material manifestations of small-scale tribal groups which may belong to larger tribal confederations. These researches help one to study human adaptation and responses to environmental changes in the region from a long-term perspective. The Chalcolithic Age and Early Bronze Age can be considered to form a climax in the desert-steppe culture relying on pastoral nomadism in this arid zone (Levy 1983: 32; Rosen 1988: 503). This era has been regarded as being the acme of the specialised pastoral nomadism in Southwest Asia, although the beginning of specialized pastoral nomadism dates already from the Neolithic period.

The few pieces of pottery from the central areas of the mountain date generally from the Chalcolithic, Early Bronze Age and the Roman period, with only some pieces coming from the Neolithic and Middle Bronze Ages. We may ask whether the earliest nomads moving on the mountain did not use much pottery but preferred perishable materials like leather instead. The grazing area is large and mobility does not encourage the carrying around of heavy burdens, although it is known that there are nomads who use pottery. From the archaeological point of view we may also venture to ask whether the scarcity of the Middle Bronze Age pottery at these sites is in turn reflecting the situation in the written sources, when nomadic groups who are largely identified with the area had moved and sedentarized widely to other areas nearer to the river zone. Along the right bank of the river beneath Jebel Bishri Early Bronze Age sites are numerous as we shall discuss in ch. 9. The relationship of these sites with the pastoral sites on Jebel Bishri will be discussed in the context.

However, the tombs recently excavated by the Syro-Japanese mission on the northwestern flanks indicate that pastoralists still were using the mountain as their burial site during the Middle Bronze Age. The evidence also accords with the Mari texts (ARM) which locate some of the nomadic and West Semitic (Amorite) tribes to the region during the early second millennium B.C. However, there is ample evidence that nomadic Amorites were on the move at the end of the third millennium B.C. (Lönnqvist 2008). We may ask: what were the major impulses for a large-scale sedentarization? Was it caused by "push" or "pull" mechanisms? Environmental explanations have been in vogue for the past decades. Land degradation, desertification and site abandonment were seen as mutual processes ending into the so-called collapse of the Akkadian state system (see, e.g., Weiss *et al.* 1993), which in turn triggered social factors into the nomadic expansion. Based on our ethnoarchaeological studies from the present situation we know that social reasons may play a concomitant role as well.

Our project has carried out ethnoarchaeological studies at the village of Shanhas and on the mountain in the Bedouins compounds and villages at Tar al-Sbai, Ash-Shujiri and Nadra (see ch. 15). We have especially concentrated on different reasons for and processes of sedentarization and abandonment of pastoral sites using *inter alia* satellite image analyses (Lönnqvist and Törmä 2004; Lönnqvist and Törmä 2006; Lönnqvist *et al.* 2006; see more in chapter 15.).

Bibliography

Akkermans, P.A., Boema, J.A.K., Clason, A. T., Hill, S. G., Lohof, E., Meiklejohn, C., Le Miere, M., Molgot, G.M.F., Roodenberg, J.J., Waterbolk-van Rooyen, W., and van Zeist, W. (1983) Bouqras Revisited: Preliminary Report on a Project in Eastern Syria, in *Proceedings of the Prehistoric Society*, Vol. 49, pp. 335-372.

Albert, R.M. and Henry, D.O. (2004) Herding and Agricultural Activities at the Early Neolithic Site of Ayn Abu Nukhayla (Wadi Rum, Jordan), The results of phytolithic and spherulite analysis, in *Paléorient*, Vol. 30, pp. 81-92.

Anati, E. (1986) *Har Karkom: The Mountain of God.* New York: Rizzoli.

ARM = *Archives Royales de Mari.*

Aveni, A. (2002) *Empires of Time: Calendars, Clocks, and Cultures.* Revised Edition. Colorado: University Press of Colorado.

Bar-Yosef, O. (1981) The Walls of Jericho: An Alternative Interpretation, in *Current Anthropology*, Vol. 27, pp. 157-162.

Bar-Yosef, O. and Khazanov, A. (eds.) (1992) *Pastoralism in the Levant: Archaeological Materials in Anthropological Perspectives.* Monographs in World Archaeology No. 10. Madison, Wisconsin: Prehistory Press.

Bradbury, J. (2009) Beyond a Landscape of Tells: Subsistence, Society and Interaction within the Basalt Environs of Homs, Syria, in *CBRL 2009, Bulletin of the Council for British Research in the Levant 4*, pp. 38-41.

Buccellati, G. and Kelly-Buccellati, M. (1967) Annual Meeting of the Archaeological Institute of America (Archaeological Survey of the Palmyrene and the Jebel Bishri), in *Archaeology*, Vol. 20, pp. 305-306.

Cauvin, J. (1981) L'occupation néolithique de la region d'El Kowm: résultats 1978-1979, in *Préhistoire du Levant: zone de contacts et d'échanges*, ed. by Cauvin, J. and Sanlaville, P., Paris: Editions du CNRS, pp. 471-483.

Cauvin, J. (1990) Nomadisme néolithique en zone aride: l'oasis d'El Kowm (Syrie), in *Resurrecting the Past: A Joint Tribute to Adnan Bounni*, ed. by Matthiae, P., Van Loon,

M. and Weiss, H., Istanbul: Nederlands Historisch-Archaeologisch Instituut te Istanbul, pp. 41-47.

Cauvin, J. (2000) *The Birth of the Gods and the Origins of Agriculture.* Trans. Watkins, T. Cambridge: Cambridge University Press.

Chang, C. (2006) The Grass is Greener on the Other Side: A Study of Pastoral Mobility on the Eurasian Steppe of Southeastern Kazakhstan, in *Archaeology and Ethnoarchaeology of Mobility*, ed. by Sellet, F., Greaves, R. and Yu, P.-L., Gainesville: University Press of Florida, pp. 184-200.

Cribb, R. (1991) *Nomads in Archaeology.* New Studies in Archaeology. Cambridge: Cambridge University Press.

Dornemann, R.H. (1986) *A Neolithic Village at Tell el Kowm in the Syrian Desert.* Studies in Ancient Oriental Civilization No. 43. Chicago: The Oriental Institute of the University of Chicago.

Durkheim, E. (1912) *Les formes élémentaires de la vie religieuse: Le système totémique en Australie.* Paris: Presses universitaires de France.

Dussaud, R. (1929) Les relevés du captaine Rees dans le désert de Syrie, in *Syria*, Vol. 10, pp. 144-163.

Eddy, F.W. and Wendorf, F. *et al.* (1999) *An Archaeological Investigation of the Central Sinai, Egypt.* Colorado: The American Research Center in Egypt, Inc. and The University Press of Colorado.

Edwards, Y.H. and Martin, L. (2007) Fauna from the Natufian and PPNA Cave Site of Iraq ed-Dubb in Highland Jordan, in *Paléorient,* Vol. 33, pp. 143-174.

Fujii, S. and Adachi, T. (2010) Archaeological Investigations of Bronze Age Cairn Fields on the Northwestern Flank of Mt. Bishri, in *International Symposium, Formation of Tribal Communities, Integrated Research in the Middle Euphrates, Syria, November 21-23, 2009, Al-Rafidan*, Special issue, pp. 61-77.

Fujii, S., Akazawa, T., Nishiaki, Y., and Wada, H. (1987) Thaniyyet Wnker: A Pre-Pottery Neolithic B Site on the Lacustirne Terrace of Paleo-Palmyra Lake, in *Paleolithic Site of Douara Cave and Paleogeography of Palmyra Basin in Syria, Part IV: 1984 Excavations*, ed. by Akazawa, T. and Sakaguchi, Y., The University Museum, The University of Tokyo, Bulletin No. 29, Tokyo.

Geyer, B. and Besançon, J. (1996) Environnement et occupation du sol dans la vallée de l'Euphrate Syrien durant le Néolithique et le Chalcolithique, in *Paléorient,* Vol. 22/2, pp. 5-15.

Goring-Morris, N. (1991) The Harifian of the Southern Levant, in *The Natufian Culture in the Levant,* ed. by Bar-Yosef, O. and Valla, F.R., International Monographs in Prehistory, Archaeological Series 1, Ann Arbor, Michigan: International Monographs in Prehistory, pp. 173-216.

Grigson, C. (1995) Plough and Pasture in the Early Economy of the Southern Levant, in *The Archaeology of Society in the Holy Land,* ed. by Levy, T.E., London: Leicester University Press, pp. 245-268.

Haiman, M. (1992) Sedentarism and Pastoralism in the the Negev Highlands Emergency Survey, in *Pastoralism in the Levant: Archaeological Materials in Anthropological Perspectives.* Monographs in World Archaeology No. 10. Madison, Wisconsin: Prehistory Press, pp. 93-104.

Haiman, M. (2002) An Unpublished Lecture at the American Centre of the Oriental Research Conference in Toronto. Nov. 2002.

Henry, D.O. (1995) *Prehistoric Cultural Ecology and Evolution: Insights from Southern Jordan.* Interdisciplinary Contributions to Archaeology. New York: Plenum Press.

Horwitz, L.K. (2003) Temporal and Spatial Variation in Neolithic Caprine Exploitation Strategies: A Case Study of Fauna from the Site of Yiftah'el (Israel), in *Paléorient,* Vol. 29/1, pp. 19-58.

Ingold, T. (1980) *Hunters, Pastoralists and Ranchers: Reindeer Economies and their Transformations.* New York: Cambridge University Press.

Jenkins, E. and Rosen, A. (2007) Phytoliths, in *The Early Prehistory of Wadi Faynan, Southern Jordan, Archaeological Surveys of Wadis Faynan, Ghuwayr and Al-Bustan and evaluation of the Pre-Pottery Neolithic A Site of WF 16*, ed. by Finlayson, B. and Mithen, S., Wadi Faynan Series 1, Levant Supplementary Series 4, Oxford: Oxbow Books, pp. 429-436.

Kansa, S.W., Kennedy, A., Campbell, S. and Carter, E. (2009) Resource Exploitation at Late Neolithic Domuztepe: Faunal and Botanical Evidence, in *Current Anthropology*, Vol. 50, pp. 897-914.

Kent, S. (1989) Cross-cultural perceptions of farmers as hunters and the value of meat, in *Farmers as Hunters*, ed. by Kent, S., New Directions in Archaeology, Cambridge: Cambridge University Press, pp. 1-17.

Kenyon, K.M. (1981) ed. by Holland, T. *Excavations at Jericho.* Vol. III. The Architecture and Stratigraphy of the Tell. Text. British School of Archaeology in Jerusalem/London: Oxford University Press.

Kochavi, M. com. (1991) Notes and News: Rogem Hiri, in *Israel Exploration Journal*, Vol. 41, pp. 182-183.

Köhler-Rollefson, I. (1993) Camels and Camel Pastoralism in Arabia, in *Biblical Archaeologist*, Vol. 56, pp. 180-188.

Legge, A.J. and Rowley-Conwy, P. (1987) Gazelle Killing in Stone Age Syria, in *Scientific American,* Vol. 248, pp. 88-95.

Lees, S. H. and Bates, D. G. (1974) The Origins of Specialised Nomadic Pastoralism: a Systemic Model, in *American Antiquity*, Vol. 39, No. 2, pp. 187-193.

Levy, T.E. (1983) The Emergence of the Specialised Pastoral Nomadism in the Levant, in *World Archaeology*, Vol. 15, pp. 15-36.

Lönnqvist, M. (2008) Were Nomadic Amorites on the Move? Migration, Invasion and Gradual Infiltration as Mechanisms for Cultural Transition, in *Proceedings of the 4th International Congress of the Archaeology of the Near East*, 29 March – 3 April 2004, Freie Universität Berlin, Vol. 2, *Social and Cultural Transformation: The Archaeology of Transitional Periods and Dark Ages, Excavation reports*, ed. by Kühne, H., Czichon, R., Kreppner, F.J., Wiesbaden: Harrassowitz Verlag, pp. 195-215.

Lönnqvist, M. and Törmä, M. (2003) SYGIS – The Finnish Archaeological Project in Syria, in *Proceedings of the XIXth International Symposium CIPA 2003 (The ICOMOS & ISPRS Committee for Documentation of Cultural Heritage), New Perspectives to Save Cultural Heritage, Antalya, Turkey*, ed. by Altan, M.O., *The ISPRS International Archives of the Photogrammetry, Remote Sensing and Spatial Information Sciences,* Vol. XXXIV-5/C15, Istanbul, pp. 609-614.

Lönnqvist, M. and Törmä, M. (2004) Different Implications of a Spatial Boundary: Jebel Bishri between the Desert and the Sown in Syria, in The ISPRS XXth Congress Proceedings, ed. by Altan, M.O., *The ISPRS International Archives of the Photogrammetry, Remote Sensing and Spatial Information Sciences,* Vol. XXXV, Part B, Istanbul, pp. 897-902

Lönnqvist, M. and Törmä, M. (2006) Observing Changes and their Effects in Nomadic Environment: Remote-sensing and Archaeology for Sustainable Development on Jebel Bishri in Syria, in *Environmental Geoinformatics and Modelling, Proceedings of ICEM '05: International Conference on Environmental Management (28-30 Oct. 2005)*, ed. by Reddy, M.A., Vol. IV, Hyderabad.

Lönnqvist, M. with the contributions by Törmä, M., Nuñez, M., Lönnqvist, K., Stout Whiting, M., Okkonen, J., Riihiaho, H. and Nissinen, M. (2006) Archaeological Surveys of Jebel Bishri, The Preliminary Report of the Finnish Mission to Syria 2000-2004, in *KASKAL*, Vol. 3, pp. 205-242.

Lönnqvist, M. with contributions by Lönnqvist, K., Törmä, M., Nuñez, M., Stout Whiting, M. (2010) Tracing Tribal Implications among the Bronze Age Tomb Types in the Region of Jebel Bishri in Syria, in *International Symposium: Formation of Tribal Communities, Integrated Research in the Middle Euphrates, Syria, November 21-23, 2009, Al-Rāfidān*, Special issue, pp. 165-173.

MacDonald, B. with major contributions by 'Amr. K., Broeder, N.H., Skinner, H.C.W., Meyer, C., Neeley, M.P., Reese, D.S. and Whitcomb, D.S. (1992) *The Southern Ghors and Northeast 'Arabah Archaeological Survey.* Sheffield Archaeological Monographs 5. Sheffield: J.R. Collis Publications.

Mizrachi, Y. and Zohar, M. (1993) Rogem Hiri, in *The New Encyclopedia of Archaeological Excavations in the Holy Land*, Vol. 4, Jerusalem: Israel Exploration Society & Carta, pp. 1286-1287.

Moore, A.M.T. (1982) A Four-Stage Sequence for the Levantine Neolithic, ca. 8500-3750 B.C., in *Bulletin of the American Schools of Oriental Research*, No. 246, pp. 1-34.

Moore, A.M.T., Hillman, G.C. and Legge, A.J. (2000) *Village on the Euphrates: From Foraging to Farming at Abu Hureyra.* Oxford: Oxford University Press.

Neumann, F., Schölzel, C., Litt, T., Hense, A. and Stein, M. (2007) Holocene vegetation and climate history of the northern Golan Heights (Near East), in *Vegetation History and Archaeobotany*, Vol. 16, pp. 329-346.

Oates, D. and Oates, J. (1976) *The Rise of Civilization.* The Making of the Past. New York: Elsevier, Phaidon.

Perkins, D. Jr. (1964) Prehistoric Fauna from Shanidar, Iraq, in *Science*, Vol. 144, pp. 1565-1566.

Peters, J., Helmer, D., Von den Driesch, A, and Saña Segui, M. (1999) Early Animal Husbandry in the Northern Levant, in *Paléorient*, Vol. 25, pp. 27-48.

Pirie, A. (2007) The Chipped Stone, in *The Early Prehistory of Wadi Faynan, Southern Jordan: Archaeological Surveys of Wadis Faynana, Ghuwayr and Al-Bustan and Evaluation of the Pre-Pottery Neolithic A Site of WF 16*, ed. by Finlayson, B. and Mithen, S., Wadi Faynan Series 1, Levant Supplementary Series 4, Oxford: Oxbow Books, pp. 227-283.

Rosen, S. (1984) Kvish Harif: Preliminary Investigations at a Late Neolithic Site in the Central Negev, in *Paléorient*, Vol. 10/2, pp. 111-121.

Rosen, S. (1988) Notes on the Origins of Pastoral Nomadism: A Case Study from the Negev and Sinai, in *Current Anthropology*, Vol. 29, pp. 498-506.

Rothenberg, B. and Weyer, H. (1979) *Sinai: Pharaohs, Miners, Pilgrims and Soldiers*. Washington – New York: Joseph J. Binns.

Rothenberg, B. and Glass, J. (1992) The Beginning and the Development of Early Metallurgy and the Settlement and Chronology of the Western Arabah, from the Chalcolithic Period to Early Bronze Age IV, in *Levant*, Vol. XXIV, pp. 141-157.

Sahlins, M.D. (1968) *Tribesmen*. Foundations of Modern Anthropology Series. Englewood Cliffs, New Jersey: Prentice-Hall, Inc.

Sherratt, A. (1981) Plough and Pastoralism: aspects of the secondary products revolution, in *Patterns of the Past: Studies in Honour of David Clarke*, ed. by Hodder, I., Isaac, G., Hammond, N., Cambridge: Cambridge University Press, pp. 261-305.

Sherratt, A. (1983) The Secondary Exploitation in the Old World, in *World Archaeology*, Vol. XV, pp. 90-104.

Smith, A.B. (2005) Desert Solitude: The Evolution of Ideologies Among Pastoralists and Hunter-Gatherers in Arid North Africa, in *Desert Peoples: Archaeological Perspectives*, ed. by Veth, P., Smith, M. and Hiscock, P., Malden, Oxford and Carlton: Blackwell Publishing, pp. 261-275.

Stordeur, D. (1993) Sédentaires et nomades du PPNB final dans le desert de Palmyre (Syrie), in *Paléorient,* Vol. 19/1, pp. 187-204.

Stordeur, D. (2000) *El Kowm 2. Une île dans le désert: La fin du Néolithique précéramique dans la steppe syrienne.* Paris: CNRS Éditions.

Stordeur, D. and Taha, A. (1996) Ressemblances et dissemblances entre les sites nomades et sédentaires de la steppe syrienne au VIe millénaire, in *Les Annales Archéologiques Arabes Syriennes*, (Numéro spécial: *Palmyra and the Silk Road*), Vol. 42, pp. 85-98.

Stordeur, D., Maréchal, C. and Molist, M. (1991) Stratigraphie générale du Tell néolithique d'El Kowm 2 – Caracol (Syrie), in *Cahiers de l'Ephrate* 5-6, pp, 33-46.

Weiss, H., Courty, M.-A., Wetterstrom, W., Guichard, F., Senior, L., Meadow, R, Curnow, A. (1993) The Genesis and Collapse of Third Millennium North Mesopotamian Civilization, in *Science*, Vol. 261, pp. 995-1004.

Zarins, J. (1989a) Jebel Bishri and the Amorite Homeland: The PPNB Phase, in *To the Euphrates and Beyond,* Archaeological Studies in Honour of Maurits N. Van Loon, ed. by Chaex, O.M., Curvers, H.H. and Akkermans, P.M.M.G., Rotterdam: Balkema, pp. 29-52.

Zarins, J. (1989b) Pastoralism in southwest Asia: the second millennium BC, in *The Walking Larder: Pattern of Domestication, Pastoralism, and Predation*, ed. by Clutton-Brock, J., One World Archaeology 2, Cambridge: Cambridge University Press, pp. 127-155.

Zarins, J. (1992) Archaeological and Chronological Problems within the Great Southwest Asian Arid Zone, 8500-1850 B.C., in *Chronologies in Old World Archaeology*, Third Edition. 2 Vols., ed. by Ehrich, R.W., Chicago: The University of Chicago Press, Vol. 1, pp. 42-62.

Zeider, M.A. (1999) Animal Domestication in the Zagros: A Review of Past and Current Research, in *Paléorient*, Vol. 25, pp. 11-25.

7. STUDYING PASTORAL GRAZING POTENTIAL BY USING CLUSTER ANALYSIS AND SITE CATCHMENT ANALYSIS

Minna Lönnqvist, Markus Törmä, Kenneth Lönnqvist and Milton Nuñez

Fig. 1. The cluster analysis of the vegetation cover of Jebel Bishri area carried out by using a Landsat-7 ETM image, taken in January 1999. The target area of the QuickBird image study is circled in the centre. Note the large area under desertification covering the southwestern part of the mountain. Data© Eurimage.

7.1. Studying Present and Past Grazing Potential

Availability of space-born data offers excellent possibilities to study the grazing potential of Jebel Bishri and execute various GIS analyses. We decided to use Landsat images to provide a clearer evidence of the present grazing potential of some areas of the mountain taking into account differences in seasons. As previously mentioned, because of their high spatial resolution, in many cases the QuickBird images (0.6 m) offer possibilities to trace, locate and record even rather flimsy structures left by mobile societies, such as pastoralists. Our aim has been to identify the earlier described stone-built animal pens, i.e., corrals, on an acquired QuickBird image, to study their position in the landscape and distribution as well as their relation to the potential of the grazing grounds in the pastoral landscape. GIS has become an efficient tool to execute predictive models and simulations in archaeology including the site catchment analysis (Savage 1990). DEM tiles from ASTER-DEM provide topographical data needed to carry out GIS-based site catchment analyses. A satellite map in the scale of 1: 100,000 constructed by the Technische Fachhochschule, Berlin (Meissner and Ripke, ed. 1995) is used here in locating elevation contours and seasonal water courses, such as *wadis*, on Jebel Bishri.

7.2. Cluster Analysis for Displaying the Present Environment and its Grazing Potential

A general overview of the present land cover was acquired using cluster analysis (Figs. 1. - 4.) with Landsat images. In the cluster analysis, or unsupervised classification, the image pixels have been divided into categories or clusters according to their spectral similarities.

The principle used is based on the idea that the spectrally similar pixels belong to the same cluster, whereas different pixels belong to different clusters. When clusters had been formed, they were interpreted by comparing them to the satellite data, or to reference images such as those presented in aerial images or maps, or ground survey data. The Landsat TM and ETM images were clustered into 30 - 60 clusters using k-means algorithm (Richards 1993).

Fig. 2. Zooming in on the study area interpreted with the cluster analysis on the Landsat-7 ETM. Grazing potential in this area appears as green areas today. Some features of the topography are revealed in the image as well. Data© Eurimage

Figs. 3. and 4. are presenting a TM image taken on 1.9.1990 and an ETM image taken on 29.1.1999, using 60 clusters. Blue indicates water. Green indicates vegetated areas, mainly agricultural fields, but also steppes, depending on the season. Light green indicates areas with lower vegetation. Brown shows non-vegetated areas. Dark blue, yellow, or grey reveals non-vegetated ground; and white or reddish yellow show sandy areas. The lower radiometric and spectral resolution of MSS image caused problems as it was difficult to separate some features such as water and some types of bare ground. It was also more complicated to separate same sets of land features when using the TM and ETM images. The main source of difference between the results in interpreting the TM and ETM images is due to the time lapse between different acquisition dates of the images concerned, causing different phonological states of vegetation. Clearly, in the TM image dating to the autumn season, dryness is visible compared to the ETM image above dating to a rainy season.

We can detect seasonal differences between winter/spring and late summer in the Figs. 3. and 4. However, the large veil of sand had expanded significantly between 1990 and 2000 due to increasing desertification (see Lönnqvist and Törmä, 2004). The veil of the sand cover is heavy and shows the progress of desertification especially to the southwest and also somewhat to the northwest.

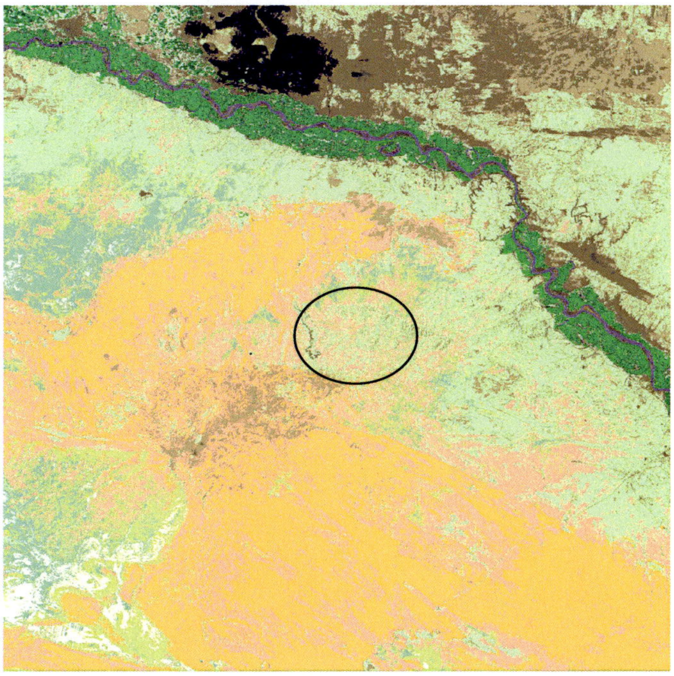

Fig. 3. The clustering result of the Landsat TM image, dated 01.09.1990, using 60 clusters. Data source: NASA.

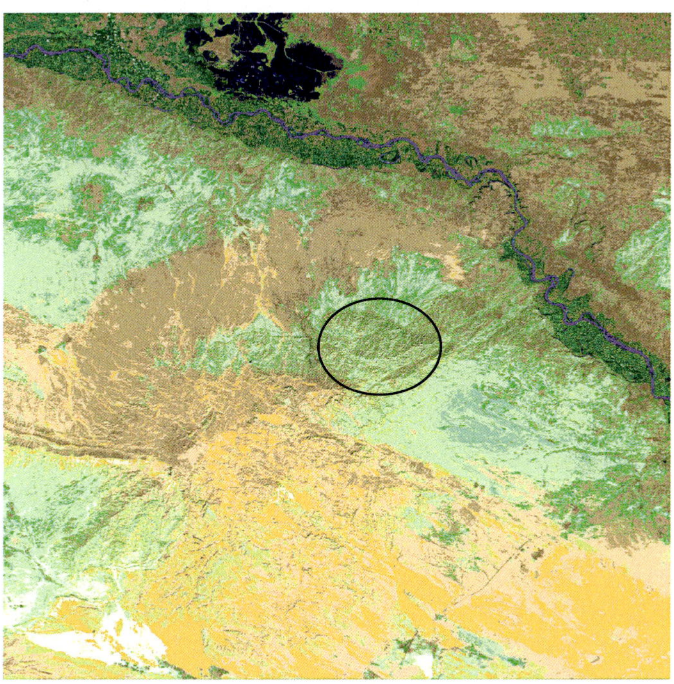

Fig. 4. The clustering result of the Landsat ETM image, dated 29.01.1999, using 60 clusters. Data source: NASA.

7.3. Locating Stone Enclosures/ Pastoral Corrals

The QuickBird satellite image data from the central district of Jebel Bishri (centre coordinates lat. 35. 422, long. 39. 4788, taken 15.7.2003, see Fig. 6.) covering 8 km x 8 km serves as a basis for prospecting more or less ancient stone enclosures/corral sites of pastoral nomads. The image area includes parts of the highest areas of the mountain, which rises ca. 867 m a.s.l. (Wirth 1971: 55). Apart from the plain of Nadra, a palaeolake, it also seems to belong to the best grazing grounds on the mountain today.

The QuickBird image was further divided into 16 smaller images in order to carry out prospecting and identifying stone enclosures/pastoral corrals by the naked eye. Each of the 16 images covered 4 km², and their order ran from the north to the south in four image sets. When a site was identified, the UTM coordinates were collected for the location information (see UTM coordinates for the corral sites identified in the QuickBird image: Tables 1.and 2.).

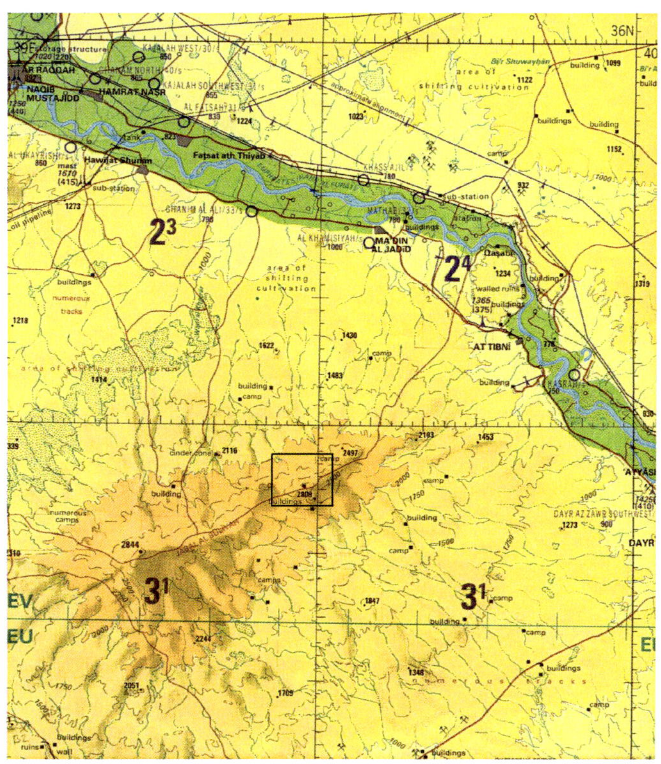

Fig. 5. The location and coverage of the QuickBird image on TPC G-4C map. Basemap: © MOD Crown Copyright, 2010 published with the permission of the Controller of HMSO.

Fig. 6. QuickBird pan-sharpened multispectral image, dated 15.07.2003.© DigitalGlobe Inc.

Image 1	Image 2	Image 3	Image 4	Image 5	Image 6	Image 7	Image 8
None	F 1 540300 3922350	F 6 540400 3920300	F 9 539920 3916700	None	F 13 542300 3921800	None	F 16 541650 3917200
	F 2 539800 3922200	F 7 540500 3920120	F 10 540080 3916800		F 14 542500 3921800		
	F 3 539850 3921950	F 8 540700 3920300	F 11 540520 3916850		F 15 543200 3921950		
	F 4 540770 3922400		F 12 540580 3916680				
	F 5 540700 3922200						
None	5 sites	3 sites	4 sites	None	3 sites	None	1 site

Table 1.

Image 9	Image 10	Image 11	Image 12	Image 13	Image 14	Image 15	Image 16
None	F 17 543800 3921700	F 19 544000 3919900	F 20 544100 3916900	F 22 547080 3924300	F 24 547530 3922400	F 28 547200 3920220	F 32 547600 3918520
	F 18 545420 3921200		F 21 544500 3916700	F 23 546650 3924350	F 25 546500 3921180	F 29 547230 3920250	F 33 547400 3918450
					F 26 546700 3920900	F 30 546800 3919930	F 34 547450 3918600
					F 27 546500 3921220	F 31 546700 3919780	F 35 547280 3918480
							F 36 546400 3918350
							F 37 546400 3918300
							F 38 54 6600 39 17200
							F 39 54 6600 39 17300
							F 40 54 6600 39 17180
							F 41 54 6550 39 17180
None	2 sites	1 site	2 sites	2 sites	4 sites	4 sites	10 sites

Table 2.

Altogether 41 sites were identified in the area covering 64 km², and they range from a single stone circle to multiple agglutinated enclosure constructions. Four images out of the 16 image set of the QuickBird area did not provide any traces of enclosures. So of the overall area coverage 16 km² remained empty from the structures corresponding to 25 % of the total area. The number of sites indicates that 3/4 of the area covering 64 km² provides one enclosure/corral site in every 1 km² on average.

Experimental visits on the ground were made only to a few sites in order to check the nature of the structures analysed in this remote sensing study, because the area in specific was outside our ground survey permission for the years 2004-2008. No associated small finds were allowed to be collected to infer dating of the structures. However, a few sites seem to include pieces of coarse handmade pottery from the Neolithic/Chalcolithic period. Based on the comparable sites found during the previous ground surveys on the mountain providing associated finds (see Lönnqvist *et al.*, 2006 and the previous ch. 6.), we know that the use of stone enclosures and corrals may vary from the Neolithic to the Medieval times, or they may even have been used by the very recent Bedouins. Some sites may be Epipalaeolithic and Neolithic hut bases as well. It needs to be emphasized that all the traced and identified stone enclosures on the image are not necessarily contemporary structures, but despite of that their location and frequency in the landscape can provide information of the preferred areas and thus the evident grazing potential in the past.

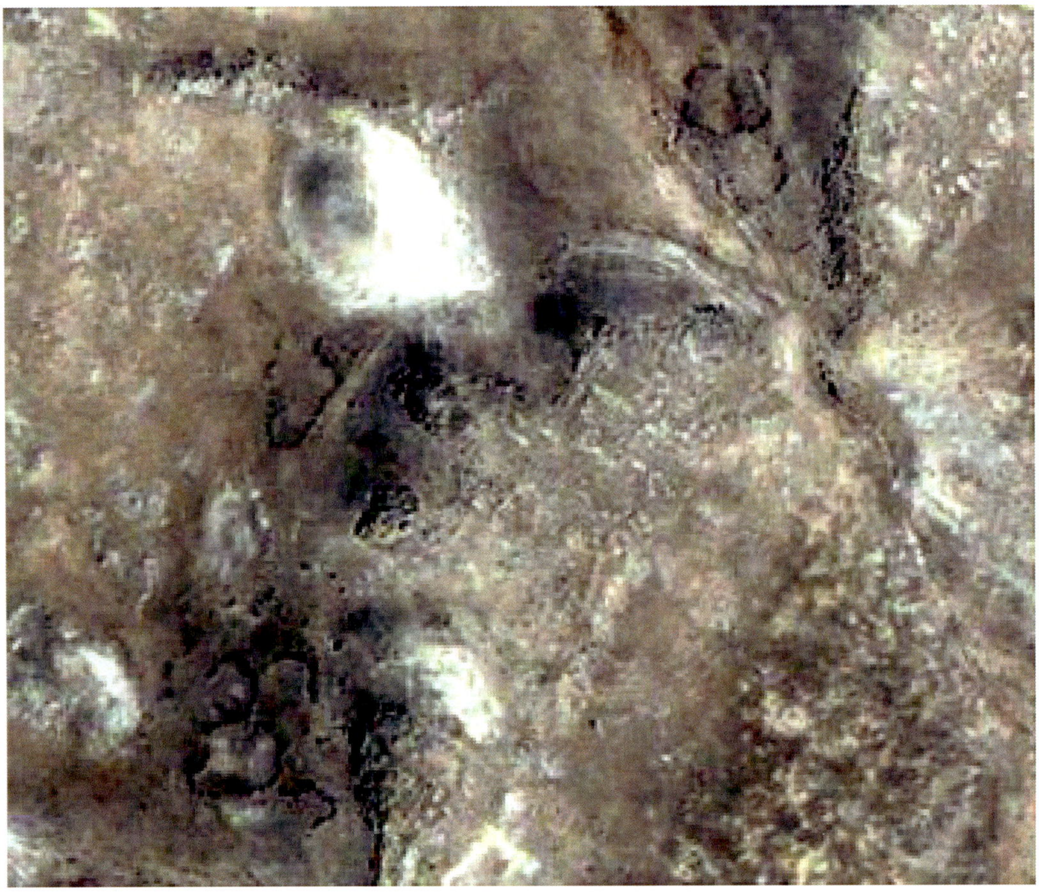

Fig. 7. Stone enclosure/corral site F 4 as visible on our QuickBird image.© Eurimage Note also the association with cairns/tumuli on nearby hillocks.

Fig. 8. Ancient stone enclosures (site F 22) identified in the QuickBird image study area of Jebel Bishri. Photo from E: Minna Lönnqvist 2008.

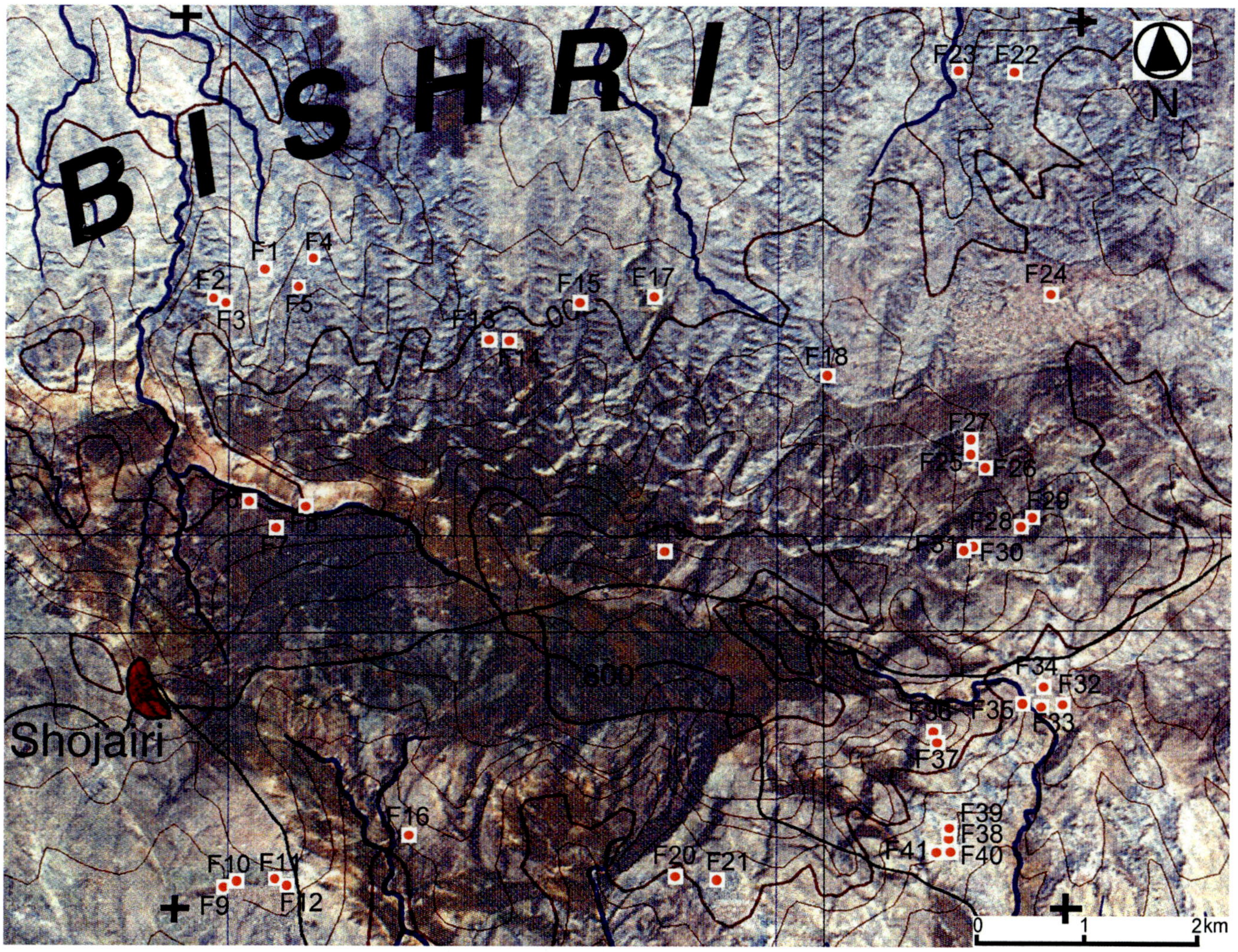

Fig. 9. Distribution of the identified corrals displayed on a plot of the satellite map with the scale of 1:100,000 (ed. by Meissner & Ripke 1995). The corners of the QuickBird image used in the present study are marked with black crosses. Mapping by Kenneth Lönnqvist 2009.

7.4. Site Catchment Analysis

"The site catchment analysis" was initially developed by C. Vita-Finzi and E. Higgs (1970) to study ancient hunter-gatherer mobility and subsistence in order to see the radius of people's movement from camps, taking the topography of the area into account. The method explores human – environment relationship. Apart from studying the hunter-gatherer economy, the site catchment analysis has been earlier applied in studying site locality in relation to ancient agricultural potential and land use for animals (Barker 1985). In our project we decided to apply the analysis to ancient curvilinear stone enclosure sites mainly known as pastoral corrals and environments surrounding them.

The site catchment analysis is based on the least-cost movement and therefore on the preference that the resources from the base (here the stone enclosures) should be within hour's or two hour's walking distance. Nowadays GIS provides a convenient tool to calculate distances taking the topography into account. (See, e.g., Savage 1990). In our study the used ASTER-DEM (ASTER Digital Elevation Model) data covering the target area of the QuickBird image (see Figs. 5. and 6; the central coordinates for the corner pixels of the ASTER-DEM: UTM 539510, 3924752 and UTM 547700, 3916562) was acquired by downloading it from the EOS data archive provided by NASA (http://asterweb.jpl.nasa.gov/gdem.aspwas). It was utilized in analysing the topography and amount of time needed to traverse a certain radius surrounding some of the pens.

Fig. 10. The ASTER-DEM image covering the study area. Data source: NASA.

The ASTER-DEM image was reprojected into the same coordinate system as the QuickBird image (UTM zone 37, > WGS84) using 30 m pixel size. The used ASTER-DEM performance follows the 20 m accuracy in height and the accuracy of ca. 50 m in location (Fujisada *et al.* 2005). Travel time was estimated using Tobler's hiking distance (Tobler 1993). The walking velocity in a hilly terrain can be estimated as W = 6 x exp {-3.5 x ABS (S +0.05)}, where W corresponds to the walking velocity in km/h, whereas S denotes a slope. The slope is estimated as dh/dx, in which dh corresponds to the difference in height, if two places (e.g. neighbouring pixels of DEM) and dx correspond to their horizontal distance in the same units as the height. On a flat terrain the walking velocity is ca. 5 km/h. The travel time is computed as distance/velocity. The results are presented in Table 3., in which each zone of colour is equivalent to 20 minutes of walk, and the maximum measured time estimated for the effort used to attain grass for flocks varies from ca. 1 hour to 2 hours according to the site catchment theory.

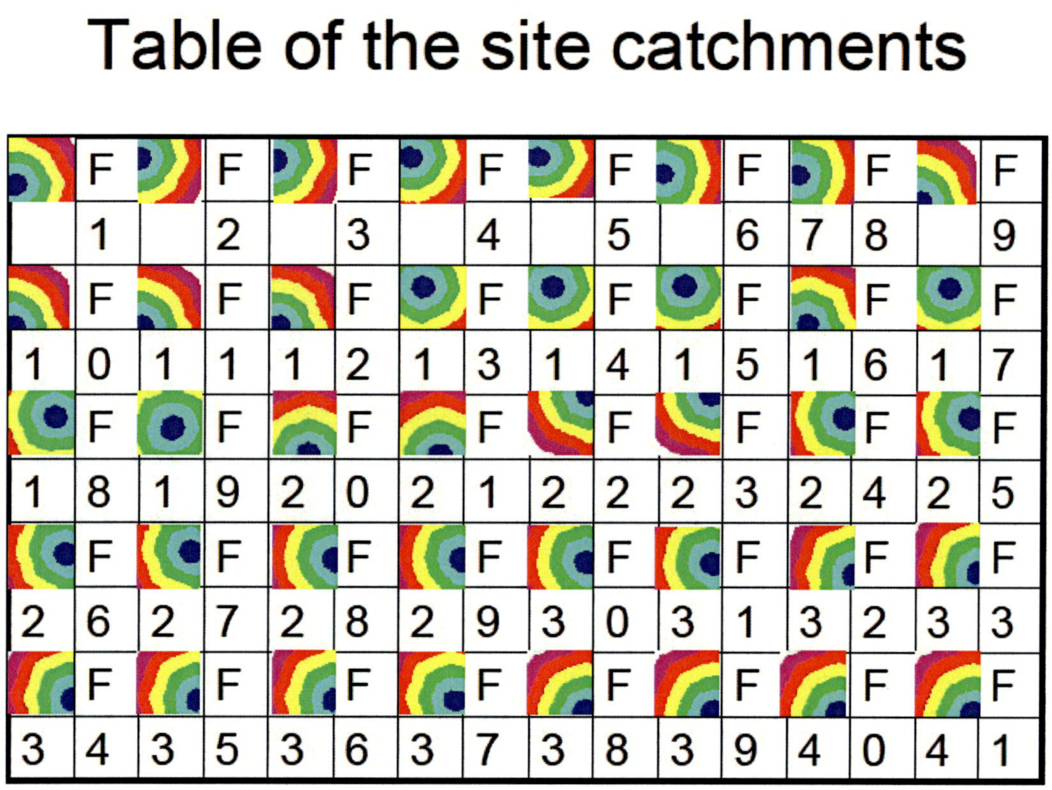

Table 3. Computer analysis by Markus Törmä 2009.

The eight sites **F 1 - F 8** analysed showed that topographical constraints for each site appeared generally to be located to the west and the northwest evidently meaning steep slopes, ridges, ravines or escarpments hindering an easy access. The direction towards the east and southeast seems to be the easiest one for moving even over 2 hours = over 10 km. However, with the flocks of sheep the pace could be considerably slower compared to a situation when a person hiking alone. The nearest *wadis* are to be found to the west, so constraints in topography do not necessarily completely hinder the progress in this case. The access to all the sites **F 9 – F 12** is most favourable to the northeast, the topographical constraints being in the southwest. However, at the sites **F 13 – F 15** there is a fairly free access in short distance to all the directions, and the terrain seems to be easiest as only an hour is needed for the distance compared to the effort of two hours for the other sites. A favourable part for movement from the site **F 16** seems to be in the northeast. The best access from the sites **F 17** to move is in the south, whereas the site **F 18** also is due to the south but has the best access to the west. The sites **F 19 – F 20** had free access for grazing to all directions nearly in the same pace. Here the altitudes are some of the highest on the mountain area. The sites **F 22** and **F 23** offer the best access to the southwest. The most favourable direction from the sites **F 24** - **F 27** is in the west, and the same applies to the sites **F 28 - F 31**. The highest number of sites, altogether 10 sites, is situated on the southeastern corner of the QuickBird image. Some vegetation is visible there unlike in other areas. Of this area the sites from **F 32** to **F 39** had topographical constraints to the east but have the best access for moving to the west and northwest while the sites **F 40** and **F 41** opened slightly better to the northwest and west. Our experience concerning the site **F 37** (Fig. 11.) on the ground confirmed the topographical analysis and restrictions of a huge ridge to the south.

Fig. 11. A stitched photograph of pens at the site F 37 taken from E: Kenneth Lönnqvist 2003.

The sites in general are clearly concentrating on particular areas as we see, for example, that the east offers the densest area of corrals: altogether 20 out of 40. This may be due to the direction of water currents and their catchment areas as visible in Fig. 9. There the grazing potential seems also to be greatest. Furthermore, the distance to the desertification line and spreading sands coming from the west and southwest was the longest.

7.5. General Observations on the Site Locations and Site Distribution

General topographical and topological observations of the sites can be carried out by the naked eye on the satellite images and maps. As previously mentioned, we used the topographical map of 1: 100,000 in scale edited by Meissner and Ripke at the Technische Fachhochschule in Berlin. The five sites **F 1** - **F 5** were located between the 600 m and 700 m elevation contours, and four of them were very close to a large *wadi,* a water course, leading from the north towards the village of Ash-Shujiri. We know that on the summit area there has been a spring or a well called 'As-Šiqri/Bîr-Šiqri (Poidebard 1934) that is close to Ash-Shujiri, and two wells are located in the plain of Nadra to the south. They are situated at the lower register of the QuickBird image, where also numerous water harvesting sites were recorded in our field survey of 2004 (Lönnqvist *et al.* 2006; see ch. 4. in this volume).

The sites **F 6** - **F 8** are in the vicinity of the 700 m elevation contour, and here again they are all situated near a *wadi.* The sites **F 9** – **F 12** are located even higher, between the 700 m and 800 m elevation contours, southeast from the village of Ash-Shujiri in slightly over 0.5 km's radius from a large *wadi* which however is not visible in the map. The sites **F 13** – **F 15** are again close to the 700 m elevation contour, but there are no visible large *wadis* nearby; only in 1 km's radius there exists a large water course. The site **F 16** is situating above the 700 m elevation contour near a Roman military camp (see ch. 12.) that we have discovered, and it is also in association with a large water course. The sites **F 17** and **F 18** are located again in between the 700 m and 800 m elevation contours; both sites are in 1 km's radius of a large *wadi.* However, the site **F 19** in the area of the 800 m elevation contour is situated 2 km from a large *wadi* area.

The two sites **F 20** and **F 21** are located in the alignment of the Roman road leading southeast from the Roman camp and more or less in 1 km's radius from a large *wadi.* The sites **F 22** and **F 23** are located between the 600 m and 700 m elevation contours and are also in 1 km's radius of a couple of *wadis.* The sites **F 24** – **F 31** are located above the 700 m contour, but all have a longer distance to the nearest large *wadi* being in 2 km's radius.

However, it needs to be emphasized that in higher altitudes the grass keeps moister. As previously mentioned, the largest amount of sites, namely sites **F 32** – **F 39**, altogether 10 sites, is situated on the southeastern corner of the QuickBird image. They are all surrounded by a large *wadi* and a water catchment area situating between the 600 m and 800 m elevation contours.

Apart from the sites in the highest altitudes, more or less 1 km seems to be a maximum length to a nearest seasonal water course. This is interesting as our prospecting has revealed that generally there is one stone enclosure/corral site in every 1 km^2. As previously mentioned, the grazing potential seems to be the best in the east where the majority of sites appear. This is not only in accordance with direction of the large seasonal currents but also due to the present desertification line coming from the west and southwest (cf. Figs 1.-4.). The major water course is associated with the direction of the Nadra plain that used to form a palaeolake and where we recorded a number of stone enclosures/corrals situated on *wadi* banks in 2004 (see Lönnqvist *et al.* 2006 and ch. 6.). In most of the cases that we identified in the QuickBird image the eastern banks of the *wadis* are preferred which seems to indicate that the morning sun is coming from behind the hills and keeps the grass moister longer than that on the western bank or slope.

As previously indicated, three of the stone enclosure/corral sites studied in the QuickBird image have been visited on the ground. The site **F 12** is one of the corral sites defined with UTM: 0540580, 3916680 corresponding to UTM: 0540576, 3916775 on the ground and situating ca. 756 m a.s.l. The site consisted of five stone rings obviously forming corrals. Their diameters varied from 4.5 m to 17 m. The corral site **F 37** was also visited on the ground and corresponds to UTM: 546400, 3918300 the location defined in the QuickBird with a GPS UTM: 0546264, 3918376 being 764 m a.s.l. The site consists of five to six stone circles of corrals varying from 7 m to 17 m in diameter. The site was very close to a steep slope that opened on its southern side, which was an intriguing phenomenon – the reason may have been to prevent the flocks from moving to a certain direction.

7.6. Conclusions

The cluster analysis and the site catchment analysis were chosen as methods in order to study grazing potential of the surroundings of ancient pastoral structures such as animal pens, known as corrals identified by remote sensing. The cluster analysis which was carried out with Landsat images taken in the 1990s and the year 2000 reveals the present land cover of the target area in different years and seasons to visualize grazing potential. Spring shows the best greenness in the chosen area, while in the autumn the area is clearly affected by dryness, blowing sand and the expanding desertification line.

A QuickBird image offering 0.6 m spatial resolution and covering 64 km^2 in the central area of the mountain was used for identifying stone enclosures/corrals and collecting coordinate information of the sites. The area lies outside our survey permission and the intensively studied sites on the ground level. Only experimental visits were made on the ground to check and confirm the nature of the structures analysed in this GIS study. UTM coordinates provided the location information that served as a base for the site catchment analysis. Altogether 40 stone enclosures were located in the area of the QuickBird image covering 64 km^2, and in them some enclosures had multiple cells. In 3/4 of the image area there was one site located in every 1 km^2 area on average, but some areas had only one structure and others had clear concentrations.

The most visible site concentrations were on the eastern side of the QuickBird image area, offering 20 pens out of the 40 sites, which may be due to the directions of the large water courses and end-points of *wadi* catchments. The desertification line was also situated in the west, where it was spreading first affecting the western areas. However, the distribution of the pen sites itself, reveals that grazing potential existed in the target area in ancient times as it is today visible in the cluster analysis of the Landsat images. We know that on the summit area there has also been a well site called Bîr Šiqri near the village of Ash-Shujiri, and two wells are located in the plain of Nadra in the lower area of the image. It is possible that the distribution of the stone enclosures/corrals in the landscape denote some pastoral territorialism but this cannot be conclusively demonstrated in our study because it lacks evidence of the cultural reference matrix based on small finds from these enclosures.

Time and effort were calculated that was used for attaining grazing areas from individual sites. The calculation was carried out with an ASTER-DEM image data. Travel time was estimated applying Tobler's hiking distance to the ASTER-DEM data, the preferred time to catch the resources being one to two hours. According to the analysis carried out here, the uneven topography caused constraints in several areas. The best directions to movement were analysed. Most of the sites appeared to be located more or less in 1 km's hiking distance from seasonal water courses, which apparently provided water in spring and early summer seasons. The sites were usually located on the eastern slope of a *wadi* bank so that grass would be moist longer when the sun was rising behind the slope. The densest areas of the sites were located at the ends of the water catchments. In the site catchment analysis the free choice of access to different areas were topographically in the sites **F 13** - **F 15** and **F 17** as well as **F 19**, situating in the central part of the image and some of them having the longest distance to the nearest water courses. However, there were apparent choices.

According to our observations, cairns/*tumuli* which form the most common structure type on the mountain extend over the mountain area reaching to the edge of the Euphrates river, but stone enclosures/corrals, which are the second commonest structure type, are generally only met within the mountain proper. The western piedmont seems generally to be devoid both of these structure types although stone rings or cairns could have had better chances of survival than a single stone or two. This spatial difference in distribution not only indicates that stone enclosures/corrals have best survived in the otherwise desolate mountain area but also that the mountain in its highest altitudes was particularly used for herding. But whether the highland–lowland interaction was also continued from the hunter-gatherer phase to the time of the earliest pastoral nomadism between the plains and the mountain is an intriguing question. The PPNB sites in the El Kowm basin indicate the beginning of pastoral nomadism in the piedmont of Jebel Bishri. There are some indications of sedentarization during the PPNB (ch. 5.) period beneath the mountain on the Euphrates side. Whether the pastoralists of the oases or the river zone were transhumants in specific, using mountain as a grazing area from the base-villages during the earliest phase in this transition to pastoralism, is not clear in our material (see ch. 6.). However, in the later periods this concomitant relationship becomes more visible.

Bibliography

Barker, G. (1985) *Prehistoric Farming in Europe*. New Studies in Archaeology. Cambridge: Cambridge University Press.

Fujisada, H., Bailey, G.B., Kelly, G.G., Hara, S., Abrams, M.J., (2005) Aster Dem Performance, in *IEEE Transactions on Geoscience and Remote Sensing*, Vol. 43, Issue 12, pp. 2707- 2714.

Lönnqvist, M. and Törmä, M. (2004) Different Implications of a Spatial Boundary: Jebel Bishri between the Desert and the Sown in Syria, in The ISPRS XXth Congress Proceedings, ed. by Altan, M.O., *The ISPRS International Archives of the Photogrammetry, Remote Sensing and Spatial Information Sciences,* Vol. XXXV, Part B, Istanbul, pp. 897-902.

Lönnqvist, M. with the contributions by Törmä, M., Nuñez, M., Lönnqvist, K., Stout Whiting, M., Okkonen, J., Riihiaho, H., Nissinen, M. (2006) Archaeological Surveys of Jebel Bishri, The Preliminary Report of the Finnish Mission to Syria 2000-2004, in *KASKAL*, Vol. 3, pp. 205-242.

Poidebard, A. (1934) *La Trace de Rome dans le Désert de Syrie: Le Limes de Trajan a la conquête arabe.* Recherches aériennes (1925-1932). Texte. Atlas. Bibliothèque Archéologique et Historique. Tome XVIII. Paris: Geuthner.

Richards, J. (1993) *Remote Sensing Digital Image Analysis.* 2nd ed. Berlin: Springer.

Savage, S.H. (1990) Modelling the Late Archaic Social Landscape, in *Interpreting Space: GIS and Archaeology*, ed. by Allen, K.M.S., Green, S.W. and Zubrow, E.B.W. London – New York – Philadelphia: Taylor & Francis, pp. 332-355.

Tobler, W. (1993) Three Presentations on Geographical Analysis and Modeling: Non-Isotropic Geographic Modeling, Speculations on the Geometry of Geography, Global Spatial Analysis. *National Center for Geographic Information and Analysis, University of California Technical Report 93-1*. See: http://www.ncgia.ucsb.edu/Publications/Tech_Reports/93/93-1.

Vita-Finzi, C. and Higgs, E.S. (1970) Prehistoric Economy in the Mount Carmel Area of Palestine: Site Catchment Analysis, in *Proceedings of the Prehistoric Society,* Vol. XXXVI, pp. 1–37.

Wirth, E. (1971) *Syrien: Eine Geographische Landeskunde.* Wissenschaftliche Länderkunden. Band 4/5. Wissenschaftliche Buchgesellschaft: Darmstadt.

8. ANCESTRAL AND RITUAL VIEWS OF PASTORALISTS OVER AN ANCIENT "SILICON VALLEY"

Minna Lönnqvist
with Contributions by Margot Stout Whiting and Kirsi Lorentz

Fig. 1. An onlooker standing at a megalithic enclosure site at the edge of Tar al-Sbai and looking towards the western plains below. Photo: SYGIS 2000.

8.1. Mortuary Sites in the Pastoral Landscape of Tar al-Sbai

In the chapters 4. and 5. we introduced the landscape and hunter-gatherer sites of Tar al-Sbai (also identified as Jebel Sabah) at the western edge of Jebel Bishri. The area was already surveyed in the year 2000. The concentration of ancient remains of mobile people starting from the Palaeolithic Age at this semi-circular escarpment area of the mountain is striking (see ch. 5; see the Palaeolithic sites circled in red and other structures in Fig. 3.). But most of the structures in the area consist of cairns/*tumuli* and stone enclosures, including corrals, the types of remains that were dealt with in the previous chapters and belong to the so-called pastoral technocomplex (cf. Zarins 1992). The list of pastoral sites at Tar al-Sbai, their coordinates and sizes of the structures are presented here in Table 1., and the distribution of the sites is displayed in Fig. 2. in order to show the frequency and continuity in locating sites at the mountain edge from prehistoric up till the Islamic times. The area falls in between the elevation contours of 700 m and 800 m, the latter roughly following the edge. At Tar al-Sbai there are over 50 cairns/*tumuli* in the stretch of 0.21 km^2, and a Syro-Japanese team which has followed in our footsteps in the region has excavated several such tombs further to the northwest in the areas of Tor Rahum (the Rahum Plateau), Be'er Rahum (The

Well of Rahum), Wadi Hedaj and Wadi Tletuat (35°19' - 35°25' N Latitudes and 39°01' – 39°10' E Longitudes). Some of those tombs produced finds dating from the end of the Early Bronze Age to the Middle Bronze Age (Fujii and Adachi 2010).

TAR AL-SBAI PASTORAL NOMADIC SITES

Site	Type	Coordinates	Size
A 12	cairn/*tumulus*	UTM 0518060, 3905850	1.70 m in diam.
A 13	stone circles (a-c)	UTM 0518060, 3905850 ca. 696 m a.s.l.	a) 1.30-1.60 m in diam.. b) 1.70 m in diam c) 1.50 m in diam.
A 14	cairn/*tumulus*	UTM 0518039, 3905940	4.70 m in diam.
A 16	cairn/*tumulus*	UTM 0528064, 3905893	3.30 m in diam.
A 17	cairn/*tumulus*	UTM 0518065, 3905826	N/A.
A 18	cairn/*tumulus*	UTM 0518065, 3905812	1.30 m in diam
A 19	house/s	UTM 0519050, 3905748	rooms: 21 m x 13 m and 13 m x 10 m
A 20	cairn/*tumulus*	UTM 0518021, 3905688	5.40 m in diam.
A 22	cairns/*tumuli* (a-b)	UTM 0517994, 3905479	a) 2.00 m in diam. b) 6.00 m in diam.
A 23	cairn/*tumulus*	UTM 0518080, 3905418 ca. 711 m a.s.l.	5.00 m in diam.
A 24	Islamic grave	UTM 0517827, 3905445	1.30 m x 1.00 m
A 26	a) cairn/*tumulus* and b) stone circle	UTM 0517717, 3905479 ca. 613 m a.s.l.	a) 2.5 m in diam. b) 5.00 m in diam.
A 27	stone circles (a-c) cairns/*tumuli* (d-f) and a stone line (g)	UTM 0517793, 3905377 ca. 720 m a.s.l.	a) 15-20 m in diam. b) 9-13 m in diam. c)12-18 m in diam. d) 9.50 m in diam. e) 7-9 m in diam. f) 8 m in diam. g) ca. 40 m l.
A 30	cairns/*tumuli* (a-b)	UTM 0517951, 3905178 ca.709 m a.s.l.	a) 4 m in diam. b) 6 m in diam.
A 31	Bedouin tent base	UTM 0518003, 3905149 ca. 704 m a.s.l.	6.10 m x 7.10 m
A 32	cairn/*tumulus*	UTM 0518002, 3905073	4.80-5.00 m in diam.
A 35	elongated cairns (a-b?)/ Islamic graves	UTM 0518003, 3905149 ca. 702 m a.s.l.	a) 1.30 m x 2.00 m b) 1.30 m x 2.00 m
A 36	elongated cairn/Islamic grave	UTM 0518358, 3904687	1.60 m x 2.30 m
B 3	cairn/*tumulus*	UTM 0518039, 3906573	3.60-4.00 in diam.
B 4	complex of cairns/*tumuli* (a-b) and a stone circle (c)	UTM 0517997, 3906520	a) 4.10-4.60 m in diam. b) 3.00-4.40 m in diam. c) 18.0-19.7 m in diam.
B 6	complex of cairns/*tumuli* (a-d) and stone circles (e-g)	UTM 0518016, 3906375	a) max. 3.60 m in diam. b) max. 8.40 m in diam. c) 3.00 in diam. d) 3.50-4.7 m in diam. e) max. 8.40 m in diam. f) max. 2.90 m in diam. g) 4.80-5.30 m in diam.
B 7	cairns/*tumuli* (a-b)	UTM 0517974, 3906208	a) 2.50-3.80 m in diam. b) 2.60-2.90 m in diam.
B 8	undefined stone structures (a-d)	UTM 0517819, 3906239	a) 5.30 m x 12 m b) 1.60 m x 2.40 m c) 2 x 2 m

TAR AL-SBAI PASTORAL NOMADIC SITES

Site	Type	Coordinates	Size
B 9	cairn/*tumulus*	UTM 0517953, 3900172	4.20-4.50 m in diam
B 10	cairns/*tumuli* (a-b)	UTM 0517983, 3906206	a) max. 3.80 m in daim. b) max. 3.90 m in diam.
B 11	Islamic grave	UTM 0518016, 3906062	1.40 x 1.20 m
B 12	cairn/*tumulus*	UTM 0518001, 3906086	max. 8.5 m in diam.
B 13	Islamic graves (a-f)	UTM 0518001, 3906092	a) 1.90 x 0.80 m b) 1.20 m in diam. c) 0.90 m in diam. d) 2.90 x 3.50 m e) 0.40 x 0.90 m f) 7.30 x 7.40 m
B 14	stone circle	UTM 0517997, 3906112	max. 9.8 m in diam.
B 15	a)boat-like stone str. b) cairn/*tumulus* c) stone wall	UTM 0517993, 3906120	a) 5.20 x 14.90 m b) max. 12 m in diam. c) 11.30 m l.
B 16	cairn/*tumulus*	UTM 0517991, 3906157	max. 3.40 m in diam.
B 17	cairn/*tumulus*	UTM 0517978, 3906179	1.70 x 1.20 m
C 2	Bedouin tent base (a) rock shelter (b)	UTM 0517927, 3907055	a) 5.90 m x 2.50 m b) 4.50 m in length, 1.40 m in width, and 1.40 m in height
C 3	cairn/*tumulus*	UTM 0518003, 3906919	6.25 – 6.40 m in diam.
C 4	Islamic grave	UTM 0518028, 3906813	0.84 x 1.07 m
C 8	cairns/*tumuli* (a-c) incl. ring-*tumulus* (c) and stone lines	UTM 0518033, 3906677	a) 7.10-8.5 m in diam. b) 5.70-6.10 m in diam. c) 5.10 m in diam.
C 9	cairns/*tumuli* (a-b)	UTM 0518072, 3906810	a) 6.90-7.60 m in diam. b) 5.0-7.10 m in diam.
C 11	cairn/*tumulus*	UTM 0518246, 3904739	max. 12.65 m in diam.
C 12	stone complex of cairns/*tumuli* (a-b, d, f-g, i-j) stone walls (c and e) and radiating stone lines (h)	UTM 0518358, 3904633	a) 2.07 m in diam. b) 2.05 m in diam. c) corner of stone walls meeting; walls: 4.5 m in l.., 1.42 m in w., and 3.6 m in l.., 1.55 w. d) 5.40-7.40 m in diam e) 1.70 in l. and 0.85 in w. f) max. 22.70 in diam. g) 11.20 in diam. h) max. 2.00 in diam. i) max. 6.20 in diam. h) not measured
C 13	stone complex of cairns/*tumuli* (a-e), walls (f) and radiating stone lines (g)	UTM 0518358, 3904627 ca. 736 m a.s.l.	a) 7.45 m-7.65 m in diam. b) 8.48 – 9.00 m in diam. c) 24 m in diam. d) 3.90 m in diam. e) 2.70 in diam. f) 12.6 m l. and 2 m l. g) not measured
C 14	cairns/*tumuli* (a-b)	UTM 0518724, 3904434	a) 8.10 – 9.6o m in diam. b) 4 m in diam.
C 15	cairns/*tumuli* (a-b)	UTM 0520095, 3910264 ca. 624 m a.s.l.	a) max. ca. 8 m in diam. b) 4.60 m in diam.
D 4	Islamic graves (a-e)	UTM 0520209, 3903263	not measured

Table 1.

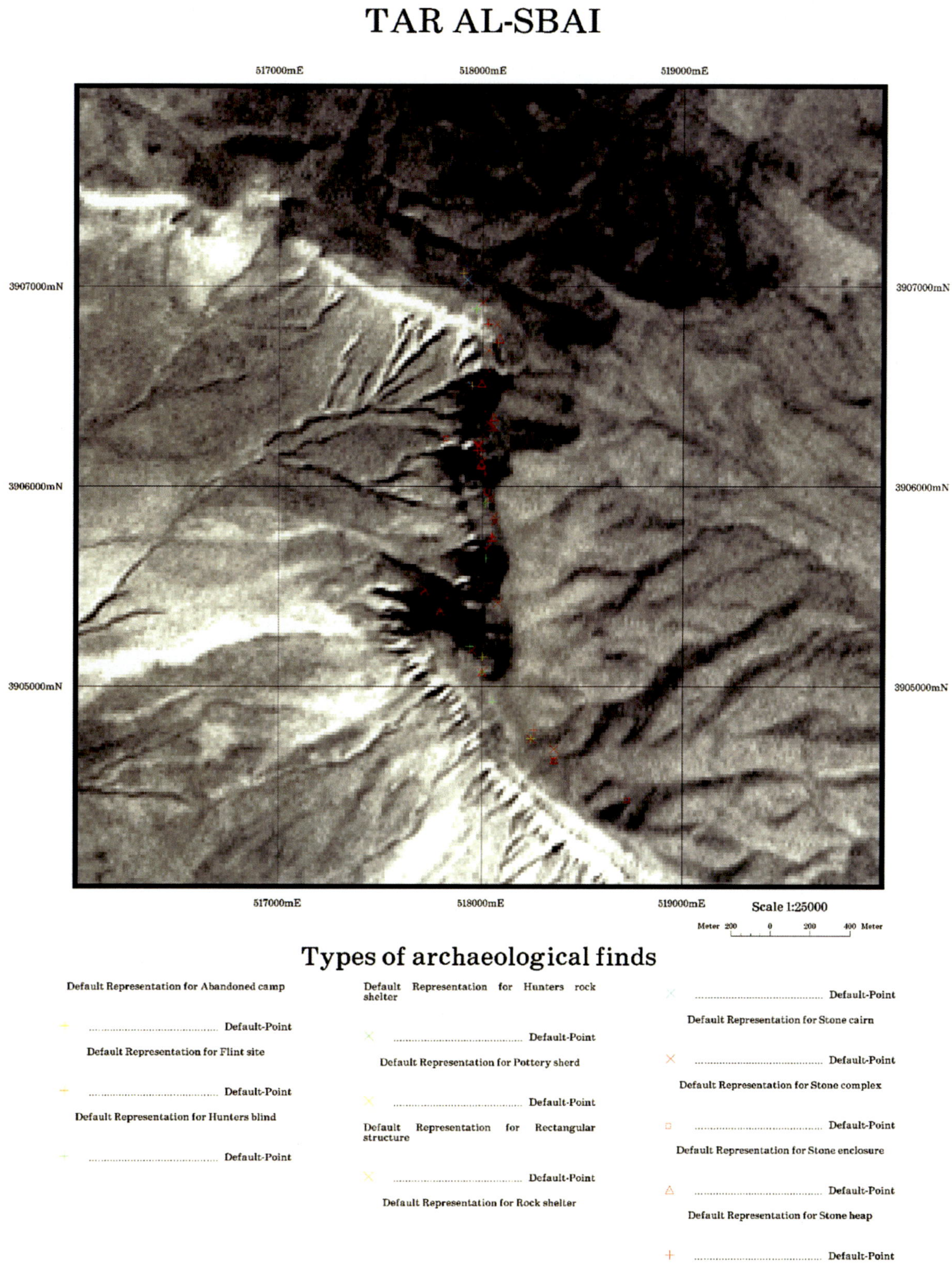

Fig. 2. The distribution of the recorded archaeological sites at Tar al-Sbai. Mapping by Markus Törmä 2004.

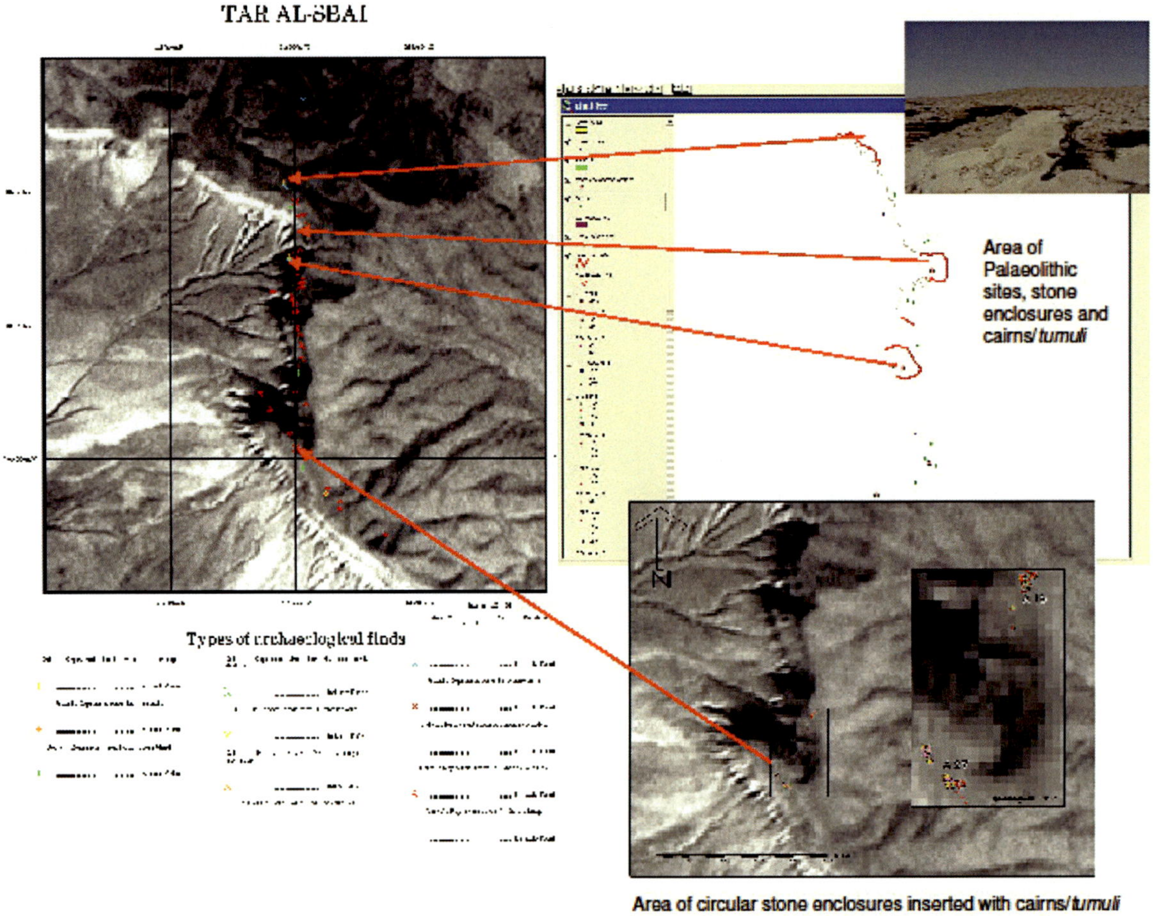

Fig. 3.

The ancient cairns/*tumuli*, which we recorded, were encountered at sites **A 12**, **A 14**, **A 16**, **A 17**, **A 18**, **A 20**, **A 22**, **A 23**, **A 26**, **A 27**, **A 30**, **A 32**, **B 3**, **B 4**, **B 6**, **B 7**, **B 9**, **B 10**, **B 12**, **B 15**, **B 16**, **B 17**, **C 3**, **C 8**, **C 9**, **C 11**, **C 12**, **C 13**, **C 14** and **C 15** (see Table 1.). They were generally ca. 2.6 – 5 m in diameter, but the so-called ring-*tumuli* at the sites **C 12** and **C 13** extended up to 22.7 m and 24 m in diameter. Some scattered structures such as met at **A 12** and **A 13** may represent cist tombs. There also were stone enclosures at the sites **A 26**, **A 27**, **A 29**, **B 4** and **B 8**, comprising structures varying from ca. 8 m to 20 m in diameter. As discussed in ch. 6. and ch. 7., comparable structures were encountered at Nadra and at Ash-Shujiri in the central eastern area of Jebel Bishri as well. At Tar al-Sbai some of the sites included both enclosures and cairns/*tumuli*, a few forming megalithic types of complexes. In these cases the question, which may only be answered by excavation, is whether the cairns/*tumuli* are secondary constructions or contemporaneous with the enclosures. Some of these megalithic type of enclosures at the edge seemed to differ from corrals, animal pens. Some of the sites included stone walls, radiating stone lines and boat–like structures in stone. Because of the tombs mortuary rituals and ancestor cults may be associated with the structures.

The landscape of Tar al-Sbai seems to have been a reason for constructing mortuary sites at the edge. Striking topographical formations around the world embody mythologies (Tilley 1994: 43). According to ethnographic studies, landscapes can be seen to contain powers offering signification of status or mythical narratives re-interpreted from generation after generation (Bradley 2000: 28). At Tar al-Sbai an impressive view opens over the western piedmont area, which we have described as an ancient "Silicon Valley" (see the description of the area in ch. 4. and 5.), full of dry *wadis* covered with shrubs, scattered grass and Palaeolithic open air sites with countless amounts of flint tools and debitage in dozens of square kilometres beside Neolithic sites of early pastoralists.

Landscape is the materialization of human memory. Memory enforces continuity in the landscape in reconstruction of sites offering an impressive view, and therefore the view is closely connected with the identity of human beings active in the area. (See Ashmore and Knapp 1999: 1-30). Nowadays archaeological landscape studies have obtained approaches from interpretive archaeology especially from cognitive archaeology, i.e., archaeology of mind. In his book *Mind´s Landscape* S. Guttenplan (2000: 7-43) mentions three main categories of mental activity: 1. experiencing, 2. attitudinizing and 3. acting. Our background, moods and feelings have an effect on experiencing and attitudinizing, and therefore the meaning of a particular landscape is personal. Landscapes are only perceived in human minds, and the perception can vary from person to person, from one human group to another. Even if we can determine the spatial dimensions of certain types of landscapes, their "deep meanings" can only be sensed in subjective ways.

The mental attachment to the view over the western plateau beneath is represented in continuous reconstruction of sites at the mountain edge transcending traditions and meanings from generation to generation. This attachment is attested in the density of remains compared to the inner areas of the mountain. J. Thomas (1999: 62-63) has also argued that monuments such as Stonehenge did not only carry one single meaning. The society which constructed it had a rich background which may have created multiple meanings contingent on deep traditions. According to Thomas, the meaning of Stonehenge has continuously altered from the Neolithic to the present, never being meaningless.

There are also more practical reasons for choosing places with good visibility, such as ridges or hilltops. From a ridge the approaching visitors can be seen and information could be spread. In the 1970s visibility was still regarded as an abstract environmental variable, whereas in the 1980s and 1990s it became to be regarded as a human act of perception. Now visibility is considered to be an issue of mobility, as visibility can be changed or enhanced, if a viewer moves through the landscape. Higuchi has observed three distances from which a view can be revealed depending on the quality and texture of the object. (Wheatley & Gillings 2000).

In theoretical studies of landscape, a *panorama* offers a wide view from a vantage-point of 360° arc over the surrounding landscape, and a *vista* is a view bound by certain margins, and normally these would be vertical or near-vertical cliff-tops or edges. Both *panoramas* and *vistas* form direct prospects, and they need to be used in order to reconnoitre from *primary vantage-points*. Natural places such as mountain tops and hilltops that offer escarpments, overhanging valley, or plateaus can be called *natural secondary vantage-points*. *Artificial secondary vantage-points* include all the human-made structures which extend the vision from the areas concerned. Lookout-points or observation-points are usually such that they provide an extension of the horizon for the field of vision in several directions. There is speculation as what may lie beyond the horizon in our perception of the whole scene, and we can better appreciate the whole panorama. (Appleton 1975: 85-90). The observation of the approaching herds of animals on the plateau beneath was experienced by the authors from the heights at Tar al-Sbai.

Especially imposing pastoral mortuary sites at Tar al-Sbai are sites C 12, A 19/C 13 and A 27. They were located so that they provided particular vistas over the desert plateau beneath.

8.2. Burials and Megalithic Stone Complexes at Tar al-Sbai

Site C 12
Cairn/Tumulus (7 tombs) - Stone Structure Complex

UTM: 0518358, 3904633
Size of the site: max. ca. 80 m x 50 m

Sizes of the cairns/*tumuli:* a) max. 2.07 m in diam., b) 2.05 m in diam.; c) rectangular corner of stone walls: 4.5 m in l., 1.42 m in w.; 3.6 m in l., 1.55 w.; d) cairn/*tumulus:* 5.4-7.4 m in diam.; e) wall: 1.70 l. and 0.85 in width; f) ring-*tumulus*: outer ring 22.7 m in diam. with a cairn/*tumulus* of 11.5 m in diam. in the middle; g) cairn/*tumulus:* max. 11.2 m in diam. with an elevation point in the middle; h) cairn/*tumulus:* max. 2.0 m in diam. and i) cairn/*tumulus:* max. 6.2 in diam.

Site **C 12** was a complex of different kinds of stone structures. There were altogether seven cairns/*tumuli* including a large ring-*tumulus* and a corner of abutting stone walls, a separate stone wall and radiating stone lines (Fig. 4.) allocated to the mountain edge. The structures appeared on both side of a desert track. The structures (a-e) were aligned from NW to SE in a row. The stone-built corner (c) was in between cairns/*tumuli* (b) and (d) at the edge of the mountain. The stone walls of this corner were aligned so that the longer one consisting of two attached stone rows was running from SW to NE, and the shorter one of three stone lines in thickness was running from SE to NW meeting the longer one in a rough right angle corner. At a cairn/*tumulus* (d) there was another wall (e), two stone rows in width, attached to the tomb in its eastern side. The ring-*tumulus* (f) was considerably larger than the other cairns/*tumuli*. Several radiating N-S oriented stone lines ran from the northern side of the ring-*tumulus* and a neighbouring cairn/*tumulus* (g). The walls may be secondary constructions or contemporaneous with the tombs or enclosures. The corridor-like stone lines may be associated with the megalithic cultures having ritual significance as pathways for spirits or they may bear some astronomical significance. The site produced worked flints (Fig. 5.) of which some, like a microburin, could represent earlier activities than the visible structures in the area. A piece of a limestone vessel was also discovered at the site (Fig. 6.).

The number of the graves at the site indicates that this is a family graveyard. The ring-*tumulus* among the cairns/*tumuli* seems to signify a pronounced social status of the deceased member of a social group: a head of a family or a tribal chief. Apart from the sites at Nadra in the central areas of Jebel Bishri (see ch. 6.), we have already referred to comparable ring-*tumuli* known from the most recent Syro-Japanese studies on the northwestern flanks the associated finds of which date from the end of the Early and the beginning of the Middle Bronze Age (Fujii and Adachi 2010). We also have discussed (ch. 6.), how these comparable megalithic structures are known from Sinai, the Negev, the Golan Heights and Arabia - from the Greater Southwest Asian arid zone in general. They are usually dating from the Chalcolithic and/or Bronze Age (see Rothenberg and Glass 1992: 142; Kochavi 2009: Fig. 3.; Kochavi 1991; Mizrahi and Zohar 1993; Zarins 1989). The corridor-like stone lines at C 12 may be associated with the megalithic tombs of the site. The stones classified among megalithic rings and rows incorporating cairns in other parts of the world also exhibit comparable structural features (cf. passim Thom *et al.* 1990) to those found at site C 12. The gravefield site C 12 seems to date from the Chalcolithic/Bronze Age.

Fig. 4. Corridor-like stone lines and a large ring-tumulus above them at site C 12. Photo from NE: SYGIS 2000.

Fig. 5. Flint implements from site C 12. Photo: Kenneth Lönnqvist 2010.

Fig. 6. Base fragment of a lime stone vessel from site C 12. Photo: Kenneth Lönnqvist 2010.

Site A 19 House, Site C 13 Cairn/Tumulus - Stone Structure Complex

Site A 19 house:
UTM: 0519050, 3905748
Size of the site: ca. 80 m x 60 m
Sizes of the rooms: 21 m x 13 m and 13 m x 10 m

Site C 13 cairns/*tumuli* (5 tombs),
UTM: 0518358, 3904627; 736 m a.s.l.
Size of the site: ca. 90 m x 25 m

Sizes of the cairns/*tumuli:* a) 7.45 m-7.65 m in diam.; b) 8.48 – 9.00 m in diam.; c) 24 m in diam.; d) 3.9 m in diam.; e) 2.7 m in diam; f) corner of meeting stone walls: 12.6 m l. and 2 m l. and g) stone lines: not measured

Site **A 19** (Fig. 7.) clearly forms a house with door entrances including thresholds. This structure was associated with the cairns/*tumuli* of **C 13** including a ring-*tumulus*. C 13, intermingling with A 19, also comprised 27 parallel radiating stone lines aligned in the N-S-axis comparable to those met with at the site C 12. The layouts of the two clearest spaces/rooms of the house at A 19 are roughly rectangular. The entrance to the larger space was from the northwest and in the smaller one from the north. A possible central courtyard was situated to the northwest of the rooms, if the site A 19 represents a house compound. Other more curvilinear structures may be animal pens attached to the possible house compound. To argue that the structures were representing a house is supported by visible thresholds. At the site there were flint tools that could be dated to the Neolithic period as well as a piece of obsidian. The flints included two fragmentary side-scrapers, a point and a burin. In addition, there was a base of a Neolithic pottery vessel (Fig. 9: 1) comparable to the El Kowm repertoire and a pottery piece from an EBIV chalice (Fig. 9: 4). The house may originally date from the Neolithic period and may represent sedentarization in the area otherwise occupied by mobile people. However, if the adjacent structures include corrals we are dealing with a pastoral site. The cairns/*tumuli* of C 13 may be contemporary or later Chalcolithic/Bronze Age additions to the site A 19, and a piece of EBIV pottery may belong to that phase. The tombs seem to represent a family structure or other kinship ties.

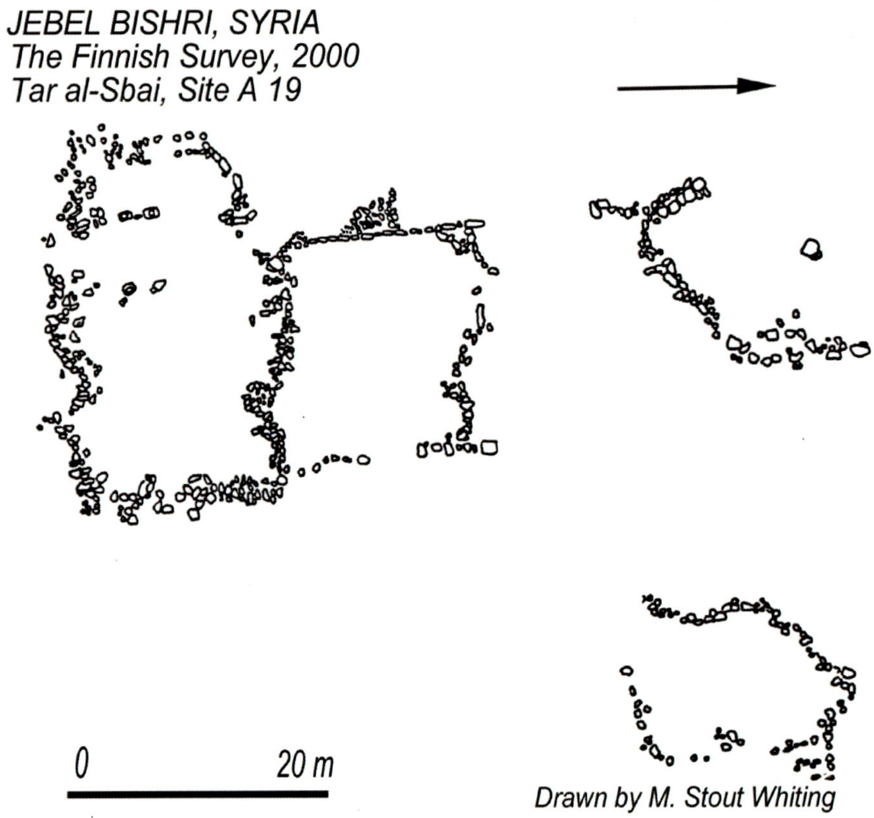

Fig. 7. A house compound A 19 at the edge of Tar al-Sbai.

Fig. 8. Worked flint finds from site A 19. Photos: Milton Nuñez 2004 and Kenneth Lönnqvist 2010.

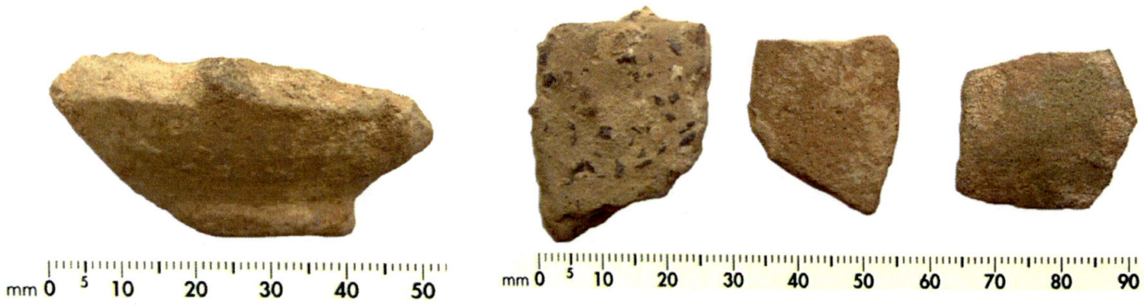

Fig. 9. Pottery finds from site A 19. Photos: Kenneth Lönnqvist 2010.

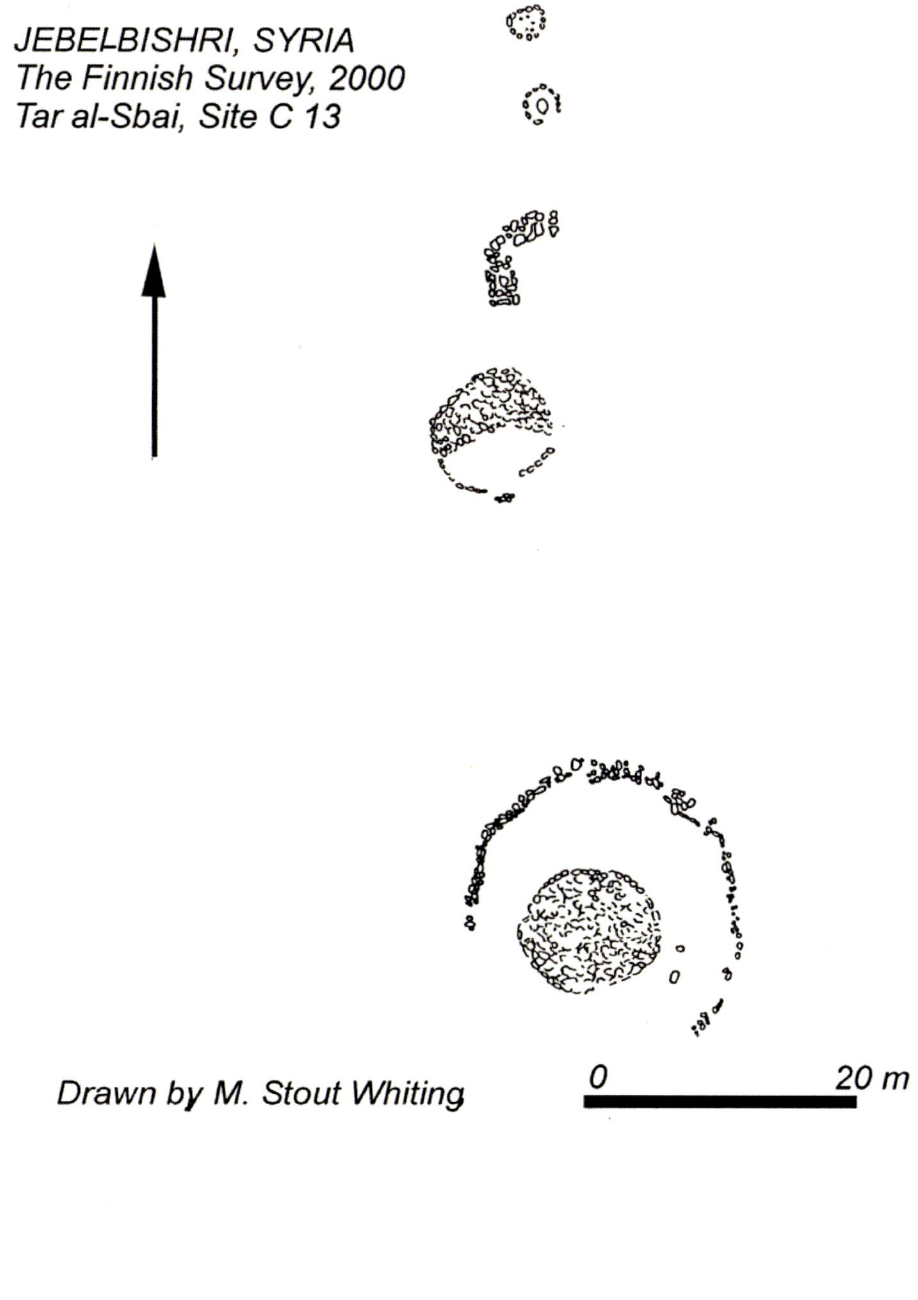

Fig. 10.

Site A 27
Cairn/Tumulus - Stone Structure Complex

UTM: 0517793, 3905377
Size of the site: ca. 90 m x 50 m

Sizes of cairns/*tumuli* : a) 9.5 m in diam., b) 7-9 m in diam., c) 8 m in diam. and a stone line d) ca. 40 m l.; the stone enclosures: e) 15-20 m in diam., f) 9-13 m in diam., g) 12-18 m in diam.

Site **A 27** (Figs. 1., 11. and 12.) consisted of large stone enclosures inserted with cairns/*tumuli*. It also incorporated a 40 m long line of hearth-like structures composed of stones in petal-like forms at the edge. The cairns/*tumuli* may also here be contemporary with the enclosures or they can be later additions. A few worked flint implements were found at the site (Fig. 14.).

Sites S-14 and S-16 near Gebel Safra in Sinai also include comparable stone enclosures inserted with cairns/*tumuli*. The associated finds in them dated from the Early and Middle Bronze Age. In addition, a comparable line of hearth-like stone structures is incorporated into the compound of curvilinear structures on a *wadi* edge at Site S-7 in Wadi Rahaya and at Site S-12 adjacent to a stone-lined ring and a *tumulus* on a terrace at Gebel Safra in Central Sinai. The former site provided finds dating from the Epipalaeolithic, Late Neolithic/Early Bronze Age. (See Eddy and Wendorf *et al.* 1999: 27-30, 33, 35, 40-41).

Fig. 11. Part of the site A 27 at Tar al Sbai. Photo: SYGIS 2000.

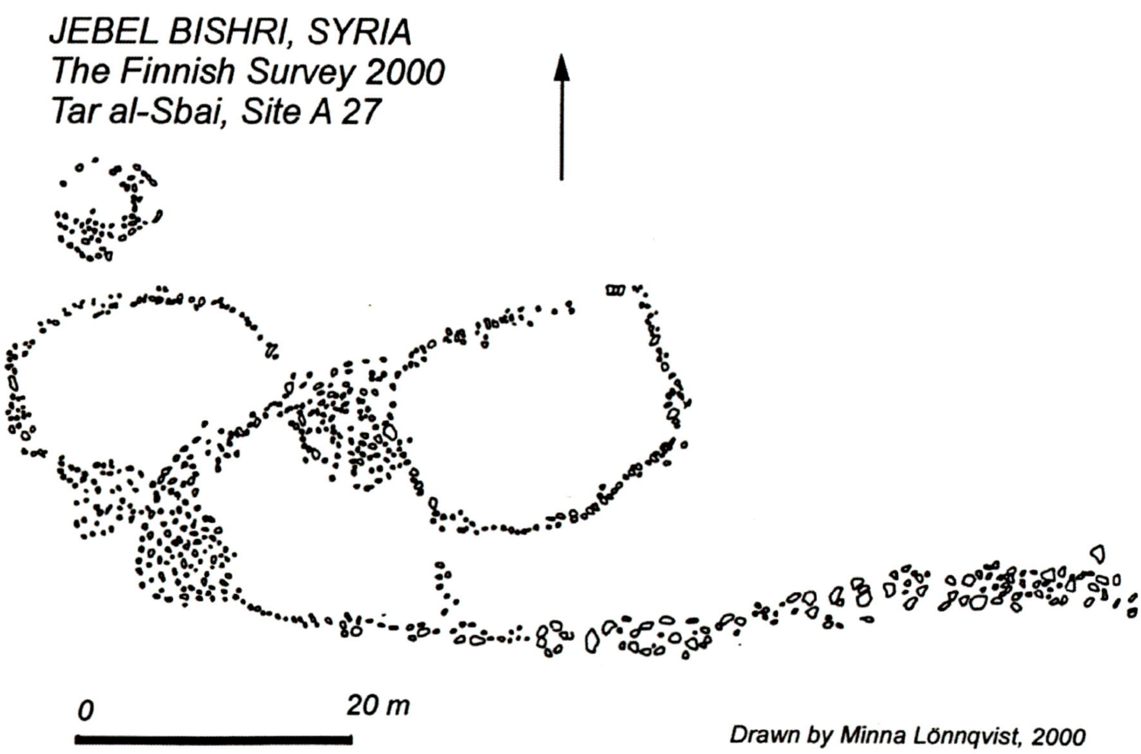

Fig. 12. Plan of site A 27.

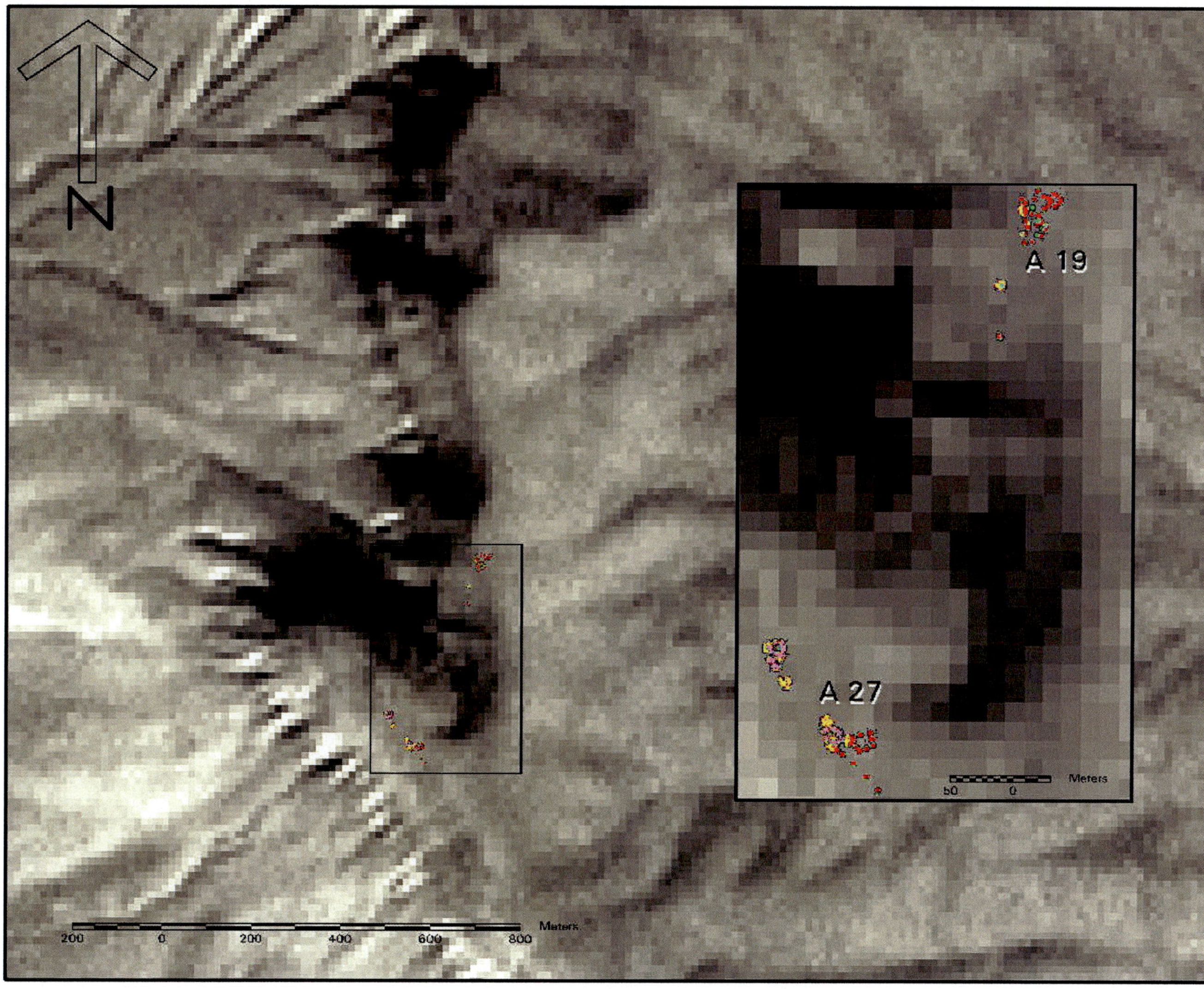

Fig. 13. Sites A 27 and A 19 at the edge of Tar al-Sbai. Mapping by Arto Vuorela 2002.

Fig. 14. A scraper and a PPN backed blade from site A 27. Photo: Kenneth Lönnqvist 2010.

Visibility from and to Sites A 19/C 13 and A 27

Sites **A 19** with **C 13** and **A 27** were all situated on spurs (Fig. 13.) so that they offered views in different directions towards the desert plateau beneath. The height of the site locations varied from 736 m a.sl. to ca. 800 m a.s.l. The structures at these sites may be interpreted as comprising composite vantage-points including natural and artificial elements (cf. Appleton 1975: 90). Already in the prehistoric times people were executing cognitive mapping, i.e., acquiring, amalgamating and storing spatial information concerning their environment (cf. Downs and Stea 1973: 13). From site A 27 a perspective model was computerized forming a 3D view ca. 13 km down towards the western plateau (Fig. 15.). The model was constructed by using Russian topographic maps (1: 50,000 in scale) and Image Drape software (see a video clip 'A View of an Ancient Silicon Valley' in the project home site: http://www.helsinki.fi/hum/arla/sygis). The entire three-dimensional geometry of the environment determines whether vision can be extended beyond the horizon. The chain of the Palmyride mountain belt that overlooks Palmyra from the north comes from the west, changing to hills near the village of Taibe (see more in ch. 14.) with its salt flats and rising again after a few kilometres towards Jebel Bishri. Beyond the horizon ca. 20 km to the west are also the oases El Kowm and Qdeir, respectively, known for their rich ancient remains of mobile and semi-sedentary cultures (see more in ch. 4 and 5.).

The stone complexes, such as A 19 including C 13 and A 27, had an extraordinarily impressive view compared to the view from several other sites situated in the central area of the mountain. It appears that sites A 19 (C 13) and A 27 at Tar al-Sbai are not only constructed at the mountain edge offering views but also to be seen from afar (cf., e.g., Martinsson-Wallin 1994: 85-107). If the line of stones in the petal-like formations at the site A 27 represents hearths, fires could have been seen from afar from the direction of the desert plateau. Traditionally, sacrificial sites have been distinguished by their striking appearance or location in the landscape (Bradley 2000: 6-8), and such may have been attached to large enclosures and burials.

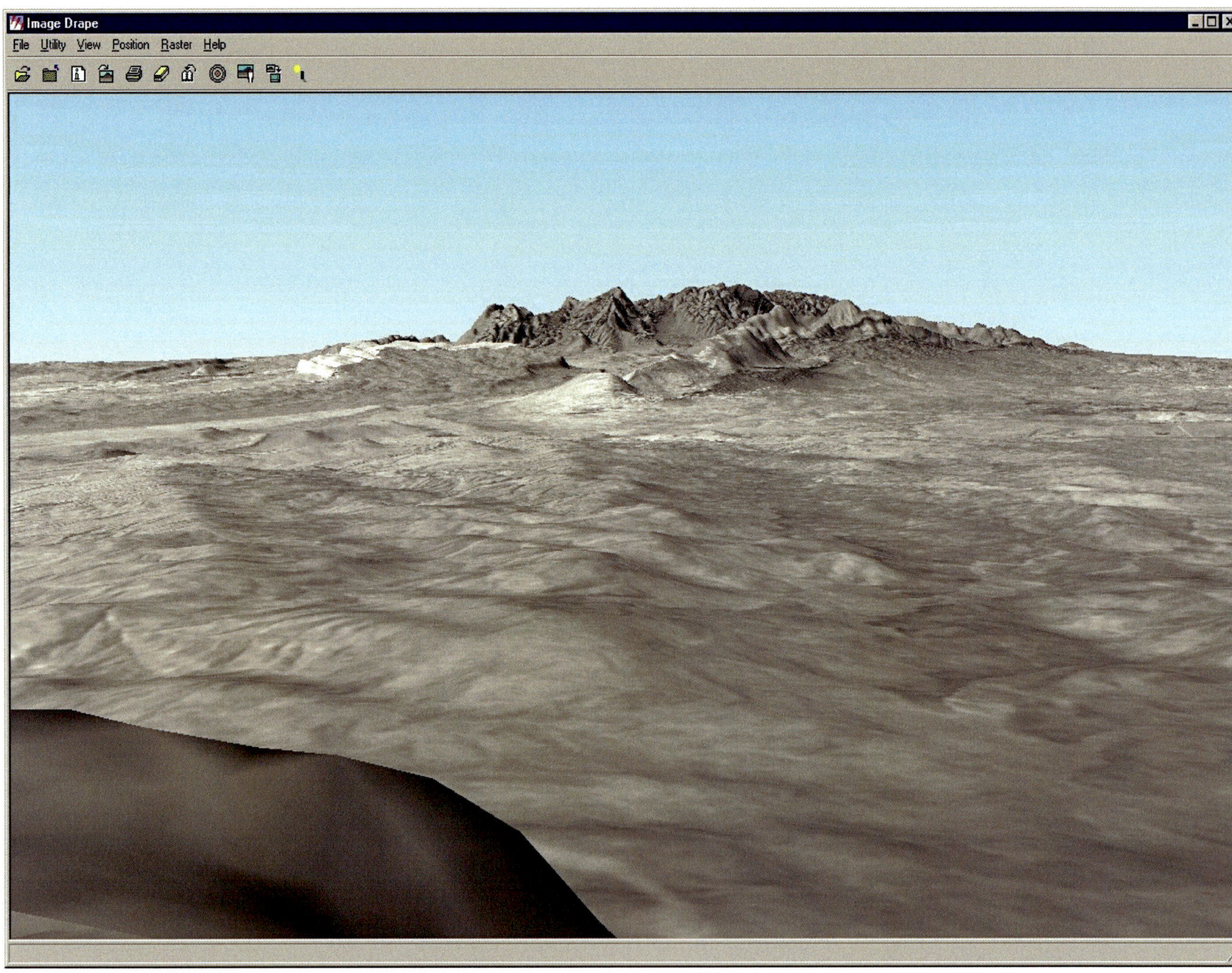

Fig. 15. A perspective view over the western piedmont seen form Tar al–Sbai. Constructed by Arto Vuorela 2002.

8.3. Cairns at Tell Tletuat

Further to the east, 3-5 km from Tar al-Sbai, stone cairns (e.g., Fig. 17.) that were not mixed with earth (like most of the other cairns/*tumuli* which we studied) were the major types of structures on three natural hills called Tell Tletuat. Despite of the name Tell Tletuat "the three tells" (Fig. 16.), these "tells" were not ancient mounds of ruined settlements but appeared to consist of natural rocky hills. We briefly visited the hills which fell outside our systematic survey area and mainly recorded the site locations (not fully documenting all the structures because of the time constraints).

Fig. 16. Tell Tletuat and the Bedouin house compound where our team stayed in 2000 as seen from Tar al-Sbai. Photo: from W Minna Lönnqvist 2000.

These three hills[1] were already passed by A. Musil (1928: 176) in his travels on the 6th of April, 1912. They are aligned from northeast to southwest. As far as the area is concerned, the hills may have been on a tribal route for centuries, if not millennia, as they clearly provide prominent landmarks in the otherwise flat desert-steppe area. Altogether fifteen cairns, which were not mixed with earth, were identified on the hills. In addition, there were stone rings and hunting blinds (the kinds of structures which were described in paragraph 5.2.5).

[1] In our preliminary report, KASKAL 3, 2006, p. 217, the site number and its coordinates are unfortunately wrong due to an error in the original documentation form. The site of the southern hill is not C 15, but D 3, and the coordinates for D 3 are given correctly here. C 15 is a cairn/*tumulus* site at Tar al-Sbai (see Table 1.).

Site D 1
Cairns (4 tombs) and Hunting Blinds

UTM: 0520981 and 3912099; 745 m a.s.l.
Sizes of the cairns: 4.4-5.0 m in diam.

The middle hill had three peaks and comprised several structures. Of the three cairns one was a small one in the middle of a hillock. One cairn was on the northern side of the peak, and the other on the southern slope. The northern slope comprised two small stone rings which were attached to each other and one *rujm* marker (1.5 m in height), which served as a path finder. Two hunting blinds were also identified on the northeastern slope of the southernmost peak.

Fig. 17. Cairns of the site D 3 at Tell Tletuat. Photo from N: Minna Lönnqvist 2000.

Site D 2
Cairns (8 tombs), Stone Circles and Hunting Blinds

UTM: 0521037, 3912210; ca.765 m a.s.l.
Sizes: not measured.

A fairly undisturbed cairn field of eight cairns was situated in the area of the northernmost hill. There also were two hunting blinds on the northeastern slope of the hill and two stone circles in the central area of the northern slope. A piece of limestone sculpture (15 cm x 10 cm x 7 cm) was discovered on the northern area of the summit, where tomb robbers' pits were identified in a cairn. The item (Fig. 18.) seems to represent female genitals and may be related to some fertility rituals at ancestral tombs. The dating of the cairns as well as the dating of the sculpture is unclear however. One cairn had a *rujm* marker on top serving as a path finder. It is possible that the site is on the route of a pilgrimage. It is evident that tribalism is enforced through rituals at the ancestral graves like these, when pilgrimage to ancestral graves has taken place (cf. Marx 1977: 42-44).

Fig. 18. A sculpture apparently representing female genitals found at Tell Tletuat. Photo: Minna Lönnqvist 2000.

Site D 3
Cairns (3 tombs), Stone Circle and Hunting Blind

UTM: 0520921, 3912020
Sizes: not measured.

This westernmost hill provided two cairns on its slopes. The larger cairn had a crater in the middle.

8.4. Sacred Landscapes and Ceremonial Sites

Ritual and cultic practices give a special way for people to act in the landscapes and attach a meaning to a place (see, e.g., Tilley 1994: 14-15). L. Kolska Horwitz (1999: 63-69) has used archaeozoological means to identify ritual sites in ancient Near Eastern contexts. Bedouins still sacrifice sheep and camels at ancestral tombs, especially at saints' tombs (like sheikhs' tombs) in the Negev. The Bedouin saints act as mediators between man and god. They are also tribe progenitors. (Klenck 1995).

In the world of ancestors the cairns/*tumuli* at the hilltops in their part were nearer to the sky and could be seen from afar (cf. ch. 6.). As earlier indicated, they could have served as signposts, as on many of them people have added piles of stones as path finders, the so-called *rujm* markers, to guide travellers through the landscape (see more in ch. 14.). One can study, how the dead are placed in the landscape in regard to tribal roads or other route ways marked as corridors for deceased spirits to move between graves and spaces. There is often inter-visibility between cairns/*tumuli* like between those on Jebel Bishri. (Cf. Parker Pearson 1991: 124; 141).

In Israel, the Negev and Jordan ritual stone enclosures dating from the Neolithic and Chalcolithic periods are known (see, e.g., Seaton 2008: 231). Apart from these previously mentioned structures comparable ritual complexes have been discovered not only in Sinai but also in Arabia consisting of *tumuli* and platforms, especially in the Nejd region (Zarins 1989: 139). In addition, S. Bourke has produced studies on the cairn-*tumulus*–cist-enclosure-ruinfields at Adeimeh, two kilometres from Teleilat Ghassul, and the ruinfield from a kilometre south of Bâb edh-Dhra', all in Jordan. At Adeimeh immediately south of the cist-*tumulus* ruinfield there is a 30 m wide rubble-mixed stone enclosure. Similarly with the cairn and rubble *tumuli* at the Bâb edh-Dhra' ruinfield there also appeared two 25 m wide enclosures. From the associated small finds and parallel structures Bourke infers that both the Adeimeh site located two kilometres southeast of Ghassul and the Bâb edh-Dhra' ruinfields are elaborate burial complexes from the Chalcolithic Age, apparently associated with the Ghassulian culture (the type site being Teleilat Ghassul in Jordan dating from the fifth and fourth millennia B.C.). An enclosure similar to those at Adeimeh and Bâb edh-Dhra' has also been identified at Nahal Mishmar, the site of the famous Ghassulian copper treasure near the Dead Sea. (See Bourke 2002: 14-17 and Bourke 2008: 109-160).

The location and the views from sites A 19 (C 13) and A 27 at Tar al-Sbai on Jebel Bishri immediately recall Chalcolithic ritual or cultic centres in the Levant, such as the Ghassulian temple at Ein Gedi (see, e.g., Ussishkin 1980: 1-44) on the mountains of the Judean Desert offering a view from the heights over the Dead Sea plateau. In the Near East, and especially in Syria, high places have also been often associated with sky gods. As indicated earlier, mountain-tops were thought to enhance observation of stars and planets and other cosmic forces. (Godwin 1981: 151). Megalithic structures are often connected with some ceremonial or ritual statuses. For example, A. Thom *et al.* (1990: 91-93, 188-189) could not study comparable structures without becoming convinced that many of such circles also had astronomical significance. In addition, astronomical observations have been associated with the megalithic ring-*tumulus* of Rogem Hiri on the Golan Heights (see, e.g., Albani and Glessmer 1997: 94).

We have suggested that tribalism as a social structure can be sought behind the cairn/*tumulus* tombs of Jebel Bishri (Lönnqvist *et al.* 2010). The tombs may also have served as tribal territorial markers. Such meaning could, for instance, be associated to the site **D 4** (UTM: 0520209, 3903263) outside our systematic survey area at the edge. The site consisted of five Islamic cairn tombs with standing capstones (Fig. 19.). One capstone bore a knot-like sign of the Fid'ān tribe. These tombs could have composed a family graveyard. The tombs clearly were located on the boundary area of the tribal territory between the Fid'ān and Sba'a tribes at Tar al-Sbai which is the eastern territorial limit of the Sba'a tribe. (See more on the tribes in ch. 14.).

Fig. 19. Continuity of the tribal burial grounds at Tar al-Sbai: the tombs of the Fid'ān Bedouin tribe seem to be serving as territorial markers at site D 4. Photo: Kirsi Lorentz 2000.

8.5. From Spatiality to Mental Landscapes

At Tar al-Sbai the relationship between the huts and houses and the burial grounds can only be studied by an excavation. It can, however, be assumed that some tombs belong to those, who also camped or dwelled more permanently at the escarpment. Bedouin tent bases at the sites A 31 and C 2 (see Table 1. and more in ch. 15.) provided evidence of camping up till recent times, and different types of cairns like the Islamic graves at the sites A 24, A 33, A 36, B 13 and B 14 could have belonged to some of those, who had more recently camped in the area.

The structures at Tar al-Sbai that can generally be associated with the ancient pastoral technocomplex, however, show the continuity from the Neolithic, Chalcolithic to the Early Bronze Age. The area has served as a site for mortuary rituals of pastoral people up till recent times. The preference of choosing Tar al-Sbai compared to the surrounding areas is striking among the mobile groups of the Chalcolithic–Early Bronze Age - during the period that can be considered to form a climax of the desert-steppe culture relying on pastoral nomadism in this arid zone.

The area of Tar al-Sbai seems to have served as a tribal burial ground for kin-based nomadic societies for a long time. The rank or the status of individuals may also be sought in the tomb types and their internal organisation at Tar al-Sbai. As shown, the head of a family or the chief of a tribe could have been given a special status signifier at a tomb in a graveyard by marking it differently. In that respect ring-*tumuli* clearly appear as signifiers being a special type of a tomb structure in some cairn/*tumulus* fields. The gravefields of these kinds may therefore be indicating the existence of a ranked social organisation, like a possible chiefdom system. (Lönnqvist *et al.* 2010).

Whether the reasons for choice of constructing sites at Tar al-Sbai are only ritual cannot be answered conclusively in every case, because different meanings could have been ascribed to the area by various groups at different times. In the ancient world vantage points were typically used for hunting, protection and ritual celebration, and it is likely that the striking landscape and views over the desert plateau had some ritual significance to ancient people at Tar al-Sbai.

Bibliography

Albani, M. and Glessmer, U. (1997) Un instrument de measures astronomiques à Qumrân, in *Revue Biblique*, Tome 104, pp. 88-115.

Appleton, J. (1975) *The Experience of Landscape.* London: Wiley.

Ashmore, W. and Knapp, A.B. (1999) Archaeological Landscapes: Constructed, Conceptualized, Ideational, in *Archaeologies of Landscape*, ed. by Ashmore, W. and Knapp, A.B., Contemporary Perspectives, Oxford: Blackwell, pp. 1-30.

Bourke, S. J. (2002) The Origins of Social Complexity in the Southern Levant: New Evidence from Teleilat Ghassul, Jordan in *Palestine Exploration Quarterly*, Vol. 134, No. 1, pp. 2-27.

Bourke, S. J. (2008) The Chalcolithic Period in *Jordan, An Archaeological Reader*, ed. by Adams, R.B., London: Equinox Publishing Ltd., pp. 109-160.
Bradley, R. (2000) *An Archaeology of Natural Places*. London: Routledge.

Downs, R.M. and Stea, D. (ed.) (1973) *Image and Environment: Cognitive Mapping and Spatial Behaviour.* Foreword by K.E. Boulding. Chicago: Aldine Publishing Comp.

Eddy, F.W., Wendorf, F. *et al.* (1999) *An Archaeological Investigation of the Central Sinai, Egypt.* Colorado: The American Research Center in Egypt, Inc. and The University Press of Colorado.

Fujii, S. and Adachi, T. (2010) Archaeological Investigations of Bronze Age Cairn Fields on the Northwestern Flank of Mt. Bishri, in *International Symposium, Formation of Tribal Communities, Integrated Research in the Middle Euphrates, Syria, November 21-23, 2009, Al-Rafidan*, Special issue, pp. 61-77.

Godwin, J. (1981) *Mystery Religions in the Ancient World.* London: Thames and Hudson.

Guttenplan, S. (2000) *Mind's Landscape: An Introduction to the Philosophy of Mind.* Oxford: Blackwell.

Klenck, J.D. (1995) Bedouin Animal Sacrifice Practices: Case Study in Israel, in *The Symbolic Role of Animals in Archaeology*, ed. by Ryen, K. and Crabtree, P.L., MASCA Research Papers in Science and Archaeology, Vol. 12, pp. 57-78.

Kochavi, M. com. (1991) Notes and News: Rogem Hiri, in *Israel Exploration Journal*, Vol. 41, pp. 182-183.

Kochavi, M. (2009) The Intermediate Bronze Age (IBA) in the Negev, Forty Years Later, in *The Levant in Transition, ed. by Parr, P.J., the Palestine Exploration Fund Annual IX,* Leeds: Maney Publishing, pp. 43-48.

Kolska Horwitz, L. (1999) The contribution of archaeozoology to the identification of ritual sites, in *Practical Impact of Science on Near Eastern and Aegean Archaeology,* ed. by Pike, S. and Gitin, S. Athens: Wiener Laboratory Publication 3.

Lönnqvist, M. with contributions by Lönnqvist, K., Törmä, M., Nuñez, M., Stout Whiting, M. (2010) Tracing Tribal Implications among the Bronze Age Tomb Types in the Region of Jebel Bishri in Syria, in *International Symposium: Formation of Tribal Communities, Integrated Research in the Middle Euphrates, Syria, November 21-23, 2009, Al-Rāfidān*, Special issue, pp. 165-173.

Martinsson-Wallin, H. (1994) *Ahu - The Ceremonial Stone Structures of Easter Island.* Analyses of Variation and Interpretation of Meanings. Uppsala: Societas Archaeologica Upsaliensis.

Marx, E. (1977) Communal and Individual Pilgrimage: the Region of Saints' Tombs in South Sinai, in *Regional Cults*, ed. by Werbner, R.P., Association of Social Anthropologists Monograph, No. 16, London: Academic Press, pp. 29-51.

Mizrachi, Y. and Zohar, M. (1993) Rogem Hiri, in *The New Encyclopedia of Archaeological Excavations in the Holy Land*, Vol. 4, Jerusalem: Israel Exploration Society and Carta, pp. 1286-1287.

Musil, A. (1928) *Palmyrena: A Topographical Itinerary.* American Geographical Society. Oriental Explorations and Studies No 4. New York: American Geographical Society.

Parker Pearson, M. (1999) *The Archaeology of Death and Burial.* Phoenix Mill: Sutton Ltd.

Seaton, P. (2008) *Chalcolithic Cult and Risk Management at Teleilat Ghassul. The Area E Sanctuary.* Monograph of the Sydney University Teleilat Ghassul Project. BAR International Series. No. 1864. Oxford: Archaeopress.

Thom, A., Thom, A.S. and Burl, A.(1990) *Stone Rows and Standing Stones: Britain, Ireland and Brittany.* Part i-ii. Collated, with archaeological notes, by Burl, A. Oxford: BAR International Series. No. 560 (i-ii).

Thomas, J. (1999) *Time, Culture and Identity.* An Interpretive Archaeology. New York: Routledge.

Tilley, C. (1994) *A Phenomenology of Landscape: Places, Paths and Monuments.* Oxford/Providence USA: Berg.

Ussishkin, D. (1980) The Ghassulian Shrine at En-Gedi, in *Tel Aviv*, Vol. 7, pp. 1-44.

Wheatley, D. and Gillings, M. (2000) Vision, Perception and GIS: developing enriched approaches to the study of archaeological visibility, in *Beyond Map*, ed. by Lock, G., Archaeology and Spatial Technology Series A: Life Sciences – Vol. 321, Amsterdam: IOS Press, pp. 1-27.

Zarins, J. (1989) Pastoralism in southwest Asia: the second millennium BC, in *The Walking Larder: Pattern of*

Domestication, Pastoralism, and Predation, ed. by Clutton-Brock, J., One World Archaeology 2, Cambridge: Cambridge University Press, pp. 127-155.

Zarins, J. (1992) Archaeological and Chronological Problems within the Great Southwest Asian Arid Zone, 8500-1850 B.C., in *Chronologies in Old World Archaeology*, Third Edition. 2 Vols, ed. by Ehrich, R.W., Chicago: The University of Chicago Press, Vol. 1, pp. 42-62.

9. OUTLOOKS TO CUNEIFORM TEXTS AND THE MESOPOTAMIAN LORE

Minna Lönnqvist

Fig. 1. Part of the victory stele of Naram-Sin. The victory stele commemorates the triumph of King Naram-Sin over the Lullubi, a mountain people. The Louvre.

9.1. Mount Basaar, Mount Tidnum and Mar.du-ki in the Third Millennium B.C.

The Basaar/Basalla, hur-sag MAR.TU, the mountain/steppe of the Amorites, mentioned in the Akkadian and Sumerian cuneiform texts from the third millennium B.C., has been identified with Jebel Bishri (Buccellati 1966: 236-237). The mountain has often been interpreted as the nomadic and tribal centre of the early West Semites,[1] especially pertaining to the Amorites and the Aramaeans. Assyriologists have been familiar with Jebel Bishri for decades. Akkadian rulers such as Naram-Sin (Fig. 1.) and Shar-kali-sharri boast about defeating Amorites in battles that took place in the mountain of *Ba-sa-ar*. Naram-Sin is also reported to have defeated a Sumerian coalition and Amorites ruled by a Sumerian city at the mountain, where he also took captives: *Naram-Šin (going) f(rom) the Euphrates River, reached Bašar, the Amorite mountain. He personally decided to fight (the two armies) made (battle) and fought one another. By the ver(dict) of the goddess Aštar (Scribal notation: it is the same) Naram-Šin, the mighty was victorious in battle over at Bašar), the Amorite mountain* ('in ba-sa-[ar] SA.DU.KI MAR.DU.KI.) (RIME 2: pp. 90-94).

[1] See, e.g., a general discussion in Gelb (1961: 29-30). Gelb equates the Old Akkadian identification of Amurru and the Gudea statue inscription MAR.TU with Jebel Bishri.

According to Gudea Statue B (Louvre AO 2; Fig. 2.), large blocks of stones were acquired and transported from the Mount Basaar, Jebel Bishri, down to the Sumerian ruler in Lower Mesopotamia (see, e.g., Buccellati 1966). These stones were of marble/alabaster, if the mountain is identical with the mountain of Tidnum, also mentioned in Cylinder A of Gudea. The blocks of marble/alabaster were namely dedicated to god Ningirsu to make lions and for the supply of door materials for his temple in Lagash. (See Lambert and Tournay 1951: 56-57).

Fig. 2. A unique headless diorite statue of Gudea: "The architect with a plan". The prince has the plan of the temple on a tablet in his lap. The Louvre.

Tid(a)num (Gudea, Cylinder A) is identified with a tribal group or an ancestral tribe, a confederation of Amorites or a closely linked nomadic group. Ur III references at the end of the Sumerian era describe *muriq Tidnim,* i.e., *the Amorite wall*: in his fourth year Šu-Sin built "the Amorite wall" to keep the raiding Tidnum nomads away (see, e.g., Wilcke 1969/1970: 9-12). The Tidnum are also mentioned in the ancestral genealogies of Shamshi-Adad, the Amorite king of the Old Assyrian Empire. Seventeen of the kings in the Assyrian king list appear to have been tent-dwellers, and in the list Di-da-a-nu is the ninth and *ha-nu-ú,* i.e., of the Hanean ancestry, appears as the tenth ruler. (Kraus 1965: 123-125.) Some lexicographers also consider the Sumerian *ti-id-nu* as being equivalent to the Akkadian *a-mur-ru-ú* (MSL III: 114, No. 209), while Old Babylonian lexicographers identify *di-ta-nu* as *su-tu-ú* (Suteans) (Heltzer 1981: 1).

During the past decades the Ebla texts have provided a stimulus for new theories and insights concerning Jebel Bishri. Emar (modern Meskéné) and Tuttul (Tall Bi'a) (see the map in Fig. 4.) have been recognized as providing evidence that the Amorites of the neighbourhood of Jebel Bishri were already known to the Eblaites in the twenty-fourth century B.C. (See, e.g., Streck 2000: 26). Where then was Mar.du-ki, the kingdom of the Amorites, mentioned in the Ebla texts and governed by a king, *lu-gal*, and a council of elders, with a famous temple? Was it somewhere in the neighbourhood of Jebel Bishri? The kingdom was located in a steppe region somewhere between the Euphrates and the Balikh. M. Astour identified its location with Jebel Bishri. (Astour 1992: 54-55; 2002: 118, note 402). Mar.du-ki mentioned in Naram-Sin's victory texts from the twenty-third century B.C. and cited above actually may refer to the same Mar.du-ki mentioned in the Ebla texts or it may just refer to an area populated by the Amorites. It could also refer to the common neighbourhood of Jebel Bishri based on the information gathered from the Naram-Sin text. At any rate, the land of the Amorites, which became to generally mean the west in the Mesopotamian point of compass, appears to be a larger entity than the mountain of the Amorites and the kingdom of the Amorites.

9.2. Trade with the Amorites in the Third Millennium B.C.

In the Ebla texts peaceful contacts with Mar.du-ki, the kingdom of the Amorites, are reflected with diplomatic gift exchange, treaties and trading with the Amorites. Garments in particular were sent from Ebla (see Fig. 4.) for the king and elders of Mar.du-ki, the types of clothing consisted of cloaks, tunics and girdles. Oil was provided for the oaths and treaties. The Amorite kingdom also received wine from Ebla, which may suggest the existence of some sort of court practices. Amorites in turn provided sheep, wool. and possibly metal weapons, such as *gír mar.tu*, i.e., the Amorite daggers, for the people of Ebla. (TM. 75.G.10079, TM.75.G.10251, ARET IV 15 rev. X 6-9). The products indicate a livelihood in pastoralism, and trade with metal objects. The Ebla texts also record that some Amorites were defeated in the neighbourhood of Emar, which seems to mean that occasional pressure was felt from that direction. (See passim Archi 1985).

G. Buccellati (1990c) also offers a reference to the "Amorite salt" (Maqlû 5: 79). As previously mentioned, the region of El Kowm on the western piedmont also provides *sabkhas,* salt plains, and a salt mine is nowadays known near Bweitiyeh (see ch. 5: Fig. 54.). The Euphrates river was a convenient means for transporting raw materials, such as marble/alabaster and salt, down to the central areas of Mesopotamia. The Ur III texts provide information that some Amorites were integrated into the Sumerian society and were trading with different materials such as textiles, wood, leather and commodities made of them. (See passim Buccellati 1966).

The alluvial zone of Lower Mesopotamia was scarce in natural resources, such as stones, metals and wood (Larsen 1979: 75-76). Those are clearly resources which Amorites had been actively providing and trading with. The goods exchanged between the people of Ebla and Sumer with the Amorites in the third millennium B.C. can be summarized as follows:

Wine
Oil?
Textiles
Leather
Wooden objects
Metal objects (including the Amorite dagger type: GIR MAR.TU?)
Stone
Salt
Lambs/goats

Later on such products as *donkeys, linen* and *silver* receive particular quality epithets as being "Amorite". (See, e.g., Vincetelli 1972: 134; Lewy 1961: 69-70; Lönnqvist 2010).

The Naram-Sin text cited above clearly states that the Akkadian ruler fought a Sumerian coalition and an Amorite city headed by a Sumerian ruler at the mountain of Jebel Bishri, which may indicate that some kind of Sumerian colonization existed in the neighbourhood. However, one needs to be cautious in drawing too far-reaching conclusions from the evidence provided with this kind of later text edition. Nevertheless, needed materials could have been transported to Sumer also from its subjected areas, and the colonization was an archaeologically traced phenomenon in the Upper Euphrates already in the Uruk period (see Algaze 1993). The Sumerian word for the king *lu-gal* appears as the head of Mar.du-ki, although sometimes the area was ruled by an Eblaite king, whose name was Amuti. (Astour 1992: 55).

9.3. The Gods Martu, i.e. *Amurru*, Ishtar Bishra and "Goat-Boy" Bazi

In the Drehem texts the area of Jebel Bishri appears as a kind of deified mountain to which lambs were sacrificed as for the god Amurru (Fig. 3.). (Buccellati 1966: 236; Astour 2002: 118, note 402). This is a reference to some kind of animism. As discussed in the previous chapter, mountains are often seen as sacred areas. They are imposing features in a landscape, i.e., archaeologically natural places to which memory and imagination are attached. They can be considered as "natural religious monuments" that are connected with myths, rituals, sacrifices, offerings and prayers. (Clack 2007: 17). The evidence of sacrificing lambs could indicate that there also was a shrine or temple in the Jebel Bishri region, but this does not imply that we have yet enough evidence to prove that the area can be identified with Mar.du-ki mentioned in the Ebla texts.

Fig. 3. God Amurru as depicted in the Mesopotamian cylinder seal impressions. Redrawn by Minna Lönnqvist from Amiet 1980.

The God Martu (i.e. *Amurru*) (see also Lönnqvist 2000: 71; Streck 2000: 68-72) occurring in the Sumerian mythology (*inter alia* in the text called *The Marriage of Martu* or *The Myth of Martu* in CBS 14061, see, e.g., Chiera 1929) is thought to be a personification of the Amorites as a people. In the myth Martu is a mountaineer, who only has a weapon as his companion. He does not know grain, house or city life. Martu is a tent-dweller (buffeted?) by wind and rain (SEM 58: iv). The habitat of Martu is a mountainous region, which accords with the areas such as Jebel Bishri. As a tent-dweller he is a nomad living a mobile life and thus cannot bury his dead companion. He is poor, as a nomad, and his belongings are restricted to a weapon. As previously noted, interestingly, in the Ebla texts there is a special type of weapon attributed to the Amorites, namely the so-called Amorite dagger. In the following ch. 10. we shall refer to some daggers recently found in the district of Jebel Bishri.

The Myth of Martu indicates that there was a plan that Martu would marry a Sumerian lady. In mythological terms it has been suggested that this is a literary metaphor of the marriage between the desert and the city life, between nomads and urban people. A symbiotic relationship between nomads and sedentary people is a well-known feature in anthropological literature. The pastoralists and farmers form complementary economies in which the farmers exploit the agricultural potential and the pastoralists the grazing lands. Both are tied into a relationship through trade and tension. The Sumerian sources such as Enmerkar and the Lord of Aratta speak, however, of times and wishes for peace and unity with the Amorites: *"Many tongued Sumer, the great land of the decrees of princeship…The land Martu, resting in security. The whole universe, the people in unison. To Enlil in one tongue (gave praise)."* (Kramer 1952, paragraphs 130-140). The Sumerian god Martu is equivalent to the god *Amurru* in the Akkadian sources. Amurru of (the) Tidnum appearing in seal inscriptions is the Lord of the Steppe (Kupper 1961: 67). He is sometimes presented on one hand as a son of the god An and on the other as a son of Ninhursaga. In some sources his wife appears to be Beletseri, Lady of the Desert (Black and Green 1992: 129-130).

In the Mari cuneiform sources (ARM), dating from the Old Babylonian period, Ishtar-Bishra appears as a special goddess attributed to the area of Jebel Bishri (ARMT 16/1, 255; ARMT 22, 64; ARMT 23, 462: 11; see also Streck 2004). In other Old Babylonian literature, which A. George has been currently publishing, there is a myth telling about a previously unattested goat-herding ram-god called Bazi. The myth describes, how he takes up residence in the Mounts Sharshar and Bashar, the latter identified with Jebel Bishri. The Mount Sharshar is the source of nomadic pastoralists, usually hostile to the peoples of the plain. The poem describes Bazi's cult-centre as an unusual mountain with high "banks" and much water inside (the waters of life and death). According to George, the imagery is so vivid that a real place must lie behind the description. In our view the "banks" may refer to the table-like character of Jebel Bishri, and the "water" to the water sources in the eastern central areas such as Nadra which has large springs. Other water sources are at Qebaqeb to the south, and as previously mentioned, there are artesian springs at El Kowm on the western piedmont. Bazi appears as a goat-boy in the poem in which the goats are taken to the uplands in the spring. Bazi left his father's territory in the south to graze his flocks at the mountain cult-centre. A. George (2008: 3) has explained, how the poem reflects the *transhumance* type of nomadism typical of the area today.

9.4. Hypothesis on the Sedentarization and the Creation of Amorite States in the Neighbourhood

In the classic wave theories the Amorites were one group of the West Semitic nomads roaming from the Arabian Desert. This theory has, however, been outdated for some decades and the role of the Kish civilization as the earliest civilization of the Semites has changed the picture (see, e.g., Gelb 1992). J.-R. Kupper (1957) did not follow the traditional wave theory but suggested a river theory instead as an explanation of the spreading Semitic nomads. They came as currents from the desert-steppes and sedentarized in the region of Syro-Mesopotamia. More recently assyriologists D. Charpin and N. Ziegler (2003: 29-32) have redefined the cradle of the Amorites as having existed between Jebel Bishri and Jebel Ansariyah, whence the people started settling in the twenty-first century B.C. According to their theories, the recurrence of certain Amorite toponyms, *toponymes en mirroir*, reflects the sedentarization process.[2] Sufficient linguistic and historical evidence has been presented to argue for a large-scale sedentarization among more or less nomadic Amorites in the late third millennium B.C., and their further expansion down to Mesopotamia has been established. However, reasons for these phenomena are not so obvious, although it is evident that environmental changes had some impact (see, e.g., Weiss 2000).

---

[2] I thank Dr. J. Eidem for mentioning this publication to me.

Archaeologically the mechanisms of this large-scale movement and sedentarization process seem to have been manifold, according to stratigraphical evidence that I have presented elsewhere: the mechanisms included gradual infiltration as well as attacks, invasions, destruction, conquests, site abandonment and reoccupation (Lönnqvist 2000; 2008; 2009). There was a clear cultural hiatus in the southern Levant in the late Early Bronze Age, and a pastoral interlude can be especially discerned in the archaeological data (Zarins 1989). However, whether one can identify the pastoralists of the southern Levant with the Amorites at the time is an open question (see passim Parr 2009).

In the following chapters we shall discuss, how the environmental evidence has become clearer in tracing some of the reasons for the nomadic movements and large-scale sedentarization. From the analogies of modern ethnographic studies and our own ethnoarchaeological observations in the Jebel Bishri region (see ch. 15.), as earlier mentioned, the environmental changes cannot necessarily always be presented as the sole deterministic factors however. According to G. Buccellati's (1992: 87; cf. Buccellati 1993: 68) later theories, the Amorites were actually originally village people who had become more nomadic in the course of time and just retained the seasonal cycle in the process of "domesticating the steppe". On the other hand, the lack of actual evidence to support this theory has been pointed out and the more or less pastoral economy and nomadism of the Amorites is supported by the earliest evidence to be retrieved in the Mesopotamian sources that deal with the Amorite habitat. The sedentary and urbanized groups that are attested in the Ur III sources (see passim Buccellati 1966) are the result of the integration of West Semitic elements in Lower Mesopotamia. Apart from the evidence of an obviously organized state like Mar.du-ki, the Amorites of the third millennium appear as rural people living on the periphery of civilization, and there are no written documents left in their own language. We only know their language through onomastics which appears in the Akkadian language (see, e.g., Gelb 1980).

In the Middle Euphrates area the change to the rule of the Amorites appears in Mari (Tell Hariri) and at Terqa (Tell Ashara) (see Fig. 4.) from the establishment of the Lim dynasty. From the King Zimri-Lim's reign the both sites belonged to the kingdom of Hana/Khana (Podany 2002: 4, 10) dating from the first half of the second millennium B.C. The situation is clearly different compared to the third millennium B.C. evidence. Archaeology has not yet revealed much of the third millennium Terqa apart from some artisan complexes and tombs. However, during the era of the *shankkanaku* governors of Mari, Terqa was partly under the influence of these rulers. O. Rouault (2000: 265-269) sees the shift of the rule as a primary result of the movements of the Amorite tribes living between Jebel Bishri and Mari. From the beginning of the Amorite occupation of Mari, during Shamshi-Adad's reign and the Lim dynasty, Terqa served as a religious centre and a provincial capital. There was the Temple of Dagan at Terqa (Buccellati and Kelly-Buccellati 1977-1978: 73-74), and the ancestors of the Amorite Lim dynasty had their burial grounds in the area (Charpin and Durand 1989: 18-19). It is no wonder either that the Amorite hegemony had religious interests in the early tribal territories, as mentioned in the Mari sources. As also previously indicated, Ishtar-Bishra appears as a special goddess attributed to the Jebel Bishri area in the Old Babylonian period. Similar material is found in the texts dating from the Sumerian renaissance: stone was quarried in the great Amorite mountain also later in the Old Babylonian period, possibly at Jebel Bishri. This is indicated by Samsuiluna's twenty-sixth- year-name (ca. 1724 B.C.) (Horsnell 1999: 217–19), when Mari was apparently under Babylonian rule after the struggles between rival Amorite dynasties.

It seems that Mari of the Lim dynasty and Terqa appear as the results of the integration of rural villagers as well as more or less nomadic people. The names of the Amorite tribal groups such as the Hana refer to tent-dwelling. Whereas the name of the Bene-sim'al (the Sons of the Left), i.e., the Simalites, is referring to the tribal territories on the left bank of the Euphrates. (Charpin and Durand 1986: 141-183). These tribes were particularly related to the Amorite Lim dynasty of Mari, and they may have consisted of more or less sedentary groups as well as nomadic ones. Some tribal disintegration was apparent in the amalgamation of the previous urban elements during the expansion and infiltration from the more steppic areas and in the creation of the state system led by the Amorites. How did then the area ruled by Mari affect the Jebel Bishri region? According to G. Buccellati's (1990b: 98) conceptual geography, Jebel Bishri was mentally much more distant to the urban Amorites of Mari compared to geographically more distant areas such as, for example, Ebla and Kish. Pastoral nomads have unfortunately left no inscribed evidence to tell about their attitudes of Mari.

Hana/Khana, as a regional state formed by tribal groups, could have enclosed other genealogically related tribal groups into its territories. Close tribal relationships built loyalty. It seems that the people of the earlier *shakkanakku* period in Mari did not have the same strong relations with the nomadic and tribal surroundings as the Amorite Lim dynasty. M. Rowton (1967; 1976) generally describes the kind of relationship between Mari and surrounding pastoral nomads as enclosed nomadism, the two groups forming the dimorphic structure of the state. During the Lim dynasty nomads became included in the city administration as a military force. The genealogy of the Hammurabi dynasty also mentions the era of the Amorite soldiers and the era of the Hanean soldiers (Finkelstein 1966: 29-31). N. Weeks (1985) assumes that the role of the Amorites as a militarily organized force enabled them to seize power in Mesopotamia. This assumption largely conforms to the second millennium B.C. evidence. But it should be stressed, as already indicated, that this is the situation reflected in the Mari texts of the second millennium B.C., when the large-scale sedentarization of the Amorites had already taken place. Like the previously mentioned Assyrian king list (Kraus 1965), Hammurabi's genealogy (Finkelstein 1966) also maintains that Amorite rulers claimed genealogical descent from tent-dwellers.

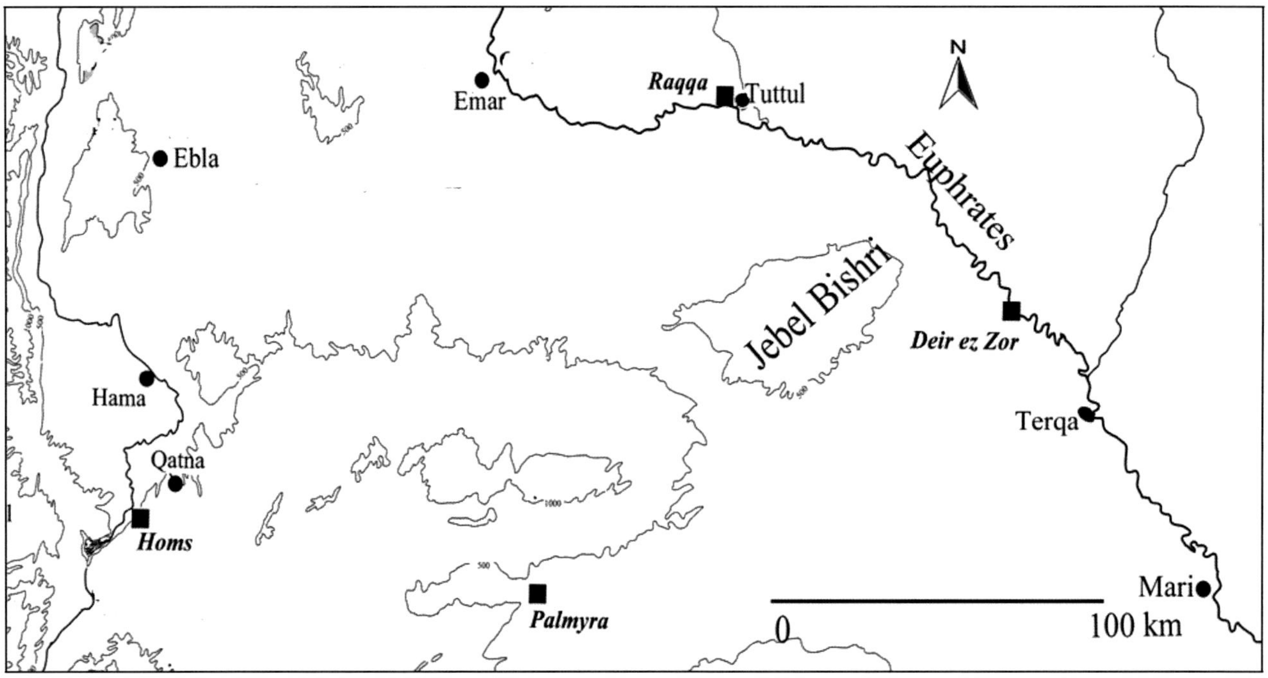

Fig. 4. Jebel Bishri between major Bronze Age kingdoms of Ebla, Qatna and Mari in Syria. Mapping by Minna Lönnqvist. (Basemap: the Syro-Japanese Tell el-Kerkh expedition).

9.5. The Mari Sources of the Second Millennium BC: Benjaminites and Suteans

The Mari documents reflect the presence of pastoral people in the area of Jebel Bishri in the eighteenth century B.C. As I have also argued elsewhere (Lönnqvist 2008), the texts cannot, however, be conclusively used for explaining the situation of the third millennium B.C., or the entire past of the local nomads/semi-nomads such as Amorites, which is manifold and complex. The symbiosis of nomadic people with agriculturalists had clearly become more pronounced in the kingdom of Mari during the second millennium B.C. compared to the third millennium B.C. evidence. (See also the introduction into the subject in Nilhamn 2008).

In addition to the previously mentioned studies by G. Buccellati and M. Rowton, several other scholars have earlier dealt with West Semitic tribal groups mentioned in the Mari texts. J.-R. Kupper's (1957) studies are classic, and J.T. Luke (1965), V.H. Matthews (1978) and M. Anbar (1991) have followed in the line. During the past decades D. Charpin and J.-M. Durand have especially continued to study the Mari archives and the tribal groups in a series of publications.

9.5.1 Yaminites, Benjaminites

In the Mari texts the Euphrates side of Jebel Bishri appears as the pasturage of the nomadic tribal confederation of Bini-Yamina (ARMT V, 27: 25-26), i.e., the Yaminites or Benjaminites = "The Sons of the Right" beside the Suteans. The name of the tribe or confederation of the Benjaminites refers to their original territory on the right bank of the Euphrates. The Benjaminites are considered to represent a more nomadic tribe than the Haneans. The Benjaminites have been identified by J.-R. Kupper as belonging to the Amorite tribal system. The Rabbeans living in the area of Yamhad (Aleppo) are identified as "brothers" of the Benjaminites in the Mari sources (ARM I, 6:9). (Kupper 1957: 47-81).

According to J.-M. Durand (2004: 158), the Benjaminites were organised in a tribe, which consisted of clans. Furthermore he lists that in the kingdom of Mari there were five Benjaminite tribes: 1. the Yahrurum, 2. the Yarîh, 3. the Amnannum, 4. the Rabbûm and 5. the Uprapûm. There were sub-clans in these sub-tribes, like the Mutiabal under the Rabbeans. Benjaminite chiefs also guided the transhumants called the *hibrum*. The Benjaminites consisted of both nomadic and settled elements, and they had their own towns. (Durand 2004: 166). A brick from the temple of Shamash at Mari reports rebellions of the Benjaminites which already took place under Yahdun-Lim, the king of Mari. From the inscription we also learn that the Benjaminite sub-tribes had kings and towns such as Samâmum, Tuttul and Abattum. (Charpin 2004: 83).

Benjaminites had their grazing grounds in the area of Dûr Yahdun-Lim (some scholars identify the site of the fortress to the neighbourhood of Deir ez-Zor), continuing beneath Jebel Bishri through Tuttul (Charpin and Durand 1986: 156) up to Harran (Kupper 1957: 48). Their encampments are also mentioned in the area of Terqa (ARM II, 90: 7-8). So their grazing areas included Mari and Terqa, but their influence even reached to the Habur, Eshnunna as well as to Yamhad (Aleppo), Qatna and Amurru. (Kupper 1957: 47-81). Their territories were also specified as being "above" in the mountainous regions identified with Jebel Bishri (see Weippert 1971: 112-113). Already the King Yahdun-Lim of Mari had conflicts with the Benjaminites. He reports that he defeated them at Abbattum, somewhere between Tuttul and Emar, achieving victory at the gates of Emar – a conflict which is associated with the chief of the Rabbeans. The problems with the Benjaminites continued during his son Zimri-Lim's reign. (Charpin and Ziegler 2003: 41; probably also the revolt of the Bedouins in Van Koppen 1997).

As previously referred to, the site of Tuttul (Tall Bi'a) beside Emar (Meskéné) (see Fig. 4.), was one of the earliest sites mentioned with special Amorite affiliations in the Ebla texts dating from the twenty-fourth century B.C. Both sites, Tall Bi'a–Tuttul and Meskéné-Emar have been under archaeological investigations in recent decades. E. Strommenger (see e.g., Strommenger *et al.* 1987; Strommenger and Kohlmeyer 1998; Strommenger *et al.* 2000; Strommenger and Miglus 2002) has led the German excavations at Tall Bi'a. The cuneiform evidence (see Krebernik 2001: 13) discovered *in situ* testifies to the Amorite presence at the site, which - as we have seen - belongs to the territories of the Benjaminites. Tall Bi'a–Tuttul provides a Middle Bronze Age palace (Strommenger *et al.* 1987: 10-17) of a type which is common in the Amorite-occupied kingdoms such as Mari and Qatna (see more in Lönnqvist 2000: 231-236).

J.-Cl. Margueron (1975a and 1975b) first led the French excavations at Emar. Up till last decade it was thought that the Middle Bronze Age strata of the site had submerged, but evidence dating from the Early and Middle Bronze Age has luckily been unearthed by the Syro-German mission led by U. Finkbeiner (Finkebeiner *et al.* 2001-2003) in this decade. Meskéné-Emar provides Early Bronze IV evidence that interests us especially due to the fact that the site is near Jebel Bishri and is relevant to those studying the Amorites of Syria. The earliest phase of the Temple of Baal at Emar was constructed in the Middle Bronze Age (Finkbeiner 1999-2000; Finkbeiner *et al.* 2003: Beilage 1) and follows the *Langraum* layout typical of several Amorite-occupied sites in Syria (cf. Lönnqvist 2000: 231-234; 239-244).

9.5.2 Suteans

The tribal people known as Suteans mainly lived between the Middle Euphrates bend and the oasis of Palmyra (Tadmor) (Kupper 1957: 98; Heltzer 1981: 57). This is the area within which Jebel Bishri is totally enclosed. The evidence of the Sutean activities in the hinterland of Jebel Bishri is attested in various cuneiform texts from the Mari period. There has been a lot of discussion whether the Suteans belonged to the Amorite tribes (see, e.g., Nilhamn 2008). According to the Mari documents, a silver treasure was discovered by the Suteans at Jebel Bishri (ARMT VI, 44, 5-6). According to D. Charpin (lecture in Tokyo, 2009), the treasure was found in a purse. The Suteans are thought to have represented even more nomadic groups than the Benjaminites, who had their grazing grounds only partly in the district. In addition, in the Mari archives the Suteans occur as pillaging in large groups (ARM I, 100: 6-7). A letter to the King Yasmah-Addu (ARM V, 23) expresses concerns and fear of the Suteans reporting that 2,000 Suteans appeared in the groups of razzia as far afield as in the pastures of the land of Qatna, and 60 pillaged the oases of Tadmor (Palmyra) and Nazala. But Suteans also acted as caravan escorts and messengers (A.3297 + A.3775, see *apud* Klengel 1996: 161; Ziegler 2004: 95-109) between Mari and Tuttul, which is the stretch in which Jebel Bishri is located. Hammi-talu appears as the principal chief of the Suteans during the reign of Yasmah-Addu and Zimri-Lim of Mari. He took part in the attack against Qatna, but he also provided guides to the caravans leading to Dilmun (modern Bahrain) (ARM I, 17) and had good relations with Zimri-Lim, whom he helped to transport silver and cattle. (See Joannès 1997: 408-409). He also guided people in the region of Tadmor (Palmyra). In the Mari sources the Suteans are also involved in slave trading, specifically of prisoners that they had acquired (ARM VIII, 9; see also Joannès 1997: 411).

Furthermore, later on the Suteans also appear in the Alalakh texts including the inscribed Statue of Idrimi, King of Alalakh, dating from the sixteenth century B.C (Smith 1949). Idrimi had to escape to Emar from the Mediterranean coastal plains, where his kingdom was established. He was forced into exile and to sojourn among his relatives in Emar and later on lived both among the *sutû* and the *hapiru*. He explains, how he took his horse and chariot *crossing to the desert country and even entering the region of the Sutian warriors. He also lived seven years among hapiru.* (ANET, 559-558). Interestingly, as we have discussed, Emar appeared as an earlier contact zone with the Amorites in the Ebla and the Mari texts. The Sutean region is presented here in association with the desert country somewhere nearby, and the *sutû* are described as warriors. The region might refer to the area of Jebel Bishri and the Palmyrene desert. According to the Idrimi text, some of the *sutû* had, however, spread out and were living as far away as the area of Alalakh. Idrimi emphasizes that he made them settle inside his kingdom and live in permanent settlements, including "those who did not want to live in settlements". So, here we have early evidence of political pressure from the side of the state to settle nomads in order to control them. Recent archaeological studies in the vicinity of Tell Atchana (Alalakh) interestingly deal with the land tenure of the kingdom incorporating small settlements (Casana 2007: 195-221).

The Suteans became to be known as (nomadic) tribal warriors in the Levant, appearing also in the El Amarna letters and in the Hittite sources (see, e.g., Masetti-Rouault 2001: 67). According to M. Heltzer (1981: 21, 80), there are specific sub-tribes inside the tribe of the Suteans such as *hibrum* and *Rabbeans*, while - as presented above - some scholars like J.-R. Kupper and J.-M. Durand include these sub-tribes to the Benjaminites.

9.6. Ahlâmu-Aramaeans and Assyrian Sources

Later in the second and first millennium B.C., the mountain of Jebel Bishri appears in the Assyrian sources as the central area of nomadic Ahlâmu-Aramaeans. For this period M.G. Masetti-Roault (2001) has written an overview of the cultures in the Middle Euphrates area.

Already in the Hittite annals Hattušili III (1275-1250 B.C.) is reported to have conquered the Mountain of Ahlâmu (Kupper 1957: 110). The Ahlamē are also mentioned in the Amarna tablets (Knudtzon 1915: 732-733). The people labelled as *ar-ma—a-ia,* Aramaeans, appear as an identifiable entity for the first time in history in connection with Jebel Bishri. (However, we do not here concentrate on the problem concerning the identification of the Ahlâmu/Ahlamē with the Aramaeans). According to a prism from Kal'at Sherkhât (K.1621a + K.13871, see Luckenbill I, § 239), the Assyrian king Tiglatpilesar I (1116-1090 B.C.) carried out a military campaign against Ahlâmu-Aramaeans ca. 1112 B.C., first passing the Euphrates in leather boats and destroying six of their towns at the foot of Jebel Bishri with fire, destruction and calamity. Despite of this boasting with the destruction, Tiglatpilesar I was unable to defeat the Aramaeans (Brinkman 1968: 278). The tribal background of the settled Aramaeans is echoed in their designations of the establishing of new states with the prefix *bīt* (e.g. Bīt Adini, Bīt Halupe) in the neighbourhood. Nomadic symbolism is present in signifying the kingship of the people and establishing new kingdoms along the Euphrates (see, e.g., Postgate 1974: 234-235). The Aramization of the region is discussed, for example, by G. Bunnens (2009) in his recent article.

According to A. Podany (2002), the Late Khana period extends to ca. 1200 B.C. The kingdom of Khana seems to disintegrate just before or simultaneously with the appearance of the Aramaean kingdom of Laqē in the region, and Khana's role as the basis of this new kingdom is possible. However, G. Buccellati (1990a: 229-253) sees that Khana's role is more obvious in the emergence of the later kingdom of Amurru further to the west towards the Mediterranean. On the other hand, there seem to have existed shared territorial interests in Khana and Laqē, as both kingdoms and their ruling classes had West Semitic and nomadic background partly associated with Jebel Bishri. The district of Jebel Bishri known as the mountain of the Amorites interestingly recurs in the sources as a possible part of the Aramaean kingdom of Laqē in the late second millennium–the first millennium B.C.

The possible affiliations between the Amorites and Aramaeans have long puzzled scholars (see, e.g., Masetti-Rouault 2001: 37). The Amorites, who appear in history as the first West Semitic and nomadic people connected with Jebel Bishri, disappear from the history of the area, becoming replaced by the nomadic Aramaeans linguistically also belonging to the West Semites. On the other hand, M. Heltzer (1981: 90-93), like K. Brinkman (1968: 280-287), identifies the matching appearance of the Suteans with the Aramaeans in the Mesopotamian sources, suggesting that the Suteans in fact merged with the Aramaeans and that the two were probably rather identical groups. Brinkman notes that they both seem to have been especially mobile in their nomadism among semi-nomadic tribes. The appearance of the Aramaeans in the Benjaminite and Sutean territory of Jebel Bishri could support the assumptions that more or less nomadic and West Semitic Aramaeans could not have emerged *ex nihilo* into the area. However, the question is complex and needs further studies.

The second major Assyrian campaign against the Aramaeans associated with Jebel Bishri took place ca. 878 B.C. and was led by Assurnasirpal II (883-859 B.C.), then king of Assyria. According to A.K. Grayson, Assurnasirpal II's reign belongs to the most significant periods in Mesopotamian history. During his reign the Aramaean city of Sūru on the right bank of the Euphrates revolted against the Assyrian ruler. The troops approached the region along the Euphrates: *Wherever the navigation of the Euphrates proved to be difficult, the boats were drawn up out of the water and dragged along the banks over rollers until they could again be safely launched; thus, partly afloat and partly on land, they passed through the gorge of Halebiyeh, landed at Kharidi, and inflicted a salutary punishment on the cities which had defied the king's wrath on his last expedition. Khindânu, Kharidi, and Kipina were reduced to ruins, and the Sukhi and the Laqi defeated, the Assyrians pursuing them for two days in the Bisuru mountains (Jebel Bishri) as far as the frontiers of Bit-Adini.* (Assurnasirpal's Annals, Rawlinson 1861-1884, see, e.g., Maspero 1903; Budge and King 1902; see also the latest translation by Grayson 1991).

The sites which were conquered were subjected to the Assyrian rule. Great fortresses were built, e.g., Dûr Karpati, i.e., Nibarti-Assur (Budge and King 1902, Assurnasirpal's Annals Col. III, lines 49-50), the location of which has been suggested to be at Halabiya, a protruding basalt peninsula belonging to Jebel Bishri area along the Euphrates (Musil 1927: 331). According to A. Musil (1927: 210-211), the neighbourhood of Rasappa (present Resafa-Sergiopolis) was a seat of an Assyrian governor after the western areas had been annexed to Assyria during the rule of Shamshi-Adad V (823-811 B.C.).

A. Musil (1927: 211) further noted that the land of Laqē situated in the vicinity of Sûru, which had earlier been the major fort of Laqē, and that the land of Suhi was also in the neighbourhood of Sûru. Sûru, according to Musil, could have been identified with modern Sûr at the foot of Jebel Bishri on the Euphrates and the site of the Roman fort Sura (cf. ch. 13.). F. Sarre and E. Herzfeld (1911) also mention that the name of Shūra is clearly of Syrian-Aramaean origin and therefore can be associated with the Assurnasirpal's Annals concerning the conflicts with the Aramaeans, but modern assyriologists do not agree with this earlier topographical identification of Sura. They locate the site of Sūra, the chief site of the tribal unit of Bīt Halupē, on the Habur river instead (see, e.g., Kühne 1980: 63, 67; Liverani 2, 1992: 32). In M. Liverani's (2, 1992: 69) topographical analysis Laqē is an important land on the Middle Euphrates extending downstream to Khindanu and Suhu, and upstream to Bit-Adini covering some northeastern areas of Jebel Bishri. However, the locations of Bit-Adini and Suhu seem topographically more difficult to assess on the ground. First of all, when Assurnasirpal built boats in Suru *he went down the Euphrates river* passing Halabiya and not upstream as currently suggested in several studies. He continued to Laqē to conquer the cities of Laqean sheikh Hemti-ili and Azi-ili. Kipinu appears as a central site in the campaign against Azi-ili, the Laqean.

In our project K. Darmark (2008) has paid special attention to the landscape at the foot of Jebel Bishri and its eventual dimensions from the view of battlefield archaeology and the Assyro-Aramaean conflicts. Assurnasirpal especially chased Azi-ilu, who was under siege at the city of Kipinu. Assurnasirpal drove him out and defeated him, but Azi-ilu finally withdrew himself to Jebel Bishri (Budge and King 1902: 357; Grayson 1991: 214-215). A. Musil (1927: 209) suggested that Kipinu could be situated somewhere near Deir ez-Zor. The nearest pre-Roman tell to Deir ez-Zor associated with Jebel Bishri is in the village of 'Ayyāsh some 12 km northwest of the city. The site has been briefly surveyed by a Syro-German mission led by M. Gschwind, but nothing else is known of the unpublished data except that the pottery is Pre-Roman. The peninsula of 'Ayyāsh provided an important strategic crossing point also in the Roman times (the bridge point of Tall al-Rum). In his expedition Assurnasirpal passed Kharidi while he was making for Sûru, where Kuduru, the prefect of the Suhi, with the help of Babylonians, fought against the Assyrians. The city of Kharidi was situated in the land of Suhi (Assurnasirpal's Annals, Col. III, line 3; Budge and King 1902: 356; Grayson 1991; see also toponyms in Parpola 1970).

The site of Kharidi has been identified with Khirbet ed-Diniye (= Old Babylonian Haradum) on the Iraqi side (see, e.g., Geyer and Monchambert 2003: 263). We have suggested that the newly discovered Tell Kharita on the Euphrates situated at the foot of Jebel Bishri could also be Kharidi (Lönnqvist *et al.* 2007 and see more in the following ch. 10.). Assurnasirpal's Annals further state: *Moving from the city of Khindanu I pitched camp (and) spent the night in a mountain by the Euphrates. Moving from the mountain by the Euphrates I pitched camp (and) spent the night among Bīt Sabi before the city of Haridu – Haridu lies on the other bank of the Euphrates* (Grayson 1991: 123). The site of Khindanu is traditionally identified at 'Anqa south of Abu Kemal (Liverani, 2, 1992; Geyer and Monchambert 2003: 140) on the right bank of the Euphrates some kilometres from the present Iraqi border. Some of the landscapes in which battles took place could be topographically nearer to the neighbourhood of Jebel Bishri than several modern studies indicate. The punishments near Kipinu also seem to take place next to the areas of the mountain. Therefore in our view some site locations in the area of Jebel Bishri could be associated with the Assyrian campaigns against the Aramaean areas. Jebel Bishri can be seen following the Euphrates at least to Raqqa or even as far as to modern town of As-Sadayan and after that we may locate the area of Bit-Adini. Therefore our view differs in 100 km's length from Liverani's (2, 1992, Fig. 2) suggestion. In our view Bit-Adini could be located to begin within the Tabqa dam area.

The finds from the Finnish survey from the seasons 2003-2006 have brought more information about the material remains of the nomadic and tribal groups in the Early and the Middle Bronze Age, and new evidence is also contributing to the presented history of the Assyrian attacks and the punishing campaigns against the Aramaeans at the foot of the mountain (Lönnqvist *et al.* 2009, Lönnqvist 2010). We shall further discuss the archaeological evidence pertaining to the area in the following ch. 10. The Aramaeans were, however, ruling the mountain well into the Roman period, and the first possible evidence of their existence discovered in the region is provided by an apparent short inscription on a pottery sherd at a domestic site found at Qseybe. Qseybe is a Roman military station on the piedmont of Jebel Bishri, where the locals such as Palmyreneans may have worked as military personnel or cooks for the Romans (see ch. 12.).

9.7. Conclusions and Discussion

The significance of bringing textual evidence of the Amorite and Aramaean activities related to the mountain of Jebel Bishri is to enable to understand the early historical framework of the area that we have surveyed for years. In order to grasp the human activities we need to take the textual evidence into account when available. The elements we can retrieve from the texts are that tribal and nomadic West Semites were connected with the area from the mid-third millennium B.C. From the socio-political point of view the basic social groups were tribes, but early on the Ebla texts elucidate that in the neighbourhood there was an organised state system already in the Early Bronze Age. The Sumerians may have ruled some of the Amorites in the area during the Akkadian period, when Naram-Sim conquered this region populated by the Amorites. Trade flourished between the Sumerian cities and Jebel Bishri during the Gudeas and probably with Ur which had an Amorite component in its society trading with different kinds of leather products during the Ur III period (see passim Buccellati 1966).

However, the Myth of Martu elucidates that there was prejudice against the Martu, i.e., the Amorite people of the steppe and desert, and these people were not totally integrated into the Sumerian society. But Jebel Bishri also appeared as a deified mountain associated with the Goat god, and sacrificial lambs were needed to be addressed in the area that was the first real highland while coming along the Euphrates from Sumer and Southern Mesopotamia. The identification of Jebel Bishri with the Mount Tidnum is re-enforced by the name of the nomadic tribe and the stone material, i.e., marble/gypsum=alabaster that is so characteristic of the area (cf. ch. 4.). We may ask whether the end of the Sumerian hegemony or colonization at the fringe of Jebel Bishri was met in a form of revenge from the side of the Tidnum at the gates of Ur, when the wall was built to keep the nomads at bay. The materials which the Amorites of the mountain traded with belonged to the typical pastoral repertoire. But also weapons and metals in general were associated with them in various sources.

There emerge intriguing questions concerning the role of Jebel Bishri in the rise of the Amorite hegemony in the kingdom of Mari and Terqa, the kingdom of Hana/Khana, in the early second millennium B.C. There is ample evidence to argue that nomadic Amorites were indeed on a move in the end of the third millennium B.C. and sedentarized widely. Environmental reasons have been seen as one of the causes, but socio-political changes could have been involved in the process as well. So, the change of the occupations of the kingdom of Mari from the earlier Sumerian-influenced rulers to the Amorite Lim dynasty apparently had its background in the large-scale movement and sedentarization of the Amorites. The tribal groups such as the Benjaminites and Suteans were especially living in the area of Jebel Bishri and specialized in caravan escorting. The former has been included in the Amorite tribes but the latter's identity is disputed. Benjaminites had both nomadic and settled elements, while the Suteans appear to have been more nomadic. Both groups were recruited as a military force in the Mari kingdom. They also caused troubles for the Mari rulers by pillaging caravans or attacking cities. The relations with the sedentary and urban Amorites varied being occasionally good and sometimes bad. In any event, during the Old Babylonian period Jebel Bishri was ascribed to the special deities such as Ishtar-Bishra or Goat-Boy Bazi – therefore the sacred character of the mountain continued from the Sumerian times onwards.

However, when the kingdom of Khana ceased to exist, another nomad-based West Semitic kingdom, namely the Aramaean kingdom of Laqē, emerged in the neighbourhood probably reaching the area of Jebel Bishri. The Assyrians boasted about their victories over Aramaean pastoralists at the foot of Jebel Bishri, like the Akkadians had done before over the Amorites. In this case no specific trade is reported as a peaceful act between the Assyrians and the Aramaeans living in the neighbourhood. There is no evidence of cultic practices that the Assyrians would have carried on in the area either.

Jebel Bishri as a natural bastion became an area for the Mesopotamian kings to boast about their victories and subject the nomadic tribes that they felt were threatening their existence from time to time. In the following chapter dealing with the Bronze and Iron Ages we shall further evaluate the evidence from the archaeological point of view and continue to discuss the role of the Amorites and Aramaeans in the history of the region.

Bibliography

Algaze, G. (1993) *The Uruk World System. The Dynamics of Expansion of Early Mesopotamian Civilization.* Chicago: University of Chicago Press.

Amiet, P. (1980) *Art of the Ancient Near East.* New York: Abrams.

Anbar, M. (1991) *Les tribus amurrites de Mari.* Orbis Biblicus et Orientalis 108. Freiburg: Universitätsverlag.

ANET = J.B. Pritchard (1969) *The Ancient Near Eastern Texts Relating to the Old Testament.* Princeton: Princeton University Press.

Archi, A. (1985) Mardu in the Ebla Texts, in *Orientalia,* Vol. 54, N.S., Fasc. 1-2, pp. 7-13.

ARM/T = *Archives Royales de Mari: Transcrite et Traduite.*

Astour, M.C. (1992) An Outline of the History of Ebla (Part 1), in *Eblaitica: Essays on the Ebla Archives and Eblaite Language,* Vol. 3, ed. by Gordon, C.H. and Rendsburg, G.A., Winona Lake Indiana: Eisenbrauns, pp. 3-82.

Astour, M.C. (2002) A Reconstruction of the History of Ebla (Part 2), in *Eblaitica: Essays on the Ebla Archives and Eblaite Language,* Vol. 4, ed. by Gordon, C.H., Winona Lake, Indiana: Eisenbrauns, pp. 57-195.

Black, J. and Green, A. (1992) *Gods, Demons and Symbols of Ancient Mesopotamia. An Illustrated Dictionary.* Avon: The Bath Press.

Brinkman, J.A. (1968) *A Political History of Post-Kassite Babylonia 1158-722 B.C. Analecta Orientalia 43.* Commentationes Scientificae de Rebus Orientis Antiqui. Roma: Pontificium Institutum Biblicum.

Buccellati, G. (1966) *The Amorites of the Ur III Period.* Pubblicazioni del seminario di semistica a cura di Giovanni Garbini. Richerche I. Naples: Istituto Orientale di Napoli.

Buccellati, G. (1990a) From Khana to Laqê: The End of Syro-Mesopotamia, in *De la Babylonie à la Syrie, en passant par Mari,* Mélanges offerts à Monseur J.-R. Kupper à l'occasion de son 70e anniversaire, Textes réunis par Ö. Tunca, Liège: Université de Liège, pp. 229-253.

Buccellati, G. (1990b) "River Bank", "High Country", and "Pasture Land": the Growth of Nomadism on the Middle Euphrates and the Khabur, in *Tall al-Hamidiya 2,*

Symposion, Recent Excavations in the Upper Khabur Region 1986, ed. by Eichler, S., Wäfler, M., Warburton, D., Orbis Biblicus et Orientalis, Series Archaeologica 6, Fribourg: Presses Universitaires Fribourg, pp. 87-117.

Buccellati, G. (1990c) Salt at the Dawn of History: the Case of the Bevelled-Rim Bowls, in *Resurrecting the Past, A Joint Tribute to Adnan Bounni*, ed. by Matthiae, P., Van Loon, M. and Weiss, H., Publications de l'Institut historique-archéologique néerlandais de Stamboul, Vol. 67, Leiden: Nederlands Historisch-Archaeologisch Insituut te Istanbul, pp. 17-40.

Buccellati, G. (1992) Ebla and the Amorites, in *Eblaitica: Essays on the Ebla Archives and the Eblaite language*, Vol. III, ed. by Gordon, C.H. and Rendsburg, G.A., Winona Lake, Indiana: Eisenbrauns, pp. 83-104.

Buccellati, G. (1993) Gli Amorrei e "l'addomesticamento" della steppa, in *Eufrate e il tempo, Le civiltà del medio Eufrate e della Gezira siriana*, A cura di Olivier Rouault et Maria Grazia Masetti-Rouault, Milano, pp. 67-69.

Buccellati, G. (2006) *The Mountain at the Core: The Geo-political Role of the Jebel Bishri in Early Historic Times*, an Abstract for the Symposium: Formation of Tribal Communities in the Bishri Region in the Prehistoric through Historical Periods, Tokyo, March 19.

Buccellati, G. and Kelly-Buccellati, M. (1977-1978) The Terqa Archaeological Project: First Preliminary Report, in *Les Annales Archéologiques Arabes Syriennes*, Vol. XXVII-XXVIII, pp. 71-96.

Budge, E.A.W. and King, L.W. (1902) *The Annals of Assyrian Kings*. Vol. I. London: Oxford University Press.

Bunnens, G. (2009) Assyrian Empire Building and Aramization of Culture as Seen from Tell Ahmar/Til Barsib, in *Syria*, Vo. 86, pp. 67-82.

Casana, J. (2007) Structural Transformations in Settlement Systems of the Northern Levant, in *American Journal of Archaeology*, Vol. 111, pp. 195-221.

Charpin, D. (2004) Nomades et Sédentaires dans l'Armée de Mari du Temps de Yahdun-Lîm, in *Nomades et sédentaires dans le Proche-Orient ancien, Compte rendu de la XLVie Rencontre Assyriologique Internationale (Paris, 10-13 juillet 2000), Amurru 3*, texts réunis par Nicolle, C., Paris: Éditions Recherche sur les Civilisations, pp. 83-94.

Charpin, D. and Durand, J.-M. (1986) "Fils de Sim'al": Les Origines Tribales des Rois de Mari, in *Revue d'assyriologie et archéologie orientale*, Vol. LXXX, pp. 141-183.
Charpin, D. and Durand, J.-M. (1989) Le tombeau de Yahdun-Lim, in *N.A.B.U.*, No. 2, pp. 18-19.

Charpin, D. and Ziegler, N. (2003) *Mari et le Proche-Orient à l'Époque Ammorite. Essai d'Histoire Politique.* Florilegium marianum. Vol. V. Mémoirs de N.A.B.U. No. 6. Antony: Société pour l'Étude du Proche-Orient Ancien.

Chiera, E. (1924) *Sumerian Religious Texts.* Crozer Theological Seminary Babylonian Publications. Vol. I. Upland. PA: Crozer Theological Seminary.

Clack, T.A.R. (2007) *Memory and the Mountain, Environmental Relations of the Wachagga of Kilimanjaro and Implications for Landscape Archaeology.* Cambridge Monographs in African Archaeology 72. Series Editors: Alexander, J., Smith, L. and Insoll, T. BAR International Series. No. 1679. Oxford: Archaeopress.

Darmark, K. (2008) The Archaeological Potential of the Assyro-Aramaean Hostility on the Euphrates Side of Jebel Bishri, Implications of Battlefield Archaeology, in *Jebel Bishri in Context, Introduction to the Archaeological Studies and the Neighbourhood of Jebel Bishri in Central Syria*, Proceedings of a Nordic Research Training Seminar in Syria, May 2004, ed. by Lönnqvist, M., BAR International Series, No. 1817, pp. 49-58.

Durand, J.-M. (2004) Peuplement et sociétés à l'époque Amorrite (I) Les Clans Bensim'alites, in *Nomades et sédentaires dans le Proche-Orient ancien, Compte rendu de la XLVie Rencontre Assyriologique Internationale (Paris, 10-13 juillet 2000), Amurru 3*, texts réunis par Nicolle, C., Paris: Éditions Recherche sur les Civilisations, pp. 111-239.

Finkbeiner, U. (1999-2000) with the Contribution by Rothmund, A., Emar & Balis 1996-1998, Preliminary Report of the Joint Syrian-German Excavations with the Collaboration of Princeton University, in *Berytus*, Vol. 44, pp. 5-33.

Finkbeiner, U. (2001) Emar 1999 – Bericht über die 3. Kampagne der syrisch-deutschen Ausgrabungen, Mit Beiträgen von H. Attoura, B. Faist, U. König, F. Sakal und F. Starke, in *Baghdader Mitteilungen*, Deutsches Archäologisches Institut Orient-Abteilung, Band 32, pp. 42-110.

Finkbeiner, U. (2002) Emar 2001 - Bericht über die 4. Kampagne der syrisch-deutschen Ausgrabungen, Mit Beiträgen von H. Attoura, W. Eixler und unter Mitarbeit

von F. Sakal, in *Baghdader Mitteilungen*, Deutsches Archäologisches Institut Orient-Abteilung, Band 33, pp. 109-146.
Finkbeiner, U. and Sakal, F. (2003) Emar 2002 – Bericht über die 5. Kampagne der syrisch-deutschen Ausgrabungen, Mit Beiträgen von H. Attoura, W. Eixler, D. Lineberry, F. Starke and S. Wilhelm, in *Baghdader Mitteilungen*, Deutsches Archäologisches Institut Orient-Abteilung, Band 34, pp. 9-100.

Finkelstein, J.J. (1966) The Genealogy of the Hammurabi Dynasty, in *Journal of Cuneiform Studies*, Vol. XX, pp. 95-118.

Gelb, I.J. (1961) The Early History of the West Semitic Peoples, in *Journal of Cuneiform Studies*, Vol. XV, pp. 27-47.

Gelb, I. (1980) *Computer-aided Analysis of Amorite.* Assyriological Studies No. 21. Chicago: The Oriental Institute of the University of Chicago.

Gelb, I.J. (1992) Mari and the Kish Civilization, in *Mari in Retrospect, Fifty Years of Mari and Mari Studies*, ed. by Young, G.D. Winona Lake, Indiana: Eisenbrauns, pp. 121-202.

George, A. (2008) Something old, something new, in *The Middle East in London*, Nov., No. 3.

Geyer, B. and Monchambert, J.-Y.(2003*) La Basse Vallée de l'Euphrate Syrien du Néolithique à l'avenement de l'Islam, Mission Archéologique de Mari.* Tome VI. Vol. 1: Texte. Vol 2: Atlas. Bibliothèque Archéologique et Historique. T. 166. Beyrouth: Institut Français du Proche-Orient.

Grayson, A.K. (1991) *Assyrian Rulers of the Early First Millennium BC.* Vol. I (1114-859 BC). The Royal Inscriptions of Mesopotamia. Assyrian Periods. Vol. 2. Toronto: University of Toronto Press.

Heltzer, M., with the contribution by Arbeli, S. (1981) *The Suteans.* Istituto Universitario Orientale. Seminario di StudiAsiatici. Series Minor XIII. Naples: Istituto Universitario Orientale.

Horsnell, M.J.A. (1999) The Year-names of the First Dynasty of Babylon, in *The Year-Names Reconstructed and Critically Annotated in Light of their Exemplars*, Vol. 2, McMaster University Press.

Joannès, F. (1997) Palmyre et les Routes du Désert au Début du Deuxième Millénaire av. J.-C., in *MARI Annales de Recherches Interdisciplinaires*, Vol. 8, pp. 393-415.

Klengel, H. (1996) Palmyra and International Trade in the Bronze Age: The Historical Background, in *Palmyra and the Silk Road, Special Issue Documenting the Activities of the International Colloquium, Les Annales Archéologiques Arabes Syriennes*, Vol. XLII, pp. 159-163.

Knudtzon, J.A. (1915) *Die El-Amarna-Tafeln.* Vorderasiatische Bibliotek Vol. I. Leipzig: Hinrichs.

Koppen van, F. (1997) L' Expédition à Tilmun et la Révolte des Bédouins, in *MARI Annales de Recherches Interdisciplinaires*, Vol. 8, pp. 417-429.

Kramer, S. (1952) *Enmerkar and the Lord of Aratta.* A Sumerian Epic Tale of Iraq and Iran. Philadelphia: University Museum, University of Pennsylvania.

Kraus, F.R. (1965) *Könige, die in Zelten wohnten. Betrachtungen über den Kern der assyrischen Königsliste.* Mededelingen der Koninklije Nederlandse Akademie van Wetenschappen. Afd. Letterkunde. Nieuwe Reeks. Deel 28. No. 2. Amsterdam.

Krebernik, M. (2001) *Tall Bi'a/Tuttul II.* Saarbrücken: Saarbrücker Druckerei und Verlag.

Kupper, J.-R. (1957) *Les nomades en Mésopotamie au temps des rois de Mari.* Bibliothèque de la Faculté de Philosophie et Lettres de l'Université de Liège. Fascicule CXLII. Paris: Les Belles Lettres.

Kupper, J.R. (1961) *L'iconographie du dieu Amurru dans la glyptique de la Ire dynastie babylonienne.* Académie Royale de Belgique. T. LV. Bruxelles: Palais des Académies.

Kühne, H. (1980) Zur Rekonstruktion der Feldzüge Adad-Nirarri II., Tukulti-Ninurta II und Aššurnasirpal II in Habur-Gebiet, in *Baghdader Mitteilungen*, Band ii, pp. 44-70.

Lambert, M. and Tournay, J.-R. (1951) Le statue B de Gudea, in *Revue d'assyriologie et archéologie orientale,* Vol. XLV, pp. 49-66.

Larsen, M.T. (1979) The Tradition of Empire in Mesopotamia, in *Power and Propaganda, A Symposium on Ancient Empires*, ed. by Larsen. M.T, Copenhagen Studies in Assyriology, Vol. 7, Copenhagen: Akademisk Forlag, pp. 75-103.

Lewy, J. (1961) Amurritica, in *The Hebrew Union College Annual*, Vol. XXXIII, pp. 31-74.

Liverani, M. (1992) *Studies on the Annals of Ashurnasirpal II. Vol. 2: Topographical Analysis. Guarderni di Geografica Storica, 4.* Università di Roma "La Sapienza". Dipartimento di Scienze storiche, archeologische e antropologiche dell' Antichità. Roma: Centro Stampa d'Ateneo.

Luckenbill, D.D. (1926) *Ancient Records of Assyria.* Vol. I. *Historical Records of Assyria: from the Earliest Times to Sargon.* Chicago: University of Chicago Press.

Luke, J.T. (1965) *Pastoralism and Politics in the Mari Period.* Unpublished Ph.D. dissertation. University of Michigan.

Lönnqvist, M. (2000) *Between Nomadism and Sedentism: Amorites from the Perspective of Contextual Archaeology.* PhD dissertation. Helsinki: Juutiprint.

Lönnqvist, M. (2008) Were Nomadic Amorites on the Move? Migration, Invasion and Gradual Infiltration as Mechanisms for Cultural Transition, in *Proceedings of the 4th International Congress of the Archaeology of the Near East,* 29 March – 3 April 2004, Freie Universität Berlin, Vol. 2, *Social and Cultural Transformation: The Archaeology of Transitional Periods and Dark Ages, Excavation reports,* ed. by Kühne, H., Czichon, R., Kreppner, F.J., Wiesbaden: Harrassowitz Verlag, pp. 195-215.

Lönnqvist, M. (2009) Jebel Bishri in Syria and the Role of Nomadism in the End of the Early Bronze Age, in *The Levant in Transition,* Proceedings of a Conference held at the British Museum on 20-21 April 2004, ed. by Parr, P.J., The Palestine Exploration Fund Annual IX, Leeds: Maney Publishing, pp. 49-55.

Lönnqvist, M. (2010) How to Control Nomads? A Case Study Associated with Jebel Bishri in Central Syria, in *City Administration in the Ancient Near East, Proceedings of the 53e Rencôntre Assyriologique Internationale,* Vol. 2, ed. by Kogan, L. et al., Orientalie et Classica, Babel + Bibel 5, Winona Lake, Indiana: Eisenbrauns, pp. 118-139

Lönnqvist, M. with the contributions by Törmä, M. Nuñez, M., Lönnqvist, K., Stout Whiting, M., Okkonen, J., Riihiaho, H., Nissinen, M. (2006) Archaeological Surveys of Jebel Bishri, The Preliminary Report of the Finnish Mission to Syria 2000-2004, in *KASKAL,* Vol. 3, pp. 205-242.

Lönnqvist, M., Törmä, M., Lönnqvist, K., Nuñez, M., Okkonen J. and Latikka, J. (2007) The Euphrates Channel Changes and Archaeology along Jebel Bishri in Syria, in *Proceedings of the XXI Symposium, CIPA 2007, Athens, Anticipating the Future of the Cultural past,* Vol. 1., ed. Georgopoulos, A., the International Archives of Photogrammetry, Remote Sensing and Spatial Information Sciences, Vol. XXXVI-5/C53, pp. 465-470.

Lönnqvist, M. with the contributions by Törmä, M., Lönnqvist, K., Okkonen, J., Herles, M., Königsdörfer, M. (2009) Archaeological Surveys of Jebel Bishri, The Preliminary Report of the Finnish Mission to Syria, 2005-2006, in *KASKAL,* Vol. 6, pp. 1-42.

MAD = Materials for the Assyrian Dictionary, Chicago 1952-.

Margueron, J.-Cl. (1975a) Rapport préliminaire sur les deux préliminaires campagnes de fouille à Meskéné-Emar (1972-1973), in *Les Annales Archéologiqes Arabes Syriennes,* Vol. XXV, pp. 73-86.

Margueron, J.-Cl. (1975b) Quatre campagnes de fouilles à Emar (1972-1974), in *Syria,* Vol. LII, pp. 53-85.

Masetti-Rouault, M.G. (2001) *Cultures locales du Moyen-Ephrate. Modèles et événements, IIe –I er mill. Av. J.-C.* Subartu VIII. Turnhout: Brepols.

Maspero, G. (1903) *History of Egypt.* Vol. 7. Part A. Chaldea, Syria, Babylonia and Assyria. Ed. by Sayce, A.H. London: The Grolier Society Publishers.

Matthews, V.H. (1978) *Pastoral Nomadism in the Mari Kingdom (ca. 1830-1760 B.C.)* American Schools of Oriental Research Dissertation Series 3. Cambridge, Mass: American Schools of Oriental Research.

Musil, A. (1927) *The Middle Euphrates: A Topographical Itinerary.* American Geographical Society. Oriental Explorations and Studies No. 3. Ed. by Wright, J.K. New York: The American Geographical Society.

Nilhamn, B. (2008) Nomadic Life in Central and Eastern Syria, A Perspective from the Present Life on Badiyah to the Amorite Nomadism in the Bronze Age, in *Jebel Bishri in Context, Introduction to the Archaeological Studies and the Neighbourhood of Jebel Bishri in Central Syria,* Proceedings of a Nordic Research Training Seminar in Syria, May 2004, ed. by Lönnqvist, M., BAR International Series 1817, pp. 15-30.

Parpola, S. (1970) *Neo-Assyrian Toponyms. Alter Orient und Altes Testament.* Band 6. Neukirchen – Vluyn: Verlag Butzon & Bercker Kevelaer.

Parr, P.J. (ed.) (2009) *The Levant in Transition.* Proceedings of a Conference held at the British Museum on 20-21

April 2004. Palestine Exploration Fund Annual IX. Leeds: Maney Publishing.

Podany, A.H. (2002) *The Land of Hana: Kings, Chronology, and Scribal Tradition.* Bethesda, Md: CDL Press.

Postgate, J.N. (1974) Some Remarks on Conditions in the Assyrian Countryside, in *Journal of the Economic and Social History of the Orient*, Vol. VII, Part 3, pp. 225-243.

Rawlinson, H. (1861-1884) *Cuneiform Inscriptions.* Vols. I-V. London.

RIME 2 = Frayne, D. (1993) *The Royal Inscriptions of Mesopotamia. Early Periods.* Vol. 2. Sargonic and Gutian Periods (2334-2113 BC). Toronto, Buffalo, London: University of Toronto Press.

Rouault, O. (2000) Quelques remarques sur la société de Terqa, in *SUBARTU*, Vol. VII, Le Djéziré et l'Euphrate Syriens, De la Protohistoire à la fin du IIe millénaire av. J.-C. Tendances dans l'interprétation historique des données nouvelles, texts réunis par Olivier Rouault et Markus Wäfler, Turnhout: Brepols, pp. 265-269.

Rowton, M. (1967) The Physical Environment and the Problem of the Nomads, in *La Civilisation de Mari, Rencôntre Assyriologique Internationale*, Vol. XV, pp. 109-121.

Rowton, M. (1976) Dimorphic Structure and the Problem of 'Apiru-'Ibrim, in *Journal of Near Eastern Studies*, Vol. XXXV, pp. 13-20.

Sahlins, M. D. (1968) *Tribesmen.* Foundations of Modern Anthropological Series. Englewood Cliffs. New Jersey: Prentice Hall.

Sarre, F. and Herzfeld, E. (1911-1920) *Archäologische Reise im Euphrat- und Tigris-Gebiet.* Band I-II. Forschungen zur islamischen Kunst I. Berlin: Dietrich Reimer (Ernst Vohsen).

Smith, S. (1949) *The Statue of Idri-mi.* Occasional Publications of the British Institute of Archaeology in Ankara No. 1. London.

Streck, M.P. (2000) *Das amurritiche Onomastikon der altbabylonischen Zeit.* Bd. 1: Die Amurriter, die onomastiche Forschung, Ortographie und Phonologie, Nominalmorphologie. Alter Orient und Altes Testament. Band 271/1. Münster: Ugarit-Verlag.

Streck, M.P. (2004) Die Religion der amurritischen Nomaden am mittleren Euphrat, in *Ofizielle Religion, locale Kulte und individuelle Religiosität, Akten des regionsgeschichtlichen Symposiums "Kleinasien und angrenzende Gebiete vom Beginn des 2. bis zur Mitte des 1. Jahrtausends v. Chr."* (Bonn, 20.-22. Februar 2003), herausgegeben von Hutter, M. und Hutter-Braunsar, S., Alter Orient und Altes Testament, Band 318, pp. 421-432.

Strommenger, E. and Kohlmeyer, K. (1998) *Tall Bi'a/ Tuttul.* Vol I. Die altorientalischen Bestattungen. Wissenschaftliche Veroffentlichungen der Deutschen Orient-Gesellschaft - WVDOG 96. Wiesbaden: Harrassowitz Verlag.

Strommenger, E. and Miglus, P.A. (2002) *Tall Bi'a – Tuttul.* Vol. VIII. *Stadtbefestigungen, Häuser und Tempel.* Saarbrucken: Saarbrucken Druckerai und Verlag.
Strommenger, E. *et al.* (1987) Ausgrabungen in Tall Bi'a 1985, in *Mitteilungen der Deutchen Orient-Gesellschaft*, Vol. 119, pp. 10-28.

Strommenger, E., Kohlmeyer, K. and Feisel, R. (2000) *Ausgrabungen in Tall Bi'a/Tuttul.* Vol. III. Die Schichten des 3. Jahrtausends v. Chr. im Zentralhugel E. Saarbrucken: Saarbrucker Druckerei und Verlag.

T.M. = Tell Mardikh (Ebla texts)

Thureau-Dangin, F. (1925) *Les Cylindres de Goudéa, découverts par Ernest de Sarzec à Tello.* Textes Copié par F. Thureau-Dangin. Musée du Louvre, Departement des antiquités orientales. Textes Cunéiformes. Tome VIII. Paris: Paul Geuthner.

Weeks, N. (1985) The Old Babylonian Amorites: Nomads or Mercenaries? in *Orientalia Lovaniensia Periodica*, Vol. XVI, pp. 49-57.

Weippert, M. (1971) *The Settlement of the Israelite Tribes in Palestine.* A Critical Survey of Recent Scholarly Debate. Studies in Biblical Theology. Second Series. No. 21. London: SCM Press Ltd.

Weiss, H. (2000) Causality and Chance: Late Third Millennium Collapse in Southwest Asia, in *Le Djéziré et l'Euphrate Syriens, De la Protohistoire à la fin du IIe millénaire av. J.-C., Tendances dans l'interprétation historique des données nouvelles*, *SUBARTU VII*, texts réunis par Rouault, O. et Wäfler, M., Turnhout: Brepols, pp. 207-217.

Wilcke, C. (1969/1970) Zur Geschichte der Amurriter in der Ur III Zeit, in *Die Welt des Orients*, Vol. V, pp. 1-31.

Vincetelli, I. (1972) Prodotti "amorrei" e prodotti "hurriti", in *Oriens Antiqvvs*, Vol. XI, pp. 133-135.
Zadok, R. (1985) Suteans and other West Semites during the Latter Half of the Second Millennium B.C., in *Orientalia Lovaniensia Periodica,* Vol. 16, pp. 59-70.

Zarins, J. (1989) Pastoralism in southwest Asia: the second millennium BC, in *The Walking Larder, Patterns of Domestication, Pastoralism, and Predation*, ed. by Clutton-Brock, J., Cambridge: Cambridge University Press, pp. 127-155.

Ziegler, N. (2004) Samsi-Addu et la Combine Sutéenne, in *Nomades et sédentaires dans le Proche-Orient ancien, Compte rendu de la XLVIe Recontre Assyriologique Internationale (Paris, 10-13 juillet 2000), Amurru 3*, Textes réunis par C. Nicolle, Paris: Éditions Rechersche sur les Civilisations, pp.95-109.

10. THE EVIDENCE FROM THE EUPHRATES SIDE OF JEBEL BISHRI: SEDENTARIZATION, SEDENTARY SITES AND SYRO-MESOPOTAMIAN CIVILIZATION

Minna Lönnqvist Kenneth Lönnqvist
with Contributions by Jari Okkonen and Michael Herles

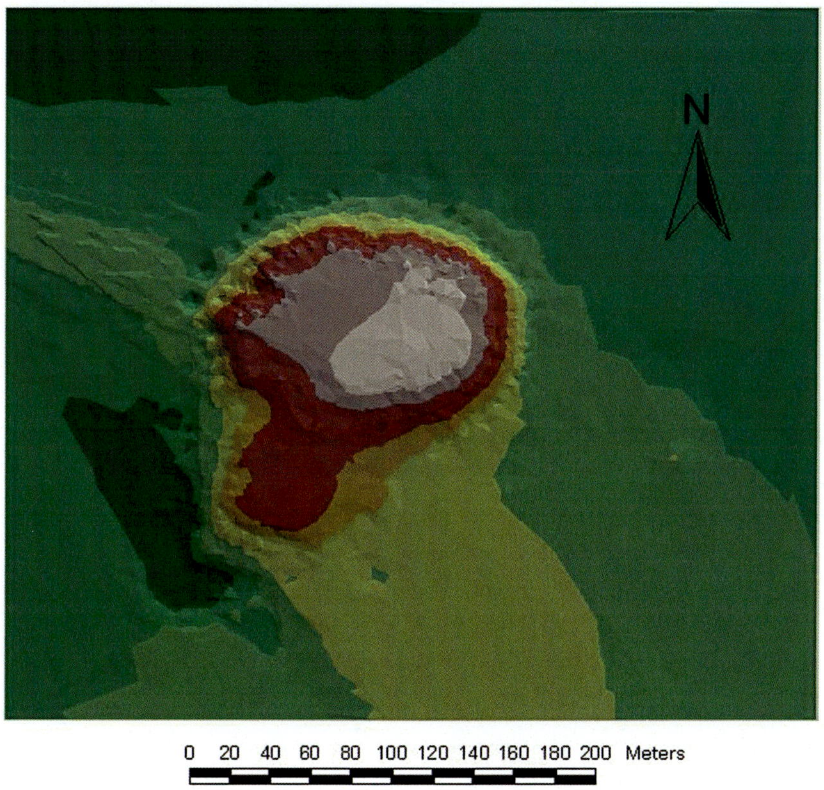

Fig. 1. The terrain model of Tell Tibne constructed by Jari Okkonen 2005.

10.1. Prospecting Sedentary Sites at the Foot of the Mountain

In this chapter we present the evidence from the surveyed and mapped sites on the Euphrates valley side of Jebel Bishri with the periods from the rise of urbanism and the earliest Syro-Mesopotamian civilization in the Chalcolithic Age, developing contacts through the Bronze Age to the Iron Age in the region. The impact of the Lower Mesopotamian cultures in this process is taken into account. The evidence of sedentarization at the foot of Jebel Bishri in specific will be closely looked at, and the role of pastoralism and tribal groups is considered in the change from movement to arrest. In C. Héraud's (1922) aerial maps several hills and possible tells between Raqqa and Deir ez-Zor can be identified in the Middle Euphrates valley, although practically most of them are sand dunes.

Our remote sensing studies and prospecting on the northern foot of Jebel Bishri, following the right bank of the Euphrates with empirical checks on the ground, have resulted in the mapping of altogether eight ancient sedentary sites of the so-called tell type of which six/seven are earlier known ones but for which no coordinates were published before. In addition, J. Lauffray (1983: Fig. 4.) mentions Tell Ma'adan which we never visited to check on the ground and which is therefore not mapped here. (Roman fortresses that we have documented on the Euphrates beside some Islamic sites will be discussed in ch. 13. and 14. dealing with the Euphrates *limes* and the changing Bedouins.)

All the tells are situated along the over 100 km long stretch of the right bank of the Euphrates following Jebel Bishri between Deir ez-Zor and Raqqa. Apart from the sites mentioned in F. Sarre's and E. Herzfeld's (1911) map, as well as C. Héraud's (1922) map, some of these tells appear in A. Poidebard's (1934), Lauffray's (1983: Fig. 4), J. Besançon's - P. Sanlaville's (1981: 10-11, Carte 1, four tells between Halabiya and Raqqa) and K. Kohlmeyer's (1984, Beilege 4) surveys and maps. Some of them are also shown in the atlas of *Syria, Archaeology from Space* (see, e.g., SAS 2002: 21). According to A.M.T. Moore (personal communication), his expedition at the Tabqa dam (Moore *et al.* 2000) also carried out some surveying on the right bank of the Euphrates as far as Halabiya but no documentation of this work has been published so far.

Several of the tells that we visited are situated in the midst of irrigated fields at the northern and northeastern foot of Jebel Bishri. In 2003 we visited Tell Hamadin (UTM: 0531832, 3960719; 216 m a.s.l.; Figs. 2., 5.) which can be found at the village of Hablé ca. 33 km from Raqqa towards Halabiya. The Syro-Japanese mission also visited the site in 2007 (Al-Maqdissi and Ohnuma 2008). The site offers marble/calcite/gypsum structures and Hellenistic-Roman pottery as well as Early and Middle Bronze Age sherds. Previous studies have discovered some Lower Palaeolithic finds at the site as well.

Fig. 2. Tell Hamadin.
Photo from NW: Kenneth Lönnqvist 2003.

Tell Grnaia, also known as Tell Ghanem Ali (UTM: 0538009, 3959311; 223 m a.s.l.; Figs. 3., 5.), is situated ca. 40 km from Raqqa to southeast towards Halabiya. We visited the site in 2003. The tell covers an area of some 400 m x 200 m and appears in the map by F. Sarre and E. Herzfeld (1911) beneath the Roman fortress of Djazlah. Tell Grnaia/Tell Ghanem Ali has been studied by a Syro-Japanese mission in 2007-2009 (see, e.g., Al-Maqdissi and Ohnuma 2008). It is essentially an Early Bronze Age site surrounded by several cemeteries stretching up to Jebel Bishri. For example, there is the Early Bronze Age cemetery of Abu Hamed which was excavated by the Syro-German team in 1990 (Falb *et al.* 2005).

Fig. 3. Tell Grnaia/Tell Ghanem Ali.
Photo from E: Kenneth Lönnqvist 2003.

Finally there is a shallow tell (UTM: 0543436, 3958166; 226 m a.s.l.; Figs. 4., 5.), situated near the village of Khamisiyah, which we called Tell Khamisiyah. The site seems to appear in Besançon's and Sanlaville's map of 1981. It is most likely to be identical with Tell Moukhla in J. Lauffray's map (1983: Fig. 4), also mentioned by Kohlmeyer (1984: 111) as Tell Mugla. This is a small tell, the diameter of which is ca. 200 m. Unlike Tell Hamadin and Tell Grnaia/Tell Ghanem Ali, which are situated on the northern side of the Deir ez-Zor–Aleppo –road, this shallow tell site (Fig. 4.) near Khamisiyah is on the southern side of the road and has Early and Middle Bronze Age pottery on its surface. This site, like Tell Grnaia/Tell Ghanem Ali, is currently occupied by an Islamic graveyard. All the mentioned tells belong to the Mouhafazat of Raqqa.

Fig. 4. Tell Khamisiyah/Tell Moukhla.
Photo from N: Kenneth Lönnqvist 2003.

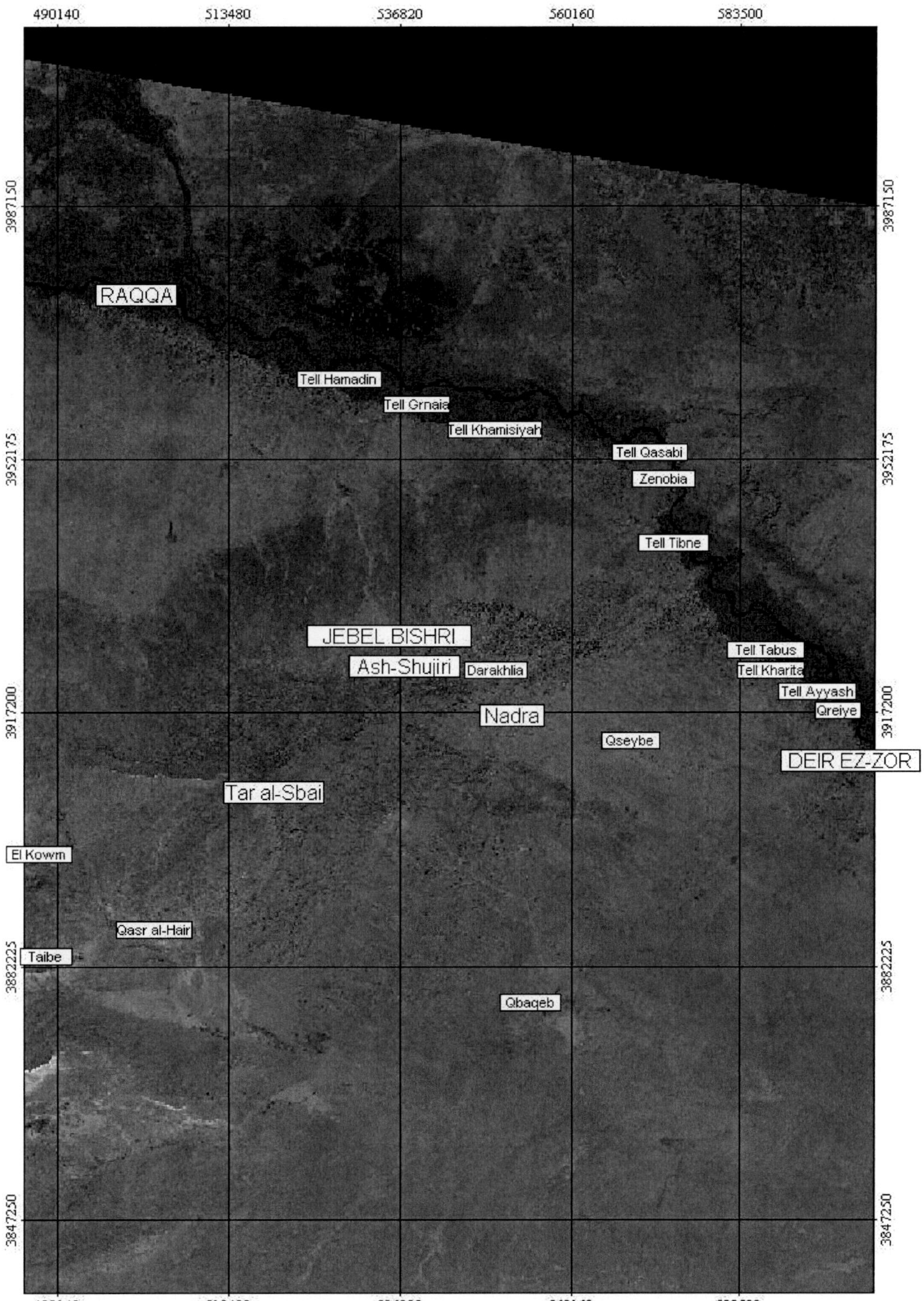

Fig. 5. Map displaying the known tell sites on the Euphrates between Raqqa and Deir ez-Zor. Mapping by Minna Lönnqvist 2007.

Before Halabiya both Lauffray (1983: Fig. 4) as well as Kohlmeyer (1984: 111) mention Tell Ma'adan, but here Sarre's and Herzfeld's map (1911) only mentions a *tumulus* (see our ch. 14., Fig. 14.). As mentioned previously, we never visited the site. At Qasabi on the peninsula of Halabiya there is a well-known tell (UTM: 0569123, 3955156; 226 m a.s.l., Figs. 5., 6. and 7.) situated northwest of Zenobia's fortress. It was already visited by F. Sarre and E. Herzfeld (1911: 165) at the beginning of the twentieth century and appears in Héraud's map (1922, XXXIII fouille). J. Lauffray (1983) and K. Kohlmeyer (1986) both mention it briefly. (The site will be further discussed in ch. 14. dealing with the coming of the Arabs into the district.) Apart from quantities of turquoise glazed pottery from the Parthian/Islamic periods, we traced Early Bronze Age pottery pieces at the site, which indicates that an earlier occupation existed at the site. To the southeast of Qasabi rise the well-preserved ruins that belong to the Byzantine fortress of Zenobia (UTM: 0574375, 3949782), the Palmyrene foundations of which have been dated to the third century A.D. But the ruins visible today represent the rebuilding of the fortress by Justinian I and date from the sixth century A.D. The site has been thoroughly studied by the French since 1944 (Lauffray 1983, 1991; see also Toivanen 2008 and more in the following chapter dealing with the Roman Euphrates *limes*). The site of Halabiya has also been tentatively identified with the location for the Assyrian fortress of Nibarti Assur/Dûr Karpati which was built on the orders of Assurnasirpal II in 877 B.C. during his campaigns against the Aramaeans, who were dwelling in the Jebel Bishri region, as was discussed in the previous chapter (see Musil 1927: 331, see also Darmark 2008).

Fig. 6 . Tell Qasabi with a modern Muslim graveyard at the foot. Photo from SE: Minna Lönnqvist 2008.

Between the peninsula of Halabiya and Deir ez-Zor there are four tells to our knowledge: they are located at Tibne (UTM: 0574179, 3941645, Figs. 1., 5.), Tabūs (UTM: 0586948, 3925414, Fig. 5.), Kharita (UTM: 0589436, 3923932, Fig. 5.) and 'Ayyāsh (UTM: 0595840, 39020927, Fig. 5.).. We have only closely studied Tell Tibne, Tell Tabūs and Tell Kharita. The tell at the peninsula of 'Ayyāsh is situated beside the Roman ruins of Qreiye and was surveyed by M. Gschwind during the years 2002-2004, as indicated in the previous chapter. According to Gschwind, the material from the tell is Pre-Roman but needs to be further studied to determine the specific periods that it represents. Tell Tabūs is known to earlier travellers, and a trial trench had been dug into its walls before the Finnish mission studied and mapped the site in 2004 (see more in ch. 13). At the village of El Kharita there is a previously unknown tell discovered by our Finnish mission in 2006 and which has not been documented before.

Fig. 7. The site of Qasabi. Mapping by Minna Lönnqvist 2006.

Fig. 8. The tell at Tibne. The Islamic graveyard can be seen on top of the tell. Photo from W: Michael Herles 2005.

10.2. Mapping a Tell at the Village of Tibne

Site G 1
Tell - Settlement

UTM: 0574179, 3941645, 231 m a.s.l
Size: ca. 200 m x 180 m

There is a tell (see Figs. 1., 5., 8, 9. and 10.) at the village of Tibne in the northern foot of Jebel Bishri situated between the Euphrates and the Deir ez-Zor–Aleppo road on the northeastern side of the road. Currently an Islamic graveyard occupies the tell. The site falls outside our systematically surveyed transects on the Euphrates. The tell already appears in F. Sarre's and E. Herzfeld's map from 1911 (see our ch. 13.). It is also mentioned and identified, e.g., by A. Poidebard (1934: 86) as the site of Diocletian's fortress of Mambri, situating 6-7.5 km southeast of Zenobia's fortress. It is further mentioned in K. Kohlmeyer's survey and located in J. Besançon's and P. Sanlaville's (1981: 10-11, Carte 1) as well as J. Lauffray's (1983: 67) map. Tibne had produced Palaeolithic finds reported by earlier missions concentrating on the prehistory. The site was first visited in 2003 and 2004 by our teams, but the more intensive surface study and mapping (see more in paragraph 13.2.) were carried out by our team in the 2005 season.

However, besides ancient structures and apparent Roman and Byzantine pottery (see more in the chapter dealing with the Euphrates *limes*) the tell has produced flints from the Palaeolithic and flint as well as pottery (Fig. 11.), including pottery from the Uruk period (Fig. 12.), from the Chalcolithic era. A number of pottery pieces, however, date from the Early and Middle Bronze Age. There are types which find parallels in the Early Bronze Age III-IV pottery repertoire of Ebla (see Fig. 13., cf., e.g., Mazzoni 1985 and Pinnock 2009) and Emar (cf. Finkbeiner 1999-202; Finkbeiner – Sakal 2003). In addition, pottery from the Middle Bronze Age II period (see Fig. 14.), typical of the Mari-Terqa district (cf., e.g., passim Rouault 1997; Geyer and Monchambert *et al.* 2003) and Baghouz (cf. passim Du Mesnil du Buisson 1948), was collected at the site. Among the Bronze Age pottery there are some Late Bronze Age pieces as well. There is also a whole jug (Figs. 13.) which finds a parallel in the graves of Mari (M 1934, T 351, see Marylou 1999: Pl. 56), the context of which has been dated to the Assyrian–Neo-Babylonian period. Apart from the Hellenistic and Roman pottery (see the details in paragraph 13.2.), some Islamic sherds (ch. 14.) were also encountered at the site.

As we briefly mentioned in the previous chapter, there also are large marble/gypsum exposures (UTM: 0578749, 3932878) at Alinbeh on the way from Deir ez-Zor just before Tibne at Jebel Bishri, and Tibne on the Euphrates could have served as a convenient site for transporting marble/gypsum = alabaster along the river to Lower Mesopotamia. This reminds us about the reference of the mountain of Tidnum (see the previous ch. 9.), obviously associated with Jebel Bishri, as a source of marble/alabaster (as alabaster is same as gypsum) in the Gudea texts dating from the late third millennium B.C. Also salt mines are currently situated near Tibne at Bweitieh (see the map in ch. 5., Fig. 53.)., and as indicated in the previous chapter, salt could have been transported from the site to Mesopotamia. Interestingly, G. Buccellati (1990) has associated certain types of Uruk vessels with salt production and incidentally or not Tell Tibne has produced Uruk pottery. Possible cultural contacts with Ebla and Emar are reflected in the Early Bronze Age pottery repertoire. This is no wonder as the site lies on the way following the river to the Mesopotamian heartland. As referred to in the previous chapter, the Sumerian cultural impact in the region is reflected in the sources describing the Sumerians as vassal rulers of the Amorites in the region. Furthermore, because of the location of Tell Tibne and the pottery associated with the tell dating from the Middle Bronze Age as well as Classical Age, J.-M. Durand (personal communication) has suggested that the site could be Dûr Yahdun-Lim's port (the port of King Yahdun-Lim's fortress that probably situated somewhere in the neighbourhood of Deir ez-Zor) Gânibatum mentioned in a recently published tablet A.859 from Mari (see Durand 2005: 64-67). Interestingly, Gânibatum served in stone transportation, which would fit well with the location of Tibne. The recovered pottery evidence at the site supports the Mari - Terqa influence of the Middle Bronze Age II period, the heyday of the Amorite kingdoms in the region. The existence of the Late Bronze Age and Iron Age pottery may connect the site to the Aramaean and Assyrian activities in the region as well (see more in ch. 9.); some maps (see, e.g., Clancier 2006: 287) even locate Nibarti-Aššur to Tibne even if the fortress is generally located to Halabiya. Because of its location and some of its Roman surface finds the site may also be identified as belonging to the fort of Mambri (Poidebard 1934: 86) built by the Emperor Diocletian and rebuilt by the Emperor Justinian I (Procopius, *De Aedificiis* II, 8.7.).

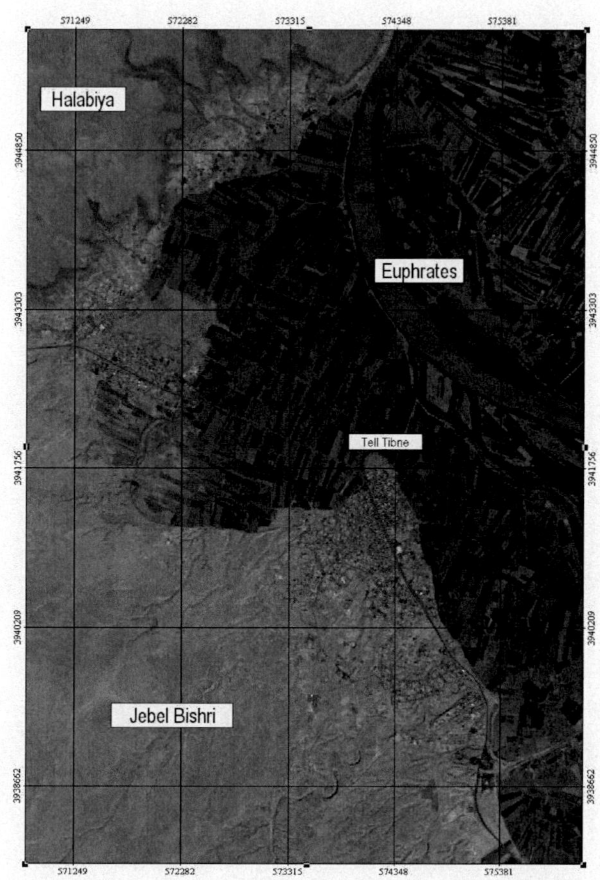

Fig. 9. The site of Tell Tibne. Mapping by Minna Lönnqvist 2008.

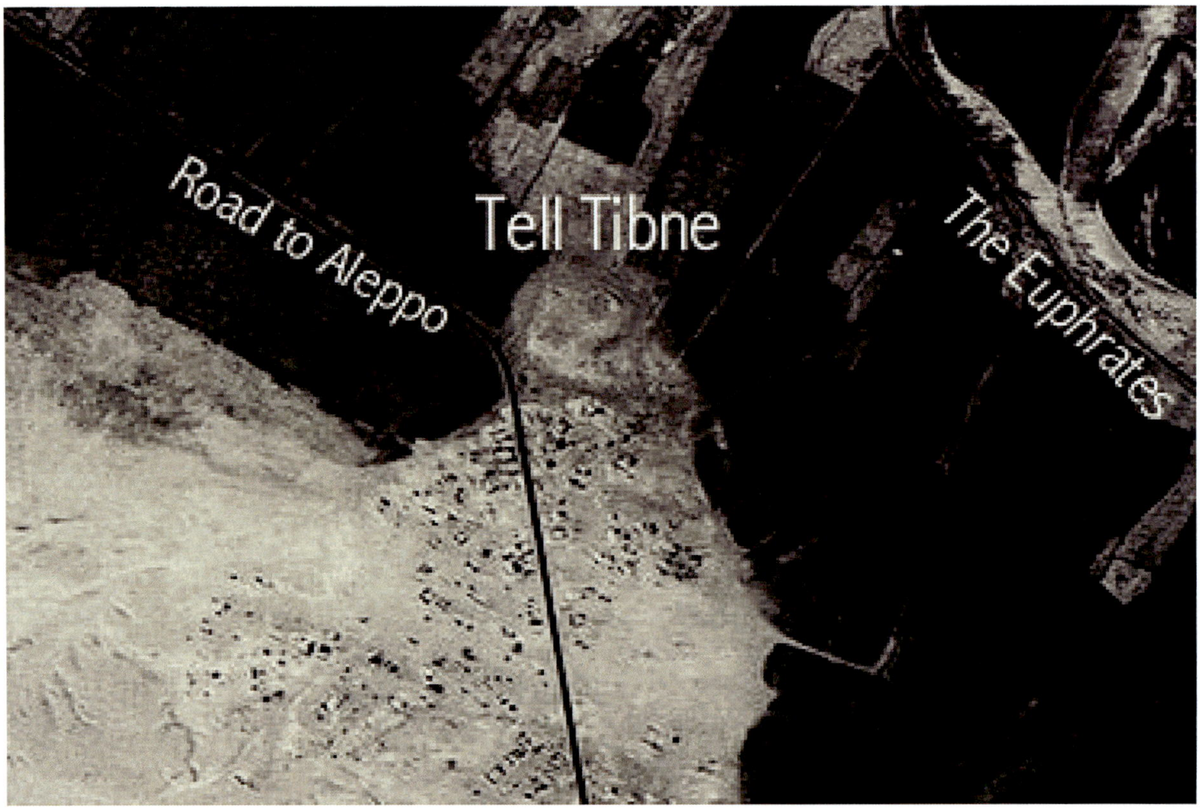

Fig. 10. Location of Tell Tibne at the edge of Jebel Bishri along the Deir ez-Zor – Aleppo road on CORONA satellite photograph from the 1960s. Mapping by Minna Lönnqvist.

Fig. 11. A piece of Chalcolithic pottery from Tell Tibne. Photo: Michael Herles 2005.

Fig. 12. Fragment of an Early Uruk pot from the Chalcolithic Age seen from the interior side. Photo: Michael Herles 2005.

Fig. 13. A ceramic jug found at Tell Tibne. Photo: Michael Herles 2005.

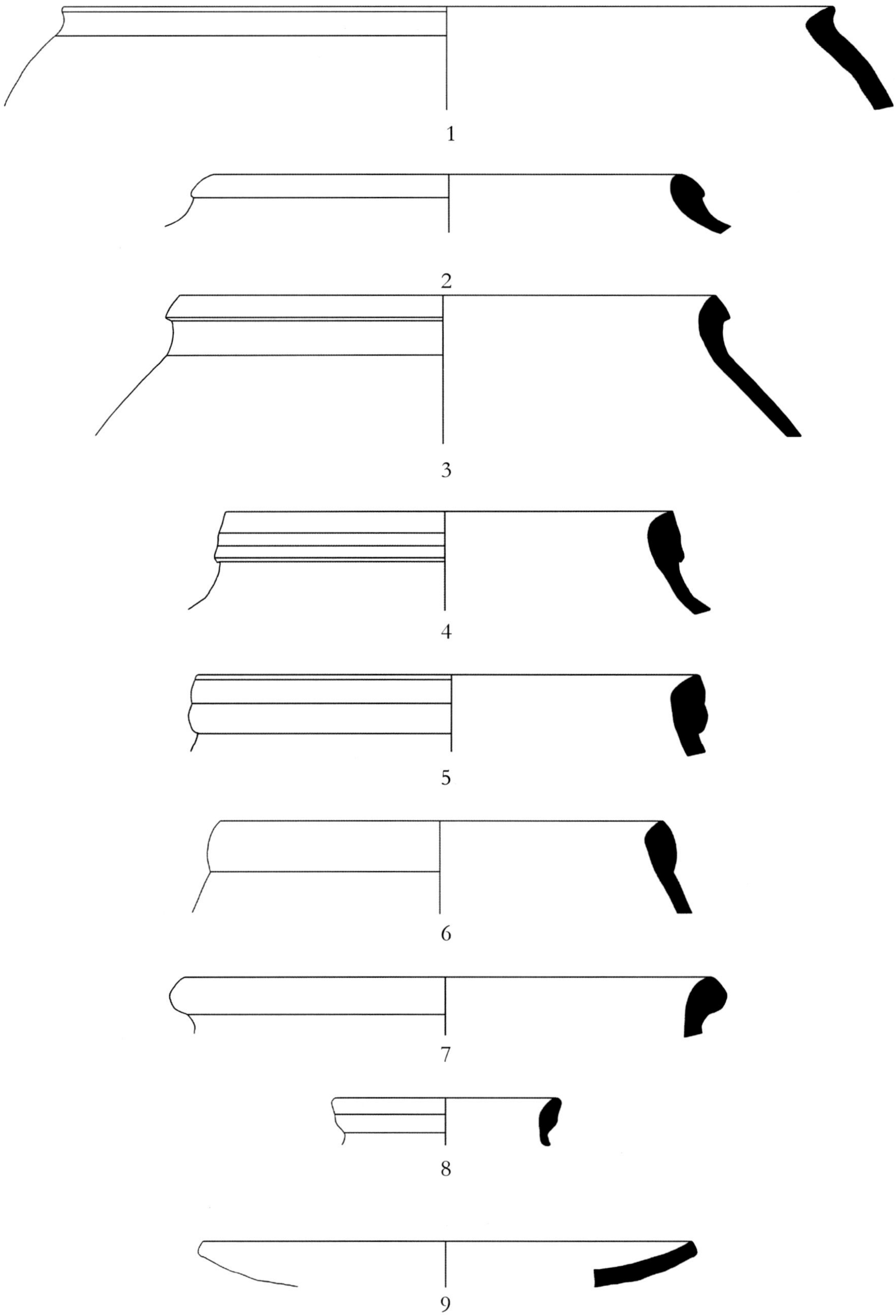

Fig. 14. Early Bronze Age pottery from Tell Tibne. Scale 1: 3.

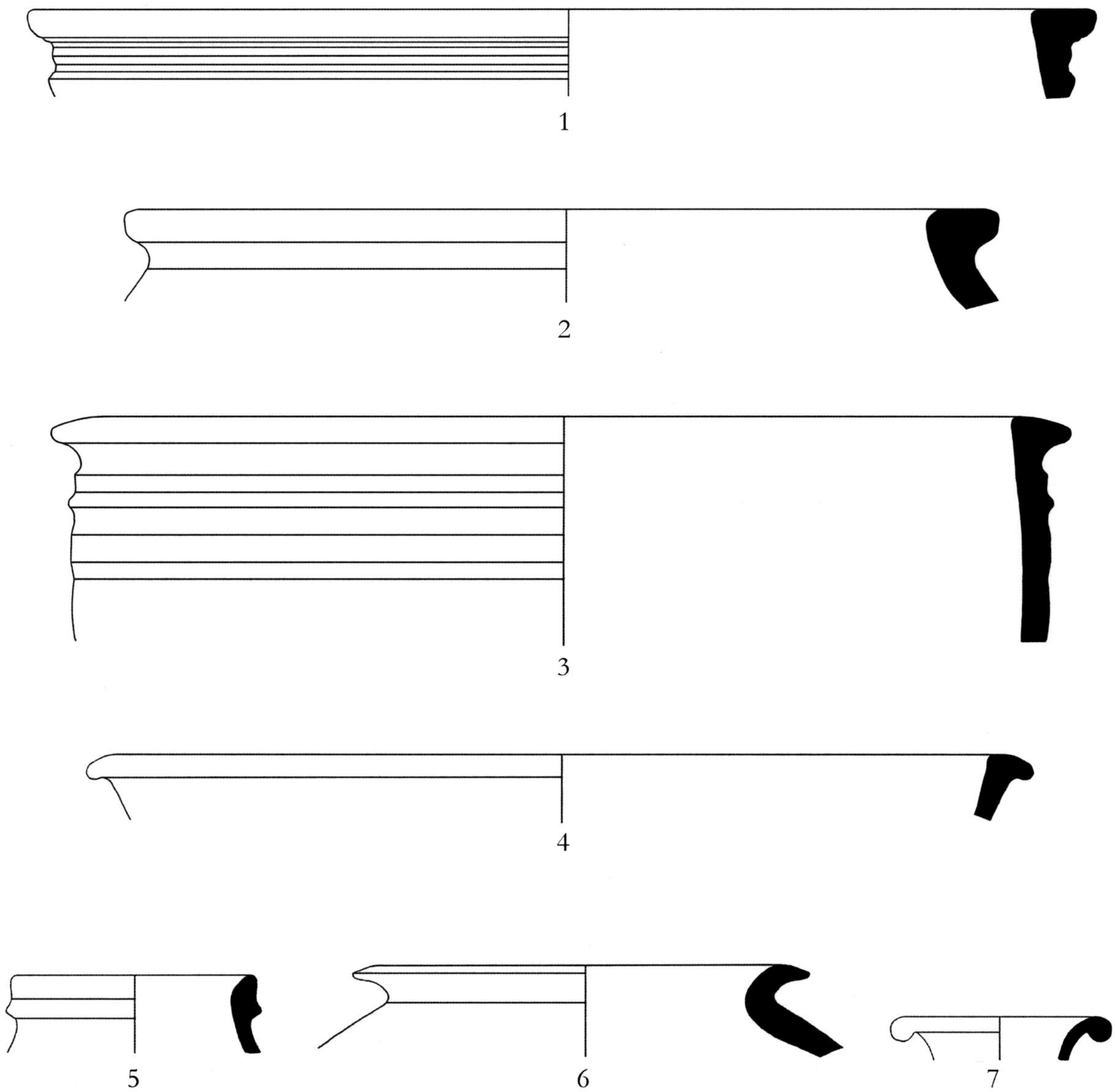

Middle Bronze Age II jars

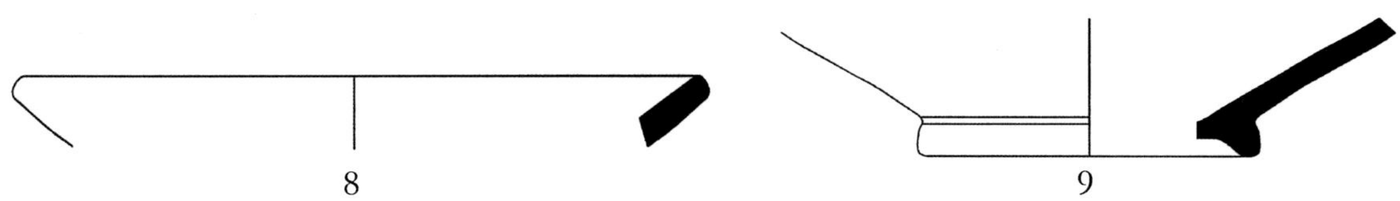

Middle Bronze Age bowls.

Fig. 15. Middle Bronze Age pottery from Tell Tibne. Scale 1: 3.

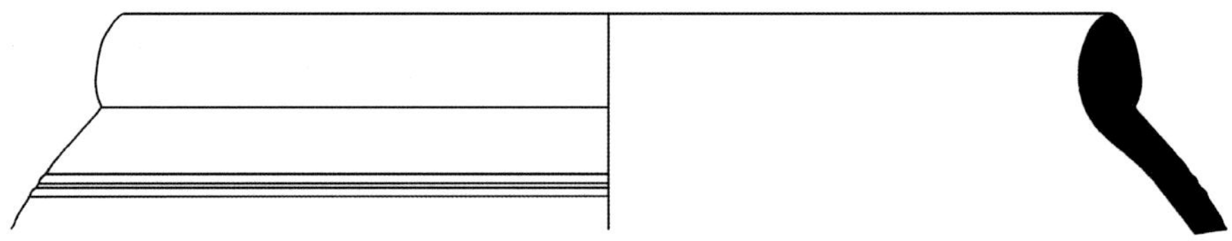

1

Late Bronze Age jar.

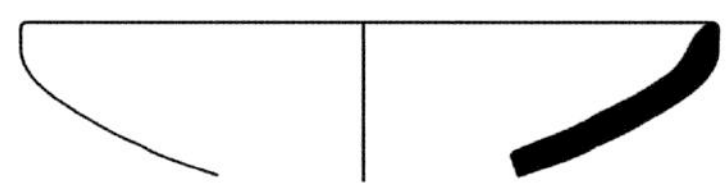

2

Late Bronze Age bowl.

Fig. 16. Late Bronze Age pottery from Tell Tibne. Scale 1: 3.

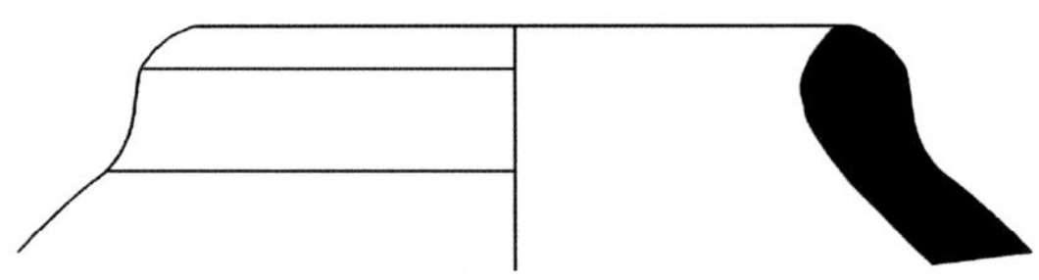

1

Iron Age jar

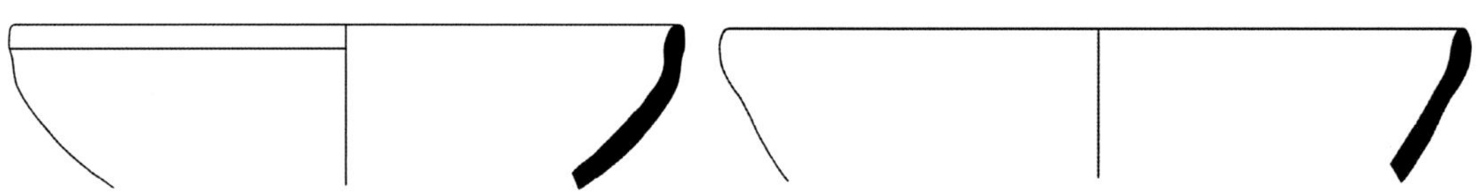

2 3

Iron Age bowls.

Fig. 17. Iron Age pottery from Tell Tibne. Scale 1: 3.

10.3. Bronze Age Evidence from Ridges and Pleistocene Terraces between Tell Tibne and Tell Kharita

The fluvial terraces were visited on the right bank of the Euphrates from Tell Tibne towards Deir ez-Zor during our unsystematic studies in 2003 by executing "window screenings". There were plateau-type terraces composed of marl and silty sands between Tell Tibne and Tell Tabūs. The first plateau **G 4** (UTM: 0580307, 3930531; 229 m a.s.l.) produced Early Bronze Age pottery, and on a similar terrace formation **G 5** (UTM: 0580484, 3929221; 224 m a.s.l.) Early Bronze Age pottery with a tabular scraper was discovered. The systematic surveys at the Roman site of Tell Tabūs **H 19** (see paragraph 13.2.2.) and its surroundings on the terrace called plateau **H 1** were carried out in 2004. The site H 1 (size: ca. 300 x 150 m) is also an ancient fluvial terrace from the Pleistocene period, the area of which was already discussed in previous chapters dealing with prehistoric finds (paragraph 5.5.). The surface finds collected on this plateau H 1 - apart from previously mentioned prehistoric flints and lime plaster bowls - offered pieces of pottery from different periods. In addition to the flints and sherds, there were two interesting finds which may be connected with the Bronze and Iron Age activities in the area. One is an eye of a cultic statue or a votive eye (Fig. 18.), and the other is a fragment of a statue made of gypsum (a wrist with a bracelet?: Fig. 19.).

The eye is carved of shell (l. 43 mm, w. 25. 48 mm and th. 2 mm.; the iris part being 11.63 mm in diam.). The eye probably dates from the Early Bronze Age. Such eyes were found inlaid in cultic statues at Mari (Temple of Ishtar and Ninni-zaza) as well as Tell Asmar, Tello,

Nippur, Ur and Ashur. They appear in the third millennium B.C. layers (see Margueron 2004: 277-285), and they are especially present in the finds dating from the Early Dynastic II and Early Dynastic III periods (see Lloyd 1984: 112-114). Tell Munbaqa in the Northern Euphrates valley has produced comparable votive eyes dating from the Bronze Age as well (Werner 1998: 41, Fig. 25.).

The arm of a statue with a possible bracelet (Fig. 19.; max. l. 13.6 cm, w. 8.2 cm and th. 1.8 cm) seems to follow the ancient Mesopotamian iconographical tradition as well. The material is gypsum/alabaster. Its precise dating is hard to judge from the fragment.

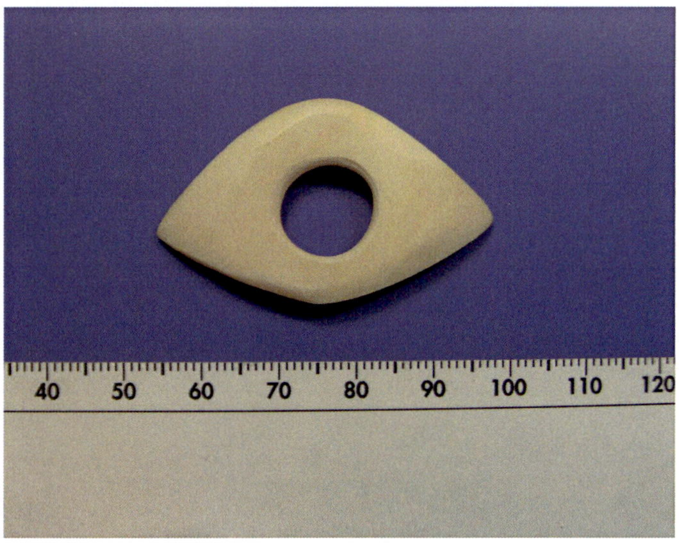

Fig. 18. An inlaid eye of a statue/a votive eye from plateau H1. Photo: Kenneth Lönnqvist 2009.

Fig. 19. A fragment of a statue from the plateau H 1. Photo: Kenneth Lönnqvist 2009.

H 1 was also systematically studied with ground penetrating radar (GPR), which could only penetrate ca. 0.5 -1 m into the clayish soil of the terrace. We especially chose to study the areas with pottery concentrations. Some possible graves and structures were identified with the GPR studies under the surface of the plateau (see ch. 4.7.2., features found in lower plateau). The pottery consisted of sherds mainly dating from the Early, Middle Bronze Age (Fig. 20. and 21.) and the Roman period. (Finds of the Roman period are discussed in ch. 13.)

Fig. 20. Early (1) and Middle Bronze Age (2) pottery from plateau H 1. Scale 1: 2.

Fig. 21. Wave-combed pottery from plateau H 1. Photo: Kenneth Lönnqvist 2004.

Several looted tombs, **H 2 - H 12** (see the plan in Fig. 39.), most of which seem to have been cist tombs, were also encountered on the mountain ridge directly above and south from the plateaus (H 1 and H 16) while walking towards Deir ez-Zor. The tombs were encountered along the ridge as far away as the border of the village of El Kharita in the east. These tombs are important because of their vicinity to Tell Kharita beneath. The tombs as listed are:

Site H 2
Cist Tomb/Looted

UTM: 0587937, 3924488; ca. 294 m a.s.l.
Size: max. 9.5 m x 3 m; depth 1.75-1.25 m

This is a cist (Figs. 22. and 23.) apparently dating from the Early Bronze Age according to associated pottery pieces (see Figs. 24. and 25.).

Fig. 22. A looted cist tomb H 2.
Photo from W: Margot Stout Whiting 2004.

Fig. 23. The looted cist tomb H 2.
Photo from W: Margot Stout Whiting 2004.

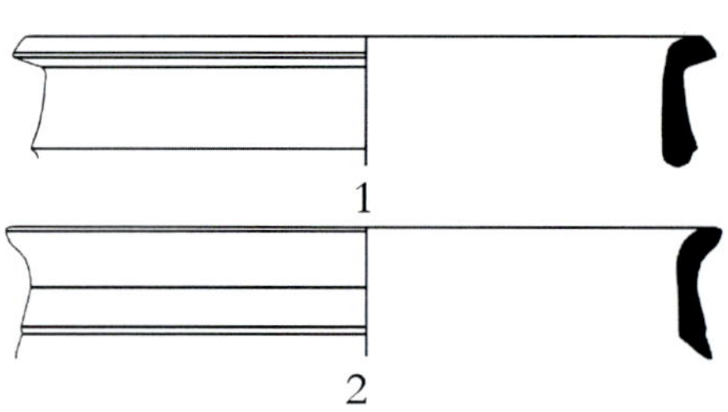

Fig. 24. Early Bronze Age pottery from the cist tomb H 2.
Scale 1: 2.

Fig. 25. Pieces of Early Bronze Age pottery from the looted tomb H 2. Photo: Kenneth Lönnqvist 2006.

Site H 3 a/b
Cist Tombs/Looted

Site H 3a UTM: 0588061, 3924426; 298.4 m a.s.l.
Size: 1.2 m in diam.

Site H 3b UTM: 0588166, 3924423; ca. 298.4 m a.s.l.
Size: 1.4 m in diam.

These are looted cists (Fig. 26.), and no finds to indicate dating were discovered at the site.

Fig. 26. A looted cist tomb H 3a.
Photo from SE: Margot Stout Whiting 2004.

Site H 4

Site H 4
Cist Tomb/Looted

UTM: 0588097, 3924407; 298.4 m a.s.l.
Size: ca. 6.8 m x 8 m, depth 1.2 m

This is another looted cist (Fig. 26.). Some associated pottery (Roman?) was discovered at the site.

Fig. 27. A looted cist tomb H 4.
Photo from SE: Margot Stout Whiting 2004.

Site H 5
Cist Tomb/Looted

UTM: 0588343, 03924358; ca. 302.7 m a.s.l.
Size: 10.4 m x 12 m, depth 1.4 m

This is a looted tomb with a stone slab and with some associated pottery (Roman?).

Site H 6
Cist Tomb/Looted

UTM: 0588678, 3923973
Size: 2.4 m x 2.8 m, depth 1.0 m

This is another looted cist (Fig. 28.). No finds to indicate dating was discovered at the site.

Site H 7
Cist Tomb/Looted

UTM: 0588709, 3923923
Size: 8.5 m x 9.6 m, depth 0.9 m

A cist tomb (Fig. 29.) with rectangular marble/gypsum slabs were found loose in the pit. No small finds to indicate dating was discovered at the site.

Fig. 28. A looted cist tomb H 6.
Photo from S: Margot Stout Whiting 2004.

Fig. 29. A looted cist tomb H 7.
Photo from SE: Margot Stout Whiting 2004.

Fig. 30. A looted cist tomb H 8.
Photo from SW: Margot Stout Whiting 2004.

Site H 8
Cist Tomb/Looted

UTM: 0588709, 3923923
Size: 8.5 m x 9.6 m, depth 0.9 m

This is a looted cist tomb (Fig. 30.) with several loose marble/gypsum slabs. No finds were made to indicate dating.

Site H 9
Cist Tomb/Looted

UTM: 0588728, 3923984
Size: 5.5 m x 5.0 m, depth: 1.30 m

Another tomb with several loose marble/gypsum slabs (Fig. 31.) was found at the edge. No small finds to indicate dating were encountered at the tomb.

Fig. 31. A looted cist tomb H 9.
Photo from NW: Margot Stout Whiting 2004.

Site H 10
Cist Tomb/Looted

UTM: 0588761, 3923953
Size: 4.8 m x 4.6 m, depth 0.7 m

This is a looted cist (Fig. 32.), which was also identified at the edge. No small finds to indicate dating were discovered at this site either.

Fig. 32. A whole in the looted cist tomb H 10. Photo from NW: Margot Stout Whiting 2004.

Site H 11
Cist Tomb/Looted

UTM: 0588796, 3923980
Size: 12 m x 12 m, depth 0.8 m

This is another looted cist tomb (Fig. 33.). No small finds to indicate dating were also discovered at the site.

Fig. 33. A looted cist tomb H 11.
Photo from NE: Margot Stout Whiting 2004.

Site H 12
Cist Tomb/Looted

UTM: 0588729, 3923848
Size: 2.8 m x 1.3 m

This looted cist tomb (Fig. 35.) is associated with several loose and scattered marble/gypsum slabs. Finds consisted of very eroded, reddish and friable pottery hard to date.

Fig. 34. A looted tomb at the graveyard H18 at Tabūs.
Photo from NW: Kenneth Lönnqvist 2004.

Fig. 35. A looted cist tomb H 12.
Photo from SE: Margot Stout Whiting 2004.

Site H 18
Ancient Graveyard/Looted

UTM: 0586868, 3925336, ca. 259 m a.s.l.
Size: ca. 100 x 100 m

This looted graveyard **H 18** is situated southwest of the Late Roman-Byzantine fortress of Tabūs ca. 25 km northwest of Deir ez-Zor. This easternmost graveyard is a rock-cut graveyard with ten plundered tombs (see an example in Fig. 35.). Some of the graves have shaft entrances (5, 6, 7) and some entrances have marble/gypsum slabs. Unlike the graveyards H 17 and H 20 which date from the Late Roman – Byzantine period, this graveyard seems to date from the Bronze Age. No outer structures are visible except bulbs of earth on the ground. The identifiable sizes of the tomb entrances are generally 50 cm x 50 cm. Further studies with ground penetrating radar showed the location of several unlooted graves. An ancient rock-cut road **H 15** winds directly below, on the southeastern side of Tabūs (see the details in ch. 13.). There is no secure evidence to determine the age of this ancient road which might originally date from very early periods, but it seems also to have been used during the Roman period. We are also reminded that Tiglathpilesar I's cylinder from Kal'at Sherkât (K. 1621a + K 13871; Col II) tells that he hewed a way with axes of bronze into a mountain to make a good passage for chariots and troops. The road H 15 leads to a valley or high plateau with a possible unlooted graveyard **J 22** (Fig. 35.) extending to both sides of the road. The area opens behind the fort of Tabūs on Jebel Bishri.

Site J 22
Ancient Graveyard?

UTM: 0586904, 3924695; 288 m a.s.l.
Size: ca. 50 m x 80 m

This gypseous area (Fig. 37.) which might be an ancient grave field consisting of dozens of bulbous formations of which five formed hollowed pits which seemed like looted tombs. Associated pottery dates *inter alia* from Bronze Age and Roman periods (Fig. 36.).

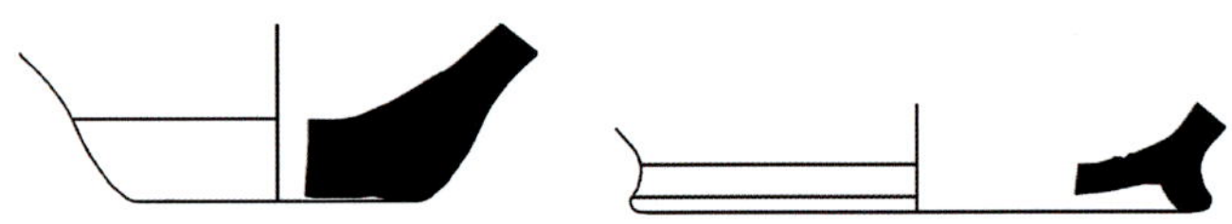

Fig. 36. Early Bronze Age (1) and Roman pottery (2) (western sigillata). Scale 1: 4.

Fig. 37. A possible ancient graveyard J 22. Photo from N: Michael Herles 2005.

Site H 16
Plateau

UTM: 0588791, 3924304
Size: ca. 50 m x 100 m in size.

Directly east of the plateau H 1, another plateau, namely H 16 (Fig. 38.) continued, beneath the graves H 2-H 12. The finds from this plateau consisted pieces of pottery from the same periods as at H 1. The terrace areas of H 1 and H 16 are situated between the fortress of Tabūs and the village of El-Kharita, both actually being inside the village of Kharita - Mustaha.

Fig. 38. A view from plateaus H 1 and H 16 towards Tell Kharita.
The tell is in the middle of the photograph, to the left of the visible school building. Photo from W: Kenneth Lönnqvist 2004.

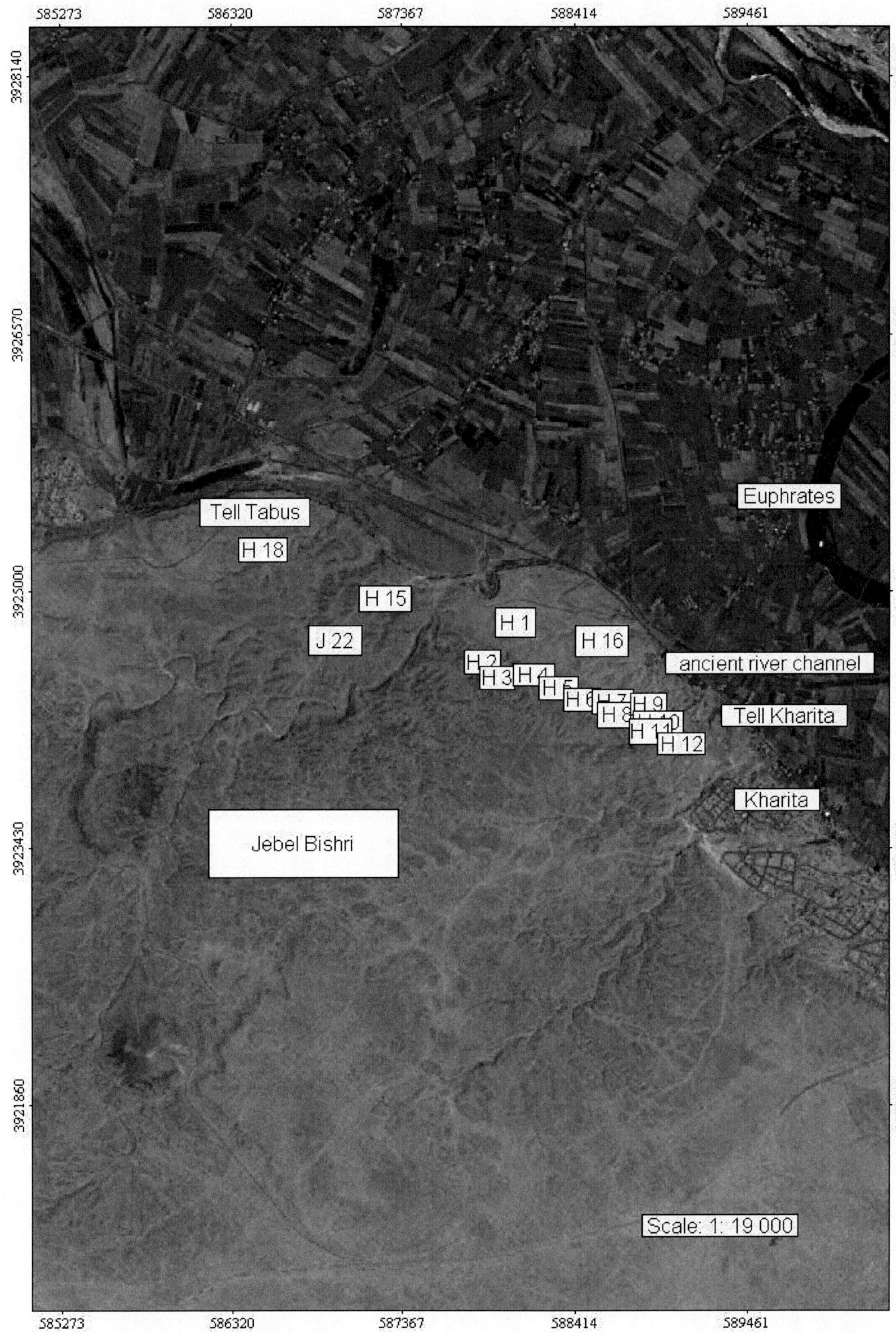

Fig. 39. Tombs, Plateaus and Tell Kharita mapped on a SPOT image. Mapping by Minna Lönnqvist 2008.

Fig. 40. A view over Tell Kharita occupied by a modern Muslim graveyard towards the Euphrates valley. Photo from S: Minna Lönnqvist 2006.

Fig. 41. A profile of Tell Kharita. Photo from S: Minna Lönnqvist 2006.

10.4. A Tell at the Village of El Kharita

Site L 14
Tell - Settlement

UTM: 0589436, 3923932; 197-207 m a.s.l.
Size of the tell on the eastern side of the road:
ca. 150 m x 150 m; according to the SPOT image overall size covers ca. 150 m x 300 m

In 2006 our intention was to reach the area of the plateaus **H 1** and **H 16** surveyed in 2004 approaching from the opposite direction from the village of Ayyāsh in the east towards the village of Kharita to the west. While we ascended onto marl terraces at the foot of Jebel Bishri at the village of Kharita we found a previously unknown tell **L 14** (see Figs 39.-41.) situated just beneath the Early Bronze Age tomb **H 12** (UTM: 0588791, 3924304) already discovered and identified in 2004. The site of the tell is over one kilometre southeast from H 1, but only a few hundred metres southeast of the eastern end of the plateau area **H 16**. The areas like these terrace plateaus and the tombs on higher ridges had provided Early Bronze Age remains, but stronger evidence and accumulation of the period became visible at the tell. Actually the tell L 14 can be seen from H 1 but we did not notice it in 2003-2004 as it is well hidden in the landscape. This newly discovered tell is situated ca. 22 km NW of the centre of Deir ez-Zor on the southern side of the Deir ez-Zor - Aleppo -road in the border zone between the villages Mustaha–Kharita and Kharita;

it is just inside the borders of the village of Kharita. At present it is used as a site for a local Islamic graveyard (see Figs. 40. and 41.) and as grounds for an elementary school. However, its stratification is not deep, two to three metres at maximum as seen in the ascending road profile (see Fig. 41.), but differences in the elevation may vary by ca. 10 m. The site has not apparently been visible to earlier archaeologists, because it is above the Euphrates valley on a marl terrace. From the satellite images it is possible to notice that the way leading from the Deir ez-Zor–Aleppo -road up to the tell and the school divides the tell into two different parts: one to the west and the other to the east, which we only recognized later from a SPOT satellite image.

Some inscriptions were found on the surface of the site. One is apparently written with a rod on a large terracotta tile (see Fig. 42.), the others are smaller unidentified scriptures, possibly potters' marks, on pottery pieces (Figs. 43. and 44.). According to G. Lanfranchi, the way in which the rod was used on the terracotta tile could indicate to the influence of Aramaean scribal techniques. Otherwise the finds from the site consist of Palaeolithic flints, Chalcolithic (for example pieces of Al-Ubaid pottery[1], see Fig. 45.), Early Bronze Age (especially EBIII and EBIV types), Middle Bronze Age and some Iron Age pottery with a few examples of Assyrian and apparently the so-called Aramaean[2] red burnished pottery. The Mesopotamian impact is especially present in Al-Ubaid (Figs. 47. and 48.) and Uruk pottery (Fig. 49.) found at the site, possibly pointing also here to the Sumerian influence in the region. As previously mentioned, the historical sources seem to hint to the Sumerian colonisation of sites along the Euphrates beneath Jebel Bishri (see the previous chapter). Later even the nomadic Amorites could have been their subjects in the region.

Interestingly, the location of the tell just at the foot of the mountain on a Pleistocene terrace and not in the river plain like the other mapped tells between Deir ez-Zor and Raqqa as well as the occurrence of the so-called hole-mouth jars types at the tell indicate that the site might have served sedentarizing nomads coming from Jebel Bishri. EBIV pottery jars are distinctively present and some types of them (Fig. 51.) are rare in the Middle Euphrates region but appear in the tombs of Jericho in Palestine instead (see Kenyon and Holland 1982: 257, Fig. 97). Large *pithoi* vessels both from the Early Bronze Age (Fig. 48.) and the Middle Bronze Age (Fig. 49.) are also among the pottery material, and one of them bear incised marks (Figs. 43. and 44.) - apparently indicating to the size of the vessels (cf., e.g., Duistermaat 2008: 563). So, storing seems to have played an important role in the life of the community. The existence of Iron Age pottery (Figs. 54.) and some inscriptions could provide archaeological evidence pertaining to the occupation of the site during the period of the Assyrian - Aramaean conflicts in the region described in the textual sources.

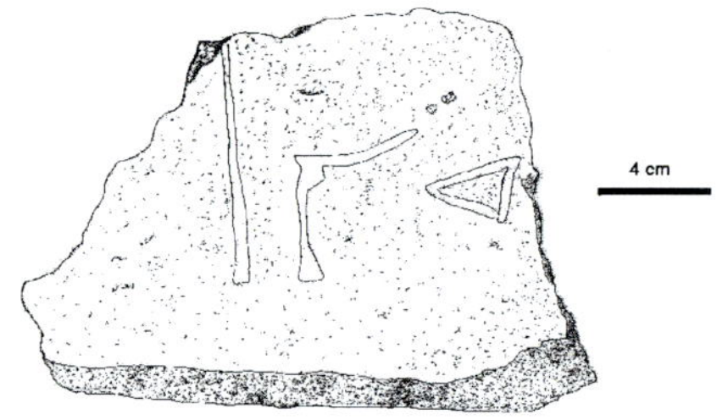

Fig. 42. A fragment of a large inscription written on a terracotta tile. Drawn by Minna Lönnqvist 2006.

Figs. 43. and 44. Inscribed pieces of pottery. Photos: Kenneth Lönnqvist 2006.

[1] See early examples of Al-Ubaid pottery finds from the Ur excavations in Hall and Woolley 1927: 163-165. One piece of Al-Ubaid pottery was also found at Site I 16 at Nadra in the central part of Jebel Bishri.

[2] Cf., e.g., Schwartz 1989: 278.

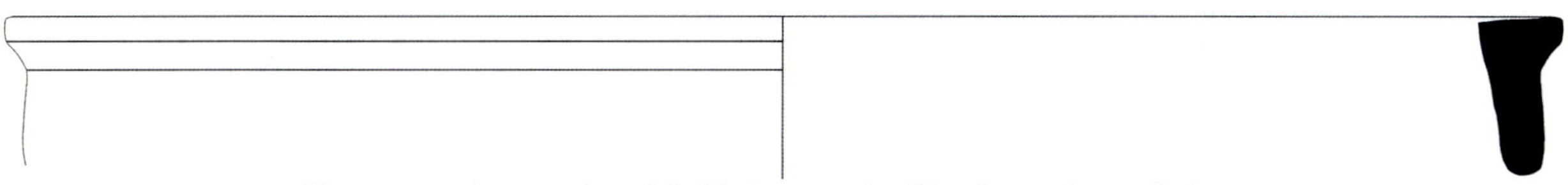

Figs. 45. and 46. A large Middle Bronze Age II pithos with inscribed signs.
Photo: Kenneth Lönnqvist 2006. Scale of the profile drawing: 1:5.

Fig. 47. Al-Ubaid pottery from Tell Kharita.
Photo: Kenneth Lönnqvist 2006.

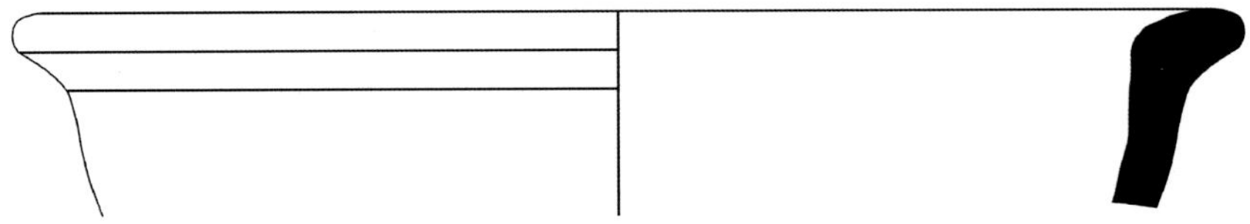

Fig. 48. Al-Ubaid bowl. Scale 1: 3.

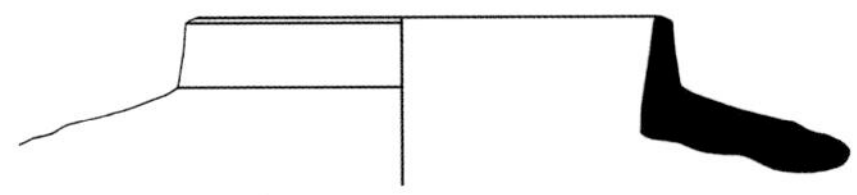

Fig. 49. Late Uruk jar. Scale 1: 4.

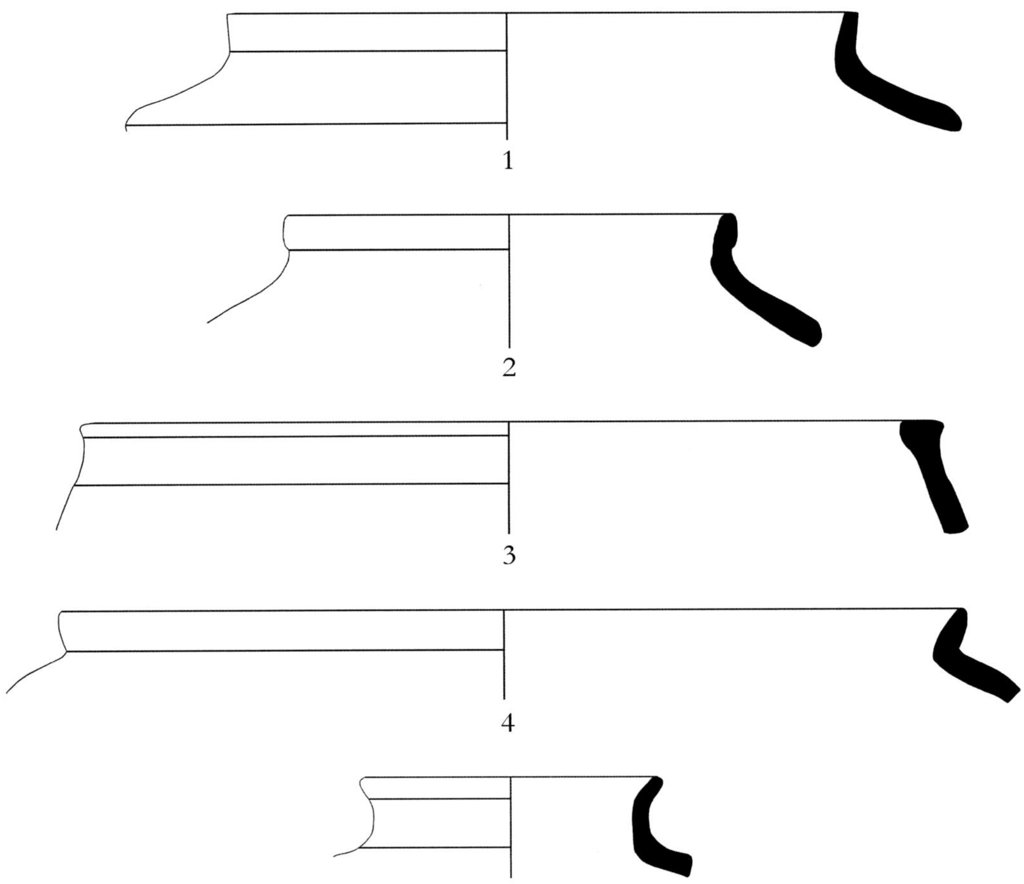

Fig. 50. Early Bronze Age jars. Scale 1:3.

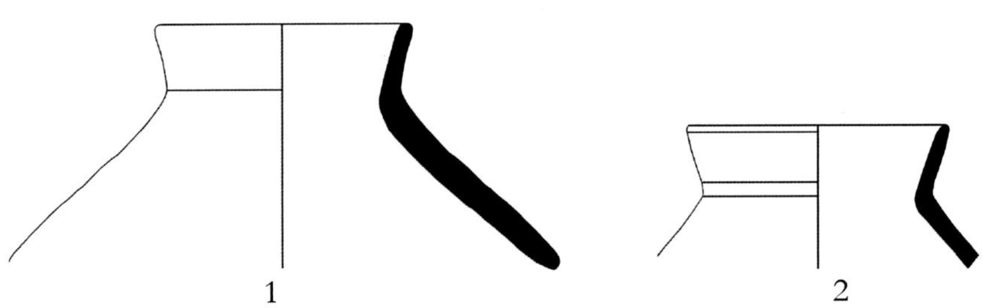

Fig. 51. Early Bronze Age IV jars. Scale 1:3.

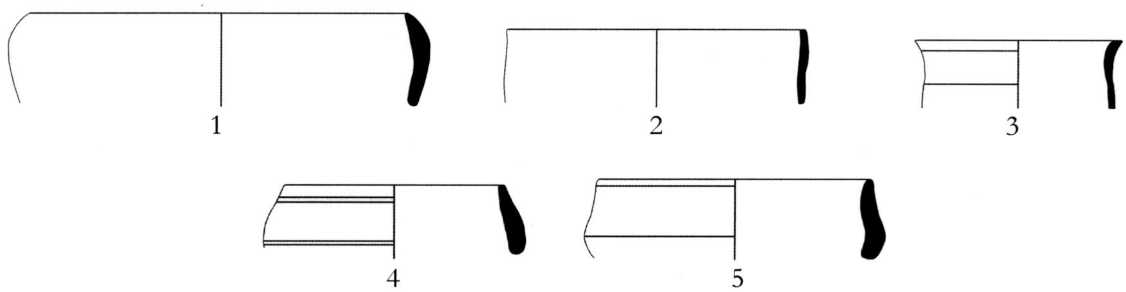

Fig. 52. Early Bronze Age III and IV bowls, cups and Early Bronze Age IV chalices. Scale 1:3.

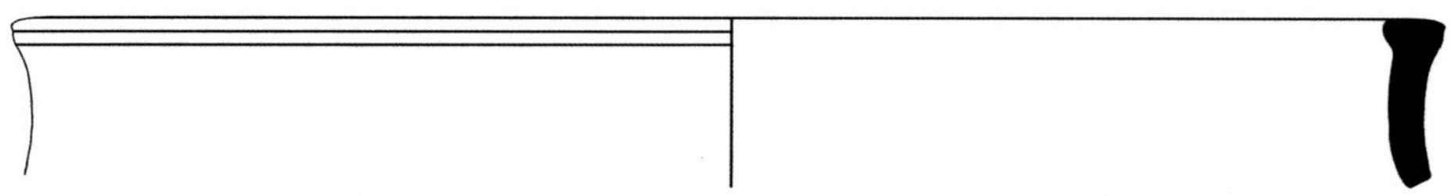

Fig. 53. Middle Bronze Age jar. Scale: 1: 3.

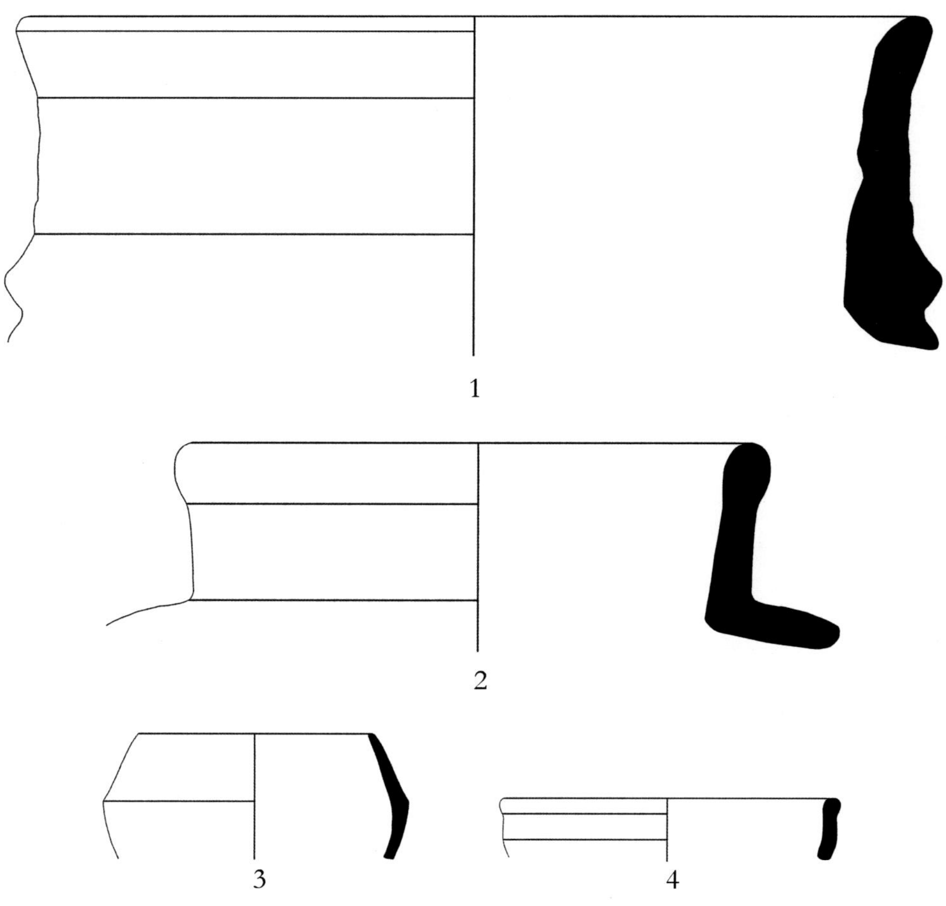

Fig. 54. Iron Age pithoi and bowls. Scale 1: 3.

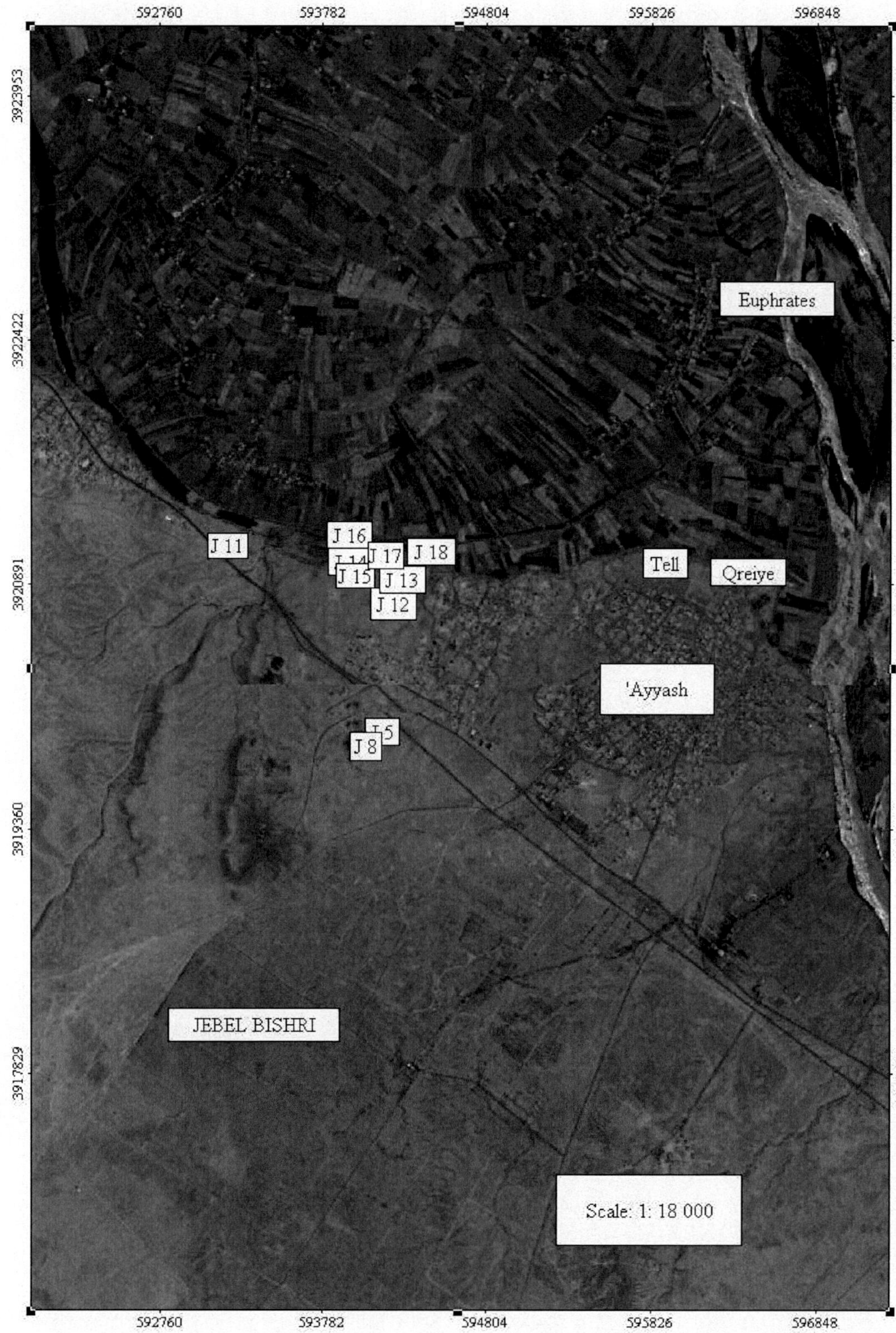

Fig. 55. Chalcolithic and Bronze Age remains near the village of 'Ayyāsh. Mapping by Minna Lönnqvist 2010.

10.5. Bronze Age Tombs and Small Sites between the Villages of 'Ayyāsh and El Kharita

In 2005 (see Fig. 55.) we studied several cairn/*tumuli* and open accumulated sites in the vicinity of 'Ayyāsh. The sites produced Chalcolithic and Bronze Age pottery comparable to the pottery discovered in the Mari–Terqa district (cf., e.g., Rouault 1997; Geyer and Monchambert *et al.* 2003) and Baghouz (cf. Du Mesnil du Buisson 1948). There were graveyards with cairns/*tumuli* **J 14 - J 18** comprising cists, in between Palaeolithic and Roman sites, and intermingling with later Islamic tombs. The tombs had been dug in clayish soil and mixed with cobbles of limestone or basalt 10-20 cm in diameter. Based on the associated pottery these tombs mainly date from the Bronze Age. The sites are as follows:

Site J 5
Plateau, Open Accumulated Site

UTM: 0593827, 3920245; 247 m a.sl.

This site formed a plateau area on the southern side of the Deir ez-Zor—Aleppo -road (Fig. 55.) in the area southwest of the village of 'Ayyāsh. It is an open accumulated site producing pottery from different periods, mainly sherds from the Roman era but also some pieces from the Bronze Age. The Bronze Age pottery had probably been washed down the mountain slope by rain and water currents, because we later observed that cairns/*tumuli* of J 8 situated on a hillock next to J 5.

Site J 8
Cairns/Tumuli

UTM: 0593324, 3920353; 255 m a.s.l.
Sizes of the cairns: ca. 8.5 m and 10.4 m in diam.

Three cairns/*tumuli* were discovered on a hillock **J 8** above the plateau **J 5** (see Fig. 55.). The area of the tombs produced Chalcolithic and Bronze Age pottery (Fig. 56.).

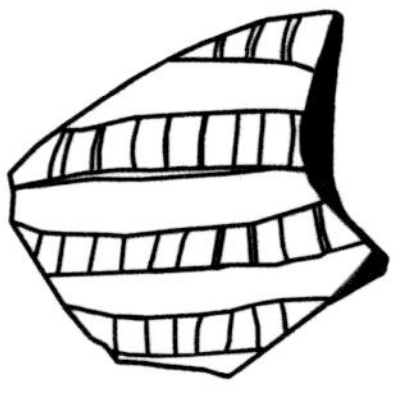

Fig. 56. A piece of Chalcolithic pottery from site J 8. Scale: 1:1. Drawing by Michael Herles 2005.

Site J 11
Open Accumulated Site and Bottom of House

UTM: 0593185, 3921140; 226 m a.sl.
Size of the site: Undefined
Size of the house: 6. 50 m x 8. 70 m

This site was discovered on a low ridge of Jebel Bishri facing the Euphrates and situated between the villages of Ayyāsh and 'Ain Abu Jimā on the northern side of the Deir ez-Zor–Aleppo –road. Beside Palaeolithic flints and Roman pottery, the remains of a rectangular stone structure, associated with some Early Bronze Age pottery (Fig. 57.), were discovered.

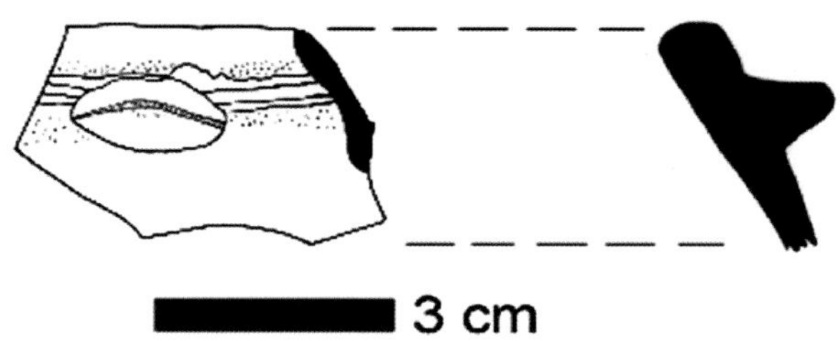

Fig. 57. A piece of an Early Bronze bowl with a ledge handle from site J 11.

Site J 12
Cairn/Tumulus Field

UTM: 0593991, 3920688; 244 m a.s.l.

This site was a gravefield consisting of elongated cairns generally orientated in SE-NW lines. These tombs may be early Islamic ones but the associated pottery dated *inter alia* from the Bronze Age (Fig. 58.). All the pottery needed not be from the time of the graves. There were more modern Islamic graves among these older cairns but the orientation of them differed.

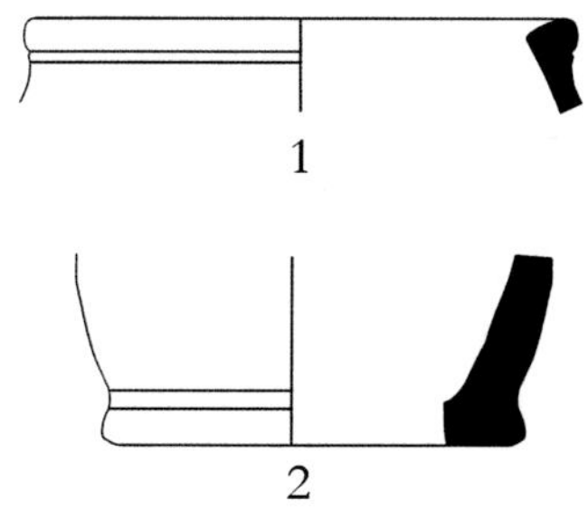

Fig. 58. Early Bronze Age and Middle Bronze Age II pottery from site J 12. Scale 1:4.

Site J 13
Open Accumulated Site

UTM: 0594016, 3920882; 256 m a.s.l.

This was an open accumulated site ca. 500 m northwest of Ayyāsh situating on the northern side of the Deir ez-Zor–Aleppo -road. It is a clayish area on the bedrock with scattered basalt boulders. The site produced Early Bronze Age pottery (Fig. 59.)

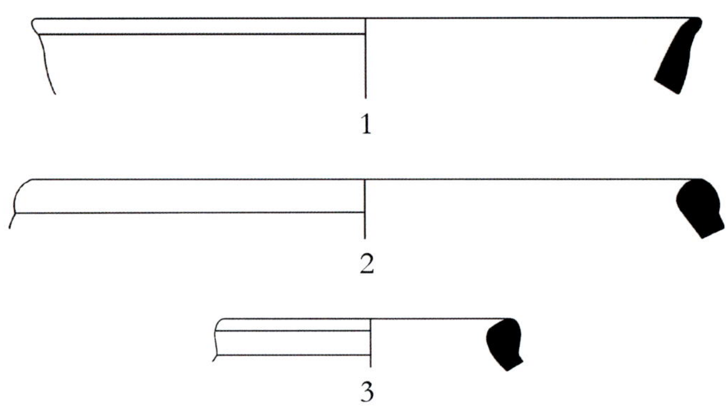

Fig. 59. Early Bronze Age pottery from site J 13. Scale 1:3.

Site J 14
Cairn/Tumulus

UTM: 0593793, 3920928; 250 m a.s.l.
Size: ca. 20 m in diam.

This is an unlooted cairn/*tumulus* tomb **J 14** (Fig. 60.) near to the site **J 13** on the northern side of the Deir ez-Zor–Aleppo -road (Fig. 55.). The area produced Bronze Age pottery (Fig. 61.).

Fig. 60. Site J 14. Photo from S: Michael Herles 2005.

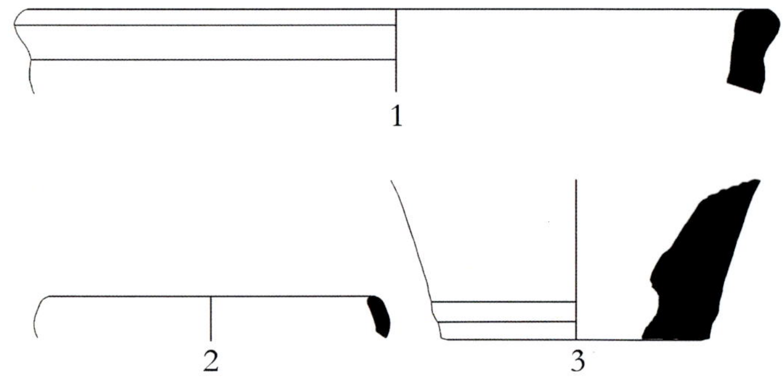

Fig. 61. Bronze Age pottery from site J 14. Scale 1: 4.

Site J 15
Cairn/Tumulus

UTM: 0593887, 3920922; 242 m a.s.l.
Size: max. 17 m in diam.

This is another unlooted cairn/*tumulus* **J 15** (Fig. 62.) next to the tomb **J 14** on the northern side of the Deir ez-Zor–Aleppo -road (Fig. 55.). The area produced Bronze Age pottery as well.

Fig. 62. Site J 15. Photo from W: Michael Herles 2005.

Fig. 63. A piece of Early Bronze Age IV pottery from site J 15. Scale 1:4.

Site J 16
Cairn/Tumulus/Looted

UTM: 0593927, 3920971; 238 m a.s.l.
Size: max. 15 m in diam.

Another cairn/*tumulus* **J 16** was found in the vicinity (Fig. 55.) of the tombs J 14 and J 15, but this tomb had been opened and looted including a large marble/gypsum cist consisting of two preserved marble/gypsum slabs (size 1.8 m x 0.3 m and 1.1 m x 0.3 m) *in situ,* and some broken fragments had been thrown out of the cist (Fig. 64.). The associated pottery dates from the transition of the Middle to the Late Bronze Age (Fig. 65.).

Fig. 64. Remains of a looted and broken cist in cairn/tumulus J 16 west of 'Ayyāsh. Photo from W: Michael Herles 2005.

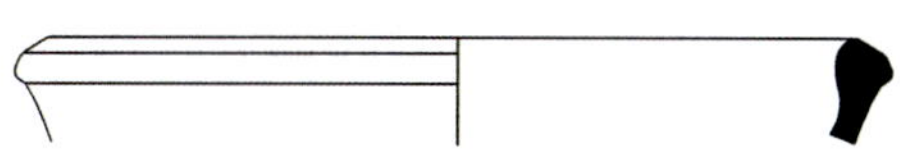

Fig. 65. Middle/Late Bronze Age pottery from tomb J 16. Scale 1: 4.

Site J 17 Cairn/Tumulus/ Looted

UTM: 0593936, 3920984; 237 m a.s.l.
Size: ca. 10 m in diam.

In the vicinity of the site **J 16** there was another cairn/*tumulus* **J 17** (Fig. 55.) which had been looted and (Fig. 66.) which may be part of the same gravefield as J 16. Associated pottery dates from the late Middle Bronze Age II period (Fig. 67.) (parallels Mardikh IIIB in Ebla, cf. also Tell Nebi Mend, see, e.g. Bourke 1993).

Fig. 66. A looted cairn/tumulus J 17 in the graveyard to the west from the village of Ayyāsh. Photo from W: Michael Herles 2005.

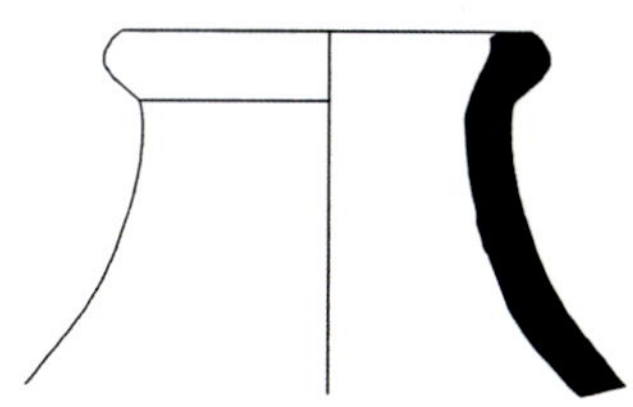

Fig. 67. A piece of a neck from a pottery jar from the Middle Bronze Age II period from cairn/tumulus Site J 17. Scale 1: 4.

Site J 18
Cairn/Tumulus Field /Looted

UTM: 0594046, 3921018
Altitude: 239 m a.s.l.

This site **J 18** (Figs. 55. and 68.) is another contemporaneous gravefield, or it belongs to the same one as J 16 and J 17. Associated pottery dates from the Middle Bronze Age.

Fig. 68. A looted graveyard J 18 west of 'Ayyāsh. Photo from SE: Michael Herles 2005.

Open accumulated sites were encountered on a limestone/gypsum terrace 14-14.5 km to the west of Deir ez-Zor on the southern side of the Deir ez-Zor – Aleppo -road. These sites are associated with the village of 'Ain Abu Jimā. The sites are **J 19** (UTM: 0593221, 3920934; 228 m a.sl.), **J 20** (UTM: 050592654, 3821375: 238 m a.sl.) and **J 21** (UTM: 0592577, 3921370: 245 m a.s.l.), and they produced pottery from the Middle Bronze Age II period.

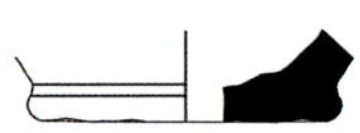

Fig. 69. A pottery piece of Middle Bronze Age II pottery from site J 19. Scale 1: 4.

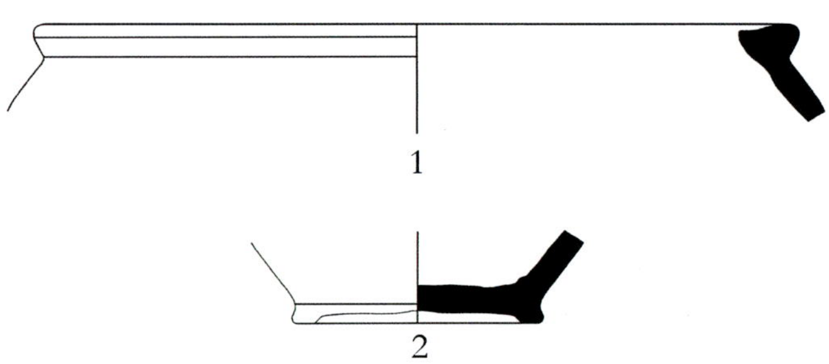

Fig. 70. Middle Bronze Age II Pottery pieces from site J 20. Scale 1: 4

10.6. The Surveyed Sites at the Foot of Jebel Bishri on the Middle Euphrates Considered

Prospecting with satellite images already shows a spatial difference in the landscape of the Euphrates compared to the mountain proper of Jebel Bishri. Also, the finds change in their character. Apart from the outer ridges, the mountain proper lacks monumental sites and clearly represents small-scale societies, the majority of the surface finds consisting of flints. There is a clear distinction between monument and non-monument areas. In the Euphrates area tells indicate sedentism and more pottery is recovered, generally concentrating on tells. Pottery production is known to be especially tied with sedentism, although pottery is not absent in pastoral nomadic cultures either. However, pastoralists often use perishable materials such as skin and basket which, as earlier indicated, do not leave much signs in the archaeological record.

The sites outside the systematic studies in transects on the Euphrates were Tell Tibne G 1, sites G 4 and G 5 and Tell Kharita L 14. They were studied as separate entities. Tell Tibne and Tell Kharita offer evidence of the spread of the Mesopotamian Al-Ubaid and early Uruk cultures to the Middle Euphrates in the area between Deir ez-Zor and Halabiya, and thus the rise of early urbanism in the region. A Chalcolithic site called Tell Zaidan (Site 16), also providing Al-Ubaid and Halaf pottery, was found on the Balikh, but during his 1983 survey K. Kohlmeyer (1984: 109) found other sites (his Sites 55, 61, 96, 99) providing Al-Ubaid evidence up the stream on the left bank of the Middle Euphrates 6 km east of Raqqa. On the left bank he also identified more Uruk sites (Site 55). In addition, El Kowm 2 – Caracol has provided evidence of occupation during the Uruk period (Cauvin and Stordeur 1985: 191-197). The Syro-Spanish team led by J.L. Montero Fenollós (see http://aidifondation.org/English/Syrian_spanish_mission.htm) has also traced considerable evidence of Uruk impact in the area of the gorge of Halabiya, namely at the site of Tell Humeida on the left bank of the Euphrates.

The Al-Ubaid pottery recovered during our surveys at Tell Kharita and Nadra (ch. 6.) as well as the Uruk pottery at Tell Kharita and Tell Tibne provide new evidence of the activities associated with the southern Mesopotamian heartland in the respective periods on the right bank of the Middle Euphrates between the Habur and the Balikh rivers. It may therefore be that Tell Tibne already served as an Uruk station while approached from Qraya and Lower Mesopotamia (cf. Algaze 1993: 48-). The transportation of stone such as marble/alabaster (= gypsum) could have taken place from nearby quarries to Sumerian centres (cf. the cuneiform evidence in ch. 9.). The site has also produced Early Bronze Age pottery which finds probable cultural contacts with Ebla as well as Emar.

The prominent impact of the Early Bronze Age (see Fig. 71.) was recognized along the Euphrates. All the recorded sedentary sites provided Early Bronze Age pottery. (See more on the early urbanization on the Middle Euphrates in Cooper 2006.) However, the Early Bronze Age sites on the Euphrates side of Jebel Bishri are still relatively small compared to the Jezira at the time. They also reveal several features which bear evidence of the continuing importance of pastoral economy. Interestingly, Early Bronze Age has been seen as the time of significant crop-livestock relationship (Levy 1983: 32 and Rosen 1988: 503). We may venture to ask whether this is the beginning of archaeologically visible symbiosis between pastoral and agricultural life beneath Jebel Bishri. Compared to the area of the Balikh (see Akkermans 1993) that attests the beginning of the symbiosis along the rise of villages already in the Neolithic period the Jebel Bishri region shows a later development dating from the Early Bronze Age. This may indicate that the area of Jebel Bishri area was under more pronounced pastoral lifestyle in the Neolithic and Chalcolithic periods.

The question whether Jebel Bishri provides evidence that enclosed nomadism or *transhumant* pastoralism already existed in the Early Bronze Age in the region - if one uses M. Rowton's nomenclature originally associated with the Middle Bronze Age data (see ch. 9.) – has been briefly touched upon in the previous chapters. It is interesting that the stone enclosures including animal pens in the central higher grounds (see ch. 6. and 7.) with good grazing potential were recorded on the mountain proper. They provided associated finds *inter alia* from the Chalcolithic and Bronze Ages. The immediate connection between Tell Kharita and Jebel Bishri seems to be unique compared to the other tells beneath Jebel Bishri on the Middle Euphrates. The position of this tell is comparable with the present habitation of village pastoralists on the marl terraces of the Euphrates. This site is topologically so closely connected with the mountain that it seems to reflect the Bronze Age highland-lowland interaction at its best, which is also visible in its material remains. There are nomadic features in the pottery, including hole-mouth jars and elongated EBIV jars. Because of the occurrence of large *pithoi* and other vessels, storage seems to have played a particular role at this site. These features reflect a gradual shift from nomadism towards sedentism or pastoral *transhumant* movement from base-villages to the mountain.

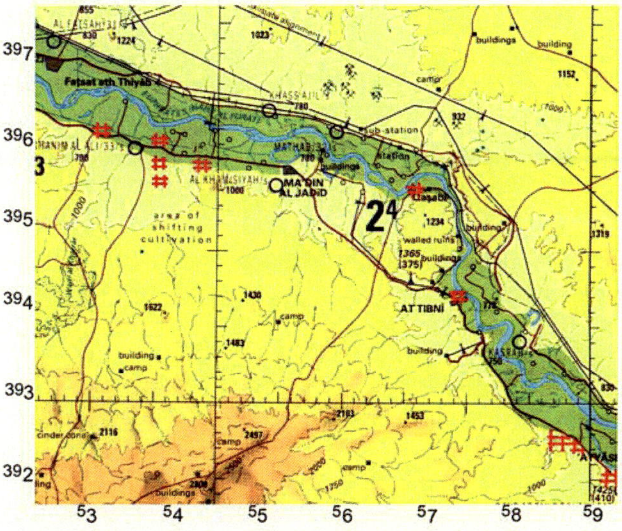

Fig. 71. Early Bronze Age sites on the Euphrates side of Jebel Bishri. Mapping Kenneth Lönnqvist 2009. Basemap: © MOD Crown Copyright, 2010 published with the permission of the Controller of HMSO.

As mentioned previously, the Mari archives from the Middle Bronze Age connect the neighbourhood of Jebel Bishri with the Benjaminite and the Sutean pastoral tribes (see ch. 9). The caravan road along the Euphrates must have needed caravanserai sites, and the Suteans, active nomadic people in the region, served as caravan escorts according to the Mari texts. (See more on the Bronze Age trade routes and contacts in the area of Jebel Bishri in Josephson Hesse 2008 and Charpin 2010).

Some of the archaeological evidence may be even more clearly connected with the Amorite impact in the area. Tell Tibne especially has produced interesting material such as the Middle Bronze Age pottery comparable with the Mari of the Amorite Age. Because of its location and the associated Middle Bronze Age as well as Classical pottery, J.-M. Durand (personal communication) has suggested that the site of Tell Tibne could be Dûr Yahdun-Lim's port Gânibatum mentioned in a recently published tablet A.859 from Mari. Interestingly Gânibatum served in stone transportation (see Durand 2005: 66-67), which accords well with the location near the apparent quarries and salt mines. Judging from the convenient location, the site may also have served as a caravan station on the route from Mari to Yamhad (Aleppo) along the Euphrates. Compared to Tell Kharita, the finds from Tell Tibne seem also to present a more organised community, which even under the nomadic influence, was serving settled people as a transition station or a port of the Mesopotamian region in historical times.

The types of cists, cairns/*tumuli* and shaft tombs on the Euphrates side of Jebel Bishri, including those from the graveyard of Abu Hamed (Falb *et al.* 2005), are typical tomb types among the EBIV sites in Syria-Palestine. This applies not only to the pottery types but also to the types of daggers as well. (Cf. Lönnqvist 2000). Furthermore, the finds from the Early Bronze Age tombs at Abu Hamed include a triangular dagger and toggle pins, the finds comparable to the Early Bronze Age/Middle Bronze Age cairns/*tumuli* discovered in the most recent excavations in the northwestern area of Jebel Bishri by a Syro-Japanese team (Fujii and Adachi 2010). Earlier it has been argued that the triangular dagger type is the type fossil of the Amorites (cf. Lönnqvist 2000: 290-297). In the Mari texts of the Amorite era there is a special designation for cairn/*tumulus* tombs namely *humûsum*/*humûsim*, which were constructed as commemorative "monuments" over the deceased persons (Durand 2005: 130). During our survey in the 'Ayyāsh–'Ain Abu-Jimā district several cairns/*tumuli* produced Middle Bronze Age II pottery which is comparable to the pottery of the Mari–Terqa district and Baghouz. The area is near the possible site of Dûr Yahdun-Lim and was under the Mari influence at the time as shown by J.-M. Durand (2010).

Cultic evidence associated with the Mesopotamian practices is testified to by the finds from plateau H 1: the inlaid eye of a cultic statue/votive eye apparently dating from the Early Bronze Age and a possible fragment of a cultic statue. There could have been a temple in the region, as indicated in the cuneiform sources. We may also reconsider the references of Jebel Bishri as a deified mountain further (see the cuneiform sources in the previous ch. 9.): the mountain could have drawn several groups of people to build their ancestral burial grounds there. This seems to have been partly the function of the island of Bahrain with its countless *tumuli*. The area has been identified with Dilmun/Tilmun in the cuneiform sources dating from the Bronze Age (2400-1700 B.C.). (Breuil 1999: 49-55.) It may be noted that some Ur traders heading from Mesopotamia to Dilmun bore Amorite names. In addition, the population of the area included Amorites, and expeditions and caravans from Mari are attested in the region (see, e.g. van Koppen 1997).

F. Højlund's *et al.* (2008) studies of the *tumulus* gravefields in Bahrain have concentrated on the question of social complexity. The tombs dating from the UR III period to the Isin-Larsa period have several features in common with our cairns that had cists inside. In the case of Bahrain, Højlund observed that the *tumulus* gravefields were reflecting kinship groups based on village structure. The division between fresh water and arable land, and non-arable land, for locating the tombs was another central feature in his analyses. Højlund associates the availability of fresh water and arable land with settlements, and *tumulus* fields associated with the desert landscape: non-arable land + cemetery *contra* arable land + water + village. As far as the arable land of the Euphrates river valley and the oases is concerned, in the case of Jebel Bishri this seems to apply to our gravefield areas as well – they were always located on the mountain ridges or slopes in the desert-steppe regions outside the arable lands. Ancient Bronze Age villages or towns appear on the river plain (Tell Hamadin, Tell Grnaia/Tell Ghanem Ali, Tell Khamisiyah/Tell Moukhla and Tell Tibne) or on mountain-based terraces (Qasabi, Tell Kharita and Tell 'Ayyāsh). However, in the case of Jebel Bishri the tombs are obviously connected with kinship based society, although not necessarily village structure; they seem to reflect tribal societies and in some cases more or less mobile chiefdoms (see Lönnqvist *et al.* 2010). In addition, the cemetery of Abu Hamed (Falb *et al.* 2005) near Tell Grnaia/Ghanem Ali may reflect a village organisation of agro-pastoralists (cf. Lönnqvist *et al.* 2010) - J.-W. Meyer (2010) suggests the groups were mere pastoralists.

The question of change and continuity in West Semitic pastoral populations in the area of Jebel Bishri is a challenging one. This area, associated in the cuneiform texts of the third and second millennium B.C. with the Amorites, appears to have been later populated by the Aramaeans (see also discussion in Zarins 1989: 133). The texts of the Akkadian rulers from the third millennium B.C. boasted about defeating Amorites at the foot of the mountain. This same landscape became a scene of the Assyrian-Aramaean hostilities: Tiglathpilesar conquered towns at the foot of Jebel Bishri. (See the cuneiform evidence in the previous ch. 9.). In the light of the present evidence there are at least eight tells that can be associated with the foot of the mountain, and in two of them, namely at Tell Tibne and Tell Kharita, we have encountered Late Bronze Age and Iron Age pottery evidence. The tell at 'Ayyāsh still awaits publication of the results from the Syro-German expedition.

Topographically Tell Kharita could match the site of Kharidi occurring in Assurnasirpal II's Annals describing the punishing campaign of the Assyrians against the Aramaeans at the foot of Jebel Bishri. The conquests of the towns of Khindanu, Kharidi and Kipinu took places downstream from the gorge of Halabiya (see, e.g. Maspero 1903; see also Musil 1927: 207-210; Grayson 1991: 214-215). However, as mentioned in the previous chapter, since A. Musil's identification Kharidi has been located in modern scholarship to Khirbet ed-Diniye (= Old Babylonian Haradum) on the Iraqi side (Liverani 2, 1992: 66; Geyer and Monchambert *et al.* 2003: 263, note 169). The reasons for our alternative suggestion are the following: 1) the location of the tell Kharita after the "Euphrates narrow", i.e., Halabiya with an old river channel beneath the tell, 2) archaeological remains from Tell Kharita, *inter alia* pottery, indicating Assyrian and Aramaean impact, 3) the name of the village "Kharita" is close to the site name "Kharidi", and modern toponyms may carry remnants from the older ones and finally 4) ancient Holocene river channels of the Euphrates can be found just below Tell Kharita, although relatively far away today, offering places for troops to go ashore easily. (See Lönnqvist *et al.* 2007 and Darmark 2008).

Bibliography

Akkermans, P.M.M.G. (1993) *Villages in the Steppe: Later Neolithic Settlement and Subsistence in the Balikh Valley, Northern Syria.* International Monographs in Prehistory. Archaeological Series 5. Ann Arbor, Michigan: International Monographs in Prehistory.

Algaze, G. (1993) *The Uruk World System: The Dynamics of Expansion of Early Mesopotamian Civilization.* Chicago: University of Chicago Press.

Al-Maqdissi, M. and Ohnuma, K. (2008) Preliminary Reports of the Syria-Japan Archaeological Joint Research in the Region of Ar-Raqqa, Syria, 2007, in *Al-Rāfidān,* Vol. XXIX, pp. 117-196.

Besançon, J. and Sanlaville, P. (1981) Aperçu Géomorphologique sur la Vallée de l' Euphrate Syrien, in *Paléorient,* vol. 7/2, pp. 5-18.

Bourke, S.J. (1993) The Transition from the Middle to the Late Bronze Age in Syria: The Evidence from Tell Nebi Mend, in *The Levant,* Vol. XXV, pp. 155-195.

Breuil, J.-Y. (1999) "The world's largest prehistoric cemetery…" in *Bahrain: The Civilisation of the Two Seas, From Dilmun to Tylos,* Ghent: Institut du Monde Arabe, Paris, pp. 49-55.

Buccellati, G. (1990) Salt at the Dawn of History: the Case of the Bevelled-Rim Bowls, in *Resurrecting the Past: A Joint Tribute to Adnan Bounni,* ed. by Matthiae, P., Van Loon, M. and Weiss, H., Publications de l'Institut historique-archéologique néerlandais de Stamboul, Vol. 67, Leiden: Nederlands Historisch-Archäologisch Instituut te Istanbul, pp. 17-40.

Cauvin, J. and Stordeur, D. (1985) Une Occupation d'Époque Uruk en Palmyrène: le Niveau Supérieur d'El Kowm 2 - Caracol, in *Cahiers de L'Euphrate,* 4, pp. 191-205.

Charpin, D. (2010) The Desert Routes around the Djebel Bishri and the Sutean Nomads according to the Mari Archives, in *International Symposium, Formation of Tribal Communities, Integrated Research in the Middle Euphrates, Syria, November 21-23, 2009, Al-Rafidan,* Special issue, pp. 239-245.

Clancier, P. (2006) Le moyen Euphrate de l'implantation des Araméens à la période romaine, in *Studia Euphratica. Le moyen Euphrate iraquien révélé par les fouilles de sauvetage de Haditha,* ed. by Ch. Kepinski, Ch., O. Lecomte, O. et Tenu, A., Paris: De Boccard, pp. 247-289.

Cooper, L. (2006) *Early Urbanism on the Syrian Euphrates.* London: Routledge.

Darmark, K. (2008) The Archaeological Potential of the Assyro-Aramaean Hostility on the Euphrates Side of Jebel Bishri, Implications of Battlefield Archaeology, in *Jebel Bishri in Context: Introduction to the Archaeological Studies and the Neighbourhood of Jebel Bishri in Central Syria,* Proceedings of a Nordic Research Training Seminar in Syria, May 2004, ed. by Lönnqvist, M., BAR International Series, No. 1817, pp. 49-58.

Du Mesnil du Buisson, R. (1948) *Baghouz: L'Ancienne Consôté.* Le Tell Archaique et la Necropole de l'Âge du Bronze. Leiden: Brill.

Duistermaat, K. (2008) *The Pots and Potters of Assyria. Technology and Organisation of Production, Ceramic Sequence and Vessel Function at Late Bronze Age Tell Sabi Abyad, Syria.* Turnhout: Brepols.

Durand, J.-M. (2005) *Le Culte des pierres et les monuments commémmoratifs en Syrie amorrite,* Florilegium manianum VIII, Mémoires de N.A.B.U. 9. Paris: SEPOA.

Durand, J.-M. (2010) The Banks of the Euphrates along the Bishri, in *International Symposium, Formation of Tribal Communities, Integrated Research in the Middle Euphrates, Syria, November 21-23, 2009, Al-Rafidan,* Special issue, pp. 253-262.

Falb, C., Krasnik, K., Meyer, J.-W. and Vila, E. (2005) *Der Friedhof vom Abu Hamed: Gräber des 3. Jahrtausends v. Chr. Im syrischen Euphrattal.* Schriften zur vorderasiatischen Archäologie. Herausgegeben von

Winfried Orthmann. Band 8. Saarwellingen: Saarländische Druckerei & Verlag.

Finkbeiner, U. (1999-2000) with the Contribution by Rothmund, A., Emar & Balis 1996-1998, Preliminary Report of the Joint Syrian-German Excavations with the Collaboration of Princeton University, in *Berytus*, Vol. 44, pp. 5-33.

Finkbeiner, U. (2001) Emar 1999 – Bericht über die 3. Kampagne der syrisch-deutschen Ausgrabungen, Mit Beiträgen von H. Attoura, B. Faist, U. König, F. Sakal und F. Starke, in *Baghdader Mitteilungen*, Deutsches Archäologisches Institut Orient-Abteilung, Band 32, pp. 42-110.

Finkbeiner, U. (2002) Emar 2001 - Bericht über die 4. Kampagne der syrisch-deutschen Ausgrabungen, Mit Beiträgen von H. Attoura, W. Eixler und unter Mitarbeit von F. Sakal, in *Baghdader Mitteilungen*, Deutsches Archäologisches Institut Orient-Abteilung, Band 33, pp. 109-146.

Finkbeiner, U. and Sakal, F. (2003) Emar 2002 – Bericht über die 5. Kampagne der syrisch-deutschen Ausgrabungen, Mit Beiträgen von H. Attoura, W. Eixler, D. Lineberry, F. Starke and S. Wilhelm, in *Baghdader Mitteilungen*, Deutsches Archäologisches Institut Orient-Abteilung, Band 34, pp. 9-100.

Fujii, S., Adachi, T. (2010) An Archaeological Investigation of Bronze Age Cairn Field on the Northern Flank of Mt. Bishri, in *International Symposium, Formation of Tribal Communities, Integrated Research in the Middle Euphrates, Syria, November 21-23, 2009, Al-Rafidan*, Special issue, pp. 61-77.

Geyer, B. and Monchambert, J.-Y. avec Besançon, J. and Cogeuegniot, É. (2003*) La Basse Vallée de l'Euphrate Syrien du Néolithique à l'avenement de l'Islam: Mission Archéologique de Mari. Tome VI.* Vol.1: Texte, Vol 2: Annexes. Institut Français du Proche-Orient. Bibliothèque Archéologique et Historique T. 166. Beyrouth: Institut Français du Proche-Orient.

Grayson, A.K. (1991) *Assyrian Rulers of the Early First Millennium BC.* Vol. I (1114-859 BC). The Royal Inscriptions of Mesopotamia. Assyrian Period. Vol. 2. Toronto: University of Toronto Press.

Hall, H.R. and Woolley, C.L. (1927) *Ur Excavations.* Vol. 1. Al-Ubaid. Publications of the Joint Expedition of the British Museum and of the Museum of the University of Pennsylvania to Mesopotamia. Oxford: Oxford University Press.

Héraud, C. (1922/1988) *Une Mission de Reconnaissance de l' Euphrate en 1922.* Damas: Institut Français de Damas.

Højlund, F. with the contributions by Johansen, K.L., Kjaerum, P., Kveiborg, J. and Lauren, S.T. (2008) *The Burial Mounds of Bahrain, Social Complexity in Early Dilmun.* Moesgaard Museum. Ministry of Information, Bahrain. Jutland Archaeological Society Publications vol. 58. Aarhus: Aarhus University Press.

Josephson Hesse, K. (2008) Aspects of Intercultural Relations in the Old Babylonian Period: The Contacts between Mari and Hazor as a Case Study, in *Jebel Bishri in Context: Introduction to the Archaeological Studies and the Neighbourhood of Jebel Bishri in Central Syria*, Proceedings of a Nordic Research Training Seminar in Syria, May 2004, ed. by Lönnqvist, M., BAR International Series, No. 1817, Oxford: Archaeopress, pp. 31-47.

Kenyon, K.M. and Holland, T.A. (1982) *Excavations at Jericho.* Vol. IV. The Pottery Type Series and Other Finds. London: British School of Archaeology in Jerusalem.

Kohlmeyer, K. (1984) Euphrat-Survey, Sie mit Mitteln der Gerda Henkel Stiftung durchgeführte archäologische Geländebegehung im syrischen Euphrattal, in *Mitteilungen der Deutschen Orient-Gesellschaft,* Vol. 116, pp. 95-118.

Kohlmeyer, K. (1986) Euphrat-Survey 1984, Zweiter Vorbericht über die Mitteln der Gerda Henkel Stiftung duchgeführte archäologische Geländebegehung im syrischen Euphrattal, *Mitteilungen der Deutschen Orient-Gesellschaft,* Vol. 118, pp. 51-64.

Lauffray, J. (1983) *Halabiyya – Zenobia: Place Forte du Limes Oriental et La Haute-Mésopotamie au VIe Siècle.* Tome I: Les duchés frontaliers de Mésopotamie et les fortifications de Zenobia. Bibliothèque Archéologique et Historique T. CXIX. Institut français d'archéologie du Proche-Orient, Beyrouth – Damas – Amman. Paris: Paul Geuthner.

Lauffray, J. (1991) *Halabiyya – Zenobia: Place Forte du Limes Oriental et La Haute-Mésopotamie au VIe Siècle.* Tome II: L'architecture publique, religieuse, privée et funéraire. Bibliothèque Archéologique et Historique T. CXXXVIII. Institut français d'archéologie du Proche-Orient, Beyrouth – Damas – Amman. Paris: Paul Geuthner.

Levy, T. (1983) The Emergence of Specialised Pastoral Nomadism in the Southern Levant, in *World Archaeology*, Vol. XV, pp. 15-36.

Liverani, M. (1992) *Studies on the Annals of Ashurnasirpal II. 2: Topographical Analysis*. Guaderni di Geografica Storica, 4. Roma: Università di Roma "La Sapienza", Dipartimento di Scienze storiche, archaeologiche e antropologiche dell' Antichità.

Lloyd, S. (1984) *The Archaeology of Mesopotamia from the Old Stone Age to the Persian Conquest*, Revised edition. The World of Archaeology. London: Thames and Hudson.

Lönnqvist, M. (2000) *Between Nomadism and Sedentism: Amorites from the Perspective of Contextual Archaeology*. Helsinki: Juutiprint.

Lönnqvist, M. with the contributions by Törmä, M., Nuñez, M., Lönnqvist, K. Stout Whiting, M., Okkonen, J., Riihiaho, H. and Nissinen, M. (2006) Archaeological Surveys of Jebel Bishri, The Preliminary Report of the Finnish Mission to Syria 2000-2004, in *KASKAL*, Vol. 3, pp. 205-242.

Lönnqvist, M., Törmä, M., Lönnqvist, K., Nuñez, M., Okkonen J. and Latikka, J. (2007) The Euphrates Channel Changes and Archaeology along Jebel Bishri in Syria, in *Proceedings of the XXI Symposium, CIPA 2007, Athens, Anticipating the Future of the Cultural past*, Vol. 1, ed. Georgopoulos, A., the International Archives of Photogrammetry, Remote Sensing and Spatial Information Sciences, Vol. XXXVI-5/C53, pp. 465-470.

Lönnqvist, M. with contributions by Lönnqvist, K., Törmä, M., Nuñez, M., Stout Whiting, M. (2010) Tracing Tribal Implications among the Bronze Age Tomb Types in the Region of Jebel Bishri in Syria, in *International Symposium: Formation of Tribal Communities, Integrated Research in the Middle Euphrates, Syria, November 21-23, 2009, Al-Rāfidān*, Special issue, pp. 165-173.

Margueron, J.Cl. (2004) *MARI: Métropole de l'Euphrate au IIIe et début du IIe millénaire av. J.-C.* Paris: Éditions A. et J. Picard.

Marylou, J.-M. (1999) *Tombes et Nécropoles de Mari.* Mission Archéologique de Mari. Tome V. Bibliothèque Archéologique et Historique. Tome CLIII. Institut Français d' Archéologique du Proche-Orient. Beyrouth – Damas – Amman. Tome V. Beyrouth: Institut Français d' Archéologique du Proche-Orient.

Maspero, G. (1903) *History of Egypt*, Vol. 7, Part A, Chaldea, Syria, Babylonia and Assyria, ed. by Sayce, A.H. London: The Grolier Society Publishers.

Mazzoni, S. (1985) Elements of the Ceramic Culture of early Syrian Ebla in Comparison with Syro-Palestinian EBIV, in *Bulletin of the American Schools of Oriental Research*, No. 257, pp. 1-18.

Moore, A.M.T., Hillman, G.C. and Legge, A.J. (2000) *Village on the Euphrates: From Foraging to Farming at Abu Hureyra*. Oxford: Oxford University Press.

Musil, A. (1927) *The Middle Euphrates: A Topographical Itinerary*. American Geographical Society. Oriental Explorations and Studies No 3. New York: American Geographical Society.

Pinnock, F. (2009) EBIVB-MBI in Northern Syria: Crisis and Change of a Mature Urban Civilisation, in *the Levant in Transition*, ed. by Parr, P.J., Palestine Exploration Fund Annual IX, Leeds: Maney Publishing, pp. 69-89.

Poidebard, A. (1934) *La Trace de Rome dans le Désert de Syrie*: Le Limes de Trajan a la conquête arabe. Recherches aériennes (1925-1932). Texte. Atlas. Bibliothèque Archéologique et Historique. Tome XVIII. Paris: Paul Geuthner.

Rosen, S. (1988) Notes on the Origins of Pastoral Nomadism: A Case Study from the Negev and Sinai, in *Current Anthropology*, Vol. XXIX, pp. 498-506.

Rouault, O. (1997) Terqa: Rapport Préliminaire (1987-1989), in *MARI Annales de Recherches Interdisciplinaires*, Vol. 8, Paris: Éditions Recherche sur les Civilisations, pp. 73-157.

Sarre, F. and Herzfeld, E. (1911) *Archäologische Reise im Euphrat- und Tigris-Gebiet.* Band I. Forschungen zur islamischen Kunst I. Berlin: Dietrich Reimer (Ernst Vohsen).

Sarre, F. and Herzfeld, E. (1920) *Archäologische Reise im Euphrat- und Tigris-Gebiet.* Band II. Forschungen zur islamischen Kunst I. Berlin: Dietrich Reimer (Ernst Vohsen).

SAS = *Syria: Archaeology from Space* (2002) Damascus: GORS.

Schwartz, G.M. (1989) The Origins of the Aramaeans in Syria and northern Mesopotamia: Research problems and potential strategies, in *To the Euphrates and Beyond,*

Archaeological studies in honour of Maurits N. van Loon, ed. by Chaex, O.M., Curvers, H.H. and Akkermans, P.M.M.G., Rotterdam: Balkema, pp. 275-291.

Toivanen, H.-R. (2008) Fortresses and Ecclesiae on the Imperial Border of Byzantium: The Architectural Influence of Constantinople on Resafa-Sergiopolis and Zenobia in Syria, in *Jebel Bishri in Context: Introduction to the Archaeological Studies and the Neighbourhood of Jebel Bishri in Central Syria*, Proceedings of a Nordic Research Training Seminar in Syria, May 2004, ed. by Lönnqvist, M., BAR International Series, No. 1817, Oxford: Archaeopress, pp. 111-126.

Van Koppen, F. (1997) L'Expédition à Tilmun et la Révolte des Bédouins, in *MARI Annales de Recherches Interdisciplinaires*, Vol. 8, pp. 417-429.

Werner, P. mit Beiträgen von Busch, R., Klengel, H. and Mayer, W. (1998) *Tall Munbāqa: Bronzezeit in Syrien.* Neumünster: Wachholtz Verlag.

Zarins, J. (1989) Pastoralism in southwest Asia: the second millennium BC, in *The Walking Larder: Pattern of Domestication, Pastoralism, and Predation*, ed. by Clutton-Brock, J., One World Archaeology 2, Cambridge: Cambridge University Press, pp. 127-155.

11. GLIMPSES TO SITES ON THE ROMAN DESERT FRONTIER

Minna Lönnqvist and Kenneth Lönnqvist

Fig. 1. The valli of the Roman fort of Qebaqeb and the heights of Jebel Bishri in the background. Photo: Minna Lönnqvist 2008.

11.1. In the Context of the Roman and Byzantine Military Policy

The city of Palmyra in the middle of the Syrian Desert, southwest of Jebel Bishri, with changing degrees of autonomy had the responsibility for controlling the desert after Syria was annexed to Rome as a province along with Pompey's conquest in 64/63 B.C. (*Plut. Pomp.*, 39). (Isaac 1990: 140-147; Sartre 2005: 238; see also more in Anadol 2008; in Lönnqvist 2008 and Seland 2008). There has been much discussion of the nature of the changing eastern border zone of Rome in Syria and whether the border can be called *limes*. At the beginning of the twentieth century V. Chapot (1907: 245-247) already described differences of the Eastern *limes* compared to the visible line of ramparts and fosses or continuous walls in Europe. Nevertheless, the definition of the "Eastern *limes*" has been generally used in dealing with the eastern border zone of the empire. The so-called *limes interior* passed Palmyra and consisted of areas directly under Rome. However, the 'desert *limes*', the *limes exterior* with its desert castles, was not so well constructed and difficulties arose in controlling the enemies (Musil 1928: 248; Chapot 1907: 245-249; Parker 1986: 643). There were two major groups which were to be kept at bay: the Parthians/Persians and nomads (Isaac 1990: 140-147; Sartre 2005: 238). As we shall discuss further, the division into the *limes interior* and the *limes exterior* is hard to define, but there seem to have been two more or less fortified areas: the desert frontier and the frontier on the Euphrates which were built to support each other. Our archaeological studies at Jebel Bishri elucidate the chronology and function of these two kinds of border zones.

Based on the historical references E. Dabrowa (1986) among others has suggested that the Romans started to build the border zone on the Euphrates already in the first century A.D. The first sites in the Palmyrene district identified as towns and military posts do already appear in Pliny's *Historia Naturalis*, i.e., Natural History dating from the first century A.D. and, as mentioned, in Ptolemy's *Geographia* dating from the second century A.D., resurfacing in later sources. However, at the accession of Trajan the Roman border was quite undefined from Palmyra to the Euphrates (Luttwark 1979: 107). Some early evidence of the Roman presence in this desert zone come in the form of a milestone found at Araq dating from A.D. 75 and small finds recovered at Sukhne and Taibe (O. Grabar *et al.* 1978: 5). Some of these sites are touching our research area ands are also mentioned in Ptolemy's (A.D. 90-168) *Geographia*. Locations in *Syria* (Book V, Ch. XIV) are the towns in Palmyrene: Rhesapha, Cholle, Oriza, Putea, Adada, Palmyra (see Fig. 2.); and near the Euphrates the sites: Alalis, Sura and Alamatha. A location in *Mesopotamia* (Book V, Ch. XVII) among the towns and villages near the Euphrates is Nicephorium. Locations in the *Arabia Deserta* (Book V, Ch. XVIII) are the towns and villages near the Euphrates: Thapsacus, Birtha, Gadirtha and Auzara.

The final defeat of the Roman garrison of Dura Europos on the Euphrates by the conquering Sassanians ca. A.D. 256 (Rostovzeff *et al.* 1936: 467-470; James 2005; Børlit 2008) was followed by the struggle of Palmyra for its independence led by the Queen Zenobia against the Romans in A.D. 272. Authority in Syria was formally restored under the Emperor Aurelian after Palmyra's defeat, but we are not well informed about what this would have meant for the area east and north of Palmyra. The strengthening of the frontier against the nomads of the desert (see, e.g., Isaac 1990 and Parker 2000) became to be organized by the Emperor Diocletian. The historical narratives from the Late Roman and Byzantine periods also refer to these undertakings, which were also aimed at securing the border against the Persians.

In the desert frontier the reinforcing of the military organisation is particularly attested by archaeological remains such as forts and fortresses in Jordan (Parker 2006) and beside fortresses also by inscriptions (CIL III, Suppl. 2, 14380) in Syria dating from A.D. 306, from the time of Diocletian's reign. The *Strata Diocletiana,* marking the fortified eastern border zone against the desert in Syria, has been identified as being the continuation of the *Via Nova Traiana* (see Kennedy 1997: 71-73). The southern part of the *Via Nova Traiana* is constructed of packed fieldstone lines set in parallel with larger stones, and a central spine and milestones also appear along it (Kennedy 1997: 73-93; see recent research in Borstad 2008). The unpaved route of the *Strata Diocletiana* is archaeologically characterized by milestones, *castra* and *castella,* and its alignment has traditionally been identified as leading from 'Azraq in Jordan through Palmyra to the Euphrates in Syria covering the area of western piedmont of Jebel Bishri (see Figs. 2. and 3.). Zosimus (2.34) describes Diocletian's operations on different *limes* zones. However, the *Tetrapyrgium (=Quadriburgium)* type of fortresses erected on the desert frontier have been especially attributed to Diocletian. More recently A. Lewin (2002: 93) has argued that the milestones according to which the *Strata Diocletiana* has been identified appear only in the stretch between Khan el-Qattar and 'Araq. In any event, desert *castra* and *castella* are known to have been reinforced by Diocletian. How many of them between Sukhne (Putea?) and Sura can be attributed to Diocletian is not exactly clear. John Malalas (Chron. 12.308) mentions that the *Strata Diocletiana* even extended from Egypt to the Euphrates.

The *Strata* can be seen as a realisation, 'a lesson learnt by the Romans', that one defence line alone, though placed along the Euphrates (see ch. 13.), could not be enough to secure the strategic province. Apart from Ptolemy's *Geographia*, there are itineraries, military tabulas of the Late Roman Empire including *Notitia Dignitatum* and *Tabula Peutingeriana*, both originally dating from the fourth century A.D., which mention sites in this frontier zone. In addition, the architectural accounts by the ancient historian Procopius illustrate the history surrounding the archaeological remains that the Finnish project SYGIS has been tracing in the area of Jebel Bishri and its piedmonts. Procopius (*De Aedificiis*) describes several refortifications by Diocletian on the Euphrates zone and how the fortresses were transformed by Justinian I into even stronger ones. At the junction of the Euphrates and the Habur, according to Ammianus Marcellinus (23.5.2), Diocletian also rebuilt the fortress of Circesium (Fig. 2.). Circesium became the last border fortress against the invaders on the Euphrates (Bowman *et al.* 2005: 255). Under Justinian it became necessary to secure further the area from Palmyra and Oriza to Circesium on the right bank of the Euphrates (Musil 1928: 254).

Some of the Roman sites on the mentioned part of the Eastern *limes* zone in Syria were already studied E. Sachau, F.R. Chesney, G. Bell, V. Chapot, F. Sarre, E. Herzfeld and A. Musil from the mid-nineteenth century to the beginning of the twentieth century. Sites were more comprehensively recorded and documented from the air by A. Poidebard (1934) and by R. Mouterde and A. Poidebard (1945). In recent decades the Germans have been actively studying the desert frontier in the neighbourhood of the *Strata Diocletiana* on the western piedmont of Jebel Bishri. T. Ulbert (1986) and M. Konrad (2001) have excavated at the sites of Resafa-Sergiopolis (see introduction to Resafa-Sergiopolis in Toivanen 2008 and more in our ch. 13) and Tetrapyrgium (Qusair as-Saila) (Figs. 2. and 3.). In addition, M. Konrad (2001) has surveyed as well as remapped the site of Holle on the western piedmont and Sura on the northwestern piedmont of Jebel Bishri the latter reaching the Euphrates *limes*. We have only briefly visited some of these sites on the desert frontier, because they did not belong to our areas of intensive and systematic survey connected with Jebel Bishri. In any event, we provide some additions to the knowledge of the region with the documentation of a few tomb sites on the western piedmont and the visit to the Roman fort of Qebaqeb providing canal systems like the site of Dedjan on the southern piedmont.

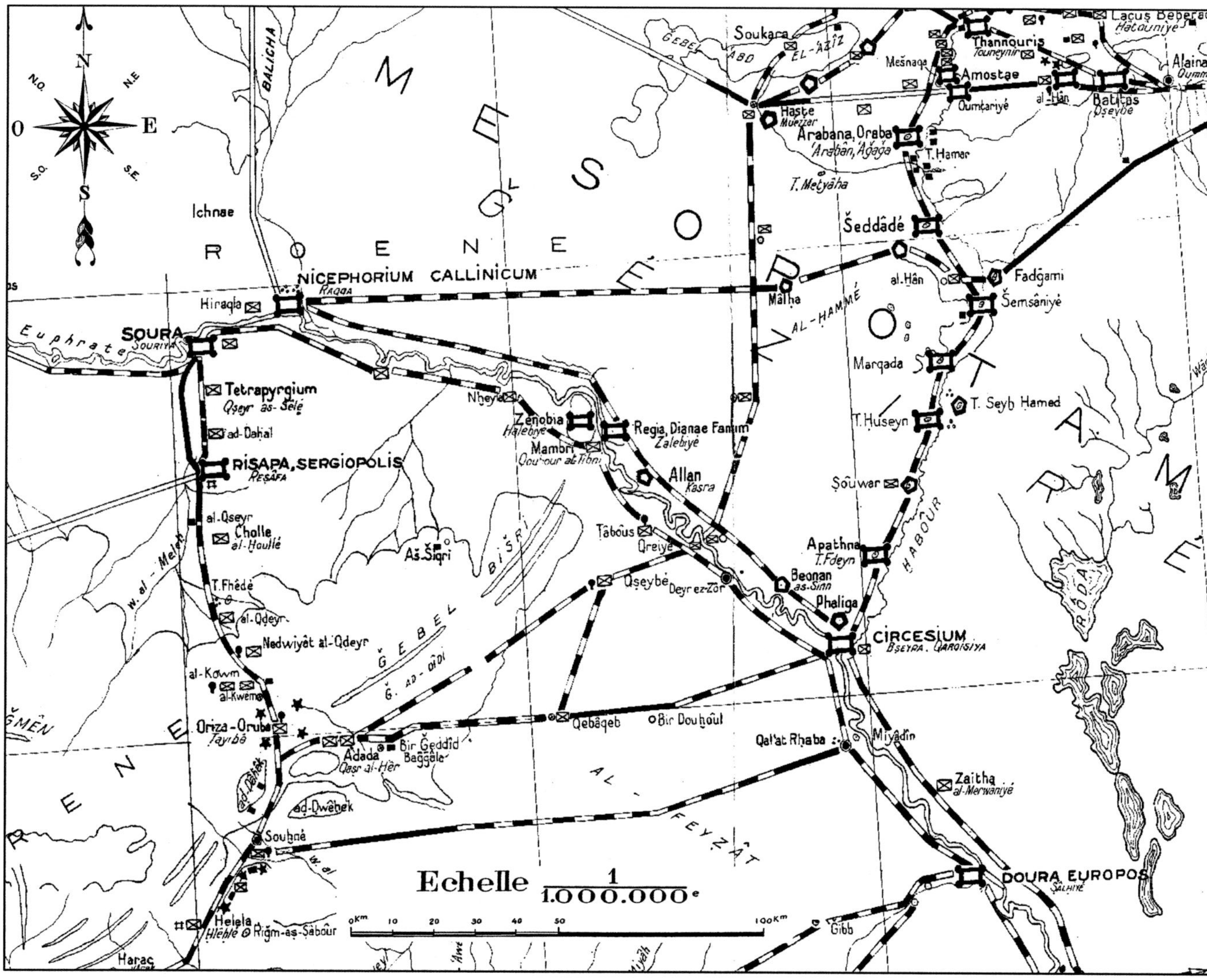

Fig. 2. Part of Poidebard's map (1934) covering Jebel Bishri and illustrating the northern part of the Strata Diocletiana extending to the Euphrates limes.

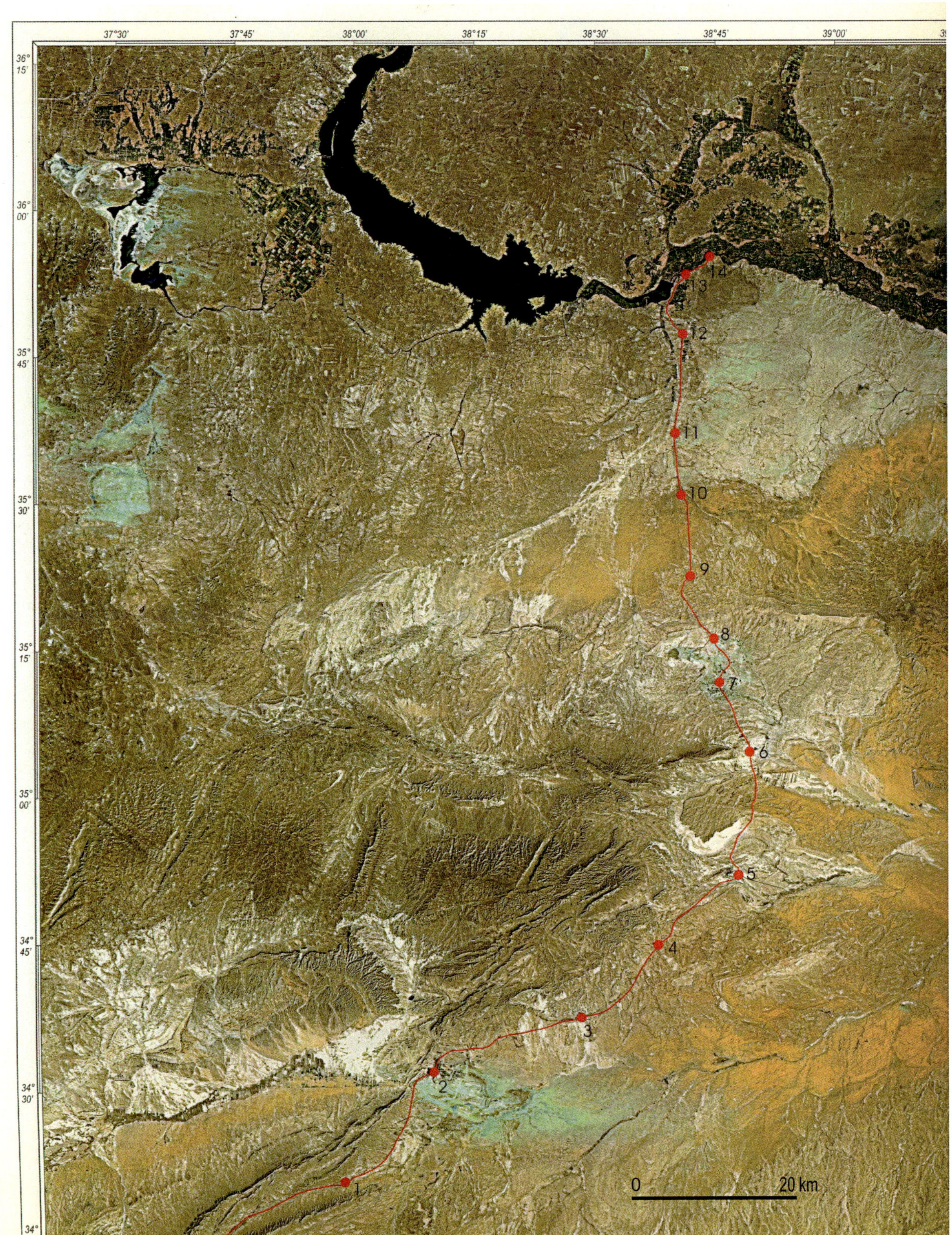

Fig. 3. A satellite map illustrating sites on the Strata Diocletiana passing Jebel Bishri: 1. Kh. al-Hallabat, 2. Palmyra, 3. 'Araq, 4. Helela, 5. Sukhne, 6. Oriza (Taibe), 7. El Kowm, 8. Qdeir, 9. Tell al-Fhada, 10. Holle, 11. Resafa-Sergiopolis, 12. Qusair al-Saila, 13. Sura I and 14. Sura II. Courtesy: GORS (SAS 2002).

11.2. Some Observations on the Western Piedmont

11.2.1 Water Resources

The operational success of the Roman army in a desert type environment - such as the one under consideration here - would have been dependent on the same factors as those in any desert-dwelling community. Therefore, to understand the disposition of the troops and military installations of the Roman army along the desert frontier requires an examination of the environment, topography, travel and duration of journeys. It is important to remind ourselves that food is not the primary issue in a hot desert, but water. In fact, a person can survive for weeks without food, but not more than two to three days without water in desert conditions. Sources mention, for instance, that Roman soldiers on a march needed a minimum of two to three litres of water per day to remain in fighting condition in addition to the same amount taken through food (Roth 1999: 36-39; Ortloff 2005: 107).

The largest of the Roman sites on the western and southern piedmont of Jebel Bishri are situated in oases which have been under human habitation from prehistoric times and which we have dealt with in the previous chapters (see ch. 4., 5. and 6.). A. Poidebard (1934) indicated - quite rightly - that many of the Roman sites, fortresses or alignments of roads in the desert frontier can be traced first by means of locating water resources. Therefore, in most places where the Romans built their military camps, stations or posts, there had to be springs, wells or reservoirs within a reasonable distance. They were either reused ones or built from scratch. A typical feature along the northern, northeastern and southeastern slopes and piedmonts of Jebel Bishri is that the Roman sites represent the collection points of some of the largest seasonal water courses - exemplified in *wadis* - producing the run-off from Jebel Bishri. In this way, it was possible to collect seasonal water into basins, cisterns, reservoirs and to build canals that brought the water to the final destination of consumption. In these places, the Romans seem to have regularly built many canals and dams, some of which we recorded in 2003-2006.

The hydraulic canal structures called *qanats* or foqqâras are a typical feature in desert and steppe regions in Syria (Lightfoot 1996a: 321-336; Lightfoot 1996b: 1-9; Mango 2008: 5-8). Such are also met within the desert between Palmyra and the Euphrates, which is in the arid zone and receives less than 200 mm annual precipitation on average. A. Poidebard (1934: 181-182) mentions that the surroundings of the Roman site of El Kowm were fed by two visible lines of ancient *foqqâras*. Later on these types of *qanats* have been especially recorded by D. R. Lightfoot (1996b: 1 and 4, Fig. 1: Distribution and status of the Syrian *qanats*, and Fig. 2: Relationship between *qanat* sites and Romano-Byzantine sites) along the northern stretch of the *Strata Diocletiana* from Sukhne to Sura (see Fig. 3.). This area overran the distance of ca. 100 km and served military installations. Examples of the *qanats* are also found at Qasr al-Hayr ash-Sharqi (Adada?), Taibe (Oriza), El Kowm, El-Qdeyr, Resafa-Sergiopolis and Raqqa (Nicephorium Callinicum). According to the topography and geology of the *Strata Diocletiana qanats*, it is likely that they tapped into shallow aquifers. They are all on the piedmont of Jebel Bishri along the converging points of some of the largest seasonal watercourses and *wadis*. (Lightfoot 1996b: 5).

In the regions where rainfall is meagre and seasonal only, the annual *transhumance* of nomadic people with their animals is normally necessary (Gichon 1991: 321), although for an occupying army it would have been much more complicated than that. However, as far as the function of the *qanats* is concerned, D.R. Lightfoot (1996b: 2) does not believe that they would have been primarily constructed to supply drinking-water or support small-scale cultivation around the sites. Regional studies by the Syrian-French-British teams in the desert areas of Ansari suggest that the *qanats* would have been used for agriculture, such as cereal production or olive plantations (Mango 2008).

11.2.2 Introduction to the Roman Sites in the Oases of the Western Piedmont

The village of Sukhne includes a tell which may be identified with ancient Putea. A tell providing Roman cooking ware, including North African imports, as well as a *qanat,* the use of which had ceased in 1981, were also encountered at Taibe by our Finnish project in 2000. Sukhne and Taibe were outside our main target areas, and therefore no actual documentation was carried out at the sites. Taibe has been identified with ancient Oriza (Figs. 2. and 3.), mentioned in Ptolemy's *Geographia*, *Notitia Dignitatum* and *Tabula Peutingeriana*. In the *Notitia* it is the legionary base of *Praefectus legionis quartae Scythicae* (OR. XXXIII, Dux Syriae). Nearby Qasr al-Hayr ash-Sharqi has been studied by an American mission led by O. Grabar *et al.* (1978) as well as a Swiss team (Genequand 2005). The site of the Qasr itself was not part of our main survey area either, although we carried out some random explorations in its surroundings (see more in ch. 5., 14. and 15.). A. Musil (1928) noted that at this ruined Umayyad castle, there had apparently existed an earlier Roman military site in a strategic position between the Roman sites of Sukhne, Taibe and El Kowm. The southwestern extremes of Jebel Bishri have been called Ad-Didi (Musil 1928: 76, 174), and the area is situated close to Qasr al-Hayr ash-Sharqi with its larger and smaller castle (UTM: 050 6457, 3883921; UTM: 0506422, 388975). Like A. Musil (1928: 254), A. Poidebard (1934: 91) as well as R. Mouterde and A. Poidebard (1945: 115) identified the site of the Qasr with ancient Roman site of Adada (Fig. 2.) (cf. the local toponym "Ad-Didi") mentioned in Ptolemy's *Geographia* and in *Notitia Dignitatum,* the latter as the place of the garrison of *equites promoti indigenae* (OR. XXXIII, Dux Syriae). Poidebard observed that at Qasr al-Hayr a number of Roman architectural fragments were scattered around, and they dated approximately from the fifth and sixth centuries A.D. being contemporary with the Late Roman and Byzantine layers of the nearby castle of Resafa-Sergiopolis.

O. Grabar *et al.* (1978: 10) and S. Gregory (1996: 180-181), however, have doubted the Roman origin of the site, but in 2000 our project observed *in situ* that the earliest architectural fragments as well as possibly reused columns (Fig. 4.) in the Qasr al-Hayr seem indeed to date from the Late Roman - Byzantine period. Architectural fragments from the castle dating from this earlier period had been collected at the nearby Guest House belonging to the General Directory of the Antiquities and Museums of Syria, where we stayed. These architectural members could, however, originate from Taibe (Oriza), but this is not so obvious as Taibe is over a dozen kilometres away and there are Late Roman/Byzantine tombs in the vicinity of the Qasr some of which we shall present below. In addition, M. Konrad (2006: 52, Fig. 12) notes that the Umayyad castle of Qasr al-Hayr follows the layout of the Late Roman *tetrapyrgium* form. Also, a site like Holle nearby was a Roman military station that was later used as a basis for an early Islamic fort of a tetrapyrgium type (Konrad 2001: 16).

In the oasis of El Kowm, Poidebard (1934: 80, Pl. LXII) identified a Roman military camp built of dry masonry limestone walls of 1 m in width and covering an area of 30 m x 60 m. Another Roman camp 80 m x 120 m in size and surrounded by a fosse and a double wall with four entrances was identified in the village of Qdeir (Poidebard 1934: 81, Pl. LXXIII). Flying from Qdeir, Poidebard (1934: 81-82) further recorded a small Roman post at Tell Fhede, and 17 km from Qdeir another previously mentioned post at Holle, which had earlier been visited by A. Musil (1928: 69-70). A. Poidebard only identified the Roman post of Holle from the air, and M. Konrad (2001: 15-22) notes that Musil actually mapped the Islamic palace and not the Roman site itself which is nearby. M. Konrad's project remapped the early Islamic stratum of Holle in the 1990s, and a topographic model of the Roman site was constructed (Konrad 2006).

Poidebard (1934: 81-83) mentions several Roman sites between Holle and Sura: Al Qseyr, Resafa, Ad-Dahal and Qseyr as-Sele (Tetrapyrgium). Al Qseyr comprises a Roman tower (15 m x 15 m in plan). The large Byzantine fortress of Resafa-Sergiopolis was built on a Late Roman castle of *equites promoti indigenae.* A *Praetorium* from the Ghâssanide period has been identified in the ruins (Konrad 2006: 50). The latest research prefers to date the large Byzantine fortification system to the reign of the Emperor Anastasios (Fowden 1999: 94), and further building projects to that of Justinian I (Toivanen 2008).

Fig. 4. Ancient columns still standing inside the palace of Qasr-al Hayr ash-Sharqi. Photo: Margot Stout Whiting 2000.

The latest excavations at Tetrapyrgium (Qusair as-Saila) led by M. Konrad (2001: 13-14) have also revealed Late

Roman-Byzantine layers beside the Islamic ones. The type of the castle follows the Diocletianic models but has its roots in Hellenistic fortifications and Parthian-Sassanian castles (Gschwind 2003; Konrad 2006; see the sites in Figs. 2. and 3.)

Following A. Poidebard, M. Konrad (2001) summarizes the list of the sites and their distribution on the Roman military desert border line beneath Jebel Bishri as follows: Taibe (Legionary Camp of Oriza) 12 km east from El Kowm, El Kowm (Small Castle) 16 km east from Al Qdeir, Al Qdeir (Small Castle) 8 km from Tell Fhada, Tell Fhada (Guarding Post) 7 km from Holle, Holle (Small Castle) 7 km from GP 2, GP 2 (Guarding Post/Small Castle) 8 km from GP 1, GP 1 (Guarding Post) 5 km from Resafa, Resafa (Cavalry Castle) 5 km from GP 3? (ad-Dahal?), GP 3? 10 km from Tetrapyrgium (Qusair as-Saila), Tetrapyrgium (Qusair as-Saila, Small Castle) 12 km from Sura and finally Sura (Legionary Camp) on the Euphrates. (See the sites in Figs. 2. and 3.)

A Roman-Byzantine site of Juwal was found during the French studies at El Kowm during the past decades. This tell site is situated 7.5 km northeast of the present village of El Kowm and seems to have been part of the desert frontier, probably contemporary with the *Strata Diocletiana.* It is identified as *castellum,* and the earliest coin discovered at the site belongs to Licinius (A.D. 308-324), and the other coins consist of Byzantine *folles* and Islamic ones. (Taha 1991: 61). There also seems to have been a Roman and Byzantine occupation at Umm el-Tlel (see ch. 5.) in the El Kowm basin. The ceramic evidence from the area includes early Roman *terra sigillata* as well as Parthian pottery and some Late Roman and Byzantine wares. Brick structures were revealed associated with the horizons exemplified in the pottery from the first to sixth centuries A.D. Two periods of occupation were identified: the one being from the first/second century to the third century A.D. and another representing a reoccupation in the fifth and seventh centuries A.D. after the abandonment. The structures from the latter occupation were better preserved compared to the previous one. (Majcherek and Taha 1993).

11.2.3 Two Tombs in a Valley between Jebel Bishri and the Western Oases

Two rock-cut tombs dating from the Late Roman/Byzantine periods were identified and documented during the year 2000 season by our Finnish project SYGIS on the western piedmont of Jebel Bishri. A survey team working between the western oases and Jebel Bishri discovered site **B 1** northeast of Qasr al-Hayr ash-Sharqi in the area called Shaharidz al-Wahab and site **B 2** in the *Wadi Suq* area northwest from the Qasr, on the way to the oasis of El Kowm (see Fig. 5.)

Site B 1
Rock-cut Tomb

UTM: 0507596, 3886103
Size of the plan: 5.35 m x 12.20 m

This looted or previously excavated tomb **B 1** in Shaharidz al-Wahab had been hewn into rock. It is oriented in an E-W direction. The flight of 11 steps is leading to an entrance (Fig. 6.) from the west and to an ante-room with two *loculi*, niches, hewn into the rock walls on both sides (N and S) of the porch section. The roof of the porch was missing. The main burial chamber (5.35 m x 12.20 m in plan), the roof of which had also partly collapsed, forms a cross in plan (Fig. 7.). In the middle of the chamber there is a rectangular bone collection pit cut into the floor. Altogether three *arcosolia* are set on the northern, eastern and western walls (Fig. 8.). A handle of Roman brittle ware and Parthian type blue-glazed pottery were found inside the tomb. Pieces of Roman pottery as well as Byzantine ware were also collected in dumps surrounding the tomb. According to the structure and the finds, the tomb seems to date from the Roman - Byzantine era (Fig. 10.).

Site B 2
Rock-cut Tomb

UTM: 0504815, 3884501
Size of the plan: ca. 6.5 m x 3.0 m

This tomb **B 2** situated in the *Wadi Suq* had also been hewn into rock but was filled with compact sand and dust so that only the upper part was visible to the height of 1.0-1.2 m. The tomb is oriented NW-SE. There are two roundish entrances: one is the northeastern entrance A, and the other is more northern one B. The entrance A is smaller (max. 0.5 m in diam.) and leads into a burial chamber that is connected to the main chamber (ca. 6.5 m x 3.0 m in plan) by a separate entrance, and the entrance B is larger (max. 0.85 m in diam.), leading directly into the main chamber. The plan of the main chamber is a trapezoid, being narrower at the entrance. (Fig. 9.) In the opposite far end of this main chamber there is a partly visible *arcosolium* and to the right at the far end there is a *loculus* set in the western corner. The associated pottery consisted of Late Roman/Byzantine sherds.

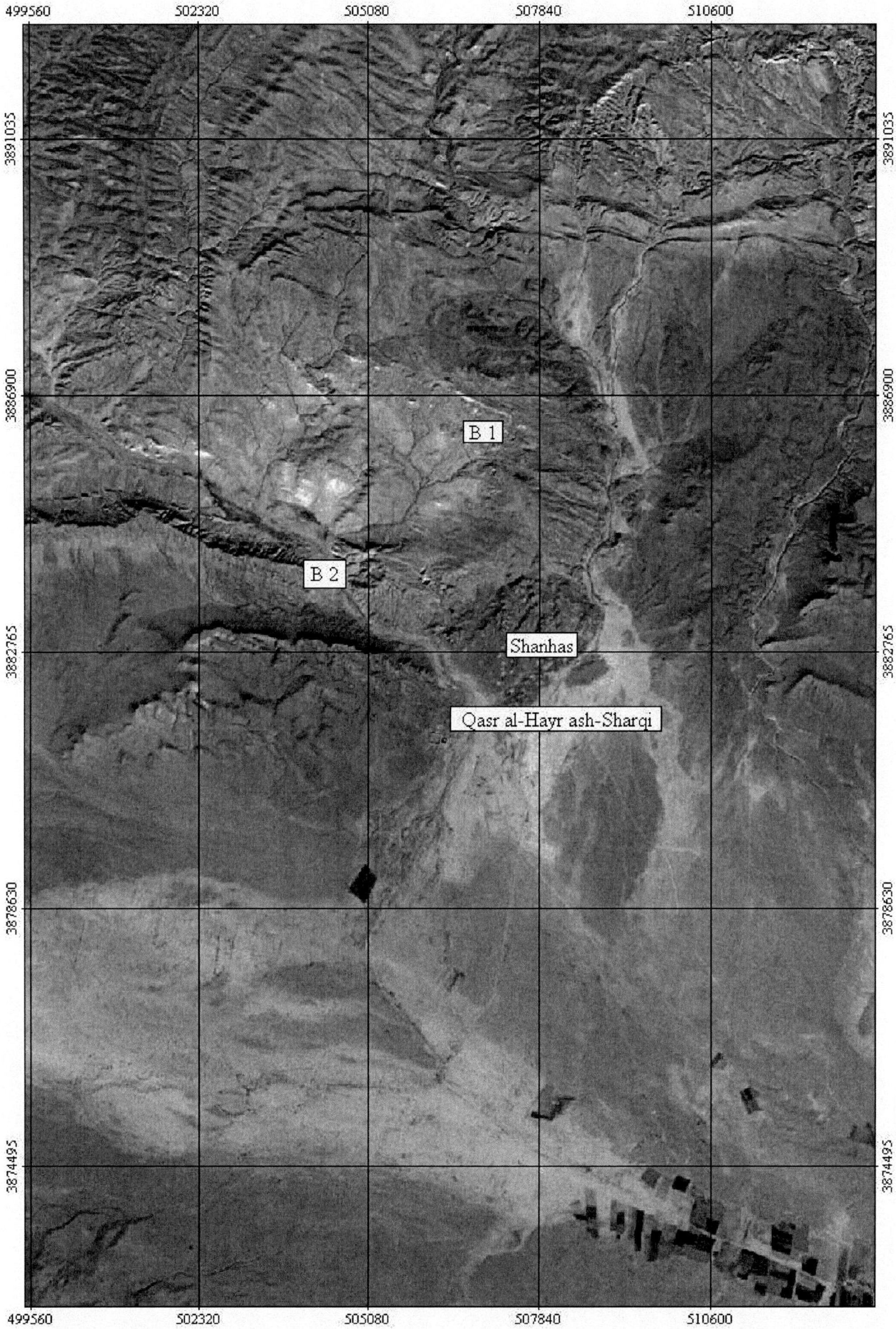

Fig. 5. Map illustrating the locations of the tombs B 1 and B 2 on a Landsat satellite image. Scale 1: 50,000. Mapping by Minna Lönnqvist 2009.

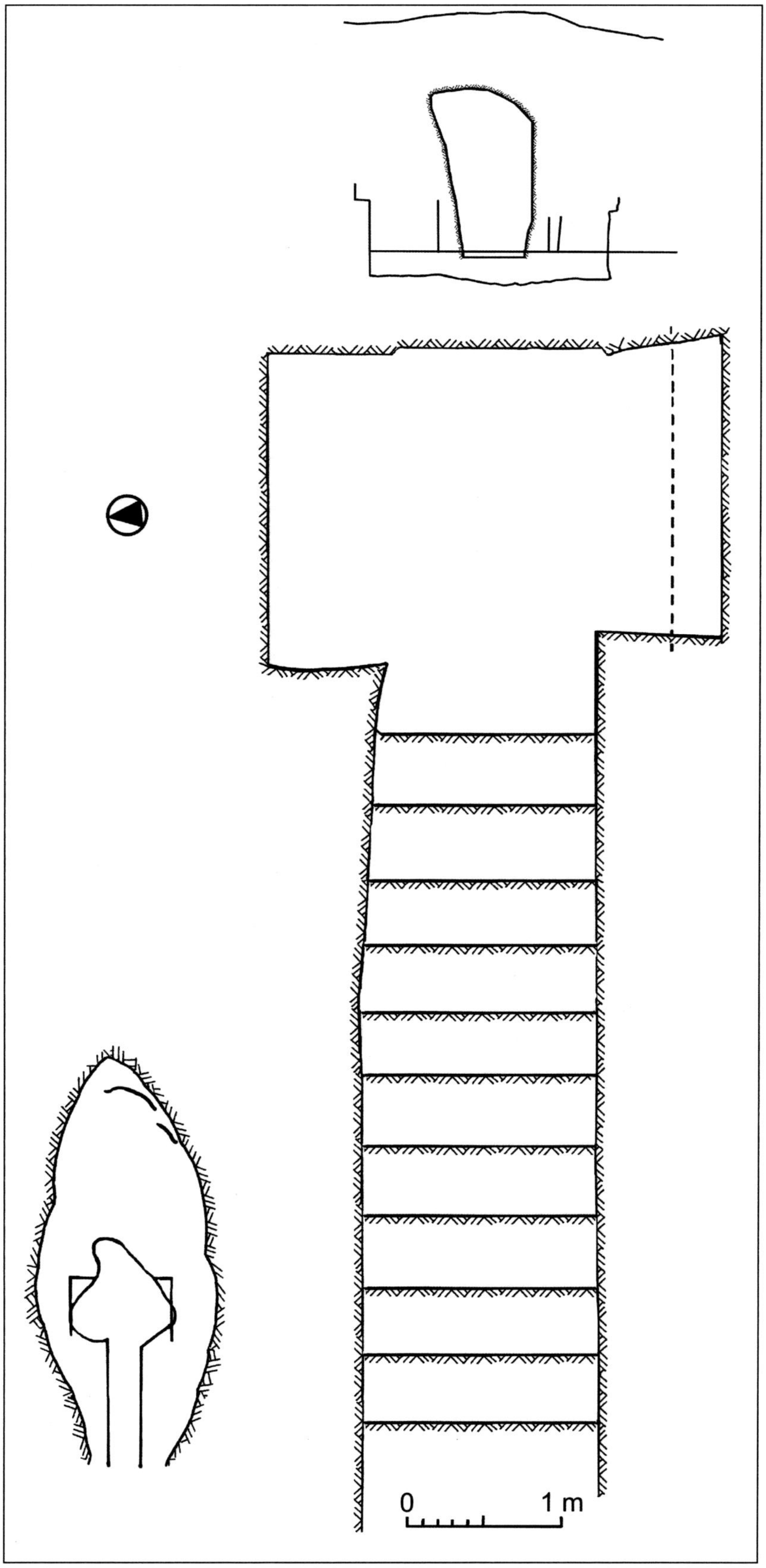

Fig. 6. The entrance and the flight of steps leading into the tomb B 1. Drawing: Martti Nissinen and Helena Riihiaho 2000 and Kenneth Lönnqvist 2009.

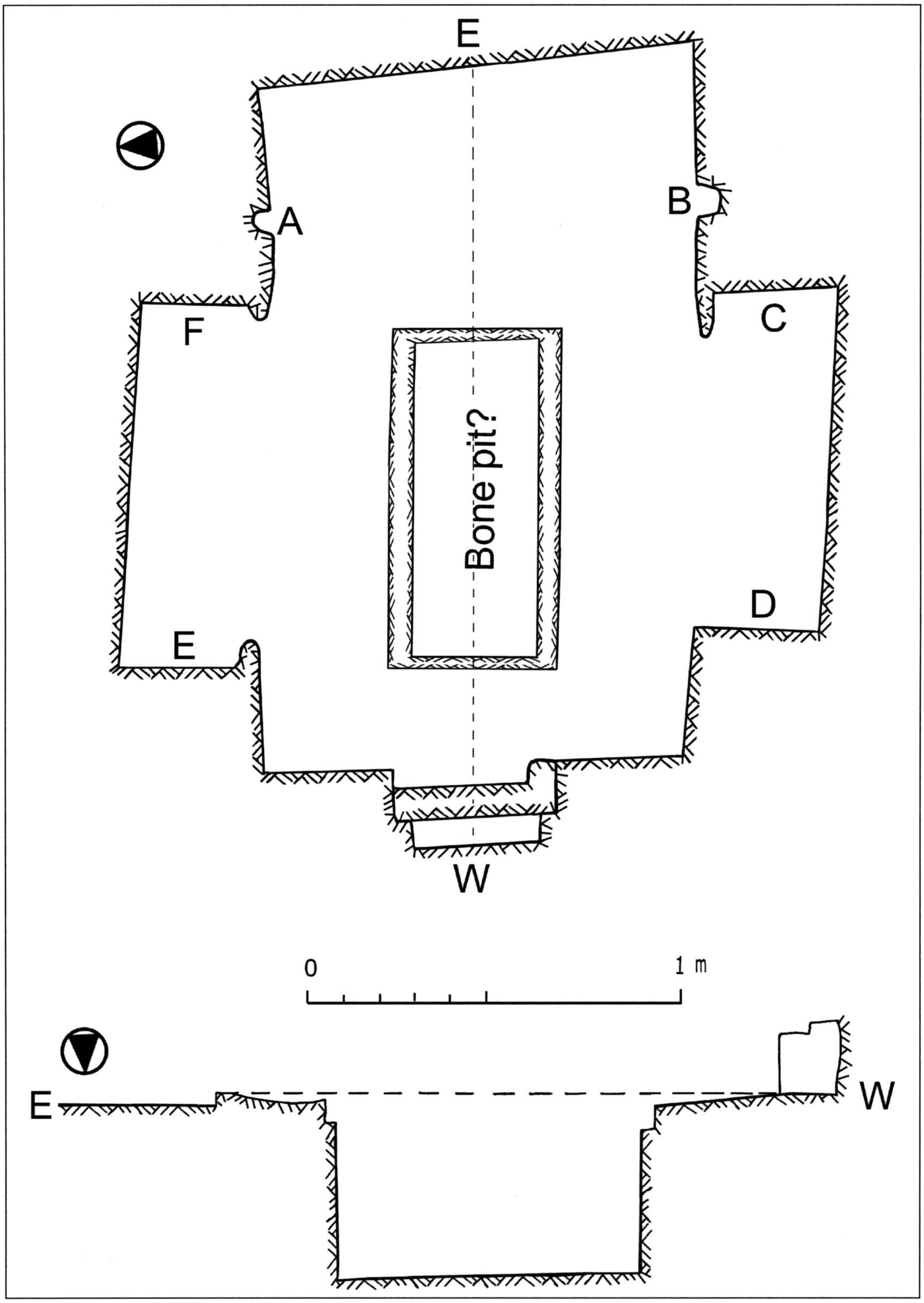

Fig. 7. The plan and the section of the main burial chamber of the tomb B 1. Drawing: Martti Nissinen, Helena Riihiaho 2000 and Kenneth Lönnqvist 2009.

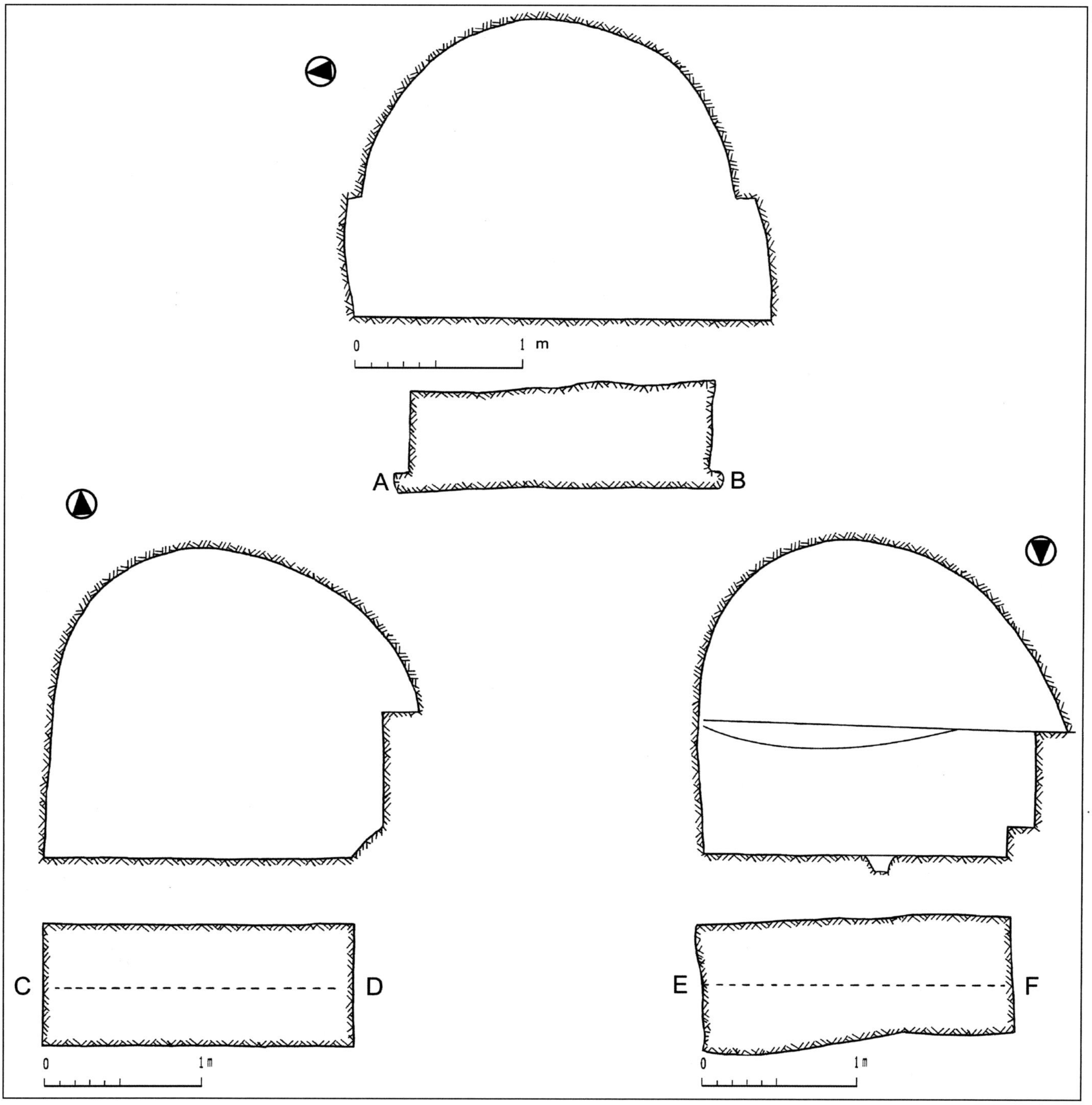

Fig. 8. The arcosolia of the main burial chamber in the tomb B 1.
Drawing: Martti Nissinen, Helena Riihiaho 2000 and Kenneth Lönnqvist 2009.

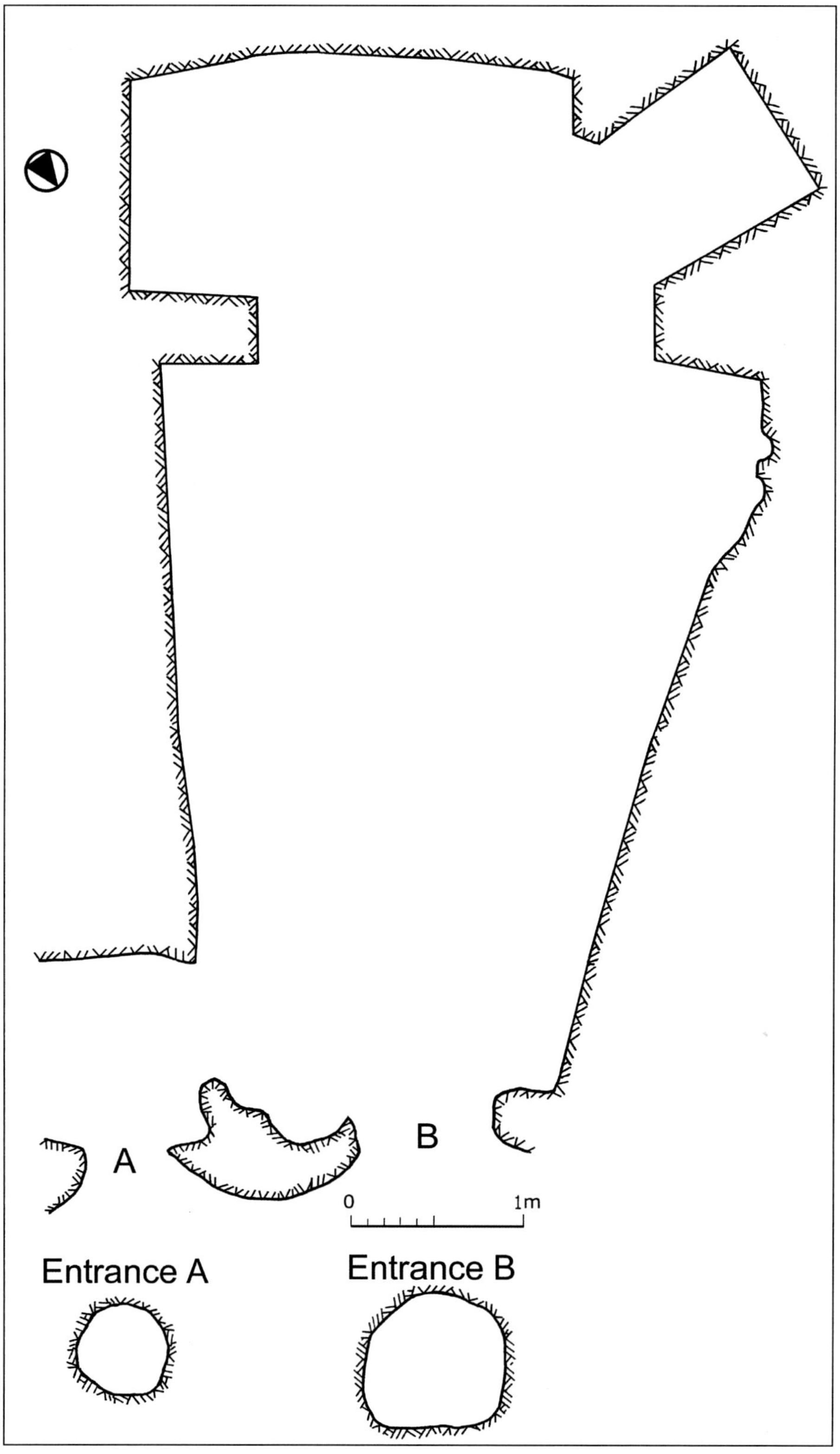

Fig. 9. The plan of the tomb B 2. Drawing: Martti Nissinen, Helena Riihiaho 2000 and Kenneth Lönnqvist 2009.

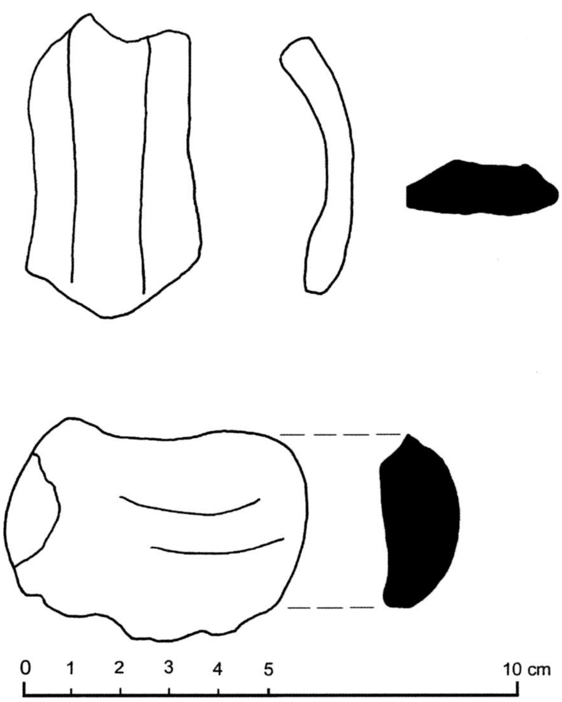

Fig. 10. Pottery associated with tomb B 1.

The tombs B 1 and B 2 are similar to those that we have documented at the cemetery of the Late Roman-Byzantine fort of Tabūs on the Euphrates side of Jebel Bishri to be described in ch. 13 dealing with the Euphrates *limes*. In the cemetery of the fortress of Zenobia at Halabiya (Lauffray 1983) there are also comparable *hypogeae* with *arcosolia* that are met within other Roman–Byzantine sites such as Kifrin in Greater Mesopotamia (Invernizzi 1986: 367). In Palmyra there some *hypogeae* have clearly comparable features. They date from the second and third centuries A.D. Especially Tomb C in the Southeastern necropolis of Palmyra comprises a rock-cut flight of stairs and an entrance which resembles our tomb B 1, but the orientation and the plan are not identical, although *loculi* and *arcosolia* are present in both (cf. Saito 2004-2005: 137-149).

Both tombs B1 and B 2, according to their types, are obviously connected with the Roman military and Byzantine pilgrim sites in the neighbourhood, possibly with the site of Qasr al-Hayr and Resafa-Sergiopolis. The plan in the form of a cross at the tomb B 1 is the same as in *hypogeae* Nos. 29 and 30 in the graveyard of Zenobia on the peninsula of Halabiya at the northern ridge of Jebel Bishri (cf. Lauffray 1991: 218-219). The plan and the W-E orientation indicate that this is an early Christian tomb. It is known that there were Christians in the neighbourhood already in the first centuries A.D. The Christian pilgrimage to Resafa-Sergiopolis especially emerged in the Late Roman and Byzantine times following the martyrdom of Sergius around A.D. 303-312. St. Sergius's relics at Resafa-Sergiopolis became an important target for pilgrimage. The building of Byzantine churches took place at the site as well as the establishing the seat of a bishop in the region of Resafa-Sergiopolis and Zenobia. (See Fowden 1999; Toivanen 2008). A. Musil (1928: 177) mentions that east of Qebaqeb, which is situated nearby on the southwestern piedmont of Jebel Bishri, there in the valley of Azaman there is a tomb called Kabr an-Nusrani, referring to the last-resting place of a Christian pilgrim. It is likely that here was a route for pilgrims going from Resafa-Sergiopolis to Mesopotamia on which at least one of the tombs that we documented may belong.

11.3. Roman Remains at Qebaqeb

A. Musil (1928: 177) mentions a gendarme station beside a well at al-Kebažeb but does not refer to the existence of any ruins. The site of the Roman fort of Qebaqeb (Kbajeb) is, however, marked on the maps by A. Poidebard (1934) and R. Mouterde and A. Poidebard (1945). The place is accessible by the Deir ez-Zor - Palmyra -road, and it is located 53.5 km from Deir ez-Zor towards Palmyra, just next to the new highway on its northern side.

Site K 7
Well and Water Canals

UTM: 0562337, 3881610; 282 m a.s.l.
Size of the well: 1.2 m in diam., depth: 12 m
Two water canals: 4 m in l.

In the present-day village of Qebaqeb, ca. 100 m to the south of the modern Palmyra - Deir ez-Zor -highway, there is a rock-cut well **K 7** with two water canals. This water system apparently dates from the Roman period. Poidebard (1934: 91) also mentions that there were two wells of the Roman date at Qebaqeb. Present-day use consisted of metal pumps and rails (Fig. 11.), and additions of stone layers on top of the originally rock-cut well. Local people also told us that there are ca. 300 wells in the village, making it a large oasis with a plentiful supply of water. The scattered rectangular houses of the village are apparently those of sedentarizing nomads from the region. The site is obviously identical with *Bir Ghabajib* in Lightfoot's (1996a: 321-336) distribution map of the Syrian *qanats*.

Fig. 11. A well and a water canal at Qebaqeb with modern pumps. Photo from NE: Kenneth Lönnqvist 2006

Fig. 12. A Roman brick at the Qebaqeb site K 8. Photo: Kenneth Lönnqvist 2006.

Fig. 13. Roman tiles from the Qebaqeb site K 8. Photo: Kenneth Lönnqvist 2006.

Site K 8
Fort

UTM: 0558858, 3878879; 270 m a.s.l.
Size of the plan: ca. 100 m x 100 m

Large earthen works of a *vallum* surround an apparent Roman camp/fort designated as Qebaqeb **K 8**. The fort is an enclosure, a square of ca. 100 m x 100 m in area (10,000 m^2 or 338 x 338 *pes Monetalis*, pM or Roman feet, or 1.0 hectare) aligned SW-NE to SE-NW (Figs. 1., 14. and 15.). The earthen works of coarse sand, gravel and stones stand high and are ca. 17 m thick in the southwest corner. At present the centre of the fort has been cleared mechanically with a bulldozer by the local Bedouins to create a natural pool for the collection of rainwater, although it may also be possible that water is actually brought to the pool by tank trucks. Roman pottery and bricks were scattered in large quantities, leaving no doubt that there had been a building of burnt brick at the site. To illustrate the abundance of material, it may be mentioned that to the northwest and southwest of the fort, an area of ca. 1 ha was literarily covered with fragments of Roman bricks, tiles (Figs. 12. and 13.) and pottery sherds (Fig. 17.). Two fragments of thin sheets from copper alloyed scale cuirasses were found at K 8, one smaller (2.2 cm x 2.1. cm in size 0.98 cm in thickness), and the other larger (measuring ca. 2 cm x 3 cm in size; ca. 1 mm in thickness). Both had one pair of holes in the short lateral edge; the other pair of holes was missing because the scales were fractured (see Fig. 16.). Associated surface finds confirm the dating of the use of the site to the Late Roman period. Outside the fort area to the northwest, modern cultivation was detected which has apparently contributed to the destruction of the ancient Roman site.

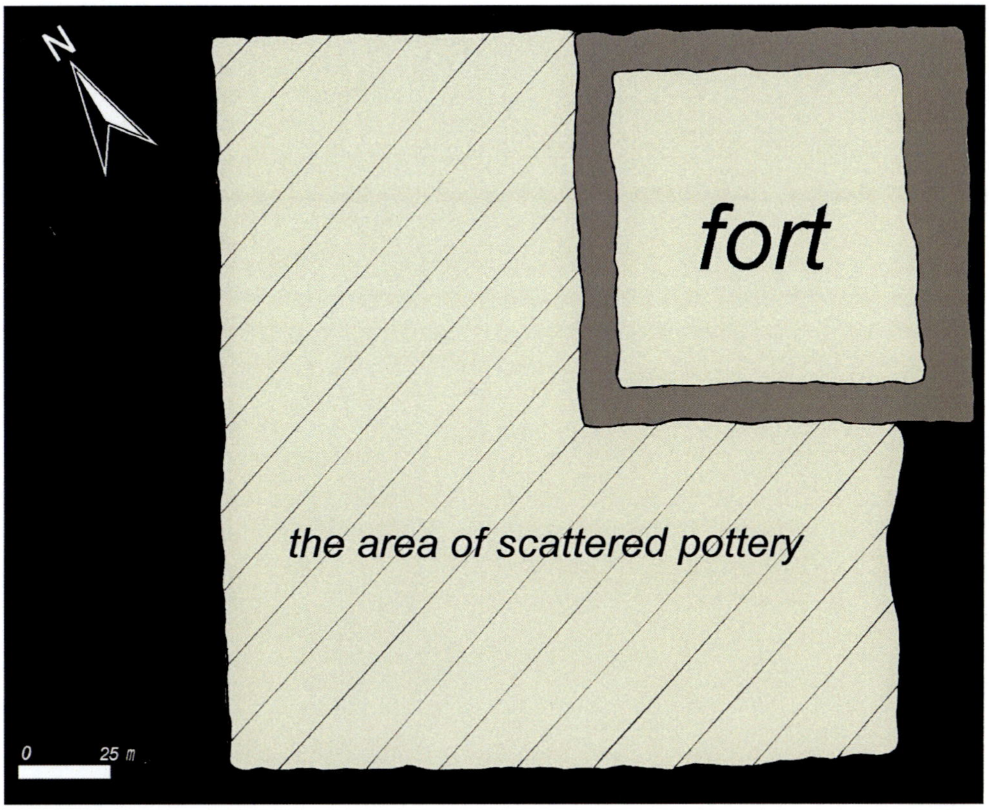

Fig. 14. The plan of the Qebaqeb fort. Drawing by Kenneth Lönnqvist 2006.

Fig. 15. The northwestern corner of the vallum at the Qebaqeb fort K 8. Photo from S: Minna Lönnqvist 2008.

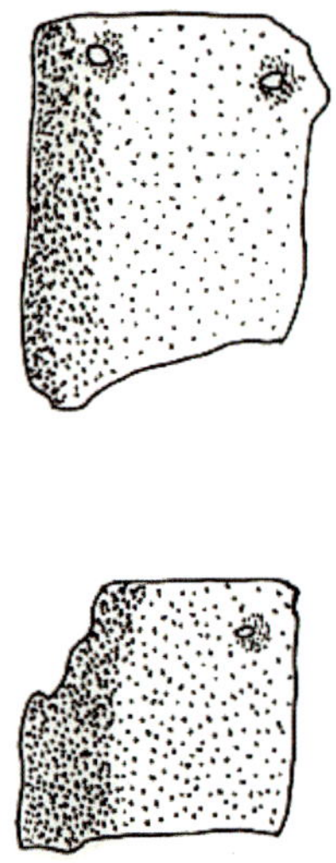

Fig. 16. Two armour scales from site K 8 at Qebaqeb. Drawn by Kenneth Lönnqvist. Scale 1:1.

Fig. 17. Roman pottery from site K 8 at Qebaqeb. Photo: Kenneth Lönnqvist 2006.

Site K 9
Ruined Post?

UTM: 0558255, 3878241
Size: 35 m x 14 m in plan, preserved 1.5 m in h.

Less than 1 km from the sites K 7 and K 8 of Qebaqeb, on the Deir ez-Zor-Palmyra -road at the distance of 54.4 km from Deir ez-Zor towards Palmyra, there is a mound **K 9** situating 50 m north of the road and covering ca. 35 m x 14 m in plan. No visible intact structures such as walls of stone, brick or earth were encountered at the site. However, broken bricks and scattered marble/limestone slabs were found in large quantities, together with associated Roman pottery. The amount of Roman pottery (see samples in Fig. 18.) was considerable taking into account the dimensions of a single small mound. Therefore, it is possible that the site represents the remains of a Roman post that was located along a Roman route. Towers and posts were placed at regular intervals along major roads, especially in the frontier zones.

Fig. 18. Pottery finds from site K 9. Photo: Kenneth Lönnqvist 2006.

11.4. Roman Water Harvesting Site at Dedjan

Site K 10
Well and Water Canal

UTM: 563635, 3894940, 352 m a.s.l.
Size of the well: max. 1.4 m in diam. and 1.8 m in depth.
Water canal: visible
230 m in l. on the surface; w. 0.70m

18 km to the northeast of Qebaqeb along a new asphalt road that runs diagonally over the piedmont of Jebel Bishri towards Qseybe and 'Ain Abu Jimā, there is another site **K 10** including a well with a water canal which we archaeologically documented. About 500 m south of the ancient site there are at present some houses occupied by local Bedouins.

The original depth of the rectangular well (1.4 m in diam.) was not explored, but it must have been much deeper in antiquity as it is now silted-up and filled with rubble. The well had not been reused, or at least it had not been rebuilt in modern times. It therefore showed no additions, such as extra layers of stone or a cement casing, suggesting perhaps that it may have been already abandoned in antiquity. From the well a roughly NW-SE oriented (300°-120°) underground water canal at least 230 m in length, had been constructed. The canal led to the southeast and disappeared in the open steppe landscape, in a field in the middle of nowhere. It is assumed that the purpose of the canal was to tap into the well from a spring or ground water, rather than the other way round.

Fig. 19. A water canal at Dedjan.
Photo from NW: Kenneth Lönnqvist 2006.

There are occasional disputes over the direction that in which the canals led, and this would obviously need to be investigated further. In the open space to the southeast some undefined heaps and earthen constructions that may have been a camp site were found. However, there were no direct surface finds to indicate either the date or place of that possible site. We were not able to determine either, for example, whether the canal would have served some irrigation purposes at the ancient site. However, Roman pottery and bricks were found in association, and the bricks obviously came from somewhere near by, although all remains have later been covered by sand. The associated pottery suggests use in the Roman period only, as no other finds from other periods were made.

The construction technique of this narrow water canal corresponds to a *qanat* or foqqâra. The building technique could be discerned from the open section of the first 30 m from the water source (i.e. the well) that was visible on the ground, before it was converted into a subterranean tunnel. The Roman water canal had been hewn into the bedrock (see Fig. 19.) in the shape of a narrow shaft with straight walls. The shaft was 70 cm wide with a minimum present-day depth of 2.5 m or more, which corresponds quite well with the shafts of Roman Imperial aqueducts in Rome, like the *Aqua Traiana* under the American Academy (Wilson 2001: 1-2). No plaster or brick constructions were seen in the shaft itself. After the first 30 m or so, the length and direction of the subterranean tunnel was indicated on the surface only by what often is called the "bomb-crater" – run, or a long, straight line of holes in the ground apparently caused by the filling up of the shafts of the tunnel. Shafts in such water systems are regular features and they were used both for cleaning purposes and ventilation.

The exact date of the *qanats* is often inconclusive, as are the dates of their possible abandonment. R. Mouterde and A. Poidebard (1945: 112-113) also deal with *foqqâras* in the area of Andarin (Androna), for example, in the Wadi of Qdeym, west of our *Strata* area. They date the initial construction of the reservoirs and canals to the second and third centuries A.D., but in their opinion some were apparently reconstructed in the Umayyad period. However, the Oxford University excavations at Andarin have expressed reservations about the earlier date of the second century. The excavators hint that the canals may be later than the reservoirs. Some were still in use in the Islamic times. (Mango 2004-2005; 2007; 2008). In our case the associated pottery seems to date the reservoirs to the Late Roman period.

Conclusions

The Roman eastern frontier zone has an organic connection to Jebel Bishri, the mountain situated between the desert frontier and the Euphrates *limes*. According to Ptolemy's *Geographia,* the desert part belonged to the Palmyrene territory at least in the second century A.D. Many of its sites are identified with the *Strata Diocletiana,* the border zone refortified by the

Emperor Diocletian from ca. A.D. 300. The Roman sites on the western piedmont of Jebel Bishri are primarily located in the desert oases and have been mainly studied by the French, Germans and Syrians. The elaborate irrigation systems known as *qanats* were apparently dug in the piedmont already by the Romans, but they were reused in the Islamic times.

The Finnish mission only made some general observations in the area of Qasr al-Hayr ash-Sharqi studying more closely two tombs and the Roman remains at Qebaqeb and Dedjan in connection with water canals, *qanats*. The neighbourhood of Qasr al-Hayr provides Late Roman–Byzantine architectural fragments which have apparently been reused in constructing the Umayyad castle. The two tombs discovered nearby clearly belonged to the Late Roman and Byzantine sites in the neighbourhood. One may be connected with the early Christian pilgrims associated with St. Sergius's site of pilgrimage at Resafa-Sergiopolis. According to our observations, Qebaqeb was full of wells, and we recorded a large Roman military camp at the site as well as a possible tower nearby, in addition to wells and canals. At Dedjan a new watering site was identified between Qebaqeb and Qseybe on the southern piedmont of the mountain. All the features testified to an elaborate organisation in the Roman desert frontier zone. The type of the fort and the finds at Qebaqeb indicated possible Roman use already in the second century A.D., when Rome extended its power during the Severan period towards the Euphrates, and the garrison of Dura Europos needed supplies through the desert. The distance to Circesium is ca. 70 km, and the sites could have served that fortress on the Euphrates as well.

Bibliography

Ammianus Marcellinus, transl. Rolfe, J.C. (1963) Loeb Classical Library. Cambridge Massachusetts: Harvard University Press.

Anadol, S. (2008) Palmyra – Identity Expressed Through Architecture and Art, in *Jebel Bishri in Context: Introduction to the Archaeological Studies and the Neighbourhood of Jebel Bishri in Central Syria*, Proceedings of a Nordic research Training Seminar in Syria, May 2004, ed. by Lönnqvist, M., BAR International Series, No. 1817, Oxford: Archaeopress, pp. 59-72.

Borstad, K.A. (2008) History from Geography: The Initial Route of the Via Nova Traiana in Jordan, in *Levant*, Vol. 40, pp. 55-70.

Bowman, A. K., Garnsey, P. and Cameron, A. (2005) *The Cambridge Ancient History*. Vol. 12. The Crisis of Empire, A. D. 193-337. Cambridge: Cambridge University Press.

Børlit, C.S. (2008) Dura Europos: The Final Siege and Abandonment, in *Jebel Bishri in Context, Introduction to the Archaeological Studies and the Neighbourhood of Jebel Bishri in Central Syria,* Proceedings of a Nordic Research Training Seminar in Syria, May 2004, ed. by Lönnqvist, M., BAR International Series 1817, Archaeopress: Oxford, pp. 99-110.

Chapot, V. (1907) *La Frontiére de l'Euphrate de Pompée a la conquête arabe.* Paris: Ancienne Libraire Thorin et Fils. Bibliothéque des Écoles Françaises d'Athènes et de Rome. Fascicule Quatre-Vingt-Dix-Neuf.

CIL = Corpus Inscriptionum Latinarum.

Dąbrowa, E. (1986) The Frontier in Syria in the First Century AD, in *The Defence of the Roman and Byzantine East, Proceedings of a colloquium held at the University of Sheffield in April 1986*, ed. by Freeman, P. and Kennedy, D., British Institute of Archaeology at Ankara Monograph No. 8, BAR International Series, No. 297, Oxford: BAR International Series, pp. 93-108.

Fowden, E.K. (1999) *The Barbarian Plain: St. Sergius between Rome and Iran.* Berkeley/Los Angeles: University of California Press.

Genequand, D. (2005) From 'desert castle' to medieval town: Qasr al-Hayr al-Sharqi (Syria), in *Antiquity*, Vol. 79, pp. 350-361.

Gichon, M. (1991) When and Why Did the Romans Commence the Defence of Southern Palestine?, in *Roman Frontier Studies 1989*, Proceedings of the XVth International Congress of Roman Frontier Studies ed. by by Maxfield, V.A. and Dobson, M.J., Exeter, pp. 318-325.

Grabar, O., Holod, H., Knustad, J. and Trousdale, W., *et al.* (1978) *City in the Desert: Qasr al-Hayr East.* An Account of the Excavations Carried out at Qasr al-Hayr East on Behalf of the Kelsey Museum of Archaeology at the University of Michigan, with the Help of Harvard University and the Oriental Institute of Chicago. Harvard Middle Eastern Monographs 23-24. Cambridge, Mass: Harvard University Press.

Gregory, S. (1996) Was there an eastern origin for the design of late Roman fortifications? Some problems for research on forts of Rome's eastern frontier in *The*

Roman Army in the East, ed. by Kennedy, D.L., Journal of Roman Archaeology, Supplementary Series, Number 18, general ed. by Foss, P. and Humphrey, J.H., pp. 169-209.

Gschwind, M. (2003) S.v. Quadriburgium in *Reallexicon der Germanischen Altertumskunde*, Band 24, Berlin –New York: Walter de Gruyter.

Invernizzi, A. (1986) Kifrin and the Euphrates Limes, in *The Defence of the Roman and Byzantine East, Proceedings of a Colloquium Held at the University of Sheffield in April 1986*, ed. by Freeman, P. and Kennedy, D., Part i, British Institute of Archaeology at Ankara Monograph No. 8, BAR International Series, No. 297 (i), Oxford, pp. 357-381.

Isaac, B. (1990) *The Limits of Empire: The Roman Army in the East.* Oxford: Clarendon Press.

James, S. (2004) *The Excavations at Dura-Europos conducted by Yale University and the French Academy of Inscriptions and Letters 1928 to 1937. Final Report VII. The Arms and Armour and other Military Equipment.* The British Museum Press. Printed by Cambridge University Press.

James, S. (2005) The Deposition of Military Equipment during the Final Siege at Dura-Europos, with Particular Regard to the Tower 19 Countermine, in *Carnuntum Jahrbuch,* pp. 189-206.

Kennedy, D. (1997) Roman roads and routes in north-east Jordan, in *Levant*, Vol. 29, pp. 71-93.

Kennedy, D. and Bewley R. (2004) *Ancient Jordan from the Air.* The Council for the British Academy in the Levant. Dorset: The British Academy.

Konrad, M. (2001) (with the contributions by Baldus, H.R. and Ulbert, T.) *Der Spätrömische Limes in Syrien: Archäologische Untersuchungen an den Grenzkastellen von Sura, Tetrapyrgium, Cholle und in Resafa. Resafa.* Vol. 5. Mainz: Philipp von Zabern.

Konrad, M. (2006) Sunt enim castella barbaris opposita in *Academie Aktuell*, Vol. 3, pp. 46-52.

Lauffray, J. (1983, 1991) *Halabiyya – Zenobia: Place Forte du Limes Oriental et La Haute-Mésopotamie au VIe Siècle.* Tome I: Les duchés frontaliers de Mésopotamie et les fortifications de Zenobia. Bibliothèque Archéologique et Historique T. CXIX. Tome II: **L'architecture publique, religieuse, privée et funéraire.** Bibliothèque Archéologique et Historique T. CXXXVIII. Institut français d'archéologie du Proche-Orient, Beyrouth – Damas – Amman. Paris: Paul Geuthner.

Lewin, A. (2002) Diocletian: Politics and limites in the Near East in *Limes XVIII, Proceedings of the XVIIIth International Congress of Roman Frontier Studies held in Amman, Jordan (September 2000)*, Vol. I, ed. by Freeman, P., Bennett, J., Fiema, Z. T. and Hoffman, B., BAR International Series 1084 (I), Archaeopress: Oxford, pp. 91-101.

Lightfoot, D.R. (1996a) Syrian Qanat Romani: History, Ecology, Abandonment, in *Journal of Arid Environments* 33, pp. 321-336.

Lightfoot, D.R. (1996b) *Syrian Qanat Romani*, at http://www.waterhistory.org/histories/syria/syrian.pdf, pp. 1-9.

Luttwark, E.N. (1979) *The Grand Strategy of the Roman Empire: From the First Century A.D. to the Third.* Baltimore, Maryland: The John Hopkins University Press.

Lönnqvist, K. (2008) The Tax Law of Palmyra and the Introduction of the Roman Monetary System to Syria – A Re-evaluation, in *Jebel Bishri in Context: Introduction to the Archaeological Studies and the Neighbourhood of Jebel Bishri in Central Syria*, Proceedings of a Nordic research Training Seminar in Syria, May 2004, ed. by Lönnqvist, M., BAR International Series, No. 1817, Oxford: Archaeopress, pp. 73-88.

Majcherek, G. and Taha, A. (1993) A Selection of Roman and Byzantine Pottery from Umm el Tlel (Syria), in *Cahiers de l'Euphrate 7*, contributions réunies par Cauvin, J., pp. 107-117.

Mango, M.M. (2004-2005) Report on Landscape Study at Andarin, Syria, September 2005, in *Les Annales Archéologiques Arabes Syriennes*, Vols. XLVII-XLVIII, pp. 117-123.

Mango, M. (2007) Landscape Study at Andarin, Syria, in *CBRL Bulletin of the Council for the British Research in the Levant*, No. 2, pp. 78-82.

Mango, M.M. (2008) Baths, reservoirs and water use at Androna in late antiquity and the early Islamic period, in *Residences, castles, settlements. Transformation processes from late antiquity to early Islam in Bilad al Sham,* ed. by Bartl, K., German Archaeological Institute, Damascus, pp. 73-81.

Mouterde, R. and Poidebard, A. (1945) *Le Limes de Chalcis, Organisation de la Steppe en Haute Syrie Romaine,*

Documents Aériens et Épigraphiques, Plans et Relevés de Lauffray, J. et Mazloum, S. Texte et Atlas. Bibliothèque Archéologique et Historique. Tome XXXVIII. Paris: Geuthner.

Musil, A. (1928) *Palmyrena: A Topographical Itinerary*. American Geographical Society. Oriental Explorations and Studies No 4. New York.

Ortloff, C.R. (2005) The Water Supply and Distribution System of the Nabataean City of Petra (Jordan), 300 BC–AD 300, in *Cambridge Archaeological Journal,* Vol. 15, pp. 93–109.

Parker, S.T. (1986) Retrospective on the Arabian Frontier after a Decade of Research, in *The Defence of the Roman and Byzantine East, Proceedings of a Colloquium held at the University of Sheffield in April 1986,* ed. by Freeman, P. and Kennedy, D., Part ii, British Institute of Archaeology at Ankara Monograph No. 8, BAR International Series 297 (ii), Oxford, pp. 633-660.

Parker, S.T. (2000) Roman Legionary Fortresses in the East, in *Roman Fortresses and their Legions*, ed. by Brewer, R.J., London – Cardiff: The Society of Antiquaries of London and National Museums & Galleries of Wales, pp. 121-138.

Parker, S.T. (2006) *The Roman Frontier in Central Jordan: Final Report on the Limes Arabicus Project 1980-1989.* Vols. I and II. Dumbarton Oaks Studies XL. Washington D.C./Cambridge, Mass: Harvard University Press.

Poidebard, A. (1934) *La trace de Rome dans le désert de Syrie, Le limes de Trajan a la conquête arabe, Recherces aériennes (1925-1932).* Bibliothèque Archéologique et Historique Tome XVIII. Texte & Atlas. Paris: Geuthner.

Rostovtzeff, M.I, Bellinger, A.R., Hopkins C. and Welles, C.B. (1936) *The Excavations at Dura-Europos* conducted by Yale University and the French Academy of Inscriptions and Letters: Preliminary Report of Sixth Season of Work October 1932 - March 1933, Preliminary Report of Sixth Season of Work, October 1932 - March 1933. New Haven: Yale University Press.

Roth, J. (1999) *The Logistics of the Roman Army at War (264 B.C.-A.D. 235).* Columbia Studies in Classical Tradition 23. Leiden: Brill Publishers.

Saito, K. (2004-2005) Excavations at the Southeast Necropolis in Palmyra, in *Les Annales Archéologiques Arabes Syriennes*, Vols. XLVII-XLVIII, pp. 137-149.

Sartre, M. (2005) *The Middle East under Rome.* Cambridge, Mass.: Harvard University Press.

SAS 2002= *Syria, Archaeology from Space* (2002) GORS, General Organization of Remote Sensing, Damascus. Syrian Arab Republic. Damascus.

Seland, E. (2008) Trade Routes of Palmyra with Special Notes on Western Routes in the Palmyrene Trade, in *Jebel Bishri in Context: Introduction to the Archaeological Studies and the Neighbourhood of Jebel Bishri in Central Syria*, Proceedings of a Nordic Research Training Seminar in Syria, May 2004, ed. by Lönnqvist, M., BAR International Series, No. 1817, Oxford: Archaeopress, pp. 89-97.

Taha, A. (1991) Prospection du Site Romain Tardif de Juwal (Cuvette d'el Kowm, Syrie), in *Cahiers de l'Euphrate* 5-6, pp. 61-66.

Toivanen, H.-R. (2008) Fortresses and Ecclesiae on the Imperial Border of Byzantium, in *Jebel Bishri in Context: Introduction to the Archaeological Studies and the Neighbourhood of Jebel Bishri in Central Syria*, Proceedings of a Nordic Research Training Seminar in Syria, May 2004, ed. by Lönnqvist, M., BAR International Series, No. 1817, Oxford: Archaeopress, pp. 111-126.

Ulbert, T. (1986) Die Basilika des Heiligen Kreuzes in Resafa – Sergiupolis, Mit Beiträgen von Bayer, I., Gatier, P.-L., Grosser, D., Khoury, R.G., Mackensen, M. und Römer, C. *Resafa.* Teil II. Mainz am Rhein: Deutsches Archäologisches Institut.

Wilson, A. (2001) *Janiculum Mills Excavations 1998*, at http://users.ox.ac.uk/~corp0057/aquatraiana.html, pp. 1-2.

12. THE "BISHRI CORRIDOR" BETWEEN THE DESERT FRONTIER ZONE AND THE EUPHRATES LIMES

Minna Lönnqvist Kenneth Lönnqvist
with Contributions by Jari Okkonen

Fig. 1. The village of Ash-Shujiri in the central area of Jebel Bishri. Photo from NW: Minna Lönnqvist 2006.

12.1. Roman Military Installations near Ash-Shujiri

12.1.1 A Well and a Tower at Ash-Shiqri

Until now there has practically existed a triangular 'blank space' on maps in the region of Jebel Bishri between the desert frontier generally referred to as the *Strata Diocletiana* district on the western piedmont and the Euphrates *limes* following the river. Apart from A. Poidebard's (1934) identification of the Roman watchtower at Aš-Šiqri (see ch. 11, Fig. 2.), no military installations typical of the Roman eastern frontier are known from the central area of the mountain. This is at least the way in which the eastern border of the Roman Empire is depicted in several history books and atlases. This picture is due to the lack of ground studies on Jebel Bishri and the fact that before the work of the Finnish project SYGIS the mountain proper had archaeologically largely remained *terra incognita.* Based on the new findings, we here define the Roman itinerary over Jebel Bishri as "the Bishri corridor".

Fig. 2. Poidebard's photograph of Aš-Šiqri and the Roman watchtower (Poidebard 1934).

A. Musil's (1928: 168-178) itinerary of 1912 partly overlaps with the corridor area, but apart from Resafa-Sergiopolis, Musil did not mention any ancient remains that he would have observed during his travel with a caravan over Jebel Bishri. Musil (1928) visited the well site of Be'er Rehub, where goats, sheep and cows were watered. He also mentions another spring, As-Sižri, at the eastern bend of Zetab al-Bishri. The latter spring seems to refer to the same location as A. Poidebard's Aš-Šiqri, which may be identified as being somewhere in the neighbourhood of the village of Ash-Shujiri on the summit area today. The map by R. Mouterde and A. Poidebard (1945: 131) illustrates the alleged Roman itinerary punctuated by wells across Jebel Bishri leading from Resafa through Qseybe to Circesium (173 km: Resafa – Be'er Rehub 35 km, Be'er Rehub – Be'er Siqri 42 km, Be'er Siqri – Qseybe 31 km, Qseybe - Circesium 65 km). Poidebard (1934: 92) reports that when he visited the site of Aš-Šiqri he identified there a Roman tower 14 m x 16 m in plan. The tower provided views as far as to Resafa-Sergiopolis and the Euphrates (see Fig. 2.). We could not confidently locate the site of Aš-Šiqri, neither the well of Be'er/Bîr Šiqri nor the tower originally identified by Poidebard (1934: 920; Mouterde and Poidebard 1945: 134), although we later identified a landscape which parallels with Poidebard's Aš-Šiqri at Ash-Shujiri (Fig. 1.). Its location also matches with Poidebard's maps. However, we found another comparable Roman water station in the neighbourhood at a place called "Darakhlia".

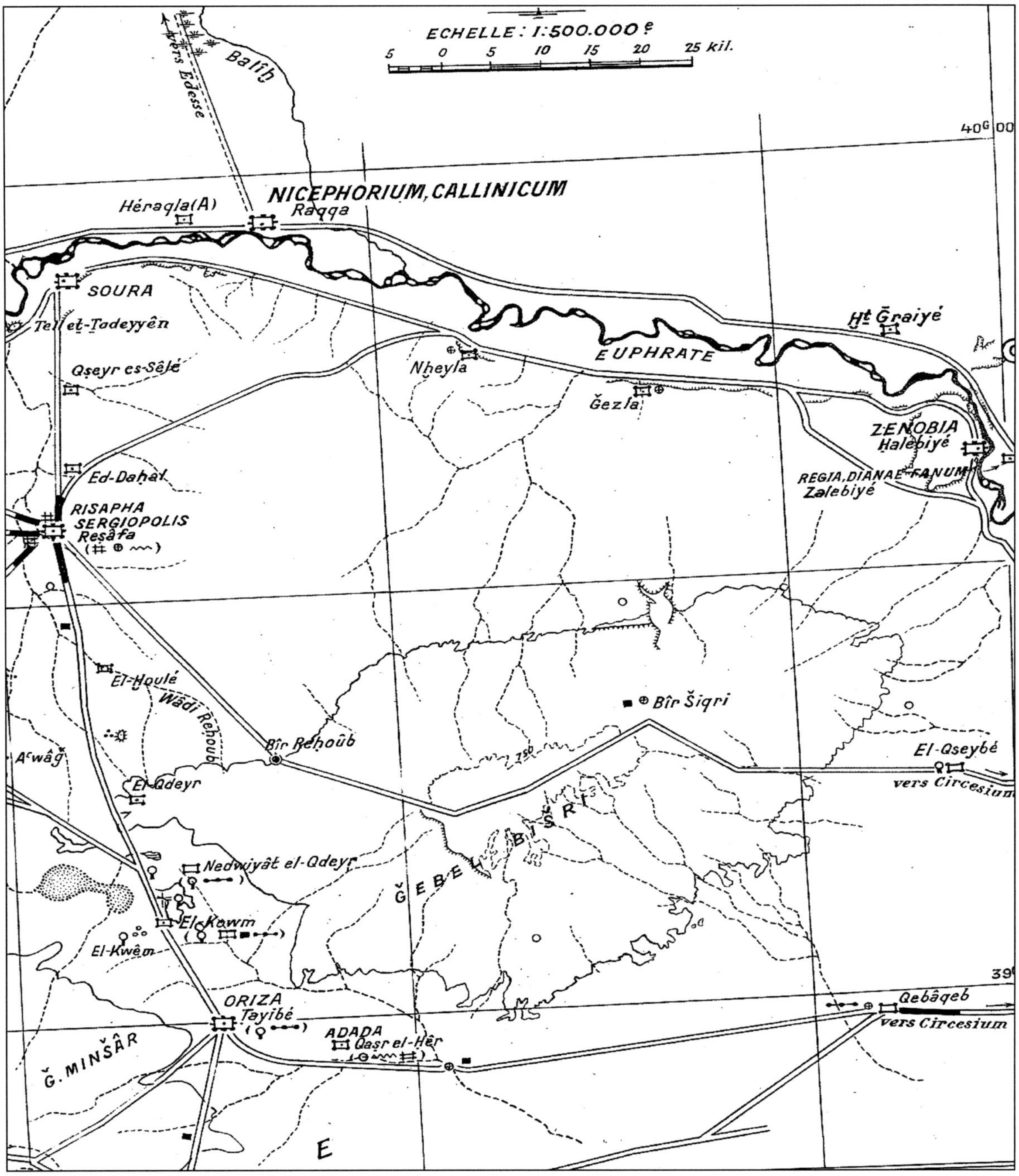

Fig. 3. A Roman itinerary over Jebel Bishri from Bîr Rehoûb to El Qseybe and towards Circesium according to Mouterde and Poidebard 1945. Courtesy by Paul Geuthner.

12.1.2 A Roman Water Reservoir, Dam and Tower at Darakhlia

Darakhlia, which was first visited in 2003 with a Bedouin guide, is situated in the eastern part of central Jebel Bishri ca. 5 km east of Ash-Shujiri (Fig. 4.). The area yielded a large water harvesting site consisting of a dam/barrage, a rectangular water reservoir and remains of a stone tower designated as site **K 3**.

Site K 3
Water Harvesting Site with Tower

UTM 0543591, 3920667, 690 m a.s.l.
Size of the site: ca. 160 m x 40 m

The size of reservoir: max. 53 m x 16.6 m in plan, preserved h. max. 3.5 m; dam/barrage: 14. 2 m in l., max. 3.8 m in h., 1.5 thick; tower: 7 m x 7 m in plan.

Ruins of a square tower are situated north of the barrage on a nearby hillock, and its apparent intention was to overlook the water installation. Because of their mutual vicinity and likely functional association the tower and the water installation are obviously contemporary. On the basis of the associated pottery from the tower, the structures can be identified as being Roman. The site **K 3** (Figs. 4., 5. and 6.) appears to be situated at the crossroads of three important routes: it can be reached from the north while crossing the mountain, from the south through the valley of Nadra, or from the west through Ash-Shujiri. K 3 is the largest one of the ancient water harvesting site encountered on Jebel Bishri so far. Some later concrete and cement constructions had appeared in 2004-2005 to the north extending the evidence of seasonal water collection activity at the site. This suggested that the system has been brought back into use by the Bedouins, meaning that it is still operative and functional at the time of seasonal rainfall.

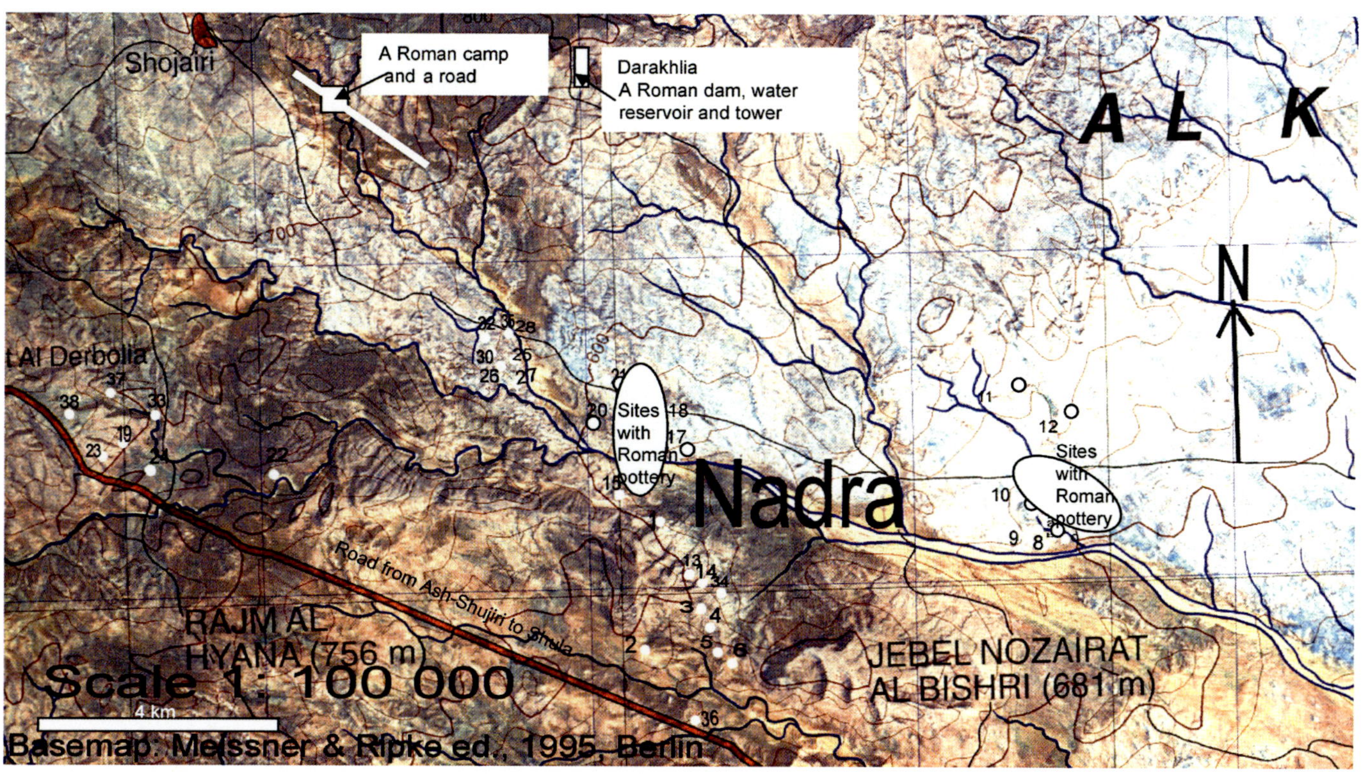

Fig. 4. Sites recorded near Ash-Shujiri (Shojairi on the map) and at Nadra during the seasons of 2004-2006.

Fig. 5. A Roman dam, water reservoir and ruined tower at Darakhlia. Photo from S: Kenneth Lönnqvist 2003.

Fig. 6. A Roman dam and water reservoir at Darakhlia. Photo from SE: Kenneth Lönnqvist 2003.

Fig. 7. The plan of the water reservoir at Darakhlia. Drawing by Minna Lönnqvist.

The rectangular reservoir is built of limestone and aligned according to the cardinal directions on the horizon. The long walls are built from the north to the south and short walls from the east to the west. The maximum length of the reservoir is 53 m on the eastern side and 46 m on the western side, the total width being 16.6 m (Fig. 7.). An original height of ca. 3.0-3.5 m is conjectured from the dam fence (Fig. 8.). All the architectonic blocks and boulders of the cistern are of local undressed limestone, fitted together with mortar. The long wall to the east is 1.10 m wide or thick, consisting of three rows of masonry that form the flight of three descending steps.

Fig. 8. Steps and the barrage of the reservoir at Darakhlia. Photo from N: Minna Lönnqvist 2003.

The long wall to the west is 2.8 m wide and consists of six rows of masonry that form six descending steps, each from the stone blocks being ca. 30-40 cm in size (Fig. 8.). The general size of the undressed rectangular step blocks is 30-40 cm x 60 cm. Some of the stones were still partially covered with plaster, so it may be suggested here that the walls of the reservoir were originally plaster-covered. On the western side, the depth of the pool from the surface to the bottom is at present ca. 2.2 m. However, this would not accurately reflect the original water capacity because of the large amount of later accumulation of soil and debris in the pool. The fact that there were steps along the long walls of the cistern suggests that these may have been used to facilitate access to the pool, both in terms of cleaning the pool and drawing of water, but also perhaps to monitor the amount of water left in the reservoir. The size of the reservoir suggests that the collection of water at the site was not a problem, and that this amount could easily be collected. The dam fence or barrage at the southern end of the water reservoir is 14.2 m in length and 1.5 m thick. Its preserved height is 3.6 m, and it is partially collapsed in the southeast (Fig. 8.).

The calculated capacity of the reservoir is ca. 2,500-3,000 m^3 of water. This suggests provision of water for large numbers of people or animals for an extended period, into the dry season. At Humayma in Jordan, a Late Roman fort had a cistern that is said to have held 1,275 m^3 of water (Kennedy and Bewley 2004: 181). The stepped reservoir or settling pool at Darakhlia was constructed to control a natural seasonal watercourse. As often indicated, the region is arid, but large quantities of water usually pass along the *wadis* on a seasonal basis. This fact led the Romans to take advantage of the water, and divert it into channels or reservoirs of this type. As we have now seen, this happened both at the upper sources, on the mountain such as here at Darakhlia, and at the final run-off on the piedmonts, in places such as Qebaqeb.

The square tower (UTM: 0548630, 3920887, 695 m a.s.l.) situating on a hillock ca. 100 m north of the dam and the water reservoir encloses an area of 7 m x 7 m in plan. According to A. Poidebard (1934: 168) this is a typical size of the small *limes* towers in Syria. The walls in the Darakhlia stone based tower have been preserved up to the height of 65-70 cm, and there were no visible 'entrances' in the foundation stone course. The construction at Darakhlia is aligned ca. 20°-180° and 110°-290°, or ca. 20° E of N. The walls are constructed of two courses of partly worked large limestone boulders (now badly weathered) up to 40-50 cm in diameter and smaller ones in the inner course with mixed fill, producing very thick walls from 1.5 m to 1.75 m thick on all sides, except for the east where the walls are thinner. Roman pottery was scattered inside and around the tower (Figs. 9. and 10.). The thickness of the walls suggests that the original structure may have been several storeys high and the entrance was higher up. There were no associated finds of building materials such as bricks at the site. Therefore, the upper storeys and structures and the roofing are matters of conjecture.

Fig. 9. Pottery from the tower at Darakhlia.
Photo: Kenneth Lönnqvist 2008

Fig. 10. Pottery from the tower at Darakhlia.
Photo: Kenneth Lönnqvist 2008.

The watch or signal towers especially appear in the *limes* zones and are often referred to in the literature as *burgus*, sometimes *turris*. During antiquity Vegetius gave even advice to build *burgi* for protecting water sources outside cities. (Isaac 1990: 178-179). The wall thickness of our tower has parallels, for instance, on the Upper Rhine, Kleiner Laufen in Switzerland (Southern and Dixon 2000: 146), with a tower measuring ca. 8 x 8 m that had walls close to 2 m in thickness. There are also examples of other watchtowers from Roman Germany, where the ground floor has no entrance and the tower is entered from a higher level by a ladder (Southern and Dixon 2000: 146). The fact that the tower was located on an E-W track road along the top of a hillock overlooking the major water reservoir on the mountain and serving the nearby Roman sites lends support to the idea that the remains were of a watchtower or signalling post. Therefore, it seems quite clear that the watchtower and the water harvesting site were part of the Roman military organisation in the inner regions of the mountain. It could have been still used in the Byzantine times as well, although no Byzantine pottery was found.

Towers of the approximate size with our tower are also found, e.g., along the Roman highways at the Dead Sea or in the Arabah, in Israel. For instance, at Migdal Tsafit a tower measuring 6.15 m x 6.15 m in plan was excavated. It was originally built in the second century A.D. and reused in the late third century A.D. (Erickson-Gini 2007: 95). The watchtowers in the Eastern *limes* area are generally of stone or brick, while in Britannia and Germania, besides stone they were often constructed of timber. A timber watchtower found in the Valkenburg-Marktveldt in Holland was 6.35 m x 6.5 m, covering an area slightly over 41 m^2. The tower is dendrochronologically dated to A.D. 123-124. Most of the references date from the second century A.D. onwards and are connected with stone towers. (Van Dierendock s.a.:75-78).

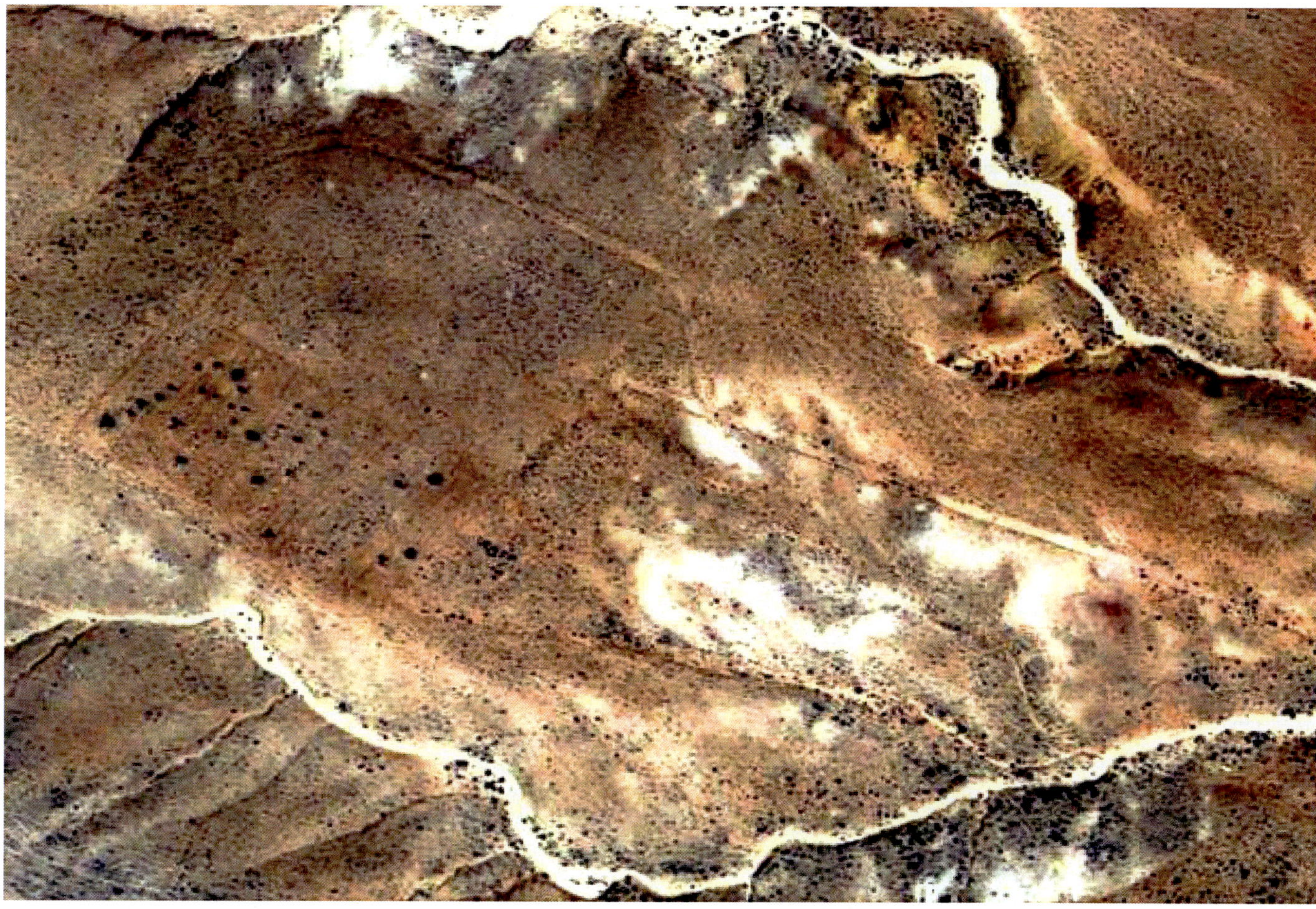

Fig. 11. Camp K 1 on Jebel Bishri visible in a QuickBird satellite image © Eurimage 2003

12.1.3 A Roman Military Camp near Ash-Shujiri

While prospecting with a QuickBird satellite image in Finland in 2003 we identified a previously unknown Roman military camp (Fig. 11.) ca. 4 km southeast of Ash-Shujiri. It is situated on the eastern slope of the central ridges of Jebel Bishri, close to the 700 m elevation contour line. In the satellite image a square pattern with rounded corners was recognized. It followed the outline of a classic military camp, the so-called playing-card shape (see Kennedy and Bewley 2004: 174, Fig. 10.2.A; Hanson and Maxwell 1983: 20), with a *via principalis* leading through the camp area. This new site was identified, recorded, documented and mapped on the ground with an EDM/tachymeter in 2005 (see Fig. 15. below).

The location of the fort in the terrain was pinpointed by means of the coordinates received from the satellite image and a GPS on the ground. The site was designated as **K 1**. It appeared to be an impressive site because of its location in the landscape. The camp or fort takes advantage of a topographical position on a small plateau that commands the water sources down into the northwest-southwest oriented *wadi* and surrounding plains (Figs. 11., 12. and 13). This is often described as one of the classical features of the Roman camps and forts. The camps and forts were situated on slightly elevated ground, preferably by a watercourse in a valley, where it was possible with a relatively small effort to keep an eye on the traffic in and out of a valley (Hanson and Maxwell 1983: 19). Another clear reason for the choice of this site is that the promontory surrounding the site on three sides protected it from the harsh winds and sandstorms coming from higher ground. In this case, the located road that traverses the camp, and which is described in more detail below, apparently represents one of the principal routes along which troops would have been travelling from the northwest to the southeast so, or from the mountain to the Nadra plain, back and forth from other Roman sites and forts. In 2005 the inner areas of the enclosure had been cultivated and ploughed in the N-S direction, the sparse plantation consisting of scattered fig trees.

Site K 1
Military Camp

UTM 0540793, 3917317, 730 m a.s.l.
Size of the site: ca. 110 m x 120 m

The camp **K 1** is a roughly square enclosure covering ca. 110 m x 120 m in plan or ca. 13, 200 m^2, 372 x 405 *pes Monetalis*, pM or Roman feet (Henderson 1991: 73) or 1.3 hectares. The short walls are aligned roughly NW-SE (115°-295°) and the long walls NE-SW (35°-215°). Remnants of what appears to be a low wall of mixed rubble of earth and stone were only found on the western and southern laterals of the square. This wall is not thicker than ca. 1 m, but it included boulders up to 100 cm x 50 cm x 10 cm in size with smaller fieldstones packed in between them.

The type of the site could be categorized as a "flimsier temporary camp" in D. Kennedy's terms. The type is frequently referred to in ancient literary sources but is not known in numbers in the Roman East. Instead over 400 examples of this type are known in Britain alone on the northwestern *limes* of the Empire, and many are found each year using remote sensing methods like aerial photography. (Kennedy 1996: 13). According to S. Gregory (1997: 40), the "playing card" type of a camp or a fort type that was in use from the first to the mid-third century A.D. Gregory further points out that the type of the fort was not primarily built to protect from or invade the enemy, but to its purpose was to establish the military organisation in hostile, more "barbaric" or recently subjected areas.

Fig. 12. General view from a mountain ridge over the camp K 1 and the wadi behind. Photo from S: Kenneth Lönnqvist 2005.

Fig. 13. The site of the camp K 1 photographed from the other side of a deep wadi. Photo from N: Kenneth Lönnqvist 2008.

As depicted in Fig. 15. below, the large stones of the fort recorded by EDM are grouped along the four sides of the square. In addition to this, inside the enclosure along the western long wall there appeared a 'squarish' area covering approximately an area of 7 m x 10 m, which showed a high concentration of large stones (0.5-1.0 m in diam.). It formed no visible shape or structure, but such concentration was not observed anywhere else in the large enclosure. For instance, at the third century A.D. camp of similar size at Oboda in Israel (Erickson-Gini 2007: 93), a *principia* was built along the western wall facing the entrance to the camp, but in the case of K 1 a surface investigation is not enough to suggest this. Many stones could also be seen down the southern slope of the camp, and these probably originated inside the enclosure, from where they had been removed by the people who are presently cultivating the plot of land. The QuickBird satellite image appears to indicate to the existence of barracks along the southern edge of the site. However, the existence of possible barracks was harder to verify on the ground because of the cultivation and the disturbances introduced by recent ploughing. No constructions that could be identified as towers were found either. It is therefore quite impossible to say anything definitive about the stone structures that may have existed inside the enclosure, other than that there appear to have been some.

Nevertheless, a clear track road, the principal road or the so-called *via principalis*, already recognizable from the satellite image enters the enclosure from the east from a direction of ca. 108°, the continuation of which can be rediscovered on the western flank of the fort, at the expected point of exit. It means that there was one entrance to the SE and one in the NW. Furthermore, before entering the enclosure, the track road divides into two, the second track following exactly the northern perimeter of the camp until it joins the principal road that traverses the square in the NW corner of the fort. The point of entry and exit of the track road into the site K 1 is one of the most typical features of Roman military architecture: namely that the *via principalis* enters a camp or a fort one third of the way along one of the long sides (Kennedy and Bewley 2004: 174-176, Figs. 10.2.A-B; Hanson and Maxwell 1983: 20, Fig. 1.5; Henderson 1991: 79; see Fig. 13.8.). We verified the existence of the track road on the ground. It is ca. 4 m wide as it continues east. No structures or pavement could be discerned up to a distance of ca. 1 km to the east of the camp. Nor were remains of bridges or other structures seen at the bottom of the *wadi*, where the road crossed the seasonal watercourse (see Fig. 14. below). The photograph below shows also clearly dominant *wadis* and seasonal streams to the left of the fort. The valley of Nadra is visible in the upper right corner.

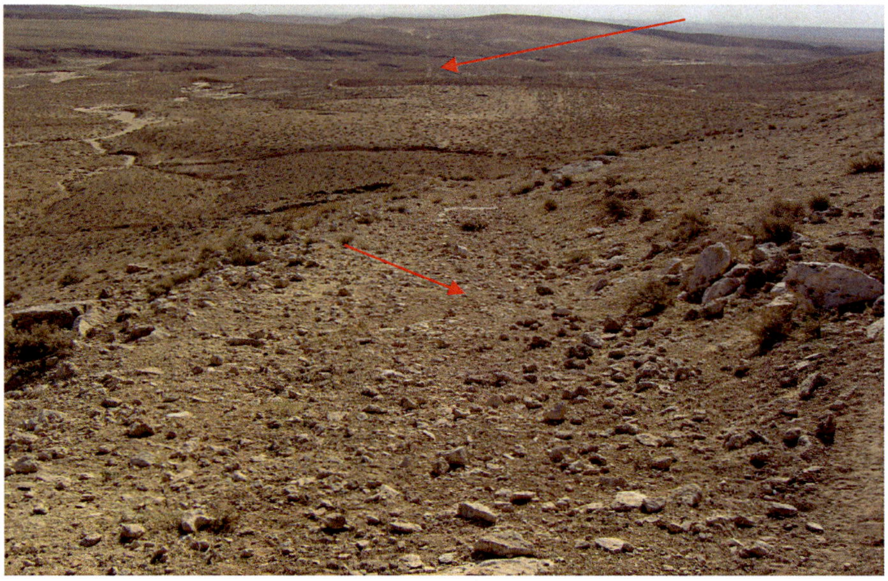

Fig. 14. A straight line of via principalis running through the camp area.
Photo from NW: Kenneth Lönnqvist 2005.

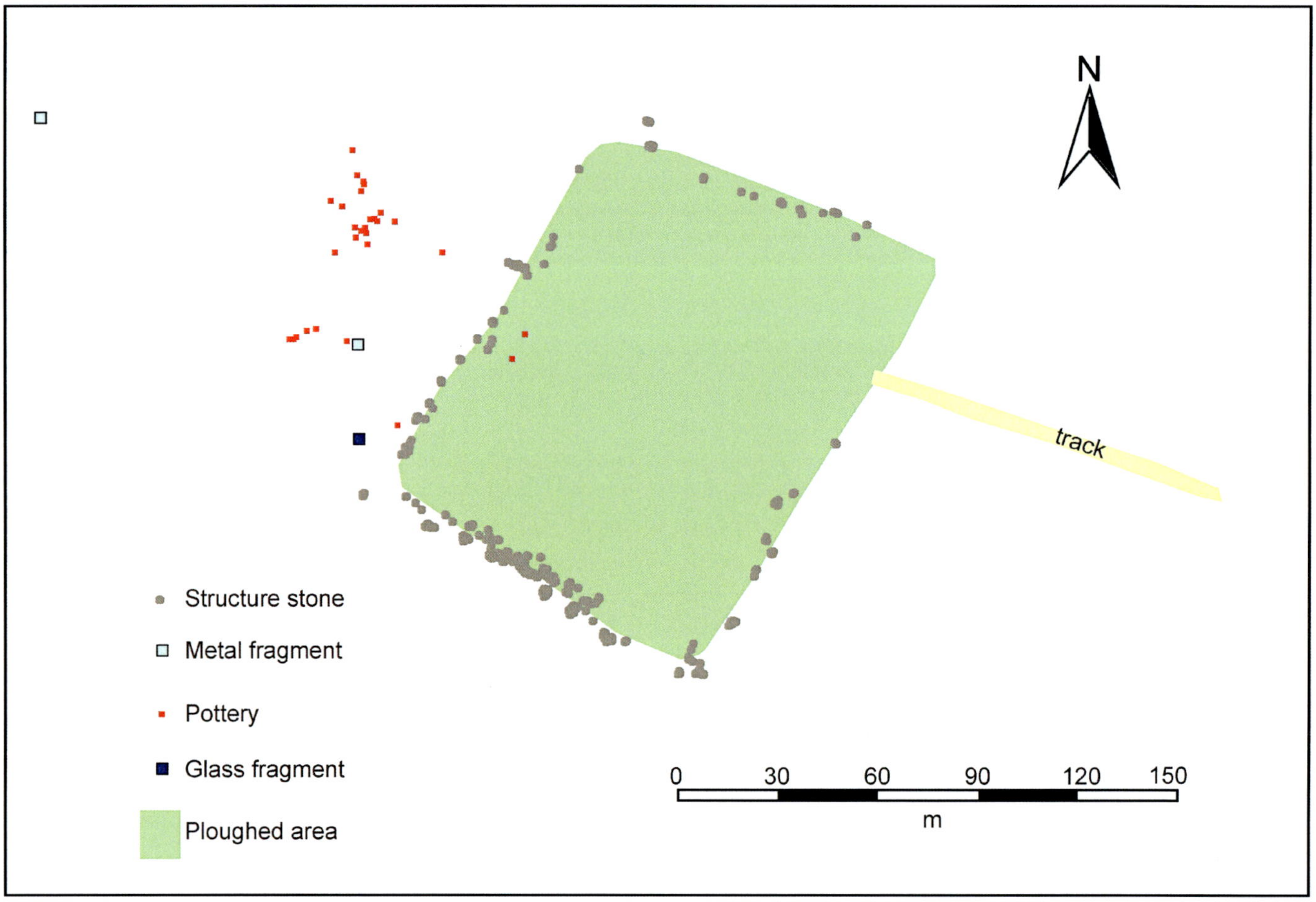

Fig. 15. The plan of the Roman camp with associated archaeological finds.
Mapping by Jari Okkonen and Kenneth Lönnqvist 2005. Data and image by Jari Okkonen.

Finds were made mainly on the western slope outside the camp area (see Fig. 15.). The site produced one fragment of copper alloyed scale armour (3 cm x 2.3 cm in size, 0.84 cm in thickness) (Fig. 16.), a fragment of Roman glass (Fig. 17.) and a number of pottery pieces including brittle ware from the Roman period (Figs. 18., 19.). This area to the west would correspond to the entrance into the camp from the west. Inside the large enclosure, only a few pieces of pottery were recovered along the western lateral side together with flint, the latter being of prehistoric origin. At the camp we found similar scale armours at the Roman fort site of Qebaqeb (ch. 11.). These finds correspond to the copper alloyed scale cuirasses discovered at Dura Europos (cf. James 2004: 121-122, 132-133) on the Euphrates providing the *terminus ante quem* A.D. 256. A recovered rim of a white glass vessel belongs to a glass jar or a bottle (Fig. 17).

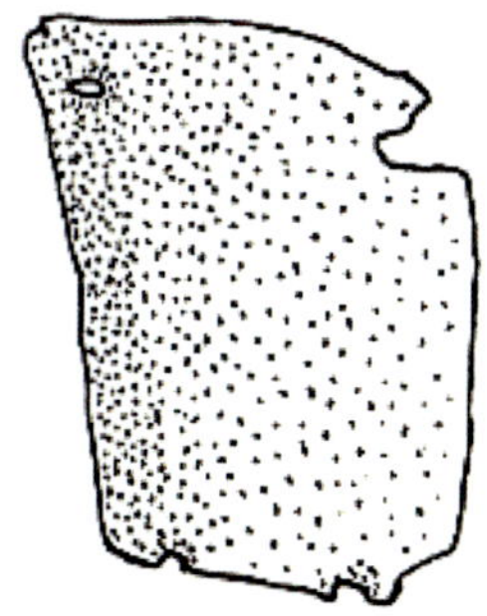

Fig. 16. An armour scale from the Roman camp K 1 (see metal fragment on the map Fig. 15.). Drawn by Kenneth Lönnqvist 2007. Scale 1: 1.

Fig. 17. A mouth-piece of a Roman glass jar or bottle from K 1. Photo: Kenneth Lönnqvist 2009.

Fig. 18. Pottery from site K 1.
Photo: Kenneth Lönnqvist 2009.

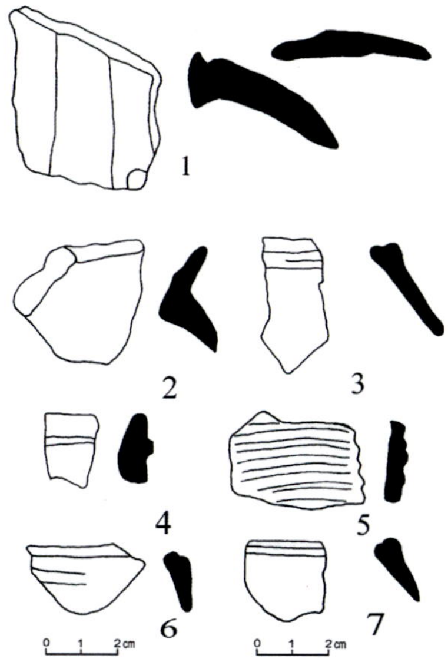

Fig. 19. Pottery profiles from site K 1.

Site K 2
Ruined Houses

UTM: 0539429, 3919109
Size: unrecorded

It needs to be noted here that the camp K 1 afforded visibility to two ruined but sturdy rectangular houses, i.e., site **K 2** (Fig. 20.) outside our target area near Ash-Shujiri ca. 2.5 km to the northwest from K 1. Therefore the site was not documented. The buildings resemble a guarding station located on a commanding hill. These rectangular houses are relatively large buildings with thick stone walls constructed in the same manner as the Roman houses at Qseybe (see paragraph 12.4. below) on a longitudinal axis towards the north, one after another, with a good view to all directions and towards the scenery in Fig. 1. On the slope of the hill, we saw the only acacia trees in the neighbourhood (see ch. 4: Fig. 16.). Surface pottery associated with the site K 2 dated from the Byzantine–Mediaeval period, but the site appears to have been later used by Bedouins as well. In the neighbouring hills there are abandoned Bedouin houses belonging to the village of Ash-Shujiri.

Fig. 20. Site K 2. Photo from SE: Kenneth Lönnqvist 2006

12.3. Roman Activities in the Plain of Nadra

According to the results of the survey in 2004, there was activity in the valley of Nadra during the Roman period (see Fig. 4.). The height of the region in the central eastern piedmont of Jebel Bishri is ca. 600-700 m a.s.l. In the transect selected for the survey, several sites such as **I 7**, **I 8**, **I 15**, **I 21** and **I 35** produced Roman pottery; reddish and blackish brittle cooking ware including examples of both handles and body parts of vessels. No architecture or structures that could be described as 'military' were identified in connection with the sites at Nadra. What was thus the possible connection of the civil activities to the military settlements, could not be determined by the surface survey, else than that the remains seemed to align along the Roman track in the southeastern piedmont. It is, therefore, apparent that the newly discovered Roman sites at Nadra are found on ancient itineraries along large and smaller *wadis* located between the military camp K 1, situated near Ash-Shujiri, and the military post of Qseybe (paragraph 12.4.) further along the piedmont to the east.

Site I 1
Bedouin Village and Ancient Roads?

UTM: 0546295, 3911090; ca. 595 m a.s.l.
Size of the roads: ca. 5 m in w., max. thickness in the profile 0.5 m.

At Nadra there are track roads (Figs. 21. and 22.) leading in a right angle through an abandoned Bedouin village site **I 1** (see more in ch. 15.). The roads are comparable in structure to the *via principalis* running through K 1, and one even follows its alignment. Because of the 5 m width, the structure of the roads and the sites providing evidence of Roman remains in the neighbourhood, the roads may date from the Roman period.

Fig. 21. A desert road near the Roman sites at Nadra. Photo: Christian Meyer 2004.

Fig. 22. The profile of the desert road at the abandoned village of I 1 at Nadra. Photo: Christian Meyer 2004.

Site I 7
House Bottom and Cairns/Tumuli

UTM: 0551944, 3910610, 530 m a.s.l.
Size of the site: ca. 120 m x 10 m
Size of the house bottom: 8 m x 3.6 m

This site, already described in ch. 6., is located on a hilltop and produced cairns/*tumuli* associated with a rectangular stone foundation of a building (8 m x 3.6 m in size), apparently the so-called *domicilium*. The site produced Roman pottery (Fig. 24.). The house was oriented roughly from the east to the west with a door entrance on the longer side facing the south. A *domicilium* consisting of one rectangular space is the simplest form of domestic buildings in the Roman and Byzantine Syria-Palestine. These structures appear in the rural village areas of the Hauran and the Golan. Most of the people lived in such houses, the size of which varied from ca. 20 m² to 200 m². (Hirschfeld 1995: 24).

Fig. 23. A bottom of a house at site I 7. Photo: Kenneth Lönnqvist 2004.

Fig. 24. Roman pottery from site I 7. Photo: Kenneth Lönnqvist 2004.

We also found houses with the same plan at site K 5 at Qseybe, a Roman fort site that we shall discuss shortly. In the vicinity of the building of I 7 there were various unidentifiable structures, the function of which is not clear.

Site I 8
Open Air Site

UTM: 0551925, 3910802, ca. 525 m a.s.l.
Size of the site: not measurable

This is an open accumulated site at the northwest bank of a *wadi* without visible structures but producing Roman pottery beside Palaeolithic flints.

Fig. 25. Roman pottery from site I 8. Photo: Kenneth Lönnqvist 2004.

Site I 15
Collapsed Bridge or Dam

15B: UTM: 0545504, 3911420; ca. 600 m a.s.l.
Size of the structure: ca. 3.5 m x 3 m

This collapsed bridge or a dam is composed of limestone boulders set 3.5 m in width on both sides of a small E-W oriented and ca. 3 m wide *wadi*. The site is associated *inter alia* with Roman pottery (Fig. 26.). However, a roofed underground water reservoir – mentioned in ch. 4 - hollowed into the marble/gypsum bedrock - was also discovered at the site.

Fig. 26. Site I 15 produced a few pieces of Roman pottery beside some Bronze Age sherds. Photo: Kenneth Lönnqvist 2004.

Site I 21
Stone Enclosures/Corrals

UTM: 0545502, 3912971- 0545511, 3913001;
582-595 m a.s.l.
Size of the site: ca. 110 m x 20 m

This is a typical pastoral stone enclosure/corral site with cairns/*tumuli* (see more in ch. 6) associated with a water barrage. The site produced Roman pottery (Fig. 27.)

Fig. 27. Roman pottery from site I 21 at Nadra. Photo: Kenneth Lönnqvist 2004.

Site I 35
Stone Enclosures/Corrals (3)

UTM: 0543344, 3913613; 616 m a.s.l.
Size of the site: 150 m x 20 m

This is also a typical pastoral enclosure corral site (see ch. 6.) Mould-made Roman/Byzantine pottery (Fig. 28.) was discovered at the site.

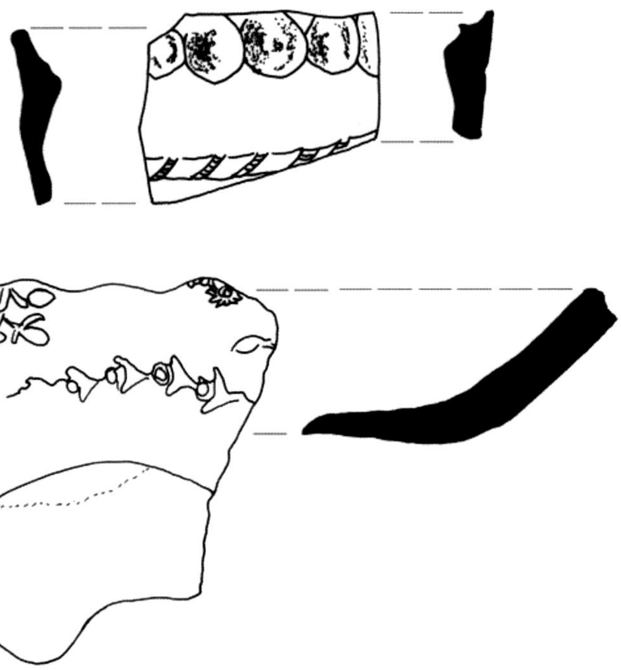

Fig. 28. Floral and other designs in the decoration of mould-made Roman/Byzantine pottery vessel from site I 35. Drawn by Helena Riihiaho 2004 and Kenneth Lönnqvist 2010.

12.4. The Military Post of Qseybe on the Southeastern Piedmont

The Roman site of Qseybe (Fig. 3.) is situated further southeast of Nadra on the piedmont of Jebel Bishri. The area of "Ksejbe" was visited by A. Musil (1928: 178), who mentions the existence of a gendarme station, khan and cultivated fields but no ruins. The Roman site was identified by A. Poidebard (1934) in his aerial studies and mapping, but he did not visit the site on the ground. The site name of Qseybe is derived from the Arabic "Qasabi" meaning a "capital". (The same site name is also associated with a village on the peninsula of Halabiya on the Euphrates.) This site on the eastern piedmont and on the desert fringe can be reached today from the Palmyra-Deir ez-Zor -highway. A road turns northwest at the crossroads of Al Kharata ca. 12 km from Deir ez-Zor towards Palmyra, continuing according to the road signs to the Al Kharata Petrol Company, which is 28 km from the modern main road. Qseybe is located ca. 1 km from the main plant of the Petrol Company. Nowadays the environment around the ancient Roman site is desert with a few shrubs, but the neighbourhood has several small green oases with springs and wells, currently occupied by local Bedouins, echoing the description of the oasis by A. Musil (1928: 178).

We identified two different archaeological sites at Qseybe on both sides of a modern track road: a Roman house compound site **K 5** and a Roman well, as well as a Roman fort or fortlet site **K 6** (Fig. 29.).

Fig. 29. The locations of sites K 5 and K 6 in between which there is a well. Mapping on a SPOT image by Minna Lönnqvist 2010.

Site K 5
Settlement

UTM 0568815, 3911211; 408 m a.s.l.
Size of the size: ca. 200 m x 200 m

Site **K 5** is a settlement obviously connected to the headquarters of the Roman fort **K 6** situated to the southwest. (This site had, however, been already occupied in the prehistoric period, as the surface area yielded numerous worked flints, among which there were many Palaeolithic and Epipalaeolithic types presented in ch. 5. This prehistoric activity was possibly due to available water sources.) The surroundings of the site K 5 described consisted of exposed bedrock and silt. There were three groups of ruined houses at K 5, of which the two easternmost ones appear to be Roman (Fig. 30.) because of their structure and the significant traces of surface pottery from the Roman period (Figs. 32. and 33.). All the houses followed the so-called simple house or *domicilium* type (Hirschfeld 1995: 24), similar to the previously mentioned house type at Nadra.

Fig. 30. Ruins of domicile type house at Qseybe. Photo from N: Minna Lönnqvist 2006.

The two obviously NE-SW aligned rectangular Roman houses had one to three rooms built on a longitudinal axis one after the other. The houses were standing 18.4 m apart. The northernmost house that was divided into three separate rooms measured 12 m x 4 m in size and it had a small annex of 2 m x 1.8 m in the southeast corner. The southern house had one single rectangular room 7.80 m x 4 m in size with an entrance facing north. The foundation walls of the houses were 20 cm thick, and the building technique involved greenish gypsum slabs (35 cm x 25 cm x 10 cm in size) fitted together with mortar. Roman bricks, like the mentioned pottery, were associated with the houses, meaning that the upper layers of the buildings had probably been built of bricks. Finally, the third rectangular house 20 m to the west had been built of limestone boulders instead and seems to have served a modern Bedouin house, or possibly being a reused Roman one. The entrance was in the northern end of the house. Rural houses from the Roman times were often still occupied in the Byzantine and Islamic times as is known in Jordan (see more on rural houses of the Roman period in Jordan, in El-Khouri 2008: 80-82).

As already indicated, significant amounts of Roman pottery pieces were discovered at site K 5. Obviously cooking/storing had taken place at the site. Interestingly one piece of pottery, apparently belonged to a locally produced cooking pot, seems to bear two Aramaic letters reading לו (Fig. 31.). If this interpretation were correct, it would be the first local proof of the existence of Aramaeans on Jebel Bishri and could indicate that locals such as Palmyreans served the Romans. Surprisingly an oyster shell was also found on the surface in the midst of the Roman pottery at K 5. Whether the oyster shell represents a trace of a culinary diet and a luxury life style in the desert fort left by the Romans is not certain. It may represent, of course, the remains of a recent picnic. On the other hand, it is not impossible that the culinary habits of the Roman soldiers in the East brought about rumours of a luxury life, as reported by ancient historians such as Tacitus and Corbulo. There is also plenty of archaeological evidence of oyster shells on many inland sites elsewhere in the Roman period, which shows that a system was in place, which enabled the transport of fresh oysters to far-away sites (Greene 1986: 79).

Fig. 31. A piece of pottery with inscribed (Aramaic?) signs from the site K 5 at Qseybe. Photo: Kenneth Lönnqvist 2006.

Fig. 32. Roman pottery pieces from site K 5 at Qseybe. Photo: Kenneth Lönnqvist 2006.

Fig. 33. Pottery pieces from various periods, inter alia from the Roman period, at site K 5 in Qseybe. Photo Kenneth Lönnqvist 2006.

Fig. 34. Pieces of Roman jars/amphorae from site K 5 at Qseybe. Photo: Kenneth Lönnqvist 2006.

Between the sites K 5 and K 6, just off the modern road on its southern side, a well was documented (UTM: 0568854, 3911118, 392 m a.s.l., see Fig. 35.). The well showed apparent signs of Roman engineering and had been cut into the marble bedrock with chisels, the marks of which could still clearly be seen in the rock. The oval shaft was slightly conical in shape, allowing for descent into the well, possibly for cleaning and inspection purposes. The oval shaped shaft had a maximum diameter of 135 cm and extended to a depth of 3.3 m. However, the present-day sand-dirt floor of the well represents later accumulation of soil, and the original depth of the well could not be determined. The thickness of the bedrock in this location was less than 3 m, and at the bottom of the well there was an open small cistern, pond or reservoir of natural (not worked) space, ca. 50-100 cm in height, it was spreading over an undefined underground area. The observation that the only inhabitants of the well were snakes may well suggest that there were hidden entrances to the underground cistern somewhere nearby. When studied in October the well was dry. Nevertheless, the fact that the well was still used in modern times - judging from its modern cement casing on top - suggests that it was tapping to groundwater, though not being particularly deep. There were no adjacent archaeological signs of any distribution or surface rainwater collection systems around the well. Indeed, the flat terrain would have been unsuitable for that. That there are still good underground water sources in the area is witnessed, as mentioned above, by the modern Bedouin wells and habitation in the oasis.

Fig. 35. A well at Qseybe. Photo: Minna Lönnqvist 2006.

Site K 6
Fortlet/Military Post

UTM: 0568664, 3911075, 392 m a.s.l.
Size of the site: ca. 100 m x 50 m

The site **K 6** represents a fortlet or a military post (Fig. 37.) situated ca. 200 m southwest of **K 5**. The fort is laid out according to the cardinal directions on the horizon, being a rectangle covering an area of 38 m x 45 m, 1,710 m^2 (128 x 152 *pes Monetalis*, pM or Roman feet, see Henderson 1991: 73) or ca. 0.2 ha. The walls are 1 m thick and are built of greenish gypsum and angular slabs (generally ca. 20-30 cm in diameter) fitted together with mortar like the above mentioned houses at K 5. The longer walls were aligned from the east to the west, and the shorter walls exactly N-S (Figs. 37. and 38.). Furthermore, there are two semicircular towers ca. 11 m in diameter protruding to the east from the wall resembling those of the Roman fort Qreiye on the Euphrates (see Gschwind and Hasan 2008). It is of importance that the towers are directed to the east, to the Euphrates frontier and the open desert. In the *Limes Arabicus* studies a military structure comparable to that of Qseybe is called *cohort castella* (Al-Khouri 2007: 149, Fig. 5.). D. Kennedy suggested for Qasr et-Telah in Jordan, a Late Roman site, that this size of military installation would in fact have garrisoned one *ala* of cavalry (Kennedy and Bewley 2004: 143).

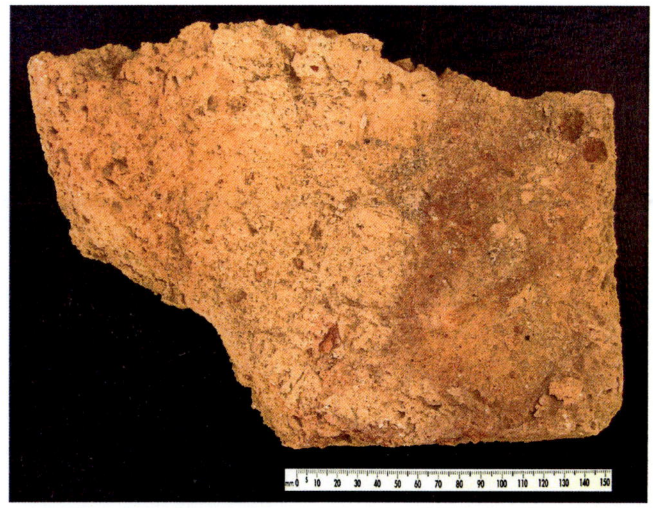

Fig. 36. A Roman tile from site K 6. Photo: Kenneth Lönnqvist 2006.

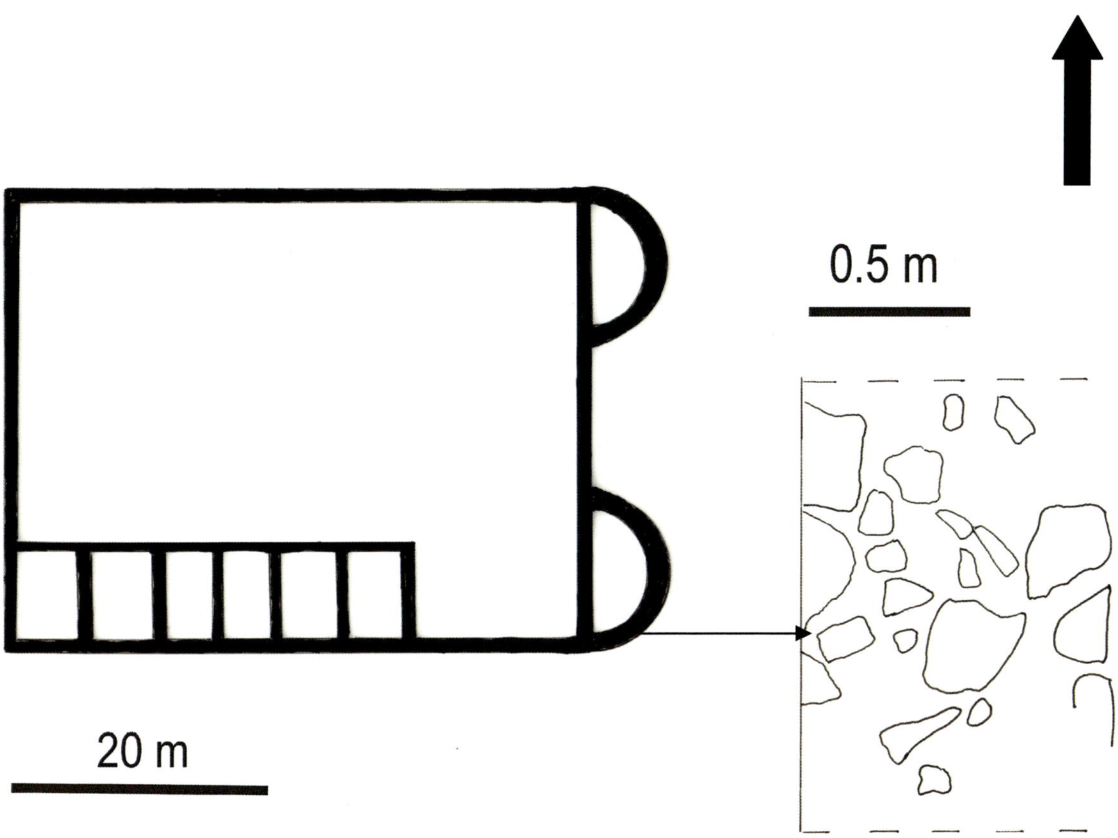

Fig. 37. Site K6. The plan of the fort at Qseybe showing the wall masonry. Drawn by Minna Lönnqvist 2006.

Fig. 38. The western wall of fortlet K 6. Photo: Minna Lönnqvist.

On the southern flank of the fort, a section of six very clearly visible rectangular rooms (each 5 m x 7.4 m in size) had been exposed, which seems to consist of barrack rooms (see above Fig. 37.). Their arrangement at Qseybe is typical of the Roman forts in which the number of *contubernia* (rooms) is reduced from 8-10 to 5-6. (Bidwell 1991: 14). *Contubernia* of the size of 5 m x 7 m, or close to, like those at Qseybe are found in the barracks of South Shields at the mouth of the Tyne, and in Antonine fortlets and Vindolanda in Britain (Hanson and Maxwell 1983: 174-175; Bidwell 1991: 14). In the case of Barburgh, the excavations confirmed that the fort had "contained sufficient accommodation for a full century of eighty men" (Hanson and Maxwell 1983: 72-73). The excavations both at South Shields and Vindolanda have shown, therefore, that the reduction of the centuries whatever it may have involved, did in fact, occur before the Emperor Diocletian's rule, perhaps before the mid-third century AD. (Bidwell 1991: 14).

There were numerous fragments of Roman bricks outside the structure on the surface. However, bricks were especially plentiful in the two mounds that stood 15 m apart to the northwest, at a distance of 33 m from the main fort. The two brick mounds were interpreted as representing the remains either of towers or a gate. The mounds now measured 10 m x 6 m and 12 m x 17 m respectively in plan, but their original size may have been somewhat smaller. The building technique appears to have been a stone slab foundation upon which a brick construction was built – similar to buildings at K 5 - a technique known also to have been employed elsewhere. In Roman military architecture gate systems also rose to form towers, and stone foundations were used for watchtowers at least in the second century A.D., for instance, along Hadrian's Wall and the Antonine Wall in Britain (Hanson and Maxwell 1983: 23, 96).

Roman ceramics were found in quantity in the area of K 6 including a few rims, handles, spouts with a stamped clay *tessera* or an invitation card (max. 2. 04 cm in diam. and 0.47 cm thick, 1.2 g) (Fig. 39.). The design of the stamp appears in the Dura Europos pottery (Welles 1968: Fig. 21: 41), which antedate the fall of the fortress ca. A.D. 256. Comparable stamped clay *tesserae* or invitation cards can also be found in the Palmyra museum dating from the first to the third century A.D. A small corroded bronze coin was found at the site as well. In addition, a fragment of thin rectangular shaped copper sheets (2.4 cm x 1.4 cm in size; 1 cm thick) with rounded corners were made at the site. One of the fragments is likely to have belonged to copper alloy scale cuirasses. The fragments are well preserved and show little signs of corrosion. Furthermore, close to one of the twin towers or a gate a fragment of a circular thin copper disc (size 2 cm x 3 cm, thickness 0.63 mm) was discovered. It may be associated with scabbards commonly found in Roman military sites. In the third century the most common form was a circular chape, and copper-alloy was used in producing the item (Southern and Dixon 2000: 104).

Fig. 39. A stamped clay tessera. Photo: Kenneth Lönnqvist 2006.

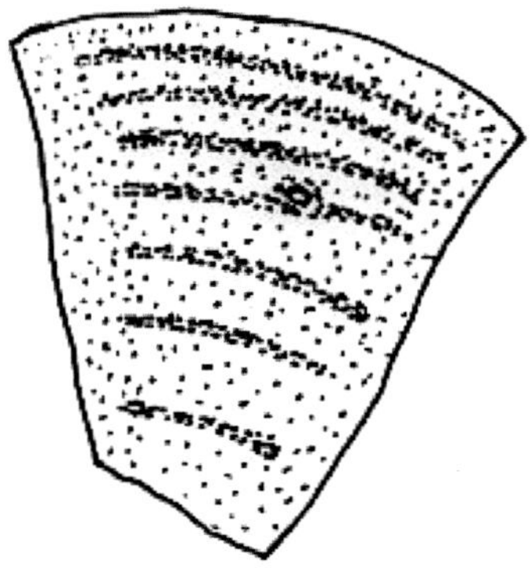

Fig. 40. A fragment of copper disc found at K 6, Qseybe. Scale: 1: 1. Drawn by Kenneth Lönnqvist 2006.

In A. Poidebard's (1934) map (visible in the previous ch. 11, Fig. 2.) a road connection between Qseybe and Qreiye is suggested, the latter fort being on the Euphrates. M. Gschwind (2005a; 2005b) has recently excavated the fortress of Qreiye (Birtha Arupan?) at the edge of Jebel Bishri 12 km northwest of Deir ez-Zor. The building of the fort by the Romans was initially associated with the Severan campaigns against the Parthians on the Euphrates *limes* (Gschwind 2005a; 2005b). The fall of the site to the Sassanians took place ca. A.D. 253 before the final blow to Dura Europos ca. A.D. 256.

On the basis of its construction type and small finds we suggest that Qseybe is contemporary with the sites of Qreiye and the garrison of Dura Europos on the Euphrates. Angular stone slabs as building material and round corner towers were met with like at Qreiye on the Euphrates (see Gschwind and Hasan 2008).

12.5. The Function and Date of the "Bishri Corridor"

R. Mouterde and A. Poidebard (1945) suggested that the route crossing Jebel Bishri through Aš-Šiqri is Roman apparently dating from Diocletian's rule, and its intention was to serve Circesium. Based on Poidebard's and our remote sensing studies the direct connection of the "Bishri corridor" to Circesium seemed logical, but the evidence gathered from the ground-level showed that this stand should be re-evaluated. Another thing is, of course, that Diocletian did not actually originally build Circesium which appears in Mouterde's and Poidebard's map (1945) as an end station of the itinerary over Jebel Bishri. According to Ammianus Marcellinus (Julianus 23, 5.2.), "*Quod Diocletianus exiguum antehoc et suspectum, muris turribusque circumdedit celcis, cum in ipsis barbarorum confiniis interiores limites ordinaret, documento recenti perterritus, ne vagarentur per Syriam Persae, ita ut paucis ante annis cum magnis provinciarum contigerat damnis.*" "This place, which was formerly small and exposed to danger, Diocletian, alarmed by a recent experience, encircled with walls and lofty towers, at the time when he was arranging the inner limes of defence on the very frontiers of the barbarians, in order to prevent the Persians from overrunning Syria, as it had happened few years before with great damage for the provinces". Therefore at Circesium Diocletian refortified an earlier constructed small Roman defence structure that possibly had fallen into disuse or may even have been destroyed or abandoned along the other eastern border forts in the Sassanian raids in the mid third century A.D.

If the minor passable roads that are connected with the Roman remains in the central eastern area of Jebel Bishri are drawn over modern maps of Jebel Bishri (Syria 1:1,000,000, Librairie International 2002), they in fact 'overlap' with Mouterde's and Poidebard's 1945 map for a route from Resafa-Sergiopolis through Aš-Šiqri across Jebel Bishri to Qseybe K 5 and K 6. As there are only a few natural choices for passable roads even today along the piedmont and the summit of the mountain, it may be assumed that the modern track roads and the 1945 map may well reflect some of the major Roman throughways, as they were arranged along the water posts. For instance, the documented water harvesting site K 3 at Darakhlia comprising a dam (barrage), reservoir and tower would have been close to the bend of the road up to the summit of Jebel Bishri. The closest parallels to the size of the watchtower near the water harvesting site K 3 at Darakhlia indicate to the second century A.D. dating.

We must further consider camp or fort K 1 in relation to Poidebard's (Pl. XC, 2) Aš-Šiqri or Bîr Šiqri and the other military arrangements that were in place on the mountain and along its piedmont. When the newly discovered fort K 1 is placed on Mouterde's and Poidepard's 1945 map, it is almost precisely under Bîr Šiqri, where the Roman road makes a sharp bend southwest. The distance between K 1 and, for instance, Qseybe K 5 and K 6 is ca. 27.8 km as the crow flies, in practice ca. 30 km or close to 20 Roman miles, which equals about a two days march. The distance is in line with Poidebard's studies of distances of ca. 23-28 km between some of the main Roman forts or fortlets on the frontier. It should be noted that the track of the *via principalis* leading through K 1 is diagonally straight in the angle leading in the direction of Qseybe. The evidence of smaller Roman sites in Nadra, such as one including a possible *domicilium* is apparently associated with this route leading over Jebel Bishri to Qseybe. Fort K 1 follows the playing card form camps that are dated from the first to the mid-third century A.D., also the pottery recovered from the camp is coherent.

Both Qebaqeb K 8 (see ch. 11.) as well as Qseybe K 5 and K 6 were Roman military forts/fortlets used when traversing from the direction of Palmyra over the desert towards the Euphrates. These sites are indicated as being either of *castellum* or *castrum* type on Poidepard's map of 1934 (see ch. 11., Fig. 2.) and Mouterde's and Poidebard's map of 1945 (see this ch., Fig. 3.). However, the road arrangements that Poidebard suggested over Jebel Bishri are quite differently portrayed on the maps of 1934 and 1945. On the earlier map, Poidebard (1934) shows no roads in the central area of Jebel Bishri. On this map, Qseybe is located on a northern E-W route from Adada (Qasr al-Hayr ash-Sharqi) leading directly to Qreiye. As previously suggested, it seems that there is a point in Poidebard's (1934) suggestion concerning a road connection between Qseybe and Qreiye. The barracks at Qseybe K 6 indicate dating to the second and third century A.D. including semicircular towers finding parallels at Qreiye that was destroyed in A.D. 253 (cf. Gschwind 2005a and 2005b). The construction date of Qreiye is from the second to the third century A.D. It functioned as a Roman military fort at the same time as Dura, which had become a Roman garrison ca. A.D. 165 and was defeated after Qreiye by the Sassanians ca. A.D. 256 (Rostovtzeff *et al.* 1936; James 2005 and Børlit 2008).

Further on the map, Qebaqeb K 8 is on another parallel southern route leading from Adada (Qasr al-Hayr ash-Sharqi) to Circesium. A N-S aligned direct road also joins Qebaqeb and Qseybe. As far as the reconstruction of the copper-alloyed scale armours recovered in our region (K 1 and K 8) are concerned, the type appears to be uniform, pointing to a single-size-type with a vertical pair of holes in each lateral edge, such as shown above in the scale fragment discovered in fort K 1. In most of the scale armours, the remains of a double pair of holes (i.e. four holes per disc) appear to have been employed and similar fragments are met at Dura Europos (James 2004: 121-122, 132-133). A copper-alloyed disc recovered at Qseybe probably belonging to a scabbard may date to the third century A.D. Scale-armours and other fittings recovered in our site are proof of the association to military activities.

Generally speaking, different types and sizes of installations belonging to the Roman military architecture correspond to the organisation of the Roman army (Hanson and Maxwell 1983: 22). Roman forts appear not normally to have been built smaller than ca. 100 m x 100 m, which appears to be the 'minimum' size, such as at the site K 1 and K 8 recorded in our surveys during 2004-2006. Examples of comparable sites along the frontier appear in the routes on the *Tabula Peutingeriana*, such as Bedoura in Numidia (Morizot 1991: 345), the Arabian frontier (Parker 1991: 498-504), the Palestinian *limes* (Negev 1993: 1159; Oboda in the Arabah) or Roman Britain (Hanson and Maxwell 1983: 90-91, 119, e.g. Polmonthill). In some *Limes Arabicus* studies this size of military structure is sometimes typologically described as a large *castellum*, with normal dimensions of 10,000-20,000 m^2, or from one to two hectares (Al-Khouri 2007: 149, Fig. 5).

D. Kennedy has suggested that a military site measuring 110 m x 75 m (0.8 hectares) would have been sufficient to house 1000 men (or a full-sized *cohors milliaria* or 'double-cohort'), meaning that a *castellum* was intended to house a legion or part of it (detachments of a legion) (Kennedy and Bewley 2004: 175; Hanson and Maxwell 1983: 175). However, this is not only true for the Roman East, but also for the West, as the size of the smallest primary and secondary forts on Rome's northwestern frontier in Britain, for instance, also focuses around three to four acres or 1.0-1.5 hectares (Hanson and Maxwell 1983: 87, Tab. 5.1). Both the fort K 1 on the mountain and K 8 at Qebaqeb on the eastern piedmont of Jebel Bishri (see ch. 11.) - which were the largest military structures documented - were typically of this size. Qseybe corresponds to what is described as a *cohort castellum*, said to have housed either a reduced-strength Roman *cohors* or perhaps from two to three mounted units (*turmae*) of an ala of cavalry. Though, this latter possibility remains hypothetical for Qseybe. It has recently been shown by Y. Tepper (2007: 63; see also Lönnqvist 2009: 49) that the Romans, for instance, in Judaea, starting from the Early Imperial period, regularly deployed detachments smaller than a full legion. Consequently, even a legionary camp does not always need to be of the assumed size of a normal legionary camp that housed a full legion in order to be 'legionary'. If an army as a rule decides to deploy smaller units in a region, the garrison sites will correspondingly be smaller too. This may be a significant factor that has not been taken into account enough, when the Roman army presence in the Near East is discussed.

The camp site of K 1, the water harvesting site of Darakhlia K 3, the sites of Qseybe K 5 and K 6 and Qebaqeb K 8 in their construction techniques and small finds altogether allow an earlier dating than the Diocletian era, although some of the sites like Darakhlia could still have served in the fourth century. Therefore the role of the sites already serving forts and fortresses on the Euphrates *limes* like those of Qreiye and Dura Europos in the second and third centuries A.D. is worth taking into account.

In our view the function of the "Bishri corridor" was primarily for supply and logistics between the larger military desert sites on the western piedmont and the Euphrates *limes*. The temporary camp like K 1 could have served the establishing of the *limes* garrisons on the Euphrates together with Qseybe. The way through Qasr al-Hayr - Adada?, Qebaqeb and Qreiye could have functioned as an additional line. The smaller and somewhat temporary nature of the sites in the "Bishri corridor" on the desert-steppe itinerary indicates that this was a line that was not constructed primarily for defence but for organisational service such as logistics of an army serving in larger compounds. It needs to be taken into account that this was an itinerary and the logistics that could not have been visually observed by the Parthians or Sassanians on the Euphrates, which the use of the Euphrates roads could have exposed. Through this line messengers could have reached troops and auxiliary troops could have been added to the larger sites without notice. A question remains, why it did not work ca. A.D. 253-256, when the Sassanians were able to conquer Qreiye and Dura Europos. As we know, Rome was in trouble not only with its outer enemies but also struggling in turmoil within during the third century A.D. until the time of Aurelian.

12.6. Concluding Notes on the Roman Military Presence

We know from previous Roman frontier or *limes* studies a few important points that are worth rehearsing here, though shortly. Firstly, there is one general feature regarding the spatial distribution of the Roman military sites that seems to hold for the Arabian *limes,* in Palestine and Syria, and in our case study. Namely, most late Roman garrisons. forts or posts that have been positively identified archaeologically or from the Late Roman military records, such as the *Notitia Dignitatum* or *Tabula Peutingeriana*, were apparently not distributed according to any major prevailing demographic patterns. In fact, these sites appear to be located along the fringe of the desert, where few people would have permanently dwelled, as mentioned, for instance, by S.T. Parker (1991: 502). There must be an obvious cause for this, which has to do with the purpose of these military structures. The surface surveys directed by D. Kennedy and R. Bewley (2004) and S.T. Parker (1991: 503), for instance, in Jordan, have uncovered a dense concentration of fortified military posts and camps in the immediate vicinity of, or overlooking major water courses or *wadis*. In Jordan towers were constructed to monitor the traffic going and coming on roads and in *wadis*.

The same observations apply for the surveys conducted by the Finnish project SYGIS to the Roman frontier in the Jebel Bishri area in Syria. Most Roman sites here are situated in the areas, where the large seasonal watercourses and *wadis* are located, or where they produce run-off of water that can be collected. From the recovered surface finds of associated Iron Age and Nabataean pottery in Jordan Parker (1991: 503) concluded that many of the Roman military sites were reoccupied by the Romans (as recorded by Ammianus 14.8.13) rather than they would have been constructed afresh. It is relatively easy to agree with Parker's view that the deployment of Roman soldiers and posts along the fringes of the desert may well correspond with paths of major nomadic migration routes. That would suggest that they could have been intended to monitor and perhaps control the *transhumance* of these people between the seasonal grazing areas and agricultural grounds (Parker 1991: 503). As far as Jebel Bishri is concerned, this is one feasible explanation for the deployment of Roman troops in the interior of the mountain. Jebel Bishri was and still is an area, where the pastoralists practice their *transhumance* between the grazing lands on the mountain to the south and the agricultural zones in the north on the Euphrates.

In classical military thinking, Poidebard's map of 1934 suggests that Roman routes and fortifications between Palmyra and the Euphrates does create a different pattern from the rest of Syria (see ch. 11: Fig. 2.), which brings about a few new thoughts. In this area we only have sparsely distributed parallel roads that form a W-E orientated grid system between the major locations on the *Strata Diocletiana* and the Euphrates. This should be understood as a foremost priority to be able to transport troops and material from the *Strata Diocletiana* to the large border fortifications on the Euphrates. At the same time, the N-S aligned major defence lines formed on the E-W axis two successive and easily defendable fortification systems that created new depth and mass to the Roman frontier. In the event that the defences of the Euphrates would have fallen, it was possible to fall back into pre-prepared positions along the *Strata Diocletiana,* which effectively blocked further advance to the coast and the large cities, such as the capital.

As will be presented in the following chapter, on the Euphrates more massive border fortifications were apparently enough to deter a direct frontal attack for a long period, and this is why the invading armies from the east circumvented in a classical flanking manoeuvre, like the Parthians/Persians (Goldsworthy 2002: 186), either from the north passing Edessa, or from the south passing Dura Europos. The shortest and easiest way was to attack first Dura Europos, which opened up the whole southern flank of the Roman army. This is the chief reason for Dura Europos' and Circesium's strategic importance on the Euphrates. At the same time, the N-S axis the W-E orientated major road system that runs in several parallel routes between the *Strata Diocletiana* and the Euphrates, would have created a considerably threat with concentrations of Roman troops along the *wadis* and water sources, along which any invading army had to march to supply itself with water. In addition, invaders crossing roads could quickly and easily be cut off and surrounded by the mobile Roman units.

In this respect, it may be that the raiding by the desert people from the south and southeast may have caused a problem as well. As far as trade and taxation are concerned, the *transhumance* on Jebel Bishri needed to be controlled also with available Roman forces and military installations, which have not previously been documented in the area. Although this is a small digression, one possible way of dealing with the 'problem' of the nomadic population in the East was always to win the locals over to the Roman view and get them enlisted in the army, which eventually could have led to Romanization and Roman citizenship. Similar ways was already practised in the Mari kingdom in controlling surrounding nomads (Lönnqvist 2010).

The impact of this would have been discernible on many levels; from the Roman point of view, it brought peace and stability, 'know-how' of the nomadic people and habits into the ranks, and it offered first-hand information on how to combat raids and security threats. For the locals, it offered a possibility to social advance and status. That this is not a mere conjecture is demonstrated by OGIS 321, a first century A.D. inscription from the northern territory of King Agrippa II. We have, for instance, the military title of Στρατηγός παρεβολών νομάδων, who - according to T. Brüggemann (2007: 278-279) - was 'captain of a troop of nomads', meaning that native nomadic people were actively converted into professional Roman soldiers and recruited into separate bands or units, and that they may have been commanded by men of their own ranks (see also Southern and Dixon 2000: 49).

Bibliography

Al-Khouri, M. (2007) Military Architecture of the *Limes Arabicus*: Strategy, Commercial Routes, Quotidian Life, in *The Late Roman Army in the Near East from Diocletian to the Arab Conquest*, Proceedings of a Colloquium held at Potenza, Acerenza and Matera, Italy (May 2005), ed. by Lewin, A.S. and Pellegrini, P. with the aid of Fiema, Z.T. and Janniard, S., BAR International Series 1717, Archaeopress: Oxford, pp. 143-151.

Bidwell, P. T. (1991) Later Roman Barracks in Britain, in *Roman Frontier Studies 1989, Proceedings of the XVth International Congress of Roman Frontier Studies*, ed. by Maxfield, V.A. and Dobson, M.J. Exeter, UK: University of Exeter Press, pp. 9-15.

Børlit, C.S. (2008) Dura Europos: The Final Siege and Abandonment, in *Jebel Bishri in Context: Introduction to the Archaeological Studies and the Neighbourhood of Jebel Bishri in Central Syria, Proceedings of a Nordic Research Training Seminar in Syria, May 2004*, ed. by Lönnqvist, M., BAR International Series 1817, pp. 99-110.

Brüggemann, T. (2007) Ἐθνάρχος, Φύλαρχος and Στρατηγός νομάδων in Roman Arabia (1st-3rd Century), Central Power, Local Administration, and Nomadic Environment, in *The Late Roman Army in the Near East from Diocletian to the Arab Conquest*, Proceedings of a colloquium held at Potenza, Acerenza and Matera, Itlay (May 2005), ed. by Lewin A. S. and Pellegrini, P. with the aid of Fiema, Z.T. and Janniard, S., BAR International Series, No. 1717, Oxford, pp. 275-284.

El-Khouri, L. (2008) The Roman Countryside in North-west Jordan (63 BC-AD 324), in *Levant*, Vol. 40, pp. 71-87.

Erickson-Gini, T. (2007) The Nabataean - Roman Negev in the Third Century CE, in *The Late Roman Army in the Near East from Diocletian to the Arab Conquest*, Proceedings of a Colloquium held at Potenza, Acerenza and Matera, Italy (May 2005), ed. by Lewin, A.S. and Pellegrini, P. with the aid of Fiema, Z.T. and Janniard, S., BAR International Series, No. 1717, Oxford: Archaeopress, pp. 91-100.

Goldsworthy, A. (2002) *Roman Warfare*. Cassel History of Warfare. London: Cassel & Co.

Greene, K. (1986) *The Archaeology of the Roman Economy*. London: Batsford Ltd.

Gregory, S. (1997) *Roman Military Architecture on the Eastern Frontier.* Vols. 1-3. Amsterdam: Adolf M. Hakkert Publisher.

Gschwind, M. (2005a) Qreiye/'Ayyāš, in *Archäologischer Anzeiger,* 2. Halbband, Deutsches Archäologisches Institut, pp. 255-257.

Gschwind, M. (2005b) The Roman Fort Qreiye/'Ayyash on the Middle Euphrates River, in *Places in Time, 25 Years of Archaeological Research in Syria 1980-2005,* Damaskus: German Archaeological Institute, pp. 123-127.

Gschwind, M. (2008) *Limes studies Qreiye, 2003-2007 excavations at Qreyie,* at Deutsches Archäologisches Institut, http://www.dainst.org/index_742_en.html.

Gschwind, M. and Hasan, H. (2008) Das römische Kastell Qreiye-'Ayyāš, Provinz Deir ez-Zor, Syrien, Ergebnisse des syrisch-deutschen Kooperations-projektes, in *Zeitschrift für Orient-Archäologie,* Vol 1, pp. 316-334.

Hanson, W.S. and Maxwell, G.S. (1983) *Rome's North West Frontier, The Antonine Wall.* Oxford: Edinburgh University Press.

Henderson, C.G. (1991) Aspects of the Planning of the Neronian Fortress of *Legio II Augusta* at Exeter, in *Roman Frontier Studies 1989, Proceedings of the XVth International Congress of Roman Frontier Studies,* ed. by Maxfield, V.A. and Dobson, M.J., Exeter, pp. 73-83.

Hirschfeld, J. (1995) *The Palestinian Dwelling in the Roman Byzantine Period.* Jerusalem: Franciscan Printing Press and Israel Exploration Society.

Isaac, B. (1990) *The Limits of Empire. The Roman Army in the East.* Oxford: Clarendon Press.

James, S. (2004) *The Excavations at Dura-Europos conducted by Yale University and the French Academy of Inscriptions and Letters 1928 to 1937. Final Report VII. The Arms and Armour and other Military Equipment.* The British Museum Press. Printed by Cambridge University Press.

James, S. (2005) The Deposition of Military Equipment during the Final Siege at Dura-Europos, with Particular Regard to the Tower 19 Countermine, in *Carnuntum Jahrbuch,* pp. 189-206.

Kennedy, D. (ed.) (1996) *The Roman Army in the East.* Journal of Roman Archaeology Supplementary Series No. 18. Ann Arbor, MI; Portsmouth: JRA.

Kennedy, D. and Bewley R. (2004) *Ancient Jordan from the Air.* The Council for the British Academy in the Levant, Dorset: The British Academy.

Lönnqvist, K.K.A. (2009) *New Perspectives on the Roman Coinage on the Eastern Limes in the Late Republican and Roman Imperial Period.* Saarbrücken: VDM Verlag.

Lönnqvist, M. (2010) How to Control Nomads? A Case Study Associated with Jebel Bishri in Central Syria, West Semitic Nomads in Reltion to the Urban World, in *City Administration in the Ancient Near East, Proceedings of the 53e Rencôntre Assyriologique Internationale,* Vol. 2, ed. by Kogan, L. *et al.*, Orientalie et Classica, Babel + Bibel 5, Winona Lake, Indiana: Eisenbrauns, pp. 118-139.

Mouterde, R. and Poidebard, A. (1945) *Le Limes de Chalcis, Organisation de la Steppe en Haute Syrie Romaine,* Documents Aériens et Épigraphiques, Plans et Relevés de Lauffray, J. et Mazloum, S. Texte et Atlas. Bibliothèque Archéologique et Historique. Tome XXXVIII. Paris: Geuthner.

Morizot, P. (1991) Les stations de la Table de Peutinger entre Lambèse et ad Calceum Herculis, in *Roman Frontier Studies 1989,* Proceedings of the XVth International Congress of Roman Frontier Studies, ed. by Maxfield, V.A. and Dobson, M.J., Exeter, pp. 337-346.

Musil, A. (1928) *Palmyrena, A Topographical Itinerary.* American Geographical Society. Oriental Explorations and Studies No 4. New York: American Geographical Society.

Negev, A. (1993) Oboda, in *The New Encyclopedia of Archaeological Excavations in the Holy Land,* Vol. 3, s.v. Jokneam-Pella, Jerusalem: Simon & Schuster, The Israel Exploration Society & Carta, pp. 1155-1165.

Parker, S.T. (1991) The Nature of Rome's Arabian Frontier, in *Roman Frontier Studies 1989,* Proceedings of the XVth International Congress of Roman Frontier Studies, ed. by Maxfield, V.A. and Dobson, M.J., Exeter, pp. 498-504.

Poidebard, A. (1934) *La trace de Rome dans le désert de Syrie, Le limes de Trajan a la conquête arabe, Recherches aériennes (1925-1932).* Bibliothèque Archéologique et Historique Tome XVIII. Texte & Atlas. Paris: Geuthner.

Rostovtzeff, M.I, Bellinger, A.R., Hopkins C. and Welles, C.B. (1936) *The Excavations of Dura-Europos* conducted by Yale University and the French Academy of Inscriptions and Letters. Preliminary Report of Sixth Season of Work

October 1932 - March 1933. New Haven: Yale University Press.

Southern, P. and Dixon, K.R. (1992) *The Roman Cavalry From the First to the Third Century AD*. London: Routledge.

Southern, P. and Dixon, K.R. (2000) *The Late Roman Army*. London: Routledge.

Tepper, Y. (2007), The Roman Legionary Camp at Legio, Israel: Results of an Archaeological Survey and Observations of the Roman Military Presence at the Site, in *The Late Roman Army in the Near East from Diocletian to the Arab Conquest, Proceedings of a colloquium held at Potenza, Acerenza and Matera, Italy (May 2005)*, ed. by Lewin, A. S. and Pellegrini, P. with the aid of Fiema, Z.T and Janniard, S., BAR International Series, No. 1717, Oxford: Archaeopress, pp. 57-71.

Van Dierendock, R.M. (s.a.) Five Postholes and a Ditch: the Valkenburg – Marktveld Timber Watch and Signal Tower, in *Archaeology in Confrontation, Aspects of Roman Military Presence in the Northwest, Studies in Honour of Prof. em. Hugo Thoen*, ed. by Vermeulen, F., Sas, K. and Dhaeze, W., Archaeological Reports Ghent University, Ghent: Ghent University Academia Press, pp. 73-102.

Welles, C.B. (ed.) (1968) *The Excavations at Dura-Europos.* conducted by Yale University and the French Academy of Inscriptions and Letters. Final Report IV: Part I, Fasc. 3. New Haven: Yale University Press.

13. DOCUMENTATION AND VIEW-SHEDS ON THE EUPHRATES *LIMES*

Minna Lönnqvist Markus Törmä Kenneth Lönnqvist
with Contributions by Jari Okkonen, Milton Nuñez and Michael Herles

Fig. 1. The ruined fort of Tabūs at the edge of Jebel Bishri facing the Euphrates valley. Photo from S: Minna Lönnqvist 2004.

13.1. Organising the Natural Border Zone of the Euphrates

The limit of the Palmyrene control is described as having extended to the Euphrates (*Appian*, the Civil Wars 5.1.9.) at least in the first century A.D. The Roman campaigns on the Middle Euphrates during the Emperor Vespasian's and Trajan's reign have not left much archaeological evidence. The first Roman military posts are thought to have been erected on the Middle Euphrates at the time when M. Ulpius Traianus (Trajan's father) was the Syrian legate (Mouterde and Poidebard 1945: 134). A short occupation of Dura Europos during Trajan's Parthian wars is visible in a ruined triumphal arch (A.D. 116) at the site. (Edwell 2008: 18-21). The establishment of more direct Roman military control at sites such as Dura Europos took place in Lucius Verus' campaign led by Marcus Aurelius (Invernizzi 1986: 359) ca. A.D. 165. Roman control and moving of the demarcation line of the *limes* to as far as Kifrin dates according to some sources from the Severan period (Invernizzi 1986: 369), but others attribute it already to Lucius Verus' campaign (Dodgeon and Lieu 2002: 1). In A.D. 194 Septimius Severus (A.D. 193-211) crossed the Euphrates, and in A.D. 197 he reconstructed the province of Mesopotamia (cf. Cary and Scullard 1984: 492; SAS 2002: 96-97).

Quite recently in the Roman frontier studies, or the so-called *limes* studies, some scholars have thought that the region of the Euphrates in Syria is insufficiently studied and known for anything definite to be said about the Roman military organization there (see, e.g., Parker 2000: 134-135; Sartre 2005). After the studies and mappings by F. Sarre and E. Herzfeld (1911, 1920), A. Musil (1928), C. Héraud (1922), A. Poidebard (1934) as well as R. Mouterde and A. Poidebard (1945) archaeological surveys and mappings of Roman and Byzantine remains have been carried out by L. Lauffray (1983, 1991) and K. Kohlmeyer (1984; 1986) on the Middle Euphrates. In addition to the surveys of our Finnish project SYGIS, the Syrians, the French, the Germans, the Spanish, the British and the Japanese have been working in the region during recent years. Our work concentrated on the less explored area between Halabiya and 'Ayyāsh at the foot of Jebel Bishri. The Roman remains of the northern part between Sura and Halabiya are first briefly introduced here in order to understand the context of the major eastern sites of Tibne and Tabūs which we were able to study more closely.

As previously mentioned, M. Konrad (2001) has restudied Sura (see the location in ch. 11, Figs. 2. and 3.) at the foot of Jebel Bishri on the Euphrates *limes*. She was following the aerial studies by A. Poidebard (1934). At Sura Konrad's survey and mapping revealed Roman occupation at least from the second and third centuries A.D. The size of the castle at Sura is 200 m x 200 m in plan enclosing a 10 m wide *vallum*. (Konrad 2001: 5-22). According to *Notitia Dignitatum*, the site was held by *Legio XVI Flavia Firma*, and its position on the promontory of the Euphrates indicates that it had been constructed to defend the border against the Parthians.

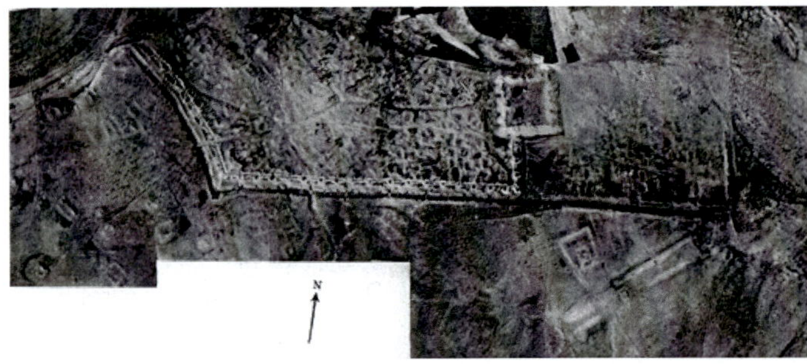

Fig. 2. Sura photographed from the air by Poidebard (1934).

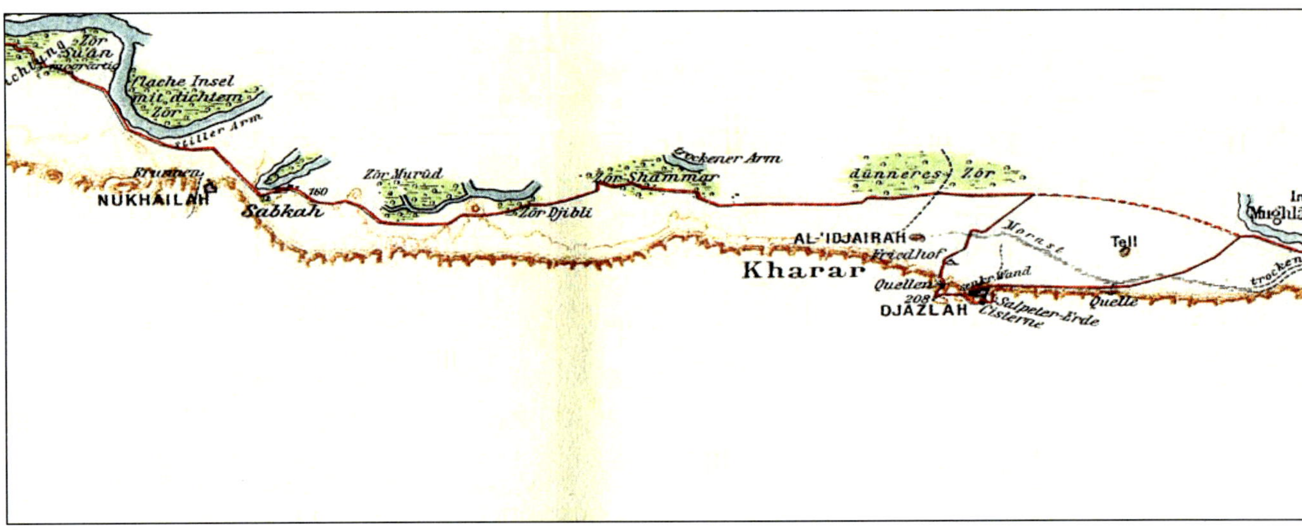

Fig. 3. An excerpt from the map by Sarre and Herzfeld (1911) illustrating the forts of Nukhailah and Djazlah at the foot of Jebel Bishri between Halabiya and Raqqa.

Qala'at Safin is a castle hanging on the cliff called "Jabal Safin" ca. 16 km southeast of Raqqa (Nicephorium-Callinicum) on the way to the peninsula of Halabiya (Zenobia). It is located on the southern side of the Deir ez-Zor–Aleppo -road at the site known as Siffin (see ch. 14, Fig. 13.). The fort provides a convenient guarding post on Jebel Bishri. It is roughly rectangular in plan covering ca. 60 m x 30 m in size and built of gypsum blocks. Some of the earlier visitors at the site date the foundation of the fort back to the Roman period. (Sarre and Herzfeld 1911: 163; Kohlmeyer 1986: 58-60 and Napoli 2000: 119). A. Musil (1927: 188), however, dated the fort to the Medieval period.

The forts of Nukhailajh or Nheyla (also known as Nouhaila, Nukhailah, Nukhaila, Naqhila) and Gezla (also known as Djazlah, Jazla, Gizler) on the Middle Euphrates are mentioned by F. Sarre and E. Herzfeld (1911: 164, Abb. 72. and 73.), and they appear on A. Poidebard's (1934) and R. Mouterde's and A. Poidebard's (1945) maps (see our ch. 12, Fig. 3.) on the northern ridges of Jebel Bishri ca. 23 km southeast of Raqqa on the way to Halabiya (Zenobia). A. Poidebard (1934: 85) further mentions that Nheyla is situated 42 km from Sura on the way to Halabiya, and Halabiya in turn is 50 km from Nheyla. There is 2 km from Sabkha to Nheyla (Sarre and Herzfeld 1911: 164). The fort has quadrilateral corner towers which even in a collapsed state still protrude from the ruins. The site clearly commands the surrounding area and was probably recognized as a good place for guarding in ancient times. The irregular plan of the fort is ca. 100 m on the north-south axis and 40 m on the east-north axis. The foundation of the fortress is laid on large stone blocks, but here baked brick walls are surmounted on the blocks instead. There are two visible gates: one in the southeast and the other in the northeast. Associated scattered potsherds date from the Roman, Byzantine and Islamic period like at Qala'at Safin (Siffin) The use of the fort is therefore vaguely dated to the Late Roman/ Byzantine/Islamic periods. (Lauffray 1983: 70; Kohlmeyer 1986: 60; Napoli 2000; Ohnuma and Al-Khabour 2007; Al-Maqdissi and Ohnuma 2008: 120).

Gezla, ca. 17 km from Sabkha, has served as a Roman fort as well. Gezla/Djazlah is situated opposite to the village of Ghanem Ali which is ca. 40 km southeast of Raqqa on the way to Halabiya. The site is on the southern side of the Deir ez-Zor - Aleppo –road, and after Sarre and Herzfeld it has been more recently studied by J. Napoli (2000). The fort has a triangular plan (see Fig. 5.). According to Napoli (2000: 122), the eastern wall measures ca. 180 m in length and the southern one 220 m. Napoli thinks that the foundations date from the Seleucid era, but the upper constructions can be dated to the third century A.D.

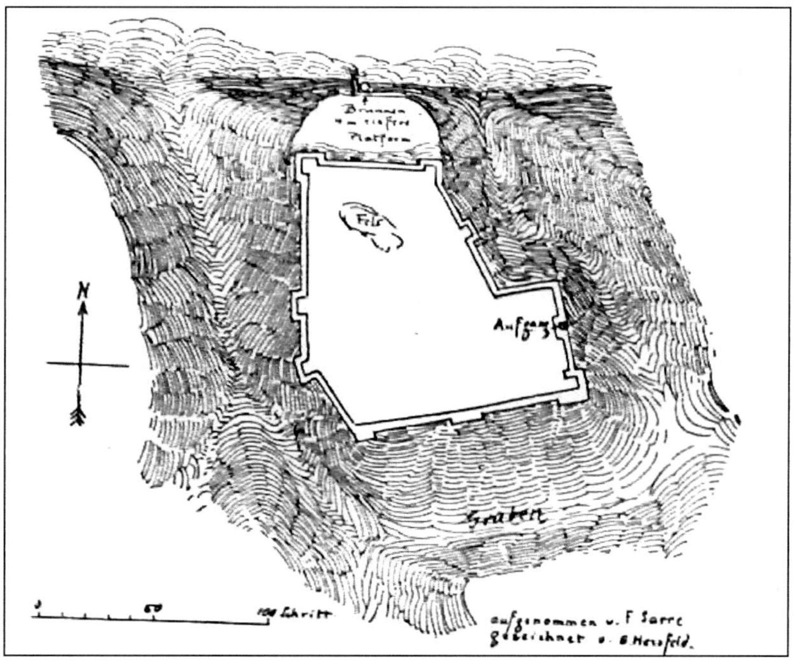

Fig. 4. The plan of the fort Nukhailah by Sarre and Herzfeld (1911).

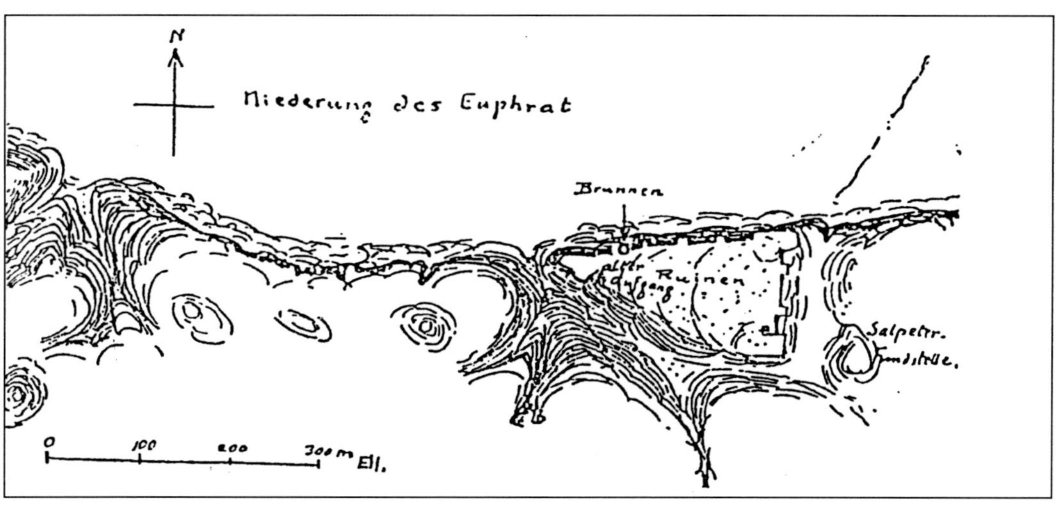

Fig. 5. The plan of the fort Djazlah (Gezla) by Sarre and Herzfeld (1911).

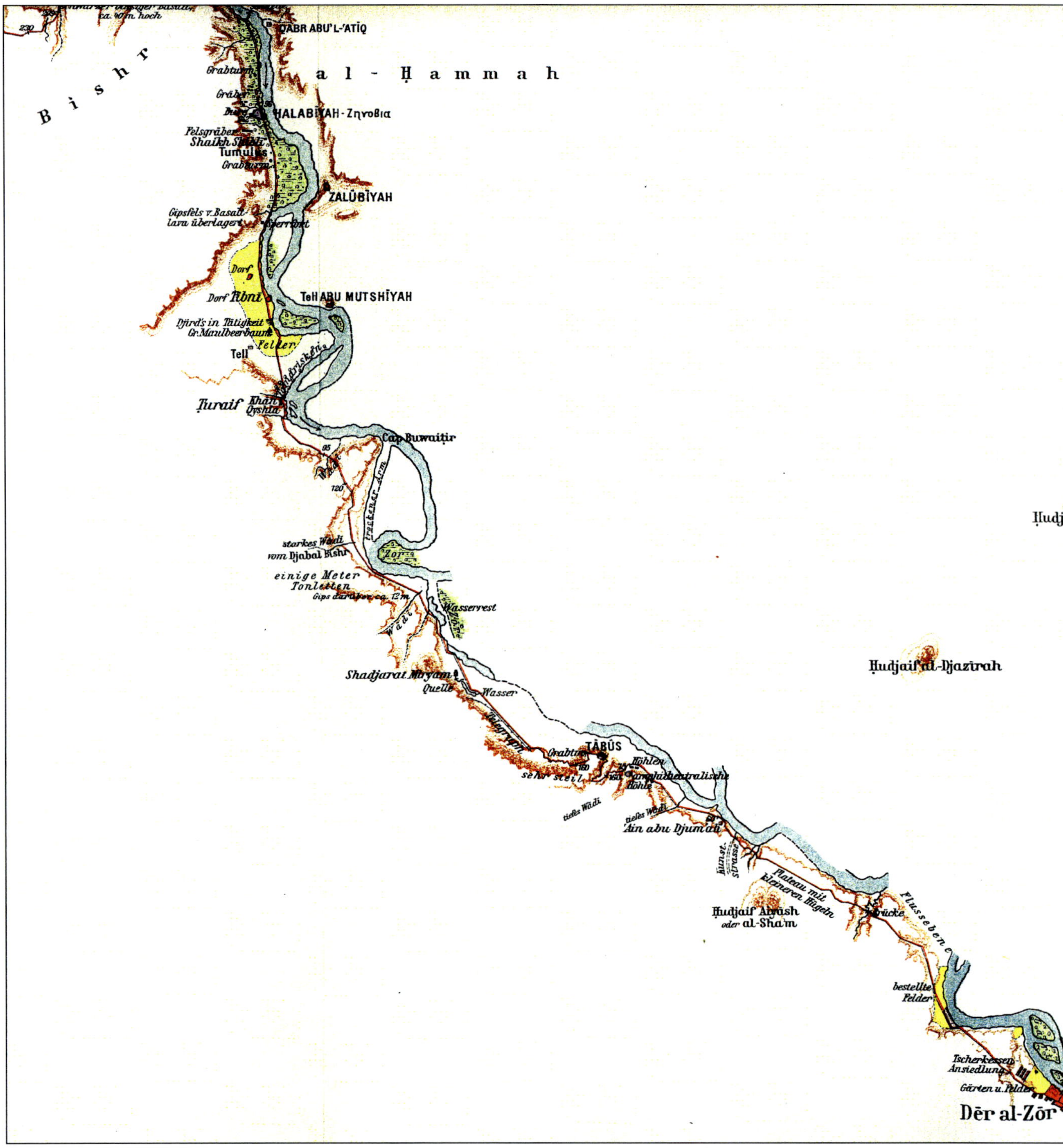

Fig. 6. A Part of the Euphrates map from Deir ez-Zor to Halabiya by Sarre and Herzfeld (1911). The map shows ancient sites such as the fortress of Zenobia as well as the tell at Tibne and the fort of Tabūs, but no fort of Qreiye, which is at the edge of the basalt peninsula of Ayyāsh, where the text "Flussebene" is visible.

The limit of the Palmyrene power in the third century A.D. was marked by two fortresses on the Euphrates: those bearing the names of Zenobia and Zalabiya. According to the historian Procopius (*De Aedificiis* II, viii, 8-9), the fortress on the peninsula of Halabiya, a basalt outcrop of Jebel Bishri (Fig. 6.), was originally built by Zenobia, the queen of Palmyra, in the third century A.D. but was rebuilt by the Emperor Justinian I in the sixth century A.D. The well-preserved fortress of Zenobia (UTM: 0574375, 3949782; see Figs. 7. and 8.) has a triangular layout (ca. the walls extending E: 380 m, N: 350 m and S: 550 m), and it is mainly built of gypsum with minor additions of brick. The well-known sites of Zenobia and Zalabiya have been mentioned by several early travellers. (See Sarre and Herzfeld 1911: 365-381, also Toivanen 2008). The Frenchmen have excavated the fortress and surrounding graveyards (Lauffray 1983; 1991). A recent restoration project carried out by Frenchmen and Syrians has concentrated on both Zenobia and Zalabiya (Blétry and Serdon 2004-2005: 197-210; Golcberg 2004-2005: 211-215).

Fig. 7. The fortress of Zenobia. Photo from E: Kenneth Lönnqvist 2004.

Fig. 8. The walls of Zenobia.
Photo from N: Kenneth Lönnqvist 2004.

13.2. Documenting Forts on the Euphrates Limes by SYGIS

13.2.1 Site G 1: The Fort of Mambri at Tibne?

The topographical details of Tell Tibne, the site **G 1**, (UTM: 574179, 3941645, ca. 231-235 m a.s.l.) were already given in ch. 10. The mapping of the site was made separately in 2003-2005, outside our systematic survey area. Tell Tibne is located on the northern side of the modern Deir ez-Zor–Aleppo -road before the crossroads leading to the right to the peninsula of Halabiya while approaching from Deir ez-Zor. The tell already appears in a map published by F. Sarre and E. Herzfeld in 1911 (Fig. 9.). In addition, C. Héraud's map from 1922 illustrates a ruined tower which we could not identify in the village area any more. The tower may have been an ancient tower tomb or/and a watchtower.

Currently an Islamic graveyard is occupying the tell. Visible structures of unidentified date had been constructed of local gypsum/marble and can still be detected on the surface of the tell (see Figs. 10. and 11.). But as indicated earlier, the first levels of the tell date from the Palaeolithic Age (see studies in the neighbourhood, e.g., Copeland 2004: 27, Fig. 2.), and the site has produced pottery from the Chalcolithic Uruk phase and the Bronze Age (see ch. 10.).

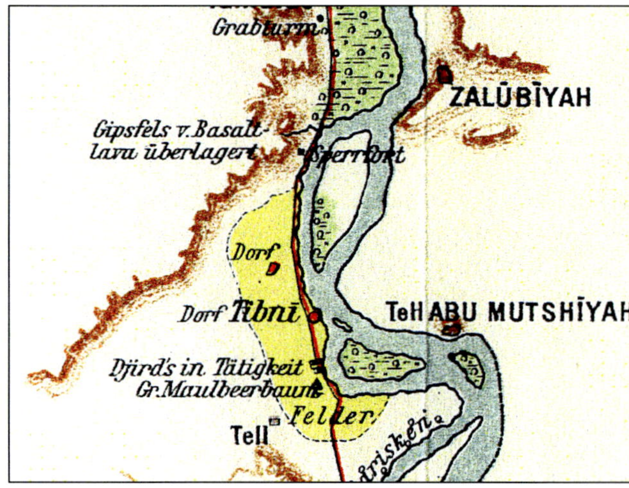

Fig. 9. Tell Tibne illustrated on the map by Sarre and Herzfeld 1911.

Fig. 10. The walls of ruins still visible at Tell Tibne. The Muslim graveyard on the top. Photo from W: Kenneth Lönnqvist 2005.

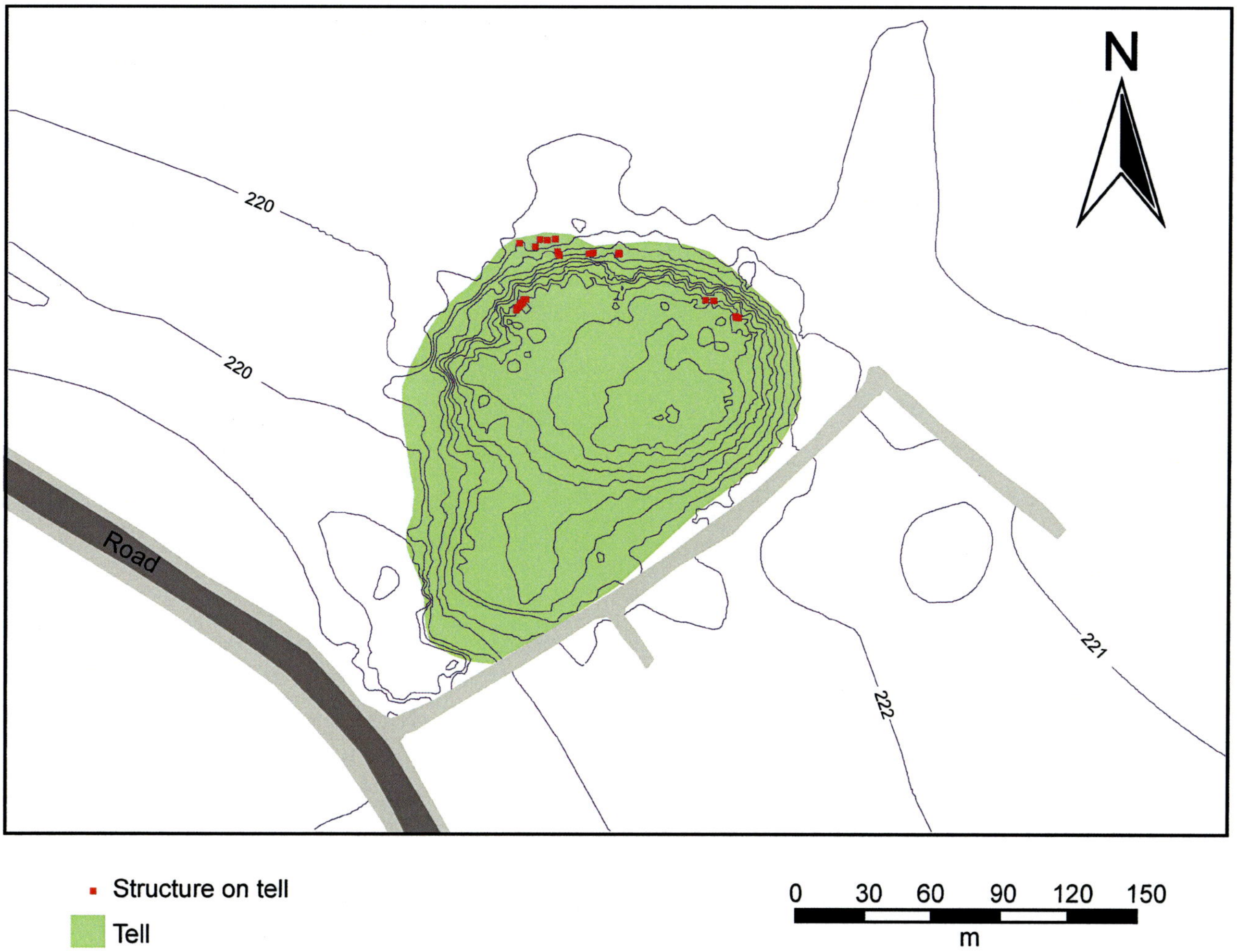

Fig. 11. A topographic map and visible structures of Tell Tibne based on the in situ captured EDM data. Mapping by Jari Okkonen 2005.

Because of its location and the exposed ruins the site may be identified as the fort of Mambri (Poidebard 1934: 86) built by the Emperor Diocletian and rebuilt by the Emperor Justinian I. The distance of 6-7.5 km southeast of the fortress of Zenobia accords with the description given by the ancient historian Procopius (*De Aedificiis* ii, 8.7.). The apparent identification of the site with Mambri can be further deduced from Late Roman and Byzantine pottery stamps and a few coins which we collected at the site. But there are possible pieces of Roman Eastern sigillata A ware (Fig. 12.) as well. The Eastern sigillata A's date ranges from the Hellenistic period of the late second century B.C. to the Augustan era. In that case we are talking about the earliest presence of the Romans in the area – actually the time when Syria was to become a Roman province. Parthian pottery and a pottery stamp depicting a swastika (Fig. 13.), comparable to the Dura Europos pottery stamps, indicate Parthian influence. The palm branch and rosette type stamps (Figs. 14. and 15.) also find parallels in the Dura Europos pottery collection (cf. Welles 1968: Fig. 21), which gives additional evidence of the Parthian and Roman influence at the site, already dating from the second and third centuries A.D. According to J.W. Hayes' (1972: 349-351) catalogue of the Late Roman pottery stamps, some of our palm leave and rosette stamps (Fig. 16.) appearing in Roman C ware could also belong to the Group I dating from ca. A.D. 360-450. It should, however, be noted that our pieces did not have any slip or gloss on their surface. Beside the Late Roman C ware, Cypriot red slip ware was collected at the site (Fig. 12.) as well. Glazed pottery and twisted pottery handles seem to date from the Byzantine era (Figs. 17. and 18.). The bronze coins are: a dupondius-sized coin (Fig. 19: Portrait of male, Emperor?, visible, facing right. w.: 14.5 g; ø 27.41 mm; th. 4.25 mm) and two smaller coins (Fig. 20: w.: 1.8 g; ø: 13.39 mm; th.: 4.2mm.w.: 1.6 g; ø: 13.16 mm; th.: 3.4 mm).

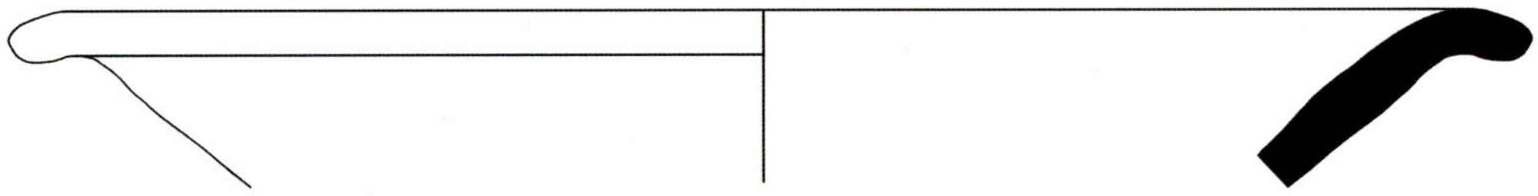

1

Roman Eastern sigillata A ware. Scale 1:3.

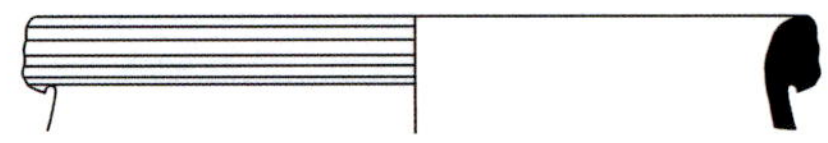

2

Parthian? ware. Scale 1: 4.

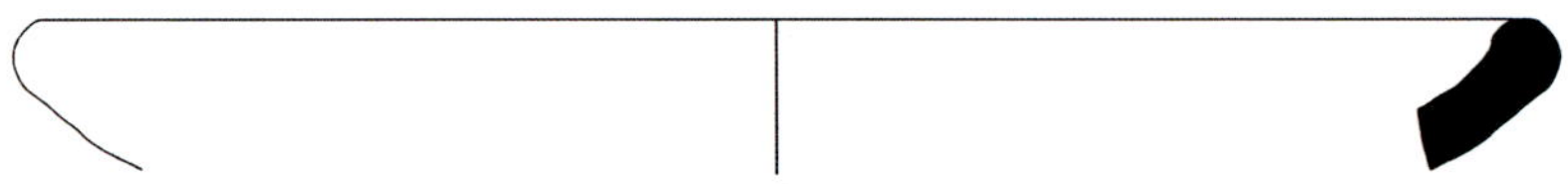

3

Cypriot red slip ware. Scale 1:3.

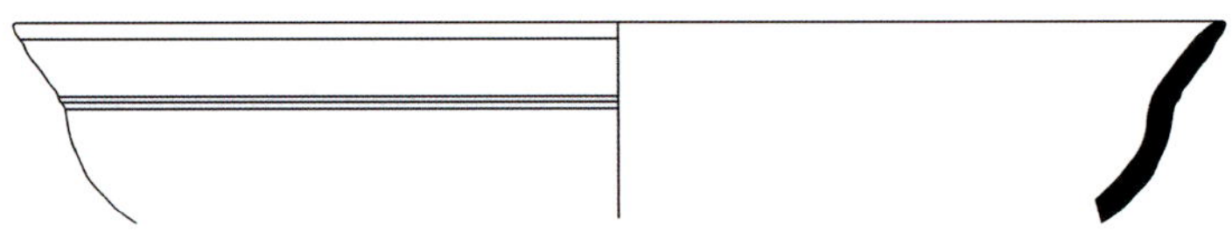

4

Late Roman C ware. Scale 1:4

Fig. 12.

Fig. 13. A pottery stamp depicting a swastika, also occurring in the pottery of Dura Europos. Photo: Michael Herles 2005.

Fig. 14. This form of a palm leave finds parallel in the relief ware decoration of the Dura Europos pottery repertoire. Photo: Michael Herles 2005.

Fig. 15. Roman pottery stamp from Tell Tibne.
Photo: Michael Herles 2005.

Fig. 16. Late Roman pottery stamp from Tell Tibne.
Photo: Michael Herles 2005.

Fig. 17. Incised decoration on a glazed pottery piece.
Photo: Michael Herles 2005.

Fig. 18. A twisted handle of a ceramic vessel.
Photo: Michael Herles 2005.

Fig. 19. A dupondius-sized coin.
Photo: Kenneth Lönnqvist 2006.

Fig. 20. Two small bronze coins.
Photo: Kenneth Lönnqvist 2006.

13.2.2 The Fort of Tabūs

Site H 19
Fort

UTM: 0586948, 3925414, northwestern tower UTM: 0586971, 3925471 and eastern tower UTM: 0587123, 3925328; ca. 266 m a.s.l.
Size of the plan: NW-SE walls: ca. 300 m and NE-SW walls: ca. 80 m - 100 m.

Ca. 20 km southeast of Tibne stands the fort of Tabūs on the cliff of Jebel Bishri. This small ruined fort guards the Euphrates valley at Tell Tabūs (Fig. 1) and rises on the southern side of the Deir ez-Zor - Aleppo -road (Fig. 25.), ca. 25 km north-west of the city of Deir ez-Zor. The fort stands at the Jebel Bishri ridge (southwestern tower) partly covered by a sandy mound ca. 90 m above the eastern alluvial terrace and the fluvial *wadi* bed (Fig. 21). The associated structural remains surrounding the fort consist of graveyards (H 17, H 18, H 22), two separate towers (H 21, H 23), a dam (H 24) and two ancient roads (H 13, H 15) at one kilometre's radius (see Fig. 25.). The documentation of the roads will be presented later in this chapter.

The corner towers can be detected in the CORONA declassified satellite photographs taken in the 1960s with a resolution of 2.7 m as well as on later SPOT images, but the remains are not very recognizable until the area is visited on the ground. In the 2004 season we recorded, documented and mapped the fort and its immediate surroundings *in situ* (Lönnqvist and Törmä 2004, under "Tabouz" in Lönnqvist *et al.* 2005; Lönnqvist *et al.* 2006). Mapping of the site was carried out from the acquired EDM field data (Fig. 27.).

Fig. 22. The ruined towers of a fort at Tabūs protruding from a mound at the edge of Jebel Bishri. Photo from SW: Minna Lönnqvist 2004.

Fig. 23. Entrance to Tabūs from west. Photo from NW: Minna Lönnqvist 2004.

Fig. 24. The ruins of the fort at Tabūs. Photo from SEE: Minna Lönnqvist 2004.

Fig. 21. The silhouette of Tabūs seen from the Deir ez-Zor - Aleppo -road. Photo from NW: Minna Lönnqvist 2008.

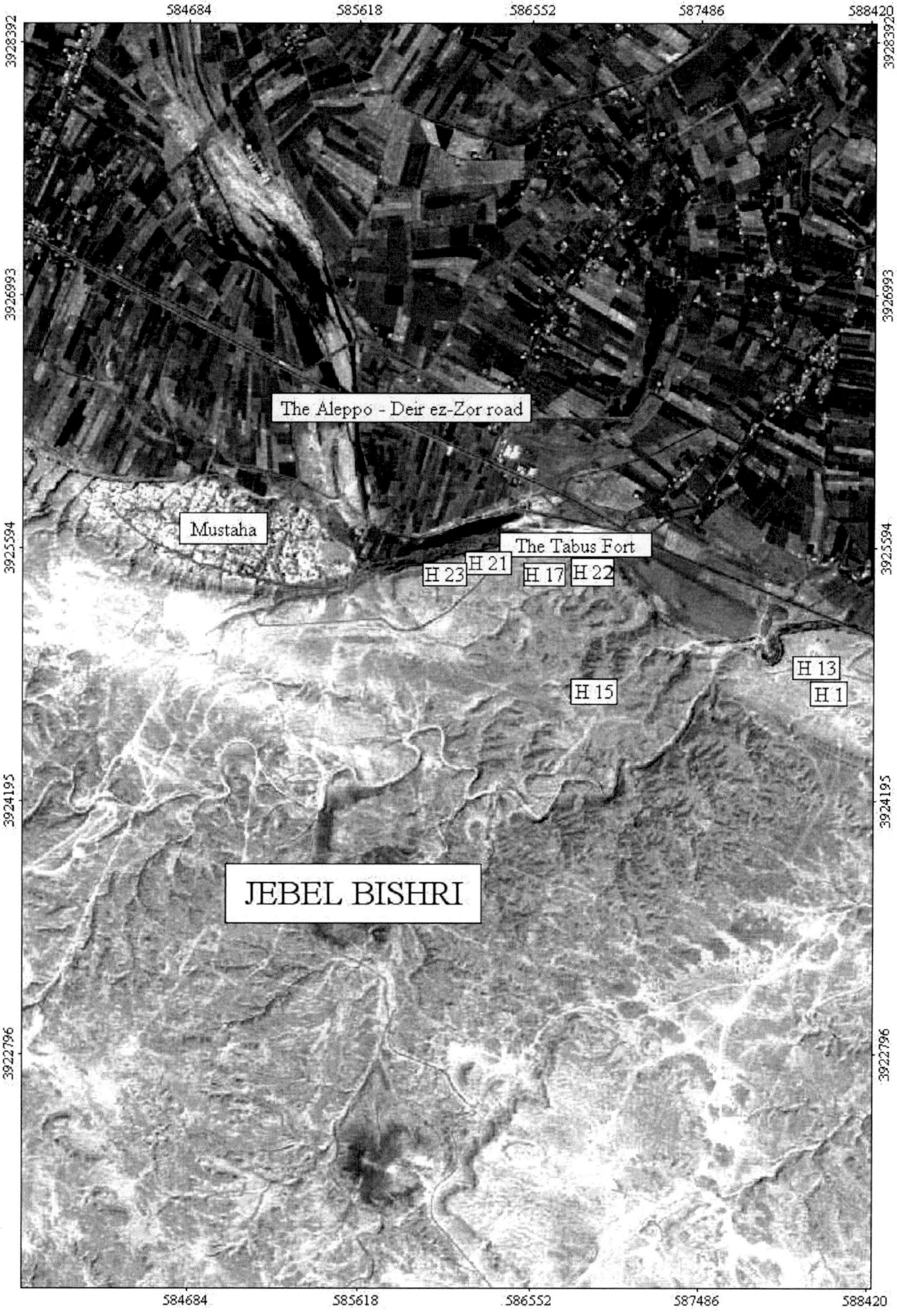

Fig. 25. The fort of Tabūs with surrounding sites/structures. Mapping by Minna Lönnqvist on a SPOT satellite image 2010.

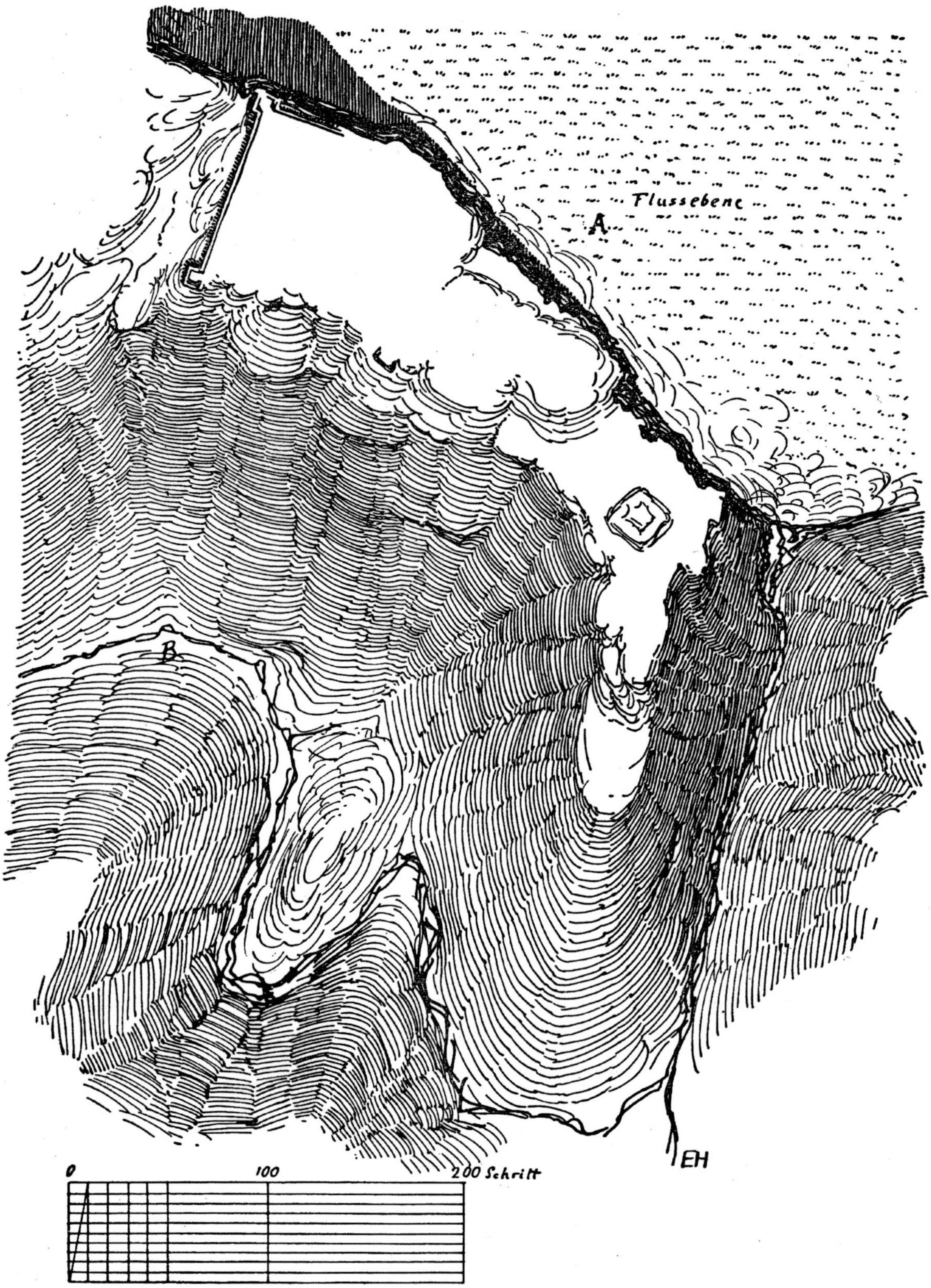

Abb. 75: Ṭābūs.

Fig. 26. The topographical drawing of Tabūs by Ernst Herzfeld from Sarre and Herzfeld 1911.

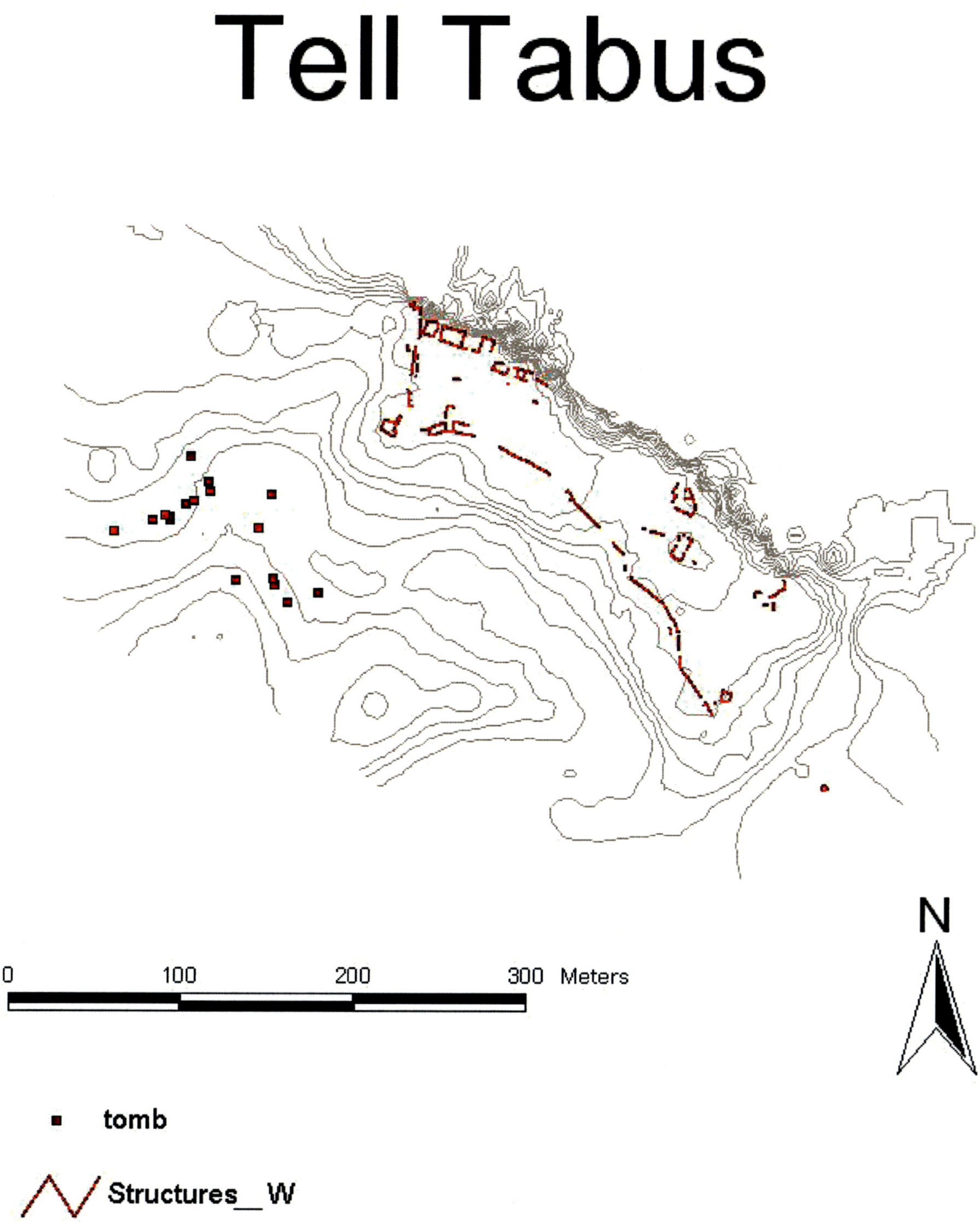

Fig. 27. A map showing the exposed ruins of the fort, the neighbouring graveyards (H 17, H 18 and H 22) and a water harvesting site to the right with topographic features at Tabūs in 2004. Map produced from the EDM field data with ArcView programme by Jari Okkonen.

The site of Tabūs has been previously described by several travellers, such as V. Chapot (1907: 293-294), F. Sarre and E. Herzfeld (1911: 168-170), as well as A. Musil (1928: 236), and it appears in C. Héraud's (1922, Fouille XXXVI) and A. Poidebard's (1934) aerial maps (see ch. 11: Fig. 2.). Apart from their general map (Fig. 6.), Sarre and Herzfeld also published a topographic map of the site (Fig. 26.). The associated flints discovered at the site indicate that there had been some habitation in prehistoric times in the area as well and some Bronze

Age activities were identified in the pottery repertoire and the types of the nearby tombs. L. Copeland (2004: 27, Fig. 2) had also earlier found a prehistoric site of 'Ain Tabūs in the neighbourhood. When we arrived at the site, we noticed that members of the General Directory of the Antiquities and Museums (GDAM) of Syria had obviously recently dug a trial trench on the western wall of the fort. The nearby graveyards to the southwest had been heavily plundered by grave robbers (Fig. 42.). However, except recently, the fort is hardly ever mentioned in the modern Roman frontier studies or encyclopaedias of the ancient world. Apart from Sarre's and Herzfeld's (1911) travel descriptions, no comprehensive archaeological documentation of the site with its surrounding ancient remains had been carried out.

The side of the fort facing the Euphrates is naturally steep and inaccessible forming a strategically convenient location. Situating on a cliff, the fort offered unhindered visibility into the plains of the Euphrates. Its eastern tower had a special command over the ancient road **H 13**, the foundations of which apparently date from the Late Roman period. The whole fortress is divided into two small hills by a shallow north-south-oriented ravine in the centre. The foundation of the fortress is laid on large marble/gypsum blocks varying ca. 0.7–0.8 m x 0.7 m x 1 m in size (see Fig. 31.).

The walls are built in a triangular/trapezoidal plan in a shell surrounding an inner courtyard (Figs. 26. and 27.). There are four ruined quadrilateral towers, one corner tower to the southwest (6.6 m x 8.6 m), another one to the northwest (6.5 m x 8.5 m) and one large keep-like tower (7 m x 14 m; Fig. 29.) to the east, in addition to one ruined quadrilateral wall tower on the southern wall. The eastern tower is still visible from the Deir ez-Zor–Aleppo -road. The narrow curtain walls (Figs. 28. and 32.) and the corner towers are built of angular and smaller crushed gypsum or calcite/marble boulders (varying ca. 0.3–0.4 m in diam.), fitted together with mortar forming a typical Roman "mortared rubble" core. The walls are covered by lime plaster with an ashlar masonry surface imitation (see Figs. 30. and 33.).

The main entrance of the fort apparently situated in the western side near the southwestern corner tower (Fig. 30.), the ruined walls of which still partly stand. The western wall of the fort has a double-walling system revealed in a trial trench. The double-walling system to the west and the southern wall tower were clearly defensive measures on the more easily accessible sides. Outside and beneath the western wall, there is a ditch which originally seems to have been the moat of the fortress but has largely been filled up with earth accumulating through the centuries. The main courtyard of the fort is filled in with room-like structures or *insulae.* However, no actual streets seem to separate them, and the site represents a citadel-type of fort rather than a *polis-castrum*-type such as the fortress of Zenobia.

Fig. 28. The southern part of the circuit wall and plundered rooms attached to the wall. Photo from E: Eivind Seland 2004.

Fig. 29. The ruins of the eastern tower or a keep. Photo from SW: Eivind Seland 2004.

Architectural fragments were documented at the site (see, e.g., Fig. 34.) and small surface finds were collected and recorded. Among the architectural fragments, there were parts of cornices and decorated masonry courses with rosettes and egg-dart-patterns dating from the Roman period. All the fragments were of greyish marble. Fragment 1 (UTM: 0587114, 3925305) is a part of a lintel 0.43 m x 0.30 m x 0.22 m in size and has "ovolo above astragal" ornament as decoration. Fragment 2 (UTM: 0587113, 3925305) is 0.43 m x 0.26 m x 0.25 m in size. It has a palm leave course in a "Palmyrene" style. Fragment 3 (UTM: 0587012, 3925429) is from an architrave (0.43 m x 0.36 m x 0.24 m in size) and has two fascia. The collected pottery mostly consists of coarse ware from different kinds of jars or bowls (Fig. 40.) But there also are pieces from globular amphoras, jugs, cooking pots, small flasks and cups. Among the pottery sherds there are Parthian blue-glazed pieces (Fig. 39.) which could be dated to the second and third centuries A.D. (cf., e.g., Fukai 1980/1981). Cooking pots include typical Late Roman dark black-greyish and grooved examples, like pieces of the third century brittle ware (Fig. 36.). There are jugs (Fig. 37.) comparable to those found at Dibsi Faraj further up the Euphrates (see, e.g., Vokaer 2009). The stamped pottery has (e.g. Fig. 38.) distant parallels in the Athenian Agora material from the fifth and sixth centuries A.D. (cf. Hayes 1972: 351-368). There are also sherds with applied "pie-crust" decorating typical of Byzantine pottery of the sixth century A.D. (cf. Fulford and Peacock 1984: 168-173) as well as twisted pottery handles from that period.

Fig. 30. The still partly standing southwestern corner tower of Tabūs. Photo from SW: Minna Lönnqvist 2004.

Fig. 31. Recording the large marble/gypsum foundation blocks of the fort, the scene beyond opens westward to the Euphrates valley. Photo from SE: Eivind Seland 2004.

Fig. 32. The visible width of the wall of Tabūs. Photo from SSW: Eivind Seland 2004.

Fig. 33. The inner face of the southern part of the circuit wall. Photo from NNE: Eivind Seland 2004.

Fig. 34. Fallen architectural fragments in situ at Tabūs. Photos: Eivind Seland 2004.

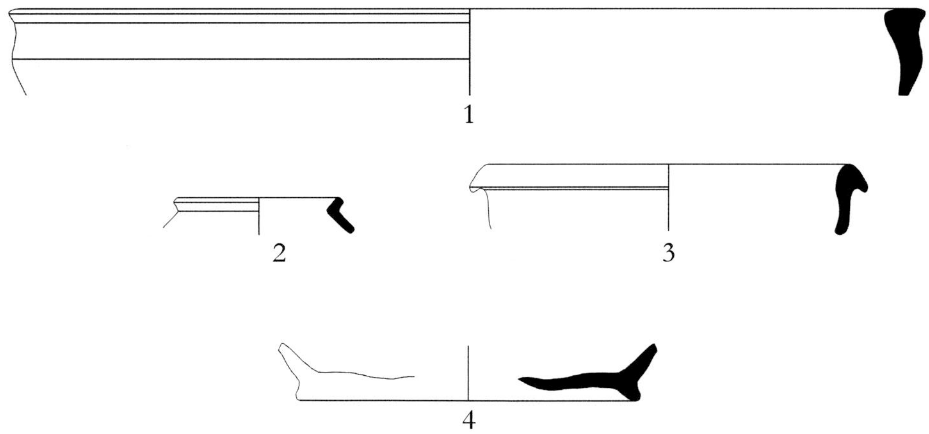

Fig. 35. Roman - Byzantine pottery. Scale: 1:3.

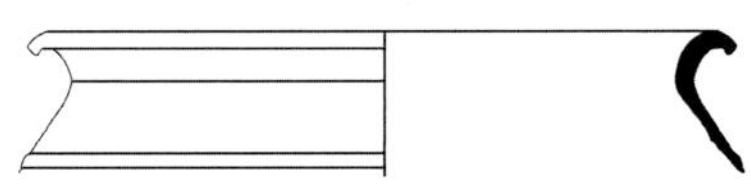

Scale 1:3

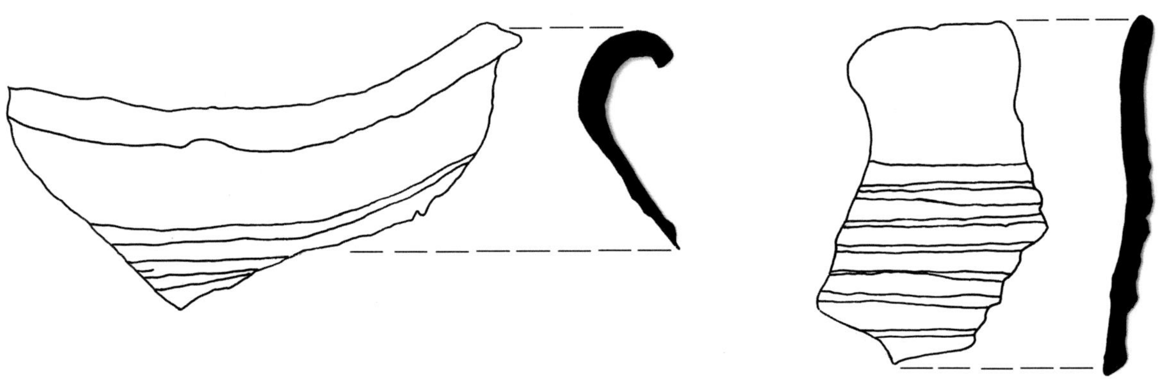

Scale: 1:1.
Fig. 36. Roman ribbed brittle ware.

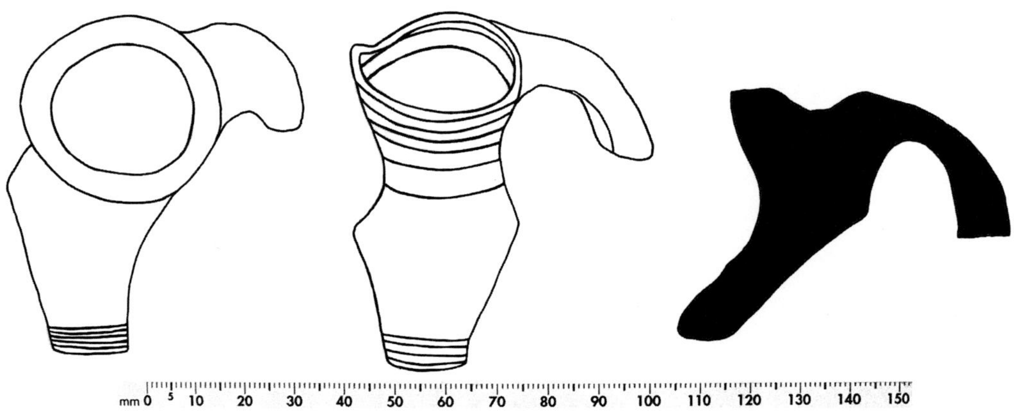

Fig. 37. A fragment of a Late Roman/Byzantine ceramic jug.

Fig. 38. Late Roman/Byzantine stamped and decorated pieces of pottery from Tabūs. Photo: Kenneth Lönnqvist 2006.

Fig. 39. Pieces of pottery from Tabūs, including a piece of turquoise and glazed Parthian pottery. Photo: Kenneth Lönnqvist 2006.

a)

b)

c)

d)

e)

f)

Figs. 40a-f.
Pieces of coarse ware from Tabūs. Photos: Kenneth Lönnqvist 2006.

Site H 22
A Water Harvesting Site

UTM: 0587192, 3925188
Size: dam: l. 10.75 m, w. 3 m and h. 2 m; rock-cut water reservoir/basin max: 6.1 m x 5.4 m x 4.2 m.

A water harvesting site **H 22** consisting of a dam and a water reservoir/basin (Fig. 43.) was identified in the *wadi* less than 100 m to the east beneath the fort of Tabūs (H 19) (Fig. 27.). The *wadi* was leading from the south to the north, being to the northwest and partly beneath the rock hewn ancient road H 15. The dam was constructed of angular gypseous stones fitted together with cement type plaster (Fig. 41.). The dam seems to be a water collecting site for the fortress of Tabūs. The reservoir is apparently formed by human hand but smoothed by water in the natural rocky *wadi* bed.

Fig. 41. A water harvesting site in the eastern wadi of Tabūs. Photo from NE: Eivind Seland 2004.

Fig. 42. A panorama over the looted graveyards of Tabūs. Photo from NE: Minna Lönnqvist 2004.

Tombs at Tabūs

Several graveyards **H 17**, **H 18** and **H 20** (Figs. 27. and 42.) with plundered graves were identified to the southwest beneath the fort on the nearby hills. They were recorded, but only one graveyard, namely **H 17,** could be securely dated to the period of the fort. H 18 was mentioned in association of the Bronze Age remains (ch. 10.). Further to the northwest there were two separate towers, **H 21** and **H 23,** stand at the edge of the mountain. They are apparently associated with the fort.

Site H 17
Graveyard/Looted

UTM: 0586819, 3925363
264 m a.s.l.

Size: ca. 100 m x 25 m

The graveyard H 17 is on a southwestern hill surrounded by *wadis* to the south, east and west. Altogether ten tombs were identified in the graveyard, eight of them nearly side by side stretching from the northwest to the southwest. The entrances were 0.5 m x 0.9 m in size on average. Two tombs, 8 and 9, in the northwest were recorded and documented. All the opened and looted graves were cut into bedrock.

Tomb 8

Size of the burial chamber: plan 6 m x 6 m, h: 1.4 m

One family tomb (H 17: 8) with a shaft entrance (Fig. 43.) had a central burial chamber, a square in plan, flanked with altogether nine niches or *loculi* on each wall and two *arcosolia,* typical of (Figs. 44.-46.) the Late Roman and Byzantine rock-cut tombs, flanking the western and northern walls. The wall surfaces of the tomb had faded traces of paint. The entrances of the *loculi* varied in shape from pointed, round to more square ones (Figs. 44.-46.). (Sizes of the entrances for the *loculi*: a) 0.70 m x 0.48 m, b) 0.85 m x 0.50 m, c) 0.85 m x 0.50 m, d) 1.08 m x 0.50 m, e) 1 m x 0.45 m, f) 1 m x 0.43 m, g) 1.05 m x 0.55 m, h) 0.80 x 0.50 m, i) 0.70 m x 0.50 m and the depths were 1.40-1.50 m; sizes of the entrances for the *arcosolia*: a) 1.60 m x 0.75 m, depth 0.80 m, b) 1 m x 1.65 m, depth 0.60 m.) In addition, in the centre of this burial chamber (Figs. 44. and 45.) there stood an empty sarcophagus (outer size: 2.25 m l. x 0.92 m w.; thickness of the walls: 0.12 m and max. preserved depth 0.77 m). The sarcophagus had been cast of concrete and was partly covered with bitumen. No associated pottery was encountered at the tomb.

Tomb 9

Size of the plan: 2.0 m x 1.8 m, h. 0.7 m

This tomb has a central burial chamber with two *arcosolia* (w. 1.8 m; depth: 0.8 m). A possible black painted *graffito* appears above one *arcosolium.* No associated pottery was found in the tomb. As previously described in ch. 11, during the 2000 survey we identified two comparable Late Roman/Byzantine tombs from the western piedmont area of Jebel Bishri. They are B 1 (UTM: 0507596, 3886103) and B 2 (UTM: 0504815, 3884501). B 1 in particular with three *arcosolia* resembles the family tomb 17: 8 recorded in the graveyard of Tabūs. These tombs in the west were clearly related to the structures and sites along the *Strata Diocletiana* while Tabūs is on the Euphrates *limes.* The tombs in the area of the fortress of Zenobia also offer parallels for these Late Roman/Byzantine tomb of Tabūs.

Fig. 43. Entrance of the tomb H 17:8. Photo from N: Kenneth Lönnqvist 2004.

Fig. 44. Sarcophagus, loculi and a part of an arcosolium visible inside the burial chamber of the tomb H 17: 8. Photo: Kenneth Lönnqvist 2004.

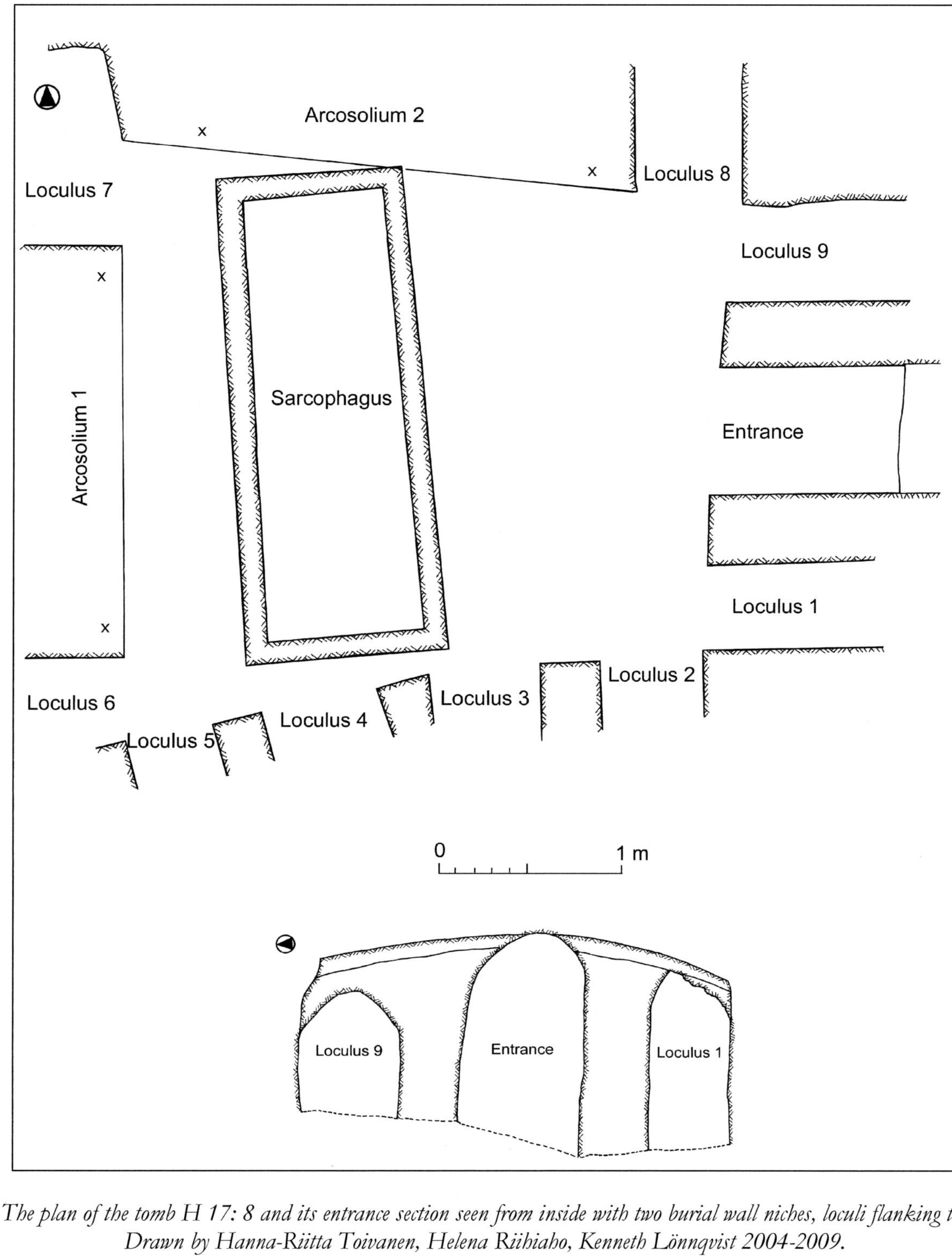

Fig. 45. The plan of the tomb H 17: 8 and its entrance section seen from inside with two burial wall niches, loculi flanking the entrance. Drawn by Hanna-Riitta Toivanen, Helena Riihiaho, Kenneth Lönnqvist 2004-2009.

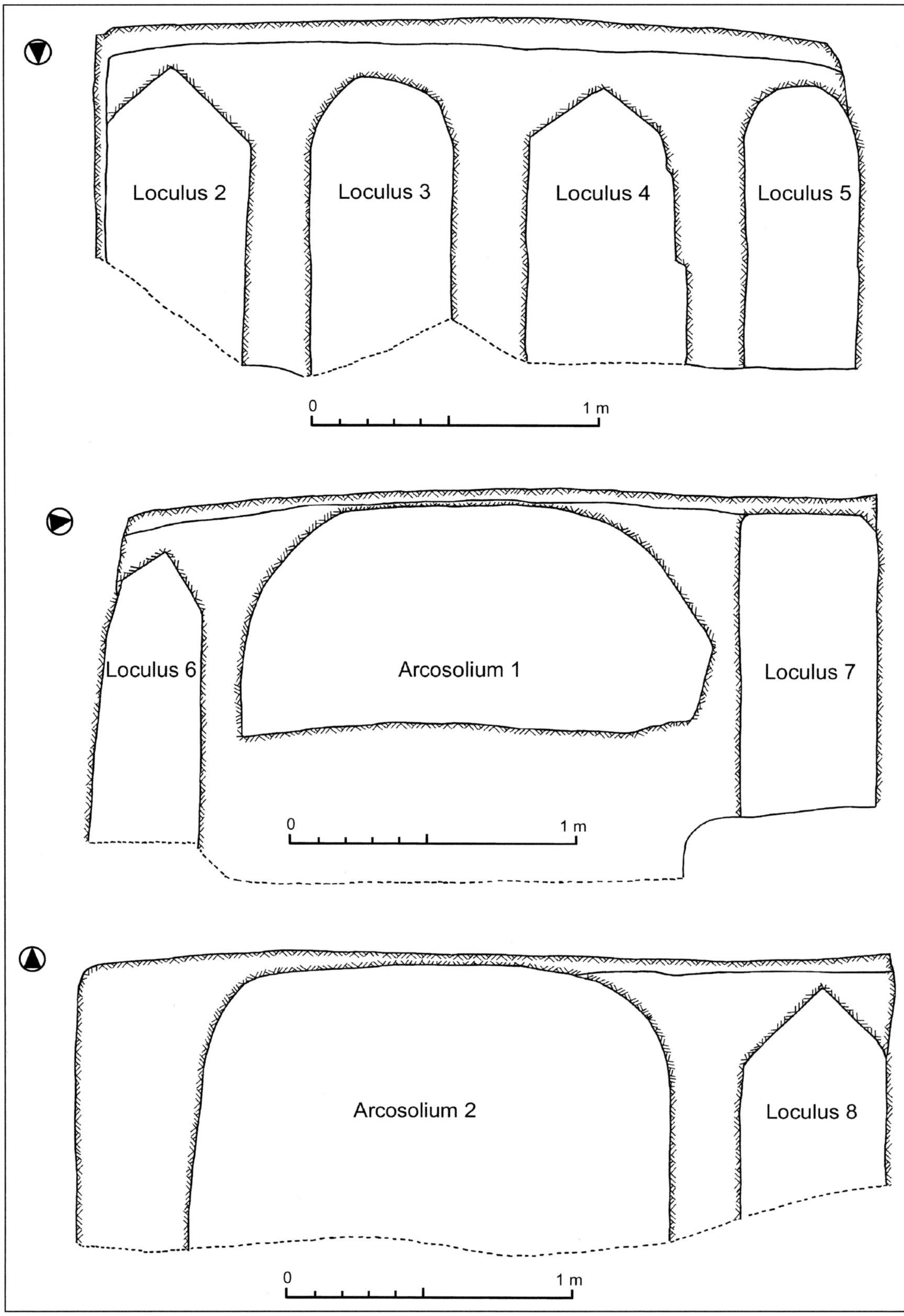

Fig. 46. Burial wall niches, loculi, and arcosolia of the tomb H 17: 8, cf. their location in the plan above. Drawn by Hanna-Riitta Toivanen, Helena. Riihiaho, Kenneth Lönnqvist 2004-2009.

Site H 20
Graveyard/Looted

UTM: 0586710, 3925444; 267 m a.s.l.
Size: undefined

This is another looted graveyard which had three rock-cut graves facing east. The openings of the graves varied from 0.4 m to 0.9 m in diameter. The depths varied from 1.0 m to 1.90 m.

Site H 21
Tower

UTM: 0586457, 3925596
Size of the outer plan: 4 m x 5 m;
preserved h. ca. 5 m

This ruined tower is situating ca. 500 m westward from the fortress of Tabūs. It has a roughly square plan, and it is constructed of angular gypsum cobbles the size of which was ca. 50 cm in diameter. The tower has collapsed and there is a robbers pit penetrating the southern entrance. There is a corner pilaster and a flight of steps covered by plaster (Fig. 47.) as well as a rectangular chamber.

Site H 23
Tower

UTM: 0586109, 3925467

Size of the outer plan: 2.0 m x 2.60 m;
inside: 1.15 m x 1.25 m

Another ruined tower, square in plan, appears on a natural hill formation ca. 800 m westward from the fortress of Tabūs (Fig. 48.). The tower is constructed from angular pieces of gypsum, the size of the stones being ca. 50 cm in diameter. Apart from this tower there also was another looted architectural construction on top of the hill and large pieces of *pithoi* were discovered on the northern slope.

F. Sarre and E. Herzfeld (1911: 367) suggested that these free-standing towers at Tabūs are tower tombs similar to those of Palmyra and Zenobia. Sarre and Herzfeld also drew a plan of one tower and dated it to the third century A.D. The ruined state of the towers does not allow very secure identification, but the remains of pilasters and signs of a plastered chamber indicate to a funerary use. The locations, however, strongly suggest that the towers could have served for monitoring the Euphrates valley at some point of time as well. It is likely that some tower tombs in Palmyra and Zenobia at the entrance or gateway sections of the cities also served to guard the traffic along the roads.[1] In addition, B. MacDonald (1992: 83) found at Rujm Umruq (Site 94) in Jordan a Hellenistic-Roman tower/tomb, which had originally served as a tower and later as a place of burial.

Fig. 47. Site H 21of a ruined tower with the visible flight of steps. Photo from W: Eivind Seland 2004.

Fig. 48. Site H 23 visible to the right on a hill formation. Photo from S: Minna Lönnqvist 2004

[1] The tower tombs in Palmyra are at the entrance of the road coming from Antioch, and at Zenobia one stands at the entrance of the southern gate like guarding the area. It needs to be noted that towers in the Graeco-Roman Near East often had multipurpose functions. The tower tombs in Palmyra had flights of steps to the roof, where the surroundings could be observed.

13.2.3 Architectural and Archaeological Comparisons

As previously mentioned, further ca. 13 km to the northeast of Tabūs M. Gschwind (2005a; 2005b) has excavated the fortress of Qreiye (UTM: 596177, 3920938; ca. 228 m a.s.l.) at 'Ayyāsh. The building of the fort of Qreiye (Birtha Arupan?) by the Romans was initially associated with the Severan campaigns against the Parthians on the Euphrates *limes* (Gschwind 2005a; 2005b). M. Gschwind and H. Hasan (2008) have vaguely dated the establishment of the site to the second or third century A.D.

Fig. 49. Aerial photograph of Qreiye (Poidebard 1934).

Qreiye follows the plan of a square Roman military camp (see Fig. 49.). The visible ruins are of basalt, and there are round corner towers like at Qseybe (see ch. 12.), thus differing from the visible ruins at Tabūs. However, the closest parallels for the fort at Tabūs in the neighbouring fortresses on the Euphrates *limes* region are the fortresses of Zenobia and Djazlah with their triangular/trapezoidal layouts and protruding quadrilateral wall and corner towers (see Poidebard 1934, Atlas, Pl. LXXXIV; Napoli 2000; Sarre and Herzfeld 1911: Fig. 73). One fortress that also can be compared with the structures of Tabūs is Dibsi Faraj to the west from Sura on the Euphrates, and its rebuilding is dated shortly after A.D. 294, the date based on a reliable stratigraphic sequence (Harper 1975: 326). Some architectural fragments at Tabūs could also indicate that the site was already in use during the heyday of Palmyra in the second or third century A.D. serving as a border station against Parthia, and it could later have been rebuilt by Byzantine rulers against the Persians. The pottery dating to the third century also supports that there already was Palmyrean/Roman activity at the site in the third century A.D.

Diocletian, who fortified and refortified the eastern desert frontier ca. A.D. 300, built *castella*, such as Qasr Bshir in Jordan, with quadrilateral towers protruding from the walls (see, e.g., Unesco World Heritage Sites UTM: 783000, 3470880, Zone 36). The Late Roman forts, *castella*, compared to those of the Principate are reduced in their size and internal structure, and the number of gates from several to one or two. The towers are also distinctive as the towers of the Principate are inserted into the curtain wall serving chiefly as observation points while in the Late Roman forts – at least in Arabia – towers are protruding from the walls, the feature which has been attributed to the change from offensive bases to defensive sites in the third century A.D. Soldiers could have more effectively used artillery and hand-held weapons against enemies from projecting towers, especially near a gate or at a curtain wall. (Parker 2006: 548). Because there are no clear barracks at Tabūs as in the Roman forts from the pre-Diocletianic and Diocletianic era, the inner structure indicates that the soldiers were living there with their families (cf. Daniels 1980 *apud* Bidwell 1991). In the case of Tabūs the graveyard with family graves also suggests to the existence of families at the site. The site could have also served as the final resort for local inhabitants. Trapezoidal forts had many tactical advantages and were often laid out against natural barriers and are frequently attributed according to their plan to the Valentinian types (Gregory 1997, vol. 1: 129). In the Late Roman forts the internal buildings were no longer separated from the curtain wall as in the case of the forts during the Principate, and the size of the garrisons was diminished (Parker 2006: 548-549).

At Tabūs there is only one major gate and two projecting towers are defending it. Apart from the Diocletianic fortresses, quadrilateral towers were especially preferred in Byzantine structures. Towers had several functions: guarding, signalling, storage and as final resorts. Towers were erected in Syrian villages in the fourth century and in border zones where active confrontations took place in the Byzantine period. (Isaac 1990: 179, 186). In Byzantine fortifications it had become common that the mortared rubble or *opus caementicium* core, faced with any available larger flattish stones, was used as surface material. This phase emphasises the distinction between the core and the facing. However, in due course in Byzantine architecture the difference between the core and the facing disappears. The rubble was preferred in fortifications, but the use without facing may date back to the fourth century A.D. in the West. An example is the Roman headquarters building at York which was rebuilt with mortared rubble in the fourth century A.D. and designed to have a lime plaster finish without ashlar stones. (Foss and Winfield 1986: 26-29). This resembles our case but being in the East and having imitation of ashlar masonry in the lime plaster surface. A good ashlar facing for a wall, often with courses of brick both on the face and through the core, were used in fortifications until the sixth century A.D. Philo of Byzantium recommends to use gypsum in fortifications, which is obviously the case at the fort of Tabūs, the fortress of Zenobia and the large *polis-castrum* of Resafa-Sergiopolis. However, lime is also commonly found in Byzantine works. It is translated as plaster which may be both lime as well as plaster of Paris, but in our case it may actually refer to the facing with lime. (See Foss and Winfield 1986: 26-29).

The Roman fort of Khirbet el-Mahatta in Jordan is also comparable in size and structure to Tabūs. The site in Jordan is mentioned by Eusebius and probably by the pilgrim Egeria. The location is reminiscent to Tabūs as the fort is situated on a ridge overlooking the Jordan valley, and at Tabūs the Euphrates valley instead. The plan of this fort is triangular and covers an area of 300 m x 150 m. The arrangement of square towers is very similar, even down to the southwestern corner tower (6.5 m x 6.5 m), which has a comparable layout. But no exact dating or identification of the builders is given to this fort either. The pottery at the site in Jordan consisted of Roman and Byzantine sherds. (Ibach 1987: 15-16). Outside the eastern *limes*, the Byzantine castle of Redina built by Justinian I is another close parallel in design to Tabūs. Redina is also characterized by a triangular/trapezoidal plan and quadrilateral towers. Its location on a ridge overlooking and protecting the *Via Egnatia*, a major highway in Greek Macedonia (see Moutsopoulos 1983: 86-99), is also comparable to the position of Tabūs. The Castle of Saône has a rectangular citadel or shell keep of Byzantine date on the summit of a hill within the castle like the one at Tabūs. This effectively exemplifies the difference between the strong points: one is in the form of an independent tower fort which may be on the outer wall, and the other is in the centre of the fortification and serves as a last resort of defence. (Cf. Foss and Winfield 1986: 26-29).

The location and structure of the fort at Tabūs indicate to a guarding post and defensive military base rather than a fortified town, *oppidum*. The fortress of Zenobia, however, as rebuilt by Justinian, is much larger (ca. 385 m x 350 m x 550 m) than the fort at Tabūs. As mentioned, it is actually a fortified town with streets, a *forum*, a *praetorium*, basilicas and baths (Lauffray 1983-1991; Kennedy and Riley 1992: 117-118 and Toivanen 2008). A much larger Byzantine building operation along the same lines is Resafa-Sergiopolis, ca. 147 km to the west and yet another is Circesium, ca. 56 km to the southeast. As previously mentioned, Circesium was rebuilt by Diocletian and later refortified by Justinian. Procopius (*De Aedificiis* II, vi, 6-8) mentions that Circesium had foundations of large, hard stones as is also the case at Tabūs (see Fig. 31.). At the fortresses of Resafa-Sergiopolis and Zenobia, mainly gypsum and some bricks are used as building material, but at Tabūs we only found calcite/gypsum rubble core walls with lime plaster facing. The double-walling system, present at Tabūs, is a typical feature of the Justinianic fortresses.

13.3. Testing the Connection of the Fortresses: Tibne – Tabūs – Qreiye

13.3.1 A Viewshed Analysis between Tibne, Tabūs and Qreiye

Already A. Poidebard (1934: 168) noted the construction of Roman watchtowers at certain intervals for the purpose of signalling and warning of coming enemies. At the Late Roman and Early Byzantine forts visibility was emphasized (Parker 2006: 549), and tests between the signalling posts have been carried out, e.g., in the *Limes Arabicus* project (Parker 2006: Fig. 2.7). From strategic points, the hill-tops and ridges offer the best commanding views over the landscape. Jebel Bishri has been a site for mobile people, such as hunter-gatherers and pastoral nomads to whom natural visibility was important (see ch. 5. and 8.), and latterly for the Romans and the Byzantine power which countered expanding empires. On the Euphrates frontier the defensive structures were first constructed towards the Parthians/ the Sassanians/the Persians. The Romans tried to choose some of the best observation points, with good visibility, at strategic sites that provide optimum vantage points. The spires of the towers offered special artificial secondary vantage-points for guarding, signalling and protection. They also provided the best panoramas for military reconnaissance or defence.

The site of Tabūs clearly belongs to the network of defences and roads on the Euphrates *limes*. We therefore carried out an experimental GIS (Geographic Information Systems) analysis to elucidate possible connections and strategic functions of the forts in the military organization at the Euphrates border zone. GIS offers the possibility of carrying out different kinds of viewshed analyses to investigate inter-visibility. (See, e.g., Wheatley 1995: 171-185). The dating, function and significance of forts and fortresses can be studied and further discussed in the light of the new basic data brought forward in the present study. This viewshed analysis between Roman forts differs from general archaeological visibility models which are static. (See Wheatley and Gillings 2000: 7).

Fig. 50. Straight visibility: the fort of Tabūs seen from the fort of Qreiye at 'Ayyāsh. Photo from SE: Minna Lönnqvist 2006.

An eventual visibility between the forts at Tibne (Mambri?), Tabūs and Qreiye (see the map in Fig. 6.) could provide some indications of their contemporaneous and functional defence purposes in the Late Roman and Byzantine periods. This inter-visibility would not, necessarily, mean that the forts were originally contemporaneously erected, but that at some point in time, they could have functioned together. Both Tibne (Mambri?) (UTM: 0574179, 3941645; ca. 231-235 m a.s.l.) and Qreiye (UTM: 0596177, 3920938; ca. 228 m a.s.l.) in the valley of the Euphrates date from the Late Roman period and are much lower compared to the fort of Tabūs (ca. 266 m a.s.l., although the original towers were much higher). Tabūs, between them on a cliff, could have functioned as a relay station, if there was visibility between the forts. (Lönnqvist *et al.* 2005). The position of Tabūs on steep cliffs was much more defendable than both that of Qreiye and Tibne along the Euphrates. At Qreiye we could also observe *in situ* that there indeed was a good visibility to Tabūs (Fig. 50.).

The viewshed analysis which we carried out is based on the DEM (digital elevation model data) covering the area, heights and the coordinate information of the sites. SRTM90m DEM was used in the analysis. PCI Geomatica 9.1 SEENARE-function determines which pixels in a digital elevation model can be seen from a view point defined by X, Y and Z coordinates. Coordinates for the forts were determined using GPS, maps and higher resolution Landsat-7 ETM panchromatic images. The heights of viewpoints were defined as being 15 m higher than the DEM surface height to decrease the effect of random errors of DEM. Possible error sources of the viewshed analysis are:

- large pixel size (90 m) of SRTM DEM
- heights of the view points were 15 m above DEM surface, so the resulting view area is likely to be larger than in reality
- SRTM DEM positional and height errors, and
- coordinate match between SRTM DEM and GPS.

The results of the viewshed analysis are presented in Figs. 51.–54. Fig. 52. presents the overview visibility image of the whole area, Fig. 51. Tibne (yellow cross), Fig. 52. Tabūs (red cross) and Fig. 53. Qreiye (red cross). The background for the visualization is a Landsat-7 ETM panchromatic image taken 29.1.1999. The image is referenced to the same coordinate system as SRTM DEM and averaged to 90 m pixel size. The colours correspond to the areas seen from different fortresses, the idea being that the areas seen from Tibne (Mambri) are presented as red, Tabūs green and Qreiye blue. Other colours mean that the place can be seen from two or three fortresses (Table 1).

	Tibne/Mambri	Tabus	Qreiye
Red	X		
Green		X	
Blue			X
Yellow	X	X	
Magenta	X		X
Cyan		X	X
White	X	X	X

Table 1. The colours used in the viewshed analysis.

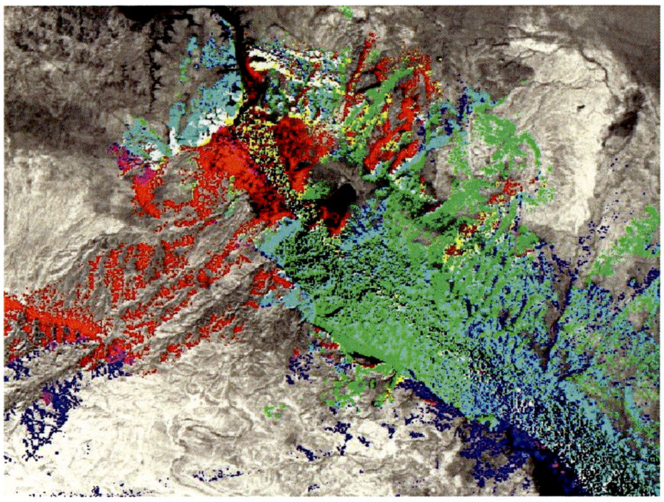

Fig. 51. The visibility image of the study area.

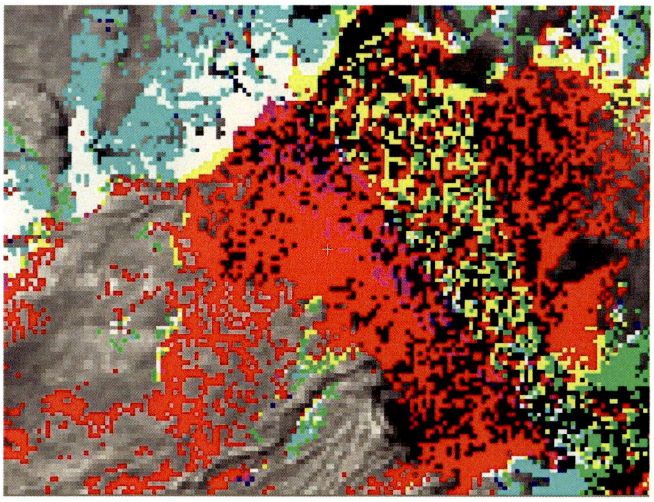

Fig. 52. The visibility image of the fort of Tibne/Mambri (yellow cross).

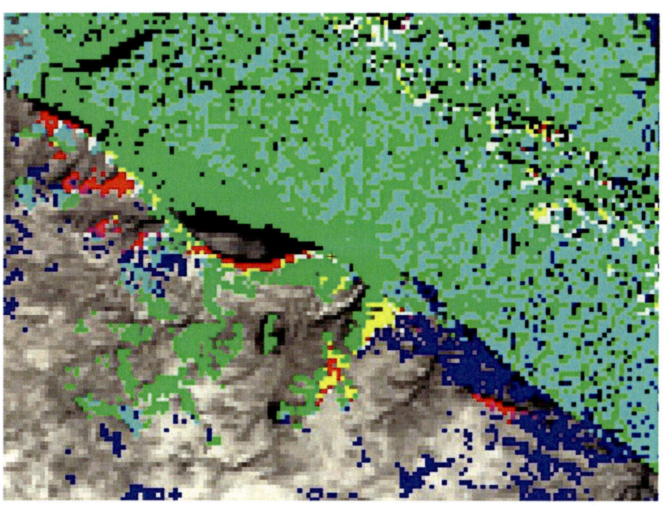

Fig. 53. The visibility image of the fort of Tabūs (red cross).

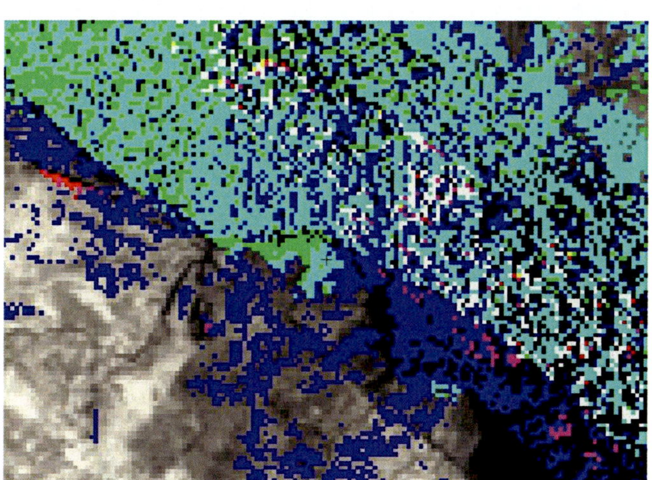

Fig 54. The visibility image of the fort of Qreiye (red cross).

According to this analysis, Tabūs could have functioned as a relay station for signaling (white) between Tibne/Mambri and Qreiye, and it offered intervisibility both with Mambri (white and yellow) and Qreiye (white and cyan). The intervisibility with Tibne (Mambri?) is dependent on the hypothesis that the structures were originally high. In the experiment, the 15 m higher level of the digital elevation model than the actual earth surface could therefore include high towers and rising smoke signals. Other means of communication could have been light-reflecting materials such as glass and metal, e.g., between Tabūs and Qreiye which are closer to each other than Tabūs and Tibne/Mambri.

As mentioned earlier, Mambri, according to Procopius, was originally built by the Emperor Diocletian. The pottery types and pottery stamps at Tibne find parallels with Dura Europos and therefore point to the beginning of the Roman presence, at least from the time of the Roman-Parthian wars, even to the earliest dates of the Roman presence in the area. Qreiye was used contemporaneously and, as mentioned, was conquered roughly at the same time (A.D. 253) as the Sassanian invasions of Dura Europos (A.D. 256) took place in the 250s A.D. Qreiye may have been used as a guarding and signalling post later on. For example, later activities are represented at Qreiye in the collected surface pottery dating from the later Roman and Byzantine times (cf. http://www.dainst.org/print.php?id=742). From the view of the accumulating evidence all the forts of Tibne (Mambri?), Tabūs and Qreiye could have originally been built during the same centuries, and after the fall of Qreiye their use could have still continued contemporaneously from the fourth century A.D. onwards.

13.3.2 The Problems in the Identification of the Fort and in its Dating

The recording, documenting and mapping of the ancient fort at Tabūs is the first step towards its protection and preservation. Looting in the area and erosion have taken their toll during the past hundred years. The identification of Tabūs has been difficult partly due to Ptolemy's *Geographia* (A.D. 90-168), which lists sites in Syria, the *Arabia Deserta* and Mesopotamia in different categories compared to our modern understanding. Tabūs has, *inter alia*, been identified with Dabausa, mentioned in *Geographia* (V, 17; see Chapot 1907: 294 reporting Sachau's original identification). Others, like A. Musil (1928: 236), identified it with Alalis, mentioned by *Geographia* as one of the towns near the Euphrates. Alalis was situated on the way from Sura to Babylon (Ptolemy's *Geographia* V, 14).

The *Tabula Peutingeriana*, originally dating to the fourth century A.D., unfortunately does not take its description as far as Tabūs on the Euphrates. The military tabulas of the *Notitia Dignitatum* mention, under *Dux Syria sub Eufratensis Syriae*, a site called Elela which is the post of the *Cohors Prima Gotthorum* (Helela). The *Notitia Dignitatum* has been linked with the military reorganization of the year A.D. 392 and the birth of Byzantium, and with the time when the barbarian Gothic legions started serving in the eastern Empire. Musil (1928: 236) suggested that the Elela mentioned in the *Notitia* should be the Alalis of Ptolemy's *Geographia.* Poidebard on his part identified Elela/Helela with Hlehle, situated between Palmyra and Shukhne, southwest of Jebel Bishri (Poidebard 1934, Atlas, Pl. V).

V. Chapot (1907: 294, note 3) suggested that Tabūs would be Diocletian's fort of Mambri rebuilt by Justinian I, and J. Lauffray agreed with this. But the distance from Tabūs to Zenobia is not in accordance with the account of Mambri by Procopius, which better matches the already mentioned Tibne. R. Dussaud (1927: 453-455) also discussed the identification of the forts of Alalis, Sura, Alamatha, Thapsaque, Birtha, Gadirtha, Auzara, Audatha, Addara, Balagea and Pharga located in Ptolemy's *Geographia* on the right river bank of the Euphrates. According to Dussaud, there were clear possibilities for the identifications of Sura, Auzara and Addara: Auzara being Deir ez-Zor and Addara being Dura Europos. Later Zenobia's fortress has been identified with Birtha Asporiscos and Qreiye with Birtha Arupan. In Ptolemy's list (V, 18) Thapsacus, i.e. Thapsaque - located obscurely in the *Arabia Deserta* but near the Euphrates - could be close to the toponym "Tabūs", and indeed its location in relation to Birtha (Zenobia/Qreiye?), Gadirtha, Auzara, Audatha and Addara would be the most convenient identification. But Thapsacus has been identified with other sites and should have offered a crossing point over the Euphrates. Alamatha, located in Syria on the Euphrates (*Geographia* V, 14), also remains obscure, but its location could be one of the sites associated with Halabiya, either later Zenobia or Mambri. The earliest Roman pottery from Tabūs dates from the second or third century A.D., to the period of Dura's Roman garrison. F. Sarre and E. Herzfeld (1911: 170) also dated the tower tombs nearby the fort to the third century A.D. As mentioned previously, it is possible that the site of Tabūs already served as a Palmyrene station during the second and third centuries A.D. We need to keep in mind that Ptolemy's *Geographia* dates from the second century A.D. According to the present evidence, there apparently were contemporaneous activities at Qreiye, Tabūs and at Tibne (Mambri) during the second and third centuries A.D. - also indicated by the viewshed analyses. In regard to the architectural remains there is more evidence of later Roman and further Byzantine activities at the site of Tabūs. These could be associated with Diocletian's and Justinian's refortification policies.

It is likely that Tabūs served during Diocletian's reign in association with Tibne (Mambri?) and was linked with Diocletian's border policy in general. The large and final rebuilding operation at Tabūs seems, however, to have taken place in the Byzantine period, apparently under Justinian I in the sixth century A.D. The large marble/gypsum foundation blocks, the triangular/trapezoidal plan of the fort, the quadrilateral and protruding corner towers, the double-walling system and plastering of the wall surfaces, seem all to follow Justinian's building projects and the Byzantine metaphysical ideals of fortifications with towers. The fort could also have not only secured the line of defence and functioned as a relay between Tibne (Mambri?) and Qreiye, but also assisted the larger fortresses of Zenobia and Circesium rebuilt by Justinian on the Euphrates.

13.4. In Search for Roman Roads and Bridge Points on the Euphrates

Tadmor/Palmyra used caravan roads over the desert already in the Bronze Age, but the "golden age" of its caravans is dated to A.D. 131-161 by inscriptions. Palmyra was an important station along the Silk Road. Palmyra used both caravans and ships in its trade networks. M. Gawlikowski (1996: 140-141) has emphasized that the caravans of Palmyra must have utilized the Euphrates waterways at least downstream. The extension of the Roman power in Syria also needed road networks to link military stations to one another. Since A. Poidebard's studies (1934: 165-168) roads have been largely neglected in archaeological surveys of the region, although they formed linear patterns and networks between monuments and sites belonging to an original cultural landscape (Gates 2006). Topography and climate largely dictate where it is most convenient to lay a road. The military troops, postal system and trade required good and fast means of communication to govern the area. The army was a central force in the strategy of organizing the Roman frontier zones. R. Chevallier (1989: 65-66) has classified Roman roads into simple categories: 1. public and military roads that were built by the state, 2. local roads that were regionally financed by town councils and 3. minor roads that were built by landowners for private use. What, however, needs to be kept in mind is that the so-called Roman roads vary in their quality, size, surface materials and techniques in different parts of the empire (Greene 1986: 35).

A. Poidebard (1934: 165-167) classified the Roman roads of the region into three categories: paved ways, track-roads and caravan roads. K. Butcher (2003: 127-128) also notes that generally in Roman Syria there were three types of roads of which the most elaborate ones were built between cities and towns, in marshy and rocky areas. The second type of roads was simply gravel-paved between towns being from 4 m to 7 m in width. The upper layers consisted of clayish soil with gravel which was laid over stones that were backed ca. 20 cm in thickness. Sometimes the foundation was on a layer of bed stones. The third type was the dirt-track typical of desert and steppe areas including the *Strata Diocletiana*. These types do not include rock-cut roads, which are met within the Levant, including some in Syria. One of the most stunning Roman rock-cut roads can be found at Wadi Barada some 35 km west of Damascus. We found one such rock-cut road beneath Tabūs, which also may date from an earlier period.

In their aerial maps and photographs A. Poidebard (1934) and R. Mouterde with A. Poidebard (1945) connected several Roman/Byzantine stations and forts in the Euphrates *limes* region with roads illustrating the sites in the neighbourhood. However, Poidebard mentions that he did not actually trace all the roads from the air or on the ground but indicated their existence and their alignments by connecting the identified ancient forts on the maps. A road which on Poidebard's map (1934) follows the right bank of the Euphrates links the fortress of Zenobia with the forts of Tibne (Mambri?), Tabūs and Qreiye.

An alluvial terrace belonging to the piedmont of Jebel Bishri formed a natural transect for our survey, delimited by the river valley and the mountain. Because of being a terrace on the hard bedrock of Jebel Bishri, the geomorphology allowed a favourable construction area above the generally flooded and cultivated river plain. The area that was systematically field-walked from plateau **H 1** in 15 m intervals in 2004, between the villages of Kharita and Mustaha, was 3 km in length and 1 km in width extending to the ancient fort of Tabūs and the surrounding hills. In 2004 we traced two ancient road alignments **H 13** and **H 15** (Fig. 55.) in the village of Kharita - Mustaha in the area 23-25 km northeast of Deir ez-Zor on the southern side of the road towards Aleppo. The two ancient roads are situated to the east and southeast from the ancient fort of Tabūs.

Site H 13
Ancient Road Alignment

UTM: W: 0587800, 3924989; E: 0588873, 3924616;
W: ca. 216 m a.s.l. and E: ca. 209 m a.s.l.
Preserved l.: ca. 1.1 km, w: 5 m.

An ancient road **H 13** (Figs. 55.-60.) can be driven up to the south by car from the side of the Deir ez-Zor-Aleppo main road ca. 23 km from Deir ez-Zor. The road bends up to the south to the Pleistocene terrace, i.e. plateau **H 1** (see details in paragraphs 5.5., 5.6, 10.3.), and it becomes parallel with the modern Deir ez-Zor - Aleppo -road. The eastern tower/keep of Tabūs provides an excellent visibility to the road (Fig. 56.), which has a rather straight alignment on the plateau. The road is laid on a silty but gypsum-based terrace that is partly the piedmont of Jebel Bishri, making a steady basis, even in winter, compared to the river plain itself. (Fig. 57.). Topographically the connection with the fort is very logical.

The road is quite well preserved for the length of ca. 1.1 km and is 5 m in width, paved with surface smoothed but angular greenish, greyish and white marble/ calcite/gypsum stones (Figs. 58. and 59.). The road had had a fourth uppermost layer of asphalt during the Ottoman or the French mandate times, but it had faded away with only some *lacunae* still visible (Fig. 60.). The road is ca. 0.35 m thick in section (Fig. 60.), now consisting of three layers from the top: in the first upper layer there are paving stones, under which there is a layer of gravel/coarse sand, and beneath there is a foundation layer of similar stones as in the pavement. Therefore the road has a similar kind of structure to Roman roads in Spain which have still been in use in modern times. Associated finds from the road area consist of several pieces of Roman pottery (Fig. 61.), including Late Roman C ware. A marble/gypsum quarry lies beneath Tabūs on the side of the Deir ez-Zor–Aleppo -road and quarries are typically connected with Roman roads for the acquisition of the paving material (cf. Chevallier 1989).

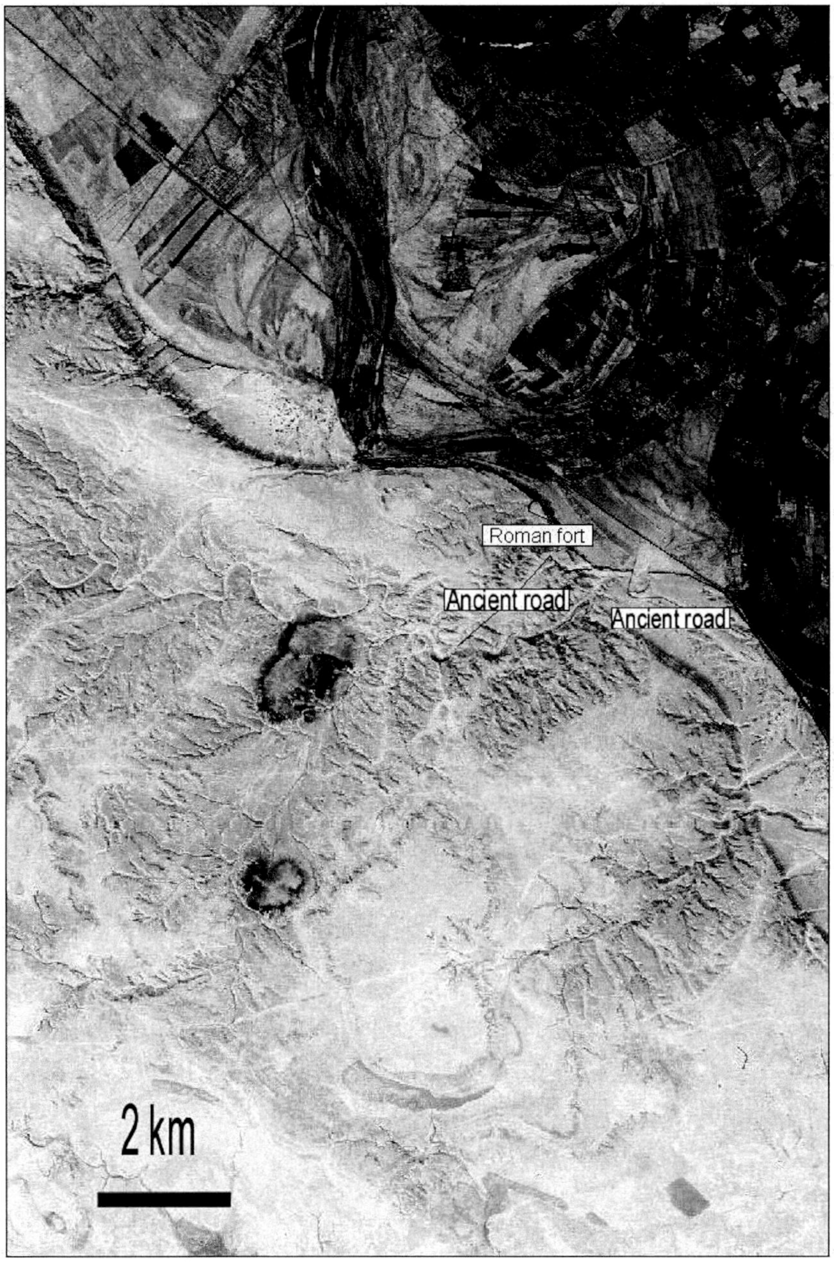

Fig. 55. The ancient Roman? roads H 13 (right) and H 15 (left) mapped on a CORONA satellite photograph. Mapping by Minna Lönnqvist 2005.

Fig. 56. The alignment of the ancient road H 13 in relation to the eastern tower/ keep of the Tabūs fort. Photo from W: Kenneth Lönnqvist 2006.

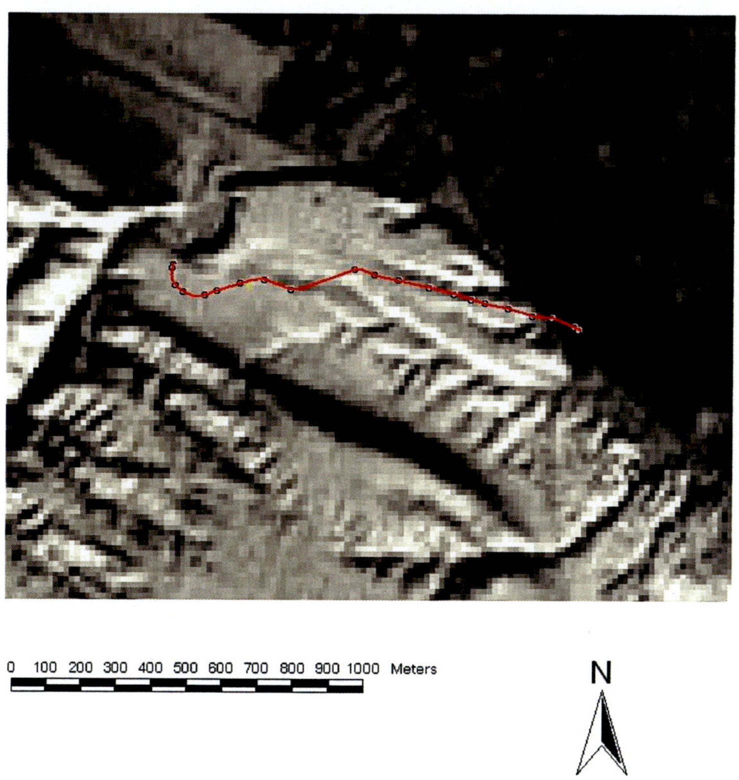

Fig. 57. The alignment of an ancient road (H 13) beneath the fort of Tabūs, GPS mapping in situ on Landsat-7 ETM satellite image. Mapping by Jari Okkonen 2004.

Fig. 58. Paving structure of the road H 13. Photo from W: Eivind Seland 2004.

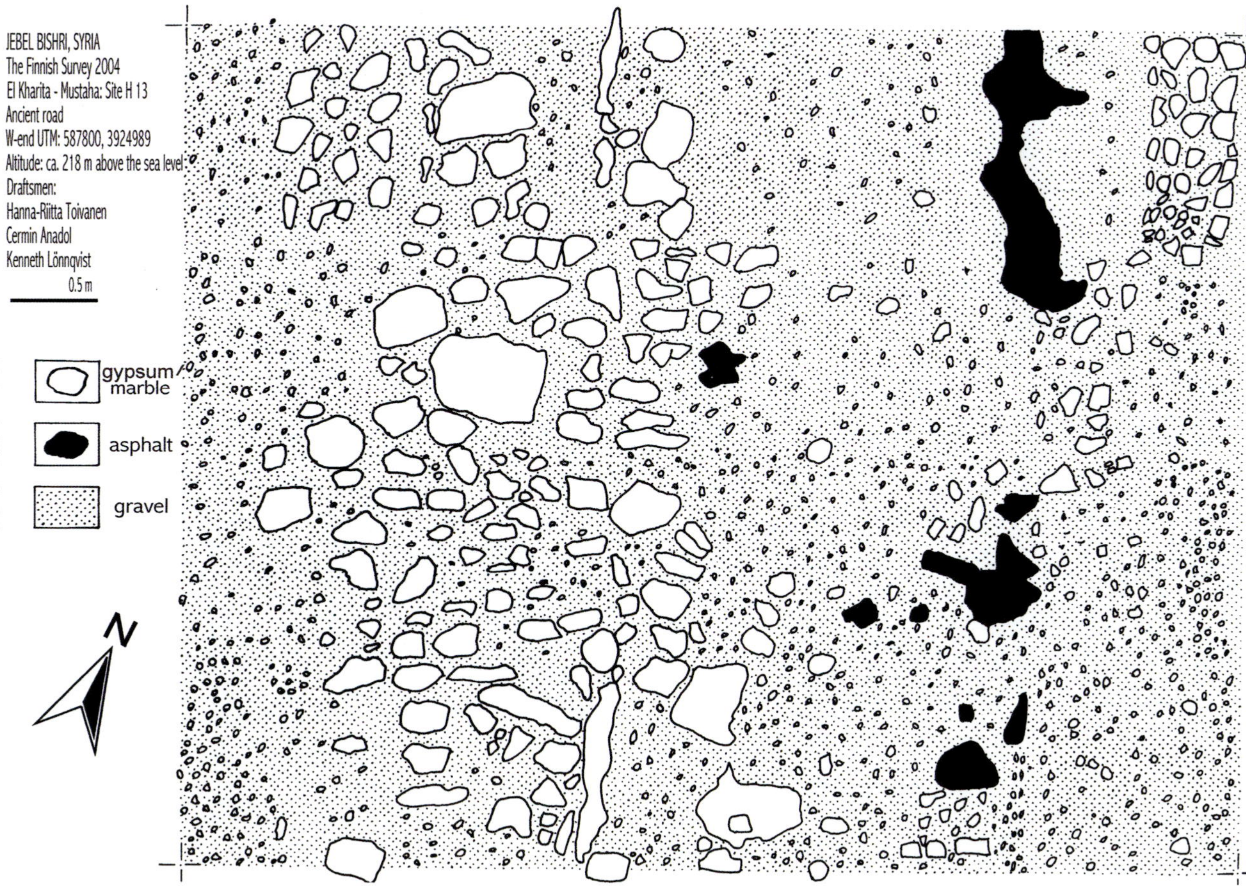

Fig. 59. Plan of the pavement structure of the road H 13 near the ruined bridge H 14.

Fig. 60. The road profile of H 13.
Photo from N: Eivind Seland 2004.

Fig. 61. Roman pottery finds associated with the road H 13.
Photo: Kenneth Lönnqvist 2006.

Fig. 62. The ruins of a rectangular stone structure H 24a. Photo from NW: Kenneth Lönnqvist 2004

Site H 24
Towers/House Bottoms (3)

UTM: 0588513, 3924837; 230 m a.s.l.
Sizes of the plans: a) 6 m x 8 m; b) 4.15 m x 6.40 m; c) 8.6 m x 5.7 m

Three ruined towers or house bottoms (Fig. 62.) were recognized on the southern side of the ancient road H 13 on the plateau area of H 1. The structures can be remains of towers, simple houses of *domicilium* type, or remains of more recent Bedouin houses. The structures consisted of only one room each. The thickness of the walls varied from 0.4 m to 0.85 m - relatively thick for being mere houses, and the vicinity of the ancient road could suggest that these were towers which were common along Late Roman/Byzantine roads. Undressed lime stone boulders had been used in construction and some had been provided an earthen wall inside. No small finds were associated with the structures, but the area of the plateau H 1 and the road produced plenty of Roman pottery.

Site H 14
Ruined Bridge

UTM: 0587835, 3925027, ca. 216 m a.s.l.
Size of the overall plan: 11 m x 5 m

The road **H 13** crosses the plateau **H 1** and bends northwestward down into a N-S oriented *wadi*, where it reaches a ruined bridge **H 14** (Figs. 63.) crossing a *wadi*. The overall plan of the bridge is ca. 11 m in length and 5 m in width. This basalt bridge has been preserved in its northern and southern parts and is seemingly later than the road alignment, apparently dating from the Ottoman era with its faceted basalt masonry of rectangular stones (0.35 m x 0.23 m in size on surface) fitted together with concrete. The foundation and the inner core are, however, made of stone and concrete rubble as in Roman bridges. No arch has preserved. In any event, the site could have served as a bridge point at the same place in antiquity.

Fig. 63. The ruined bridge H 14 of the road H 13. Photo from NE: Gullög Nordquist 2004.

Site H 15
Ancient Road

The collected UTM way points: 1) 0587778, 3925051, 2) 0587677, 3) 0587677, 3925043, 4) 0587371, 3925035, 5) 0587354, 3925029, 6) 0587330, 3925036 and 7) 0587276, 3925041, ca. 236 m a.s.l.
Width: max. 3.5 m

As mentioned previously (paragraph 10.3.), in the 2004 season we also found an ancient road **H 15** hewn into the marble/gypsum cliff directly beneath the eastern tower/keep of Tabūs, but surprisingly the road did not lead to the fort (see Fig. 55.). In 2005 we followed up its curving path over 2 km to the south into the mountain, after which it disappears perhaps due to the changes in the geomorphology or it is possible that the road was constructed for the use of a possible graveyard **J 22** (discussed in paragraph 10.3.) which we identified. The rock-hewn part of the road on the slope has wheel marks of carriages and horizontal stripes carved into rock to prevent animals such as donkeys or horses from slipping in rainy seasons. These are typical features of Roman roads, but it is possible that the road had been hewn and used much earlier. The Finnish project of Jabal Harūn in Jordan has also identified a rock-cut Nabataean/Roman road ca. 2 km in length at Petra. The road is ca. 4 m in width (see more studies of the road, e.g., Ben-David 2007).

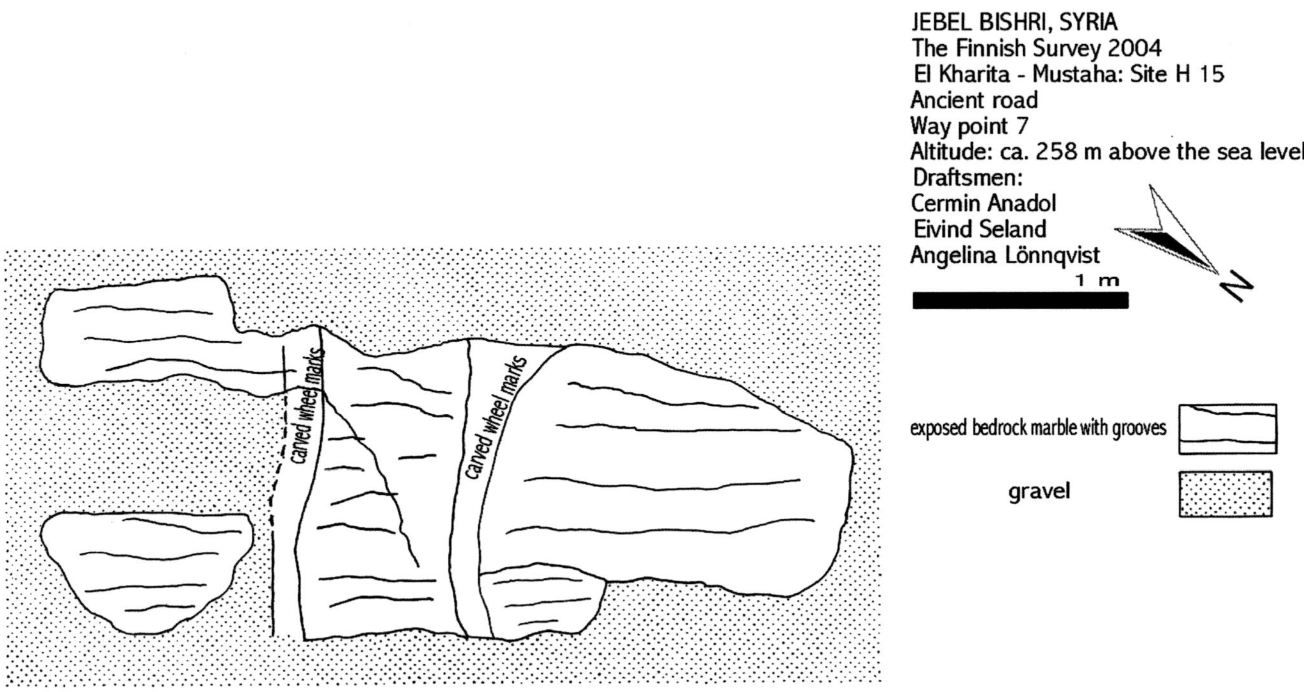

Fig. 64. Part of a rock hewn road H 15.

After the discovery of the road **H 13** on the ground we carried out remote sensing studies of its further alignments with a CORONA satellite photograph from the 1960s (see Fig. 65.; see other Finnish studies using remote sensing in tracing Roman road alignments in the Near East in Haggrén *et al.* 2002). A continuation of the road was recognized further in the direction of Deir ez-Zor near the villages of 'Ayyāsh and 'Ain Abu Jimā. These parts are not visible in later satellite images like SPOT not only because of their lower spatial resolution than the CORONA photographs, but probably because construction works in the area. We studied the traced and visible parts on the ground in 2005-2006. The survey of the year 2005 that started on the southern side of the village of 'Ayyāsh and south from the Deir ez-Zor – Aleppo -road (UTM: 0595830, 391814; ca. 240-250 m a.s.l.) ca. 12 km northwest of the centre of Deir ez-Zor continued to the village of 'Ain Abu Jimā. In 2005 we traced a part of this ancient road southwest of 'Ayyāsh.

Fig. 65. The road parallel to the Euphrates is apparently originally a Late Roman road which is part of the military and public road system connecting the forts along the Euphrates. The locations of the ruined bridges H 14 (the upper northwestern one) and J 7 (the lower southeastern one) are marked with white ovals. Mapping on a CORONA photograph by Minna Lönnqvist 2005.

Site J 6
Ancient Road Alignment

UTM: 0593719, 3920320
Width: 5 m, max. thickness 0.75 m.

The ancient road alignment **J 6** can be found outside the village of 'Ayyāsh, the site of Qreiye, on the southern side of the Deir ez-Zor–Aleppo road (Figs. 65., 66.). The width of the road is again 5 m like that of **H 13** on plateau **H 1**, but the thickness reaches as much as 0.75 m and in some parts consist of as many as 10 asphalt and Macadam layers with crushed gypsum and marble (Fig. 67.), clearly differing from the structure of the road H 13 passing through the plateau H 1 in the village of Kharita - Mustaha. The Macadam structuring indicates that the road has been reconstructed in recent centuries, possibly in the Ottoman era, but it seems to follow the alignment of an ancient road that has preserved parts elsewhere. Apart from some Middle Bronze Age pottery, Roman pottery from the third century A.D. including Late Roman C ware and sherds from the Byzantine era followed the alignment of the road J 6 and site J 10 but did not appear outside their area (see Fig. 68.).

Fig. 66. The alignment of road J 6. Photo from E: Michael Herles 2005.

Fig. 67. Foundations of the road J 6. Photo from W: Minna Lönnqvist 2005.

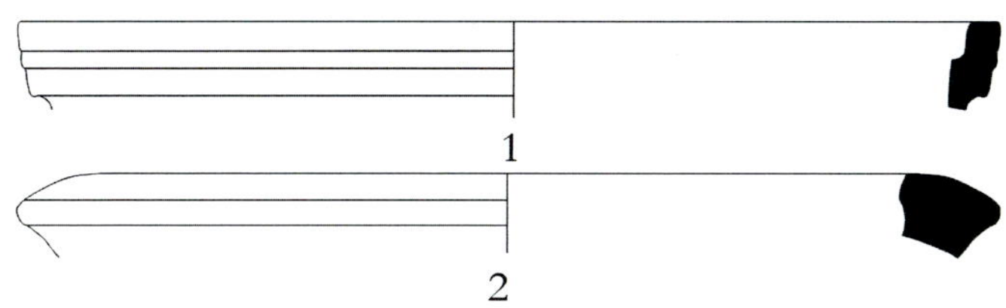

Fig. 68. Middle Bronze Age II pottery and Roman C ware from site J 6. Scale 1:4.

Site J 7 Ruined Bridge

UTM: 0593354, 3920769; ca. 226 m a.s.l.
Size of the overall plan: ca. 17 m x 13 m.

While surveying the alignment of the road **J 6** we recorded and documented a ruined bridge **J 7** (Fig. 69.) that leads over a N-S-oriented *wadi* dividing the village of 'Ayyāsh from the village of 'Ain Abu Jimā. This *wadi* poured water in rainy seasons from Jebel Bishri. The bridge is situated ca. 100 m to the south of the new bridge carrying the Deir ez-Zor–Aleppo -road over the *wadi*. In the CORONA satellite photograph (Fig. 65.), where the old road line is traced, the bridge is situated at a *wadi* where the old road crosses the modern road. The bridge has a semicircular arch. The space beneath the arch and between the bridge-ends is 7.95 m in width.

The arch rib can be formed from individual stones in different ways. In this case it is a stretcher bond (Fig. 70.), which is common in modern brick works, that is visible in the vault. This is the most common vault type in Roman arches. A study of a number of Roman bridges and aqueducts showed that 62% were constructed using stretcher bond. (Cf. O'Connor 1993: 172). Our bridge stands on a foundation layer ca. 3 m in height consisting of mixed rubble of concrete and unworked calcite and basalt stones (Fig. 69.). This is a typical structural feature of Roman foundations (see, e.g., Spence 1990, Fig. 27.). A similar kind of mixture, the so-called rubble core, also forms the inner structure of Roman bridges and triumphal arches. Here the stones in the concrete mixture of the rubble core consist of gypsum and calcite/marble. The arch is partly preserved; with four limestone layers at the top. (See Figs. 69. and 70.) The inside wall to the east is 11.9 m in width and has 16 layers of preserved limestone courses of ashlars as dressing masonry. The inside wall towards west is 13 m in width; 20 layers of ashlars have been preserved on this side. The inner side of limestone ashlar masonry consists of stones 0.65 m-0.45 m l. x 0.32 m h. in size (Figs. 71. and 72). The surface of some stones shows preserved frames.

The outer dressing and its stone course inside the bridge and the arch resemble the Roman bridge leading over the Sabu Suyu at Cyrrhus in Ancient Syria (Butcher 2003: 128, Fig. 40.). Several comparable Roman bridges can be found elsewhere in present Turkey. A resembling example is also the bridge of Segura between Spain and Portugal, as well as Puenta de la Cigarrosa at Petin and the bridge at Bibey (O'Connor 1993: Figs. 88., 92., 93.).

Fig. 69. Bridge J 7. Photo from S: Minna Lönnqvist 2005.

Fig. 70. The stretcher bond in the vault of the bridge J 7.
Photo from beneath: Minna Lönnqvist 2005.

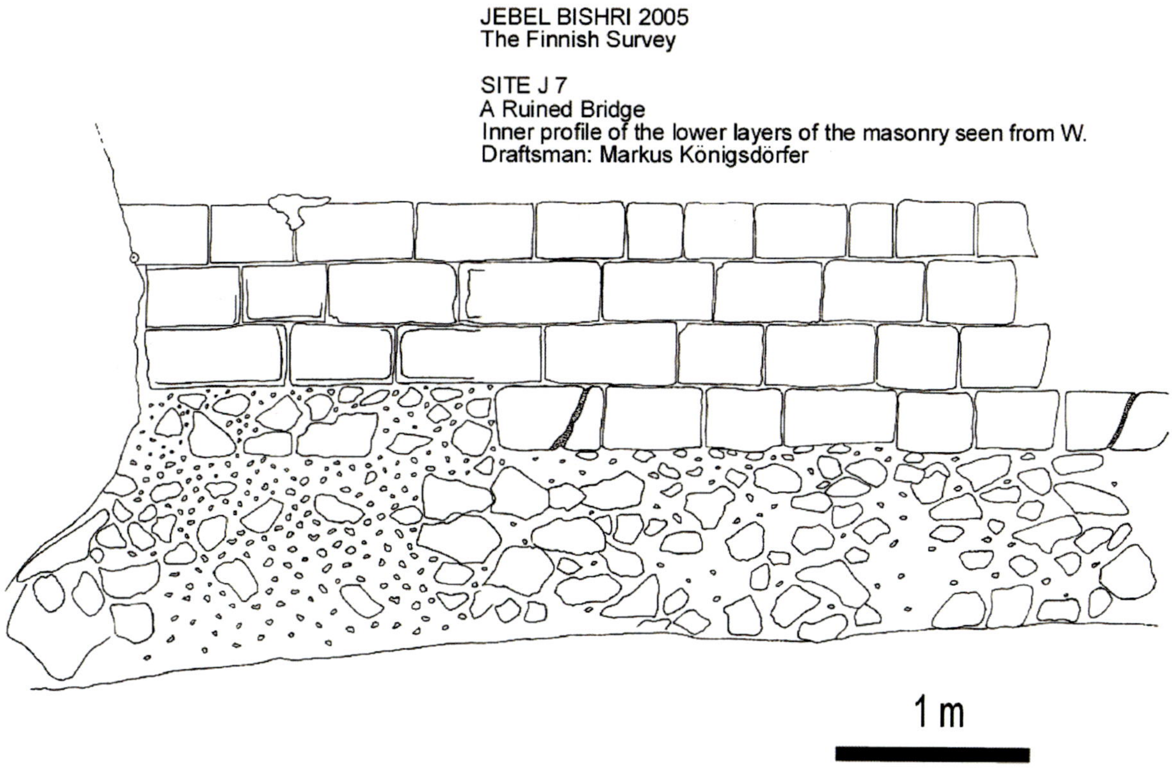

Fig. 71. Lower masonry of bridge J 7.

Fig. 72. Masonry of bridge J 7. Photo: from E: Minna Lönnqvist 2005.

Site J 10
Ancient Road Alignment

UTM: 0593109, 3921225; 237 m a.s.l.

This site **J 10** is in the continuing alignment of the road **J 6** visible in the CORONA satellite photograph. The site has only produced open accumulated finds such as pottery (Figs. 73. and 74.), but it seems to belong to the same alignment as the ancient road.

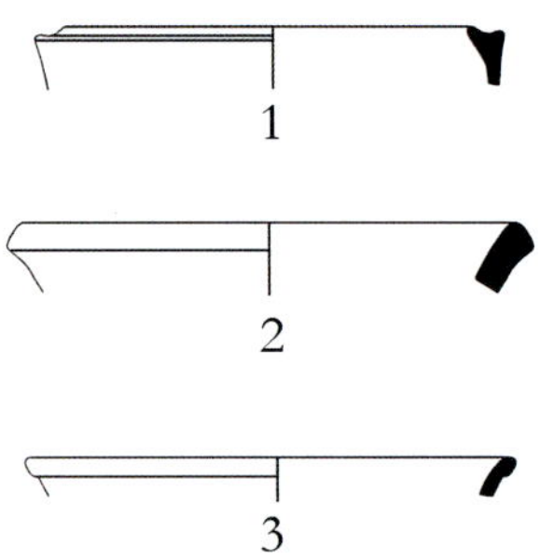

Fig. 73. Roman pottery from the terrace J 10. Scale 1:4.

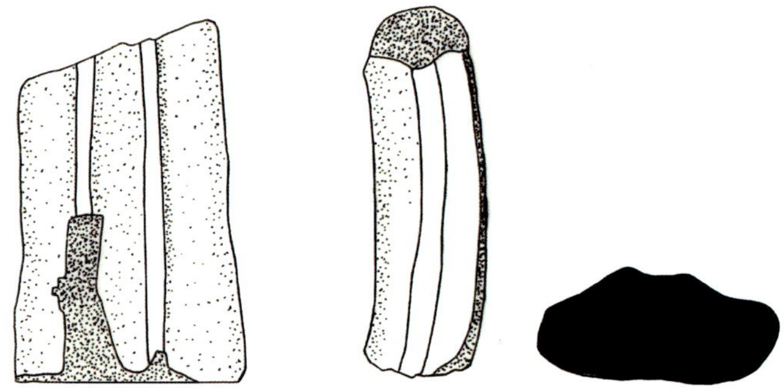

Fig. 74. A Roman pottery handle from J 10. Scale 1:2.

Site L 1
Ancient Road Alignment

UTM: 0592625, 3921364
Width: 5.4 m; thickness: 0.5 -0.6 m

In 'Ain Abu Jimā in 2006 we identified another stretch **L 1** (Figs. 55. and 75.) of the road between **H 13** and **J 6**, which is also visible in the CORONA photograph. This stretch is N-S oriented unlike the other part of the road, and its width reached 5.40 m. The road is laid on gypseous bedrock and some paving consisted of gypseous boulders varying between 20 cm x 20 cm x 15 cm or 30 cm x 30 cm in size. There were horizontal carved stripes in every 20 cm like in Roman roads (Fig. 75.). Here and there was remains of old asphalt on the top. The condition of this short stretch of road was very weathered.

Fig. 75. Part of the ancient road L 1. Photo from NWW: Minna Lönnqvist 2006.

Several features, including the common line of shadow on the satellite photographs, the location of the road **H 13** connecting **J 6** and **L 1**, its structure and associated finds consisting of Late Roman/Byzantine pottery, show that the road alignment and its preserved paved parts apparently date from the Late Roman/Byzantine era. This road differs from the Roman military unpaved desert track roads shown in Poidebard's Atlas (1934) and in our study at the Roman military camp **K 1**, and possibly at an abandoned and ruined Bedouin village **I 1** (see ch. 15.). The road following the Euphrates on its right bank must have been used for several public purposes varying from military and administrative to trading and caravan travelling. As the traffic was heavy, it would be understandable that it was built on unflooded terrace areas, partly paved and connected to the fortifications.

13.5. The Date and Structure of Roman Military Organisation along the Euphrates side of Jebel Bishri

As mentioned previously, the city of Dura Europos came under Roman control during the reign of Marcus Aurelius (A.D. 161-180), ca. A.D. 165. The site is well identified in Ptolemy's *Geographia* as well. The Roman garrisons at Dura may have been reinforced around A.D. 200, but as previously mentioned the site was already destroyed ca. A.D. 256 (Rostovtzeff *et al.* 1936: 467-470; James 2007: 29-47; Børlit 2008: 99-110). The Roman troops at Dura Europos belonged to the *Legio IV Scythica*, the *XVI Flavia Firma* or to the *Cohors Vigesima Palmyrenorum*. The *Cohors II Ulpia* occupied the town alone until the time of Septimius and Caracalla, and a discovered inscription have indicated that it still was in Dura in the 250s A.D. (Rostovtzeff *et al.* 1939: 86).

On the Euphrates, then, within a year or so in the mid-third century A.D., the whole Roman frontier from Qreiye to Dura Europos had collapsed and had been overrun by the Sassanians. The desperate situation is shown by the archaeological fact that none of these major military installations on the Euphrates were rebuilt, and the Emperor Valerian had from then on to rely on the benevolence of Odenathus of Palmyra (Bowman *et al.* 2005: 40). A temporary deathblow to Roman rule came in A.D. 260 when the Roman leaders with the Emperor Valerian were captured by Shapur, the Emperor being kept in captivity until his death. Although there is some dispute over the extent of the Sassanian conquest in Syria, space does not allow us a further discussion in this context. In practical terms the last bastions before Antioch, Carrhae and Edessa fell, and eventually the whole province of Syria (Cary and Scullard 1984: 510). It is into this power vacuum that Palmyra rocketed, first under Odenathus, and later under his widow Queen Zenobia a revolt against Rome broke out. For a few years the kingdom of Palmyra covered the entire Roman East. The Euphrates frontier was fortified anew at Halabiya/Zenobia. (Stoneman 1992).

Reasons for this above-mentioned arrangement have previously been sought from the fact that the Euphrates river is one of the longest rivers in the entire Near East with its ca. 2,800 km. Therefore, it constituted a natural border between important regional powers in antiquity. The importance of the Euphrates has been emphasized in several recent contributions to the study of the Roman - Parthian relations (Edwell 2008; Dignas and Winter 2007; Sommer 2005; Dąbrowa 1986: 93-108; 1994: 19-27; 2002: 275-279; 2007; 237-242). For example, Dąbrowa has emphasized that not only the banks of the Euphrates, but the river itself, was in fact the theatre of many military operations by the Roman army as early as the Flavian period. Furthermore, E. Dąbrowa mentions the important fact that the Roman army had the logistical capacity for large scale crossings of the river, or the capability to turn it into a direct theatre for war, which meant important strategic security for Rome.

The Roman sites, such as Tibne (Mambri?) and Tabūs that our Finnish project studied on the ground, provide indications that they had already been occupied during the time of the Palmyrene control of the area in the second - third centuries A.D. and possibly had been involved in the Roman activities during the Parthian wars. The refortifications at both sites are apparently due to the Emperor Diocletian's and Justinian's fortification policies on the Euphrates *limes*, and the nowadays largely visible structures of the fort of Tabūs are most likely from the period of Justinian I.

Our studies have further brought new light to the apparent road networks which A. Poidebard (1934) had indicated from the air as existing between major Roman forts. An ancient winding road cut into the bed rock beneath Tabūs, but not leading to the fort, could have been built earlier than the fort, although it could have been used and reworked in the Roman period. The wheel marks indicate that donkeys had been used for pulling carriages up to Jebel Bishri. Topographical and structural evidence as well as associated pottery find date the road alignment along the right bank of the Euphrates beneath the eastern tower of Tabūs from the Late Roman and Byzantine period. Its possible eastern end that was visible in a CORONA satellite photograph does not provide any traceable structural features that could be dated to the Roman or Byzantine period. However, the width and the shadow of the alignment is the same. In addition the alignment area offered Roman pottery, and an apparent Roman bridge arches over a *wadi* associated with the road. It is clear that a highway should have followed the bank, served the military troops to connect the forts and eased communication and trade of the caravans like in the earlier and later periods. For example, in Jordan practically all the Late Roman roads are associated with the military function (Kennedy 2007: 92).

The following map summarizes all the sites which provided Roman/Byzantine remains (Fig. 76.) during the surveys of the Finnish project SYGIS in the area of Jebel Bishri.

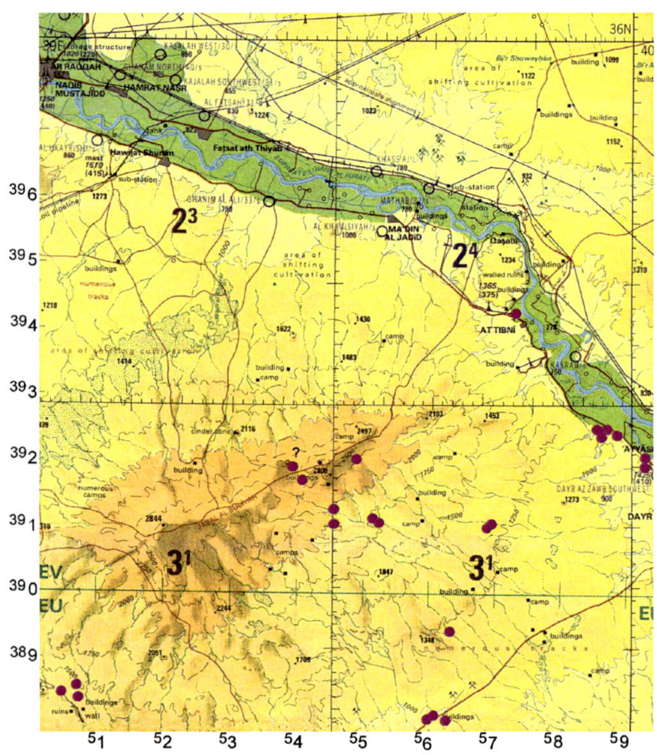

Fig. 76. The distribution of the Roman/Byzantine sites studied during the ground surveys of the Finnish project SYGIS. Basemap: © MOD Crown Copyright, 2010 published with the permission of the Controller of HMSO.

Bibliography

Al-Maqdissi, M. and Ohnuma. K. (2008) Preliminary Reports of the Syria-Japan Archaeological Joint Research in the Region of Ar-Raqqa, Syria, in *Al-Rāfidān*, Vol. XXIX, pp. 117-196.

Ammianus Marcellinus, transl. Rolfe, J.C. (1963) The Loeb Classical Library. Cambridge Massachusetts: Harvard University Press.

Bidwell, P. T. (1991) Later Roman Barracks in Britain, in *Roman Frontier Studies 1989, Proceedings of the XVth International Congress of Roman Frontier Studies*, ed. by Maxfield, V.A. and Dobson, M.J. Exeter, UK: University of Exeter Press, pp. 9-15.

Ben-David, C. (2007) The Paved Road from Petra to the 'Arabah – Commercial Nabataean or Military Roman? in *The Late Roman Army in the Near East from Diocletian to the Arab Conquest*, Proceedings of a colloquium held at Potenza, Acerenza and Matera (Italy (May 2005), ed. by Lewin, A.S. and Pellegrini, P. with the aid of Fiema, Z.T. and Janniard, S., BAR International Series, No. 1717, Oxford: Archaeopress, pp. 101-110.

Blétry, S. and Serdon, V. (2004-2005) Mission Franco-Syrienne à Halabiyé-Zalabiyé, in *Les Annales Archéologiques Arabes Syriennes*, Vols. XLVII-XLVIII, pp. 197-210.

Bowman, A. K., Garnsey, P. and Cameron, A. (2005) *The Cambridge Ancient History*. Vol. 12. The Crisis of Empire, A. D. 193-337. Cambridge: Cambridge University Press.

Butcher, K. (2003) *Roman Syria and the Near East*. London: The British Museum Press.

Børlit, C.S. (2008) Dura Europos, The Final Siege and Abandonment, in *Jebel Bishri in Context, Introduction to the Archaeological Studies and the Neighbourhood of Jebel Bishri in Central Syria*, Proceedings of a Nordic Research Training Seminar in Syria, May 2004, ed. by Lönnqvist, M., BAR International Series, No. 1817, Oxford: Archaeopress, pp. 99-110.

Cary M. and Scullard, H. H. (1984) *A History of Rome down to the reign of Constantine*. Third Edition, Hong Kong.

Chapot, V. (1907) *La Frontière de l'Euphrate de Pompée a la conquête arabe*. Paris: A. Fontemoing.

Chevallier, R. (1989) *Roman Roads*. Batsford Studies in Archaeology. London: Chrysalis Books.

Copeland, L. (2004) The Paleolithic of the Euphrates Valley in Syria, in *From the River to the Sea, The Paleolithic and the Neolithic on the Euphrates and the Northern Levant*, Studies in honour of L. Copeland, ed. by Aurenche, O., La Mière, M. and Sanlaville, P., BAR International Series, No. 1263, Oxford: Archaeopress, pp. 19-114.

Dąbrowa, E. (1986) The Frontier in Syria in the First Century AD, in *The Defence of the Roman and Byzantine East, Proceedings of a colloquium held at the University of Sheffield in April 1986*, ed. by Freeman, P. and Kennedy, D., British Institute of Archaeology at Ankara Monograph No. 8, BAR International Series, No. 297, Oxford: BAR International Series, pp. 93-108.

Dąbrowa, E. (1994) The *Bellum Commagenium* and the *Ornamenta Triumphalia* of M. Ulpius Traianus, in *The Roman and Byzantine Army in the East, Proceedings of a colloquium held at the Jagiellonian university, Kraków in September 1992*, ed. by Dąbrowa, E., Kraków: Uniwersytet Jagielloński Instytut Historii, pp. 19-27.

Dąbrowa, E. (2002) "*...ostentasse Romana arma satis...*" The military factor in Roman-Parthian relations under Augustus and Tiberius, in *Limes XVIII, Proceedings of the XVIIIth International Congress of Roman Frontier Studies held in Amman, Jordan (September 2000)*, A conference held under the auspices of the Department of Antiquities of the Hashemite Kingdom of Jordan, The Council for the British Research in the Levant and the Department of Archaeology at the University of Liverpool, Vol. I, ed. by Freeman, P., Bennet, J., Fiema, Z.T. and Hoffmann, B., BAR International Series, No. 1084 (i), Oxford: Archaeopress, pp. 275-279.

Dąbrowa, E. (2007) Naval Operations During Persian Expedition of Emperor Julian (363 AD), in *The Late Roman Army in the Near East from Diocletian to the Arab Conquest, Proceedings of a Colloquium held at Potenza, Acerenza and Matera, Italy (May 2005)*, ed. by Lewin, A.S. and Pellegrini, P. with the aid of Fiema, Z.T. and Janniard, S., BAR International Series, No. 1717, Oxford: Archaeopress, pp. 237-242.

Dignas, B. and Winter, E. (2007) *Rome and Persia in Late Antiquity: Neighbours and Rivals*. New York: Cambridge University Press.

Dodgeon, M.H. and Lieu, S.N.C. (compl. and ed.) (2002) *The Roman Frontier and the Persian Wars AD 226-363*. A Documentary History. London – New York: Routledge.

Dussaud, R. (1927) *Topographie Historique de la Syrie Antique et Médiévale*. Paris: Geuthner.

Edwell, P.M. (2008) *Between Rome and Persia: The Middle Euphrates, Mesopotamia and Palmyra under Roman Control*. Routledge Monographs in Classical Studies. London and New York: Routledge.

Foss, C. and Winfield, D. (1986) *Byzantine Fortifications: An Introduction*. The University of South Africa. Pretoria: Sigma Press.

Fukai, S. (1980/1981) *Ceramics of Ancient Persia*. New York – Tokyo: John Weatherhill Inc.

Fulford, M.G. and Peacock, D.P.S. (1984) *Excavations at Carthage. The British Mission*. Vol. I, 2. The Avenue du President Habib Bourguiba, Salammbo: the Pottery and Other Ceramic Objects from the Site. Sheffield: Published for the British Academy from the University of Sheffield Department of Prehistory and Archaeology.

Gates, J (2006) Hidden Passage: Graeco-Roman Roads in Egypt's Eastern Desert, in *Space and Spatial Analysis in Archaeology*, ed. by Robertson, E.C., Seibert, J.D., Fernandez, D.C., Zender, M.U., Calgary: The University of Calgary Press pp. 315-322.

Gawlikowski, M. (1996) Palmyra and its Caravan Trade, in *Les Annales Archeologiques Arabes Syriennes*, Vol. XLII, Special Issue Documenting the Activities of the International Colloquium: Palmyra and the Silk Road, pp. 139-144.

Golcberg, P. (2004-2005) Expertise Concernant la Restauration du Site de Halabiyé (Syrie), et Confortement de l'Eglise Nord Ouest, in *Les Annales Archéologiques Arabes Syriennes*, Vols. XLVII-XLVIII, pp. 211-215.

Greene, K. (1986) *The Archaeology of the Roman Economy*. Tiptree, Essex: B.T. Batsford Ltd.

Gregory, S. (1995-1997) *Roman Military Architecture on the Eastern Frontier. From AD 200–600*. Vols. 1-3. Amsterdam: Adolf M. Hakkert.

Gschwind, M. (2005a) Qreiye/'Ayyāš, in *Archäologischer Anzeiger*, 2. Halbband, Deutsches Archäologisches Institut, pp. 255-257.

Gschwind, M. (2005b) Qreiye/Ayyash – Ein römisches Kastell am Euphrat, in *Zwischen Kulturen und Kontinenten, 175 Jahre Forschumg am Deutschen Archäologischen Institut*. Berlin.

Gschwind, M. (2008) *Limes studies Qreiye, 2003-2007 excavations at Qreyie*, at Deutsches Archäologisches Institut, http://www.dainst.org/index_742_en.html.

Gschwind, M. and Hasan, H. (2008) Das römische Kastell Qreiye-'Ayyāš, Provinz Deir ez-Zor, Syrien,

Ergebnisse des syrisch-deutschen Kooperationsprojektes, in *Zeitschrift für Orient-Archäologie*, Vol. 1, pp. 316-334.

Haggrén, H., Nuikka, M., Junnilainen, H. and Järvinen, J. (2002) Photogrammetric Approach for Archaeological Documentation of an Ancient Road, in *CIPA Heritage Documentation*, Proceedings of the XVIIIth International Symposium of CIPA, Potsdam (Germany) Sept. 18-21, ed. Albertz, J., The ISPRS International Archives of Photogrammetry, Remote Sensing and Spatial Information Sciences, Vol. XXXIV-5/C7, Berlin: CIPA, pp. 108-113.

Harper, R.P. (1975) Excavations at Dibsi Faraj, Northern Syria, 1972-1974: A Preliminary Note on the Site and Its Monuments, in *Dumbarton Oaks Papers*, Vol. 29, pp. 319-337.

Hayes, J.W. (1972) *Late Roman Pottery*. Rome: The British School of Rome.

Héraud, C. (1922/1998) *Une Mission de Reconnaissance de l' Euphrate en 1922*. Reprinted by Institut Français de Damas 1998.

Ibach, R.D. (1987) *HESBAN 5. Archaeological Survey of the Hesban Region*. Catalogue of Sites and Characterization of Periods. Berrien Springs, MI: Andrews University Press.

Invernizzi, A. (1986) Kifrin and the Euphrates Limes, in *The Defence of the Roman and Byzantine East, Proceedings of a Colloquium Held at the University of Sheffield in April 1986*, ed. by Freeman, P. and Kennedy, D., Part i, British Institute of Archaeology at Ankara Monograph No. 8, BAR International Series, No. 297 (i), Oxford, pp. 375-381.

Isaac, B. (1990) *The Limits of Empire. The Roman Army in the East.* Oxford: Clarendon Press.

Isaac, B. and Roll, I. (1982) *Roman Roads in Judaea 1: The Legio-Scythopolis Road.* British Archaeological Reports. International Series 141. Oxford.

James, S. (2005) The Deposition of Military Equipment during the Final Siege at Dura-Europos, with Particular Regard to the Tower 19 Countermine, in *Carnuntum Jahrbuch 2005*, pp. 189-206.

Kennedy, D. (2007) *Gerasa and the Decapolis: A 'Virtual Island' in Northwest Jordan.* Duckworth Debates in Archaeology. Duckworth: London.

Kennedy, D. and Riley, D. (1992) *Rome's Desert Frontier from the Air.* London: Batsford.

Kohlmeyer, K. (1984) Euphrat-Survey, Die mit Mitteln der Gerda Henkel Stiftung durchgeführte archäologische Geländebegehung im syrischen Euphrattal, *Mitteilungen der Deutschen Orient-Gesellschaft,* Vol. 116, pp. 95-118.

Kohlmeyer, K. (1986) Euphrat-Survey 1984 Zweiter Vorbericht über die Mitteln der Gerda Henkel Stiftung durchgeführte archäologische Geländebegehung im syrischen Euphrattal, *Mitteilungen der Deutschen Orient-Gesellschaft,* Vol. 118, pp. 51-65.

Konrad, M. (2001) (with the contributions by Baldus, H.R. and Ulbert, T.) Der Spätrömische Limes in Syrien: Archäologische Untersuchungen an den Grenzkastellen von Sura, Tetrapyrgium, Cholle und in Resafa. *Resafa.* Vol. 5. Mainz: Philipp von Zabern.

Lauffray, J. (1983) *Halabiyya – Zenobia: Place Forte du Limes Oriental et La Haute-Mésopotamie au VIe Siècle.* Tome I: Les duchés frontaliers de Mésopotamie et les fortifications de Zenobia. Bibliothèque Archéologique et Historique T. CXIX. Institut français d'archéologie du Proche-Orient, Beyrouth – Damas – Amman. Paris: Paul Geuthner.

Lauffray, J. (1991) *Halabiyya – Zenobia: Place Forte du Limes Oriental et La Haute-Mésopotamie au VIe Siècle.* Tome II: L'architecture publique, religieuse, privée et funéraire. Bibliothèque Archéologique et Historique T. CXXXVIII. Institut français d'archéologie du Proche-Orient, Beyrouth – Damas – Amman. Paris: Paul Geuthner.

Lönnqvist, M. and Törmä, M. (2004) Different Implications of a Spatial Boundary: Jebel Bishri between the Desert and the Sown in Syria, The ISPRS XXth Congress Proceedings, Edited by M. Orhan Altan, *The ISPRS International Archives of the Photogrammetry, Remote Sensing and Spatial Information Sciences,* Vol. XXXV. Part B. Istanbul, pp. 897-902.

Lönnqvist, M. Lönnqvist, K., Whiting, M.S., Törmä, M., Nuñez, M. and Okkonen, J., (2005) Documenting, Identifying and Protecting a Late Roman – Byzantine Fort at Tabus on the Euphrates, in *International Cooperation to Save the World's Cultural Heritage,* Proceedings of the XX International Symposium CIPA 2005, Turin (Italy) 26 September - 01 October, 2005, Vol. 1, ed. Dequal, S., The International Archives of Photogrammetry, Remote Sensing and Spatial Information Sciences, Vol. XXXVI-5/C34, CIPA – The

ICOMOS/ISPRS Committee for Documentation of Cultural Heritage, Turin: CIPA, pp. 427-432.

Lönnqvist, M. with the contributions by Törmä, M., Nuñez, M., Lönnqvist, K. Stout Whiting, M., Okkonen, J., Riihiaho, H. and Nissinen, M. (2006) Archaeological Surveys of Jebel Bishri, The Preliminary Report of the Finnish Mission to Syria 2000-2004, in *KASKAL*, Vol. 3, pp. 205-242.

MacDonald, B. (1992) Hellenistic, Nabataean and Roman Period Sites, in *The Southern Ghors and Northeast 'Arabah Archaeological Survey* by MacDonald, B., Sheffield Archaeological Monographs 5, Sheffield: J.R. Collins Publications, pp. 83-95.

Mouterde, R. and Poidebard, A. (1945) *Le Limes de Chalcis, Organisation de la Steppe en Haute Syrie Romaine*, Documents Aériens et Epigraphiques, Plans et Relevés de Lauffray, J. et Mazloum, S. Texte et Atlas. Bibliothèque Archéologique et Historique. Tome XXXVIII. Geuthner: Paris.

Moutsopoulos, N. (1983) La section macédonienne de la Via Egnatia – Le Bourg de Redina, in *Bulletin de l'Institut International des Châteaux historiques*, Vol. 41, pp. 86-99.

Musil, A. (1927) *The Middle Euphrate:s: A Topographical Itinerary*. American Geographical Society. Oriental Explorations and Studies No 3. New York: American Geographical Society.

Musil, A. (1928) *Palmyrena: A Topographical Itinerary*. American Geographical Society. Oriental Explorations and Studies No 4. New York.

Napoli, J. (2000) Les Remparts de la Fortresse de Djazla sur le Moyen-Euphrate, in *Syria*, Vol. 77, pp. 117-136.

O'Connor, C. (1993) *Roman Bridges*. Cambridge: Cambridge University Press.

Ohnuma, K. and Al-Khabour, A. (2007) Archaeological Research in the Bishri region – report of the second working season. Unpublished manuscript.

Parker, S.T. (2000) Roman Legionary Fortresses in the East, in *Roman Fortresses and their Legions*, ed. by Brewer, R.J., The Society of Antiquaries of London and National Museums & Galleries of Wales, Trowbridge, pp. 121-138.

Parker, S.T. (2006) *The Roman Frontier in Central Jordan: Final Report on the Limes Arabicus Project*, 1980-1989, 2. vols. Dumbarton Oaks Byzantine Studies. Washington D.C: Harvard University Press.

Poidebard, A. (1934) *La trace de Rome dans le désert de Syrie: Le limes de Trajan a la conquête arabe, Recherces aériennes (1925-1932)*. Bibliothèque Archéologique et Historique. Tome XVIII. Texte. Atlas. Paris: Geuthner.

Rostovtzeff, M.I., Bellinger, A.R., Hopkins, C. and Welles, C.B. (ed.) (1936) *Excavations at Dura-Europos*, conducted by Yale University and the French Academy of Inscriptions and Letters. Preliminary Report of Sixth Season of Work, October 1932 - March 1933. New Haven: Yale University Press.

Rostovtzeff, M.I., Brown, F.E. and Welles, C.B. (ed.) (1939) *Excavations at Dura-Europos*, conducted by Yale University and French Academy of Inscriptions and Letters. Preliminary Report of the Seventh and Eight Seasons of Work 1933 - 1934 and 1934 and 1935. New Haven: Yale University Press.

Sarre, F. and Herzfeld, E. (1911) *Archäologische Reise im Euphrat- und Tigris-Gebiet.* Band I. Forschungen zur islamischen Kunst I. Berlin: Dietrich Reimer (Ernst Vohsen).

Sarre, F. and Herzfeld, E. (1920) *Archäologische Reise im Euphrat- und Tigris-Gebiet.* Band II. Forschungen zur islamischen Kunst I. Berlin: Dietrich Reimer (Ernst Vohsen).

Sartre, M. (2005) *The Middle East under Rome*. Cambridge, Mass.: Harvard University Press.

SAS = *Syria: Archaeology from Space* (2002) Damascus: GORS.

Sommer, M. (2005) *Roms orientalische Steppengrenze: Palmyra – Edessa – Dura Europos – Hatra. Eine Kulturgeschichte von Pompeius bis Diocletian.* Oriens et Occidens. Studien zu antiken Kulturkontakten und ihren Nachleben. Herausgegeben von Wiesenhöfer, J. in Zusammenarbeit mit Briant, P., Kuhrt, A., Millar, F. und Rollinger, R. Band 9. Wiesbaden: Franz Steiner Verlag.

Spence, C. (ed.) 1990 *Archaeological Site Manual.* Department of Urban Archaeology. Museum of London. London: The Museum of London.

Stoneman, R. (1992) *Palmyra and its Empire: Zenobia's Revolt against Rome.* Ann Arbor: The University of Michigan Press.

Toivanen, H.-R. (2008) Fortresses and Ecclesiae on the Imperial Border of Byzantium: The Architectural Influence of Constantinople on Resafa-Sergiopolis and Zenobia in Syria, in *Jebel Bishri in Context: Introduction to the Archaeological Studies and the Neighbourhood of Jebel Bishri in Central Syria*, Proceedings of a Nordic Research Training Seminar in Syria, May 2004, ed. by Lönnqvist, M., BAR International Series 1817, Oxford: Archaeopress, pp. 111-126.

Welles, C.B. (ed). (1968) *The Excavations at Dura-Europos conducted by Yale University and the French Academy of Inscriptions and Letters*. Final Report IV. Part I. Fasc. 3. The New Haven: Yale University Press.

Wheatley, D. (1995) Cumulative viewshed analysis: a GIS-based method for investigating intervisibility, and its archaeological application, in *Archaeology and Geographic Information Systems*, ed. by Lock, G. and Stančič, Z., London- Bristol: Taylor & Francis, pp. 171-185.

Wheatley, D. & Gillings, M. (2000) Vision, Perception and GIS: developing enriched approaches to the study of archaeological visibility, in *Beyond Map*, Archaeology and Spatial Technology Series A: Life Sciences – Vol. 321, ed. by Lock, G., Amsterdam: IOS Press, pp. 1-27.

Vokaer, A. (2009) Brittle Ware trade in Syria between the 5th and 8th centuries, in *Byzantine Trade, 4th - 12th Centuries: the Archaeology of Local, Regional and International Exchange*, Papers of the Thirty-eight Spring Symposium of Byzantine Studies, St. John's College, University of Oxford, March 2004, ed. by Mungo, M.M., Society for the Promotion of Byzantine Studies, Publication 14, Farnham: Ashgate Publishing Ltd., pp. 121-136.

14. THE COMING OF THE ARABS AND CHANGING BEDOUINS

Minna Lönnqvist

Fig. 1. Qasr al-Hayr ash-Sharqi. Photo: SYGIS- Jean-Pierre Matar 2000.

14.1. Establishing the Arab Influence through Caravan Roads and Stations

The area of Jebel Bishri provides valuable material evidence for ethnographic and ethnoarchaeological approaches among pastoral nomads. Landsat images (Fig. 2.) reveal long straight lines of old caravan routes traversing the Arabian Desert towards Jebel Bishri and its neighbourhood. These kinds of straight tracks were caused by utilizing stars in navigation. Venus in particular has been meaningful for the caravans travelling over these deserts (Madoun 1996: 69-70).

As known, Arabs originate from the Arabian Peninsula, and early Arabs appear in Syria and the Arabian Desert already in the Assyrian sources. One of the first references to the Arabs is found in Shalmaneser III's monolith inscription dealing with the battle of Qarqar which took place in 853 B.C. (ANET: 278-279). The Arabs are often associated with pastoral nomadism and caravan trade in the early sources – but also as an employed military force. In the classical sources, such as Strabo and Pliny, the area between Damascus and the Euphrates was populated by Arab nomads during the Roman period.

Strabo (Geography XVI, Ch. I, 28) writes: *The Euphrates and its eastern banks are the boundaries of the Parthian empire. The Romans and the chiefs of the Arabian tribes occupy the parts on this side the Euphrates as far as Babylonia. Some of the chiefs attach themselves in preference to the Parthians, others to the Romans, to whom they adjoin. The Scenitae nomads, who live near the river, are less friendly to the Romans than those tribes who are situated at a distance near Arabia Felix.* Strabo further reports (Geography XVI, Ch. II, 20): *Above Damascus are the two (hills) called Trachones; then, towards the parts occupied by Arabians and Itureans promiscuously, are mountains of difficult access, in which were caves extending to a great depth. One of these caves was capable of containing four thousand robbers, when the territory of Damascus was subject to incursions from various quarters. The Barbarians used to rob the merchants most generally on the side of Arabia Felix, but this happens less frequently since the destruction of the bands of the robbers under Zenodorus, by the good government of the Romans, and in consequence of the security afforded by the soldiers stationed and maintained in Syria.* (Transl. W. Falconer 1903).

Fig. 2. A Landsat- ETM satellite image reveals the straight lines of caravan roads coming from the Arabian Peninsula towards Deir ez-Zor and Jebel Bishri.

Palmyra in the Syrian Desert was known as one of the most famous caravan cities of the Near East in antiquity. The Palmyreans included nomadic tribes of Aramaeans and early Arabs in the shift to the Common Era. (See, e.g., Rostovtzeff 1932). During the Roman period Palmyra, which conducted the caravan trade in the region of Jebel Bishri, also largely controlled the nomadic Arabs in its neighbourhood.

Hatra in Mesopotamia was probably destroyed by Sassanians in A.D. 240; it also was a caravan city which had included Arabs and co-operated with Rome. As previously mentioned (e.g. ch. 12.), the Roman caravan city of Dura Europos also fell ca. A.D. 256 to a Sassanian invasion. The power of Palmyra subsequently collapsed in the defeat of Queen Zenobia's revolt by the Emperor Aurelian in A.D. 273. (See, e.g., Dignas and

Winter 2007: 163-164). After the crises Rome started directly controlling the old caravan routes in the area while extending the military control.

After Palmyrene and Roman prominence, major Arab influence in the Jebel Bishri region started with the Christian Arabs known as Ghassānids, who were Monophysites serving as *foederati* in Byzantium, and for whom Resafa-Sergiopolis on the western piedmont of Jebel Bishri became an important site for their army and centre of pilgrimage in the sixth century A.D. The Ghassānids probably repaired the large cisterns at Resafa-Sergiopolis and, as previously mentioned, built a *praetorium* outside the city. (Shahîd 2002: 125-133). In the seventh century A.D. Muslim Arabs started spreading into the area. Halid ibn al-Walid fought a battle at Jebel Bishri in A.D. 633, and the Muslims conquered Raqqa (Nichephorium-Callinicum) in A.D. 639. The Umayyads built Qasr al-Hayr ash-Sharqi (Figs. 1., 3.-7.), the already mentioned desert castle, visible on the western piedmont of Jebel Bishri today. It was constructed by Khalif Hisham b. ʻAbd al-Malik ca. A.D. 728-729 according to an inscription found at the site but now lost. (Grabar *et al.* 1978: 12; Cornu 1985: 208; Genequand 2008: 96). We have already observed (ch. 11.) that the neighbourhood of Qasr al-Hayr ash-Sharqi, although not the visible ruins themselves, provides evidence of activities already in the Roman-Byzantine period to be connected with the *Strata Diocletiana.* As discussed in connection with the Roman desert frontier, apart from the Roman military sites, the neighbourhood has belonged to the area of ancient caravan roads that led through Sukhne, Taibe and Qasr al-Hayr to El Kowm, Qdeir and Resafa-Sergiopolis. The Bazaar oasis of Sukhne and Taibe were later involved in the caravan trade of the Bedouins. The transportation involved such products as potash, wood, pistachio oil, antiquities, salt and truffles. (Wirth 1971: 442-447).

Fig. 3. Poidebard's aerial photograph of Qasr al-Hayr ash-Sharqi showing the ruins of the two main enclosures and some of the outer enclosure to the north. Source: Poidebard 1934.

Fig. 4. The main entrance of the palace, Qasr al-Hayr ash-Sharqi, with an elaborate façade. Photo: Minna Lönnqvist 2000.

Ibn Battuta is known to have travelled from Baghdad to Damascus via Sukhne and Tadmor (Palmyra) in the year 1348 (Tufa, Vol. 4, 314-315). The era of flourishing caravan trade in the oases of the western piedmont dates from the sixteenth and seventeenth centuries. Apart from early travellers such as Pietro della Valle, Texeira and Rousseau (Grabar *et al.* 1978: 12), the site of the Qasr was studied by A. Musil (1928a: 77-81). It was later excavated by O. Grabar *et al.* (1978). Therefore, we did not carry out any particular documentation inside the Qasr but concentrated on some surrounding archaeological remains (see ch. 4. and 11.) and ethnoarchaeological recording and documentation (ch. 15.). We also recorded one dumb of early Islamic pottery, **A 4,** 3.5 m x 6 m in size (UTM: 0506178, 3881506) and remains of a rectangular structure, **A 5,** 4.9 m x 3.1 m in size (UTM: 050617793, 3905377) in the neighbourhood. After our work in the surroundings of the Qasr an Italian restoration team was active in the area. Later on a Syrian-Swiss excavation team led by D. Genequand (2004-2005; 2005, 2008, 2009) has been studying the Islamic remains at the site, unearthing an Ayyubid mosque, some cemeteries and houses, such as an aristocratic residence, surrounding the main buildings.

The large complex of the Qasr is imposing in the midst of the desert and has two main buildings: *a smaller enclosure often referred to as a palace or caravanserai* and *a larger enclosure known as a madīna* (UTM: 0506457, 3883921; UTM: 0506422, 3883975), the latter identified as a city by some scholars. The ruins themselves extend over an area of ca. 6.2 km on the north-south axis. There is also an *outer enclosure* built of stone-based mud-brick which covers 7 km^2 and the periphery of which extends ca. 15 km to the north. In the northern and southern extremes there is a sluice system which directs and collects water from *Wadi Suq* (see Fig. 6.). The size of the larger enclosure of the Qasr covers an area of 167 m x 167 m. (Grabar *et al.* 1978: 9; Genequand 2005: 351). There is a reservoir in the middle and an inscription dating from 1407, and a mosque was built in the southwest corner (Musil 1928a: 77). The size of the smaller enclosure identified as a palace or caravanserai is 70 m x 70 m in plan. (See Figs. 3., 7.). Beside the palace/caravanserai there is a bathhouse following the basilica plan to the north of the main enclosures. (Genequand 2005: 351-352).

There has been much discussion in the past of the actual function of each structure. According to O. Grabar *et al.* (1978: 13), the area encompassed agricultural, palatial and urban dimensions. The medieval geographer Yaqût al-Hamawi ar-Rumi (1179-1229) reports in Mu'djam that the ridge of Bishri belonged to the Tarleb and extended from the settlement of 'Ord to the Euphrates. 'Ord also appears as Urdu in other Arabic sources and has been identified with Qasr al-Hayr ash-Sharqi. O. Grabar *et al.* (1978: 11-13) suggested that even if the identification of 'Ord may have changed in the Medieval times from Taibe to the Qasr, the two formed parts of the entity of 'Urd-Oriza, in which the former is a Muslim one and the latter the classical one.

In Grabar's *et al.* (1978: 13) view the site of the Qasr could also be Hisham's al-Zaytunah with olive plantations. Grabar follows Sauget's original hypothesis supporting the agricultural potential of the site. The outer enclosure was understood to form an agricultural area, a kind of paradise, where animals were also hunted. With an elaborate irrigation system the site may have been used as an oasis with gardens (Fairchild Ruggles 2005: 277-278). Like in A. Poidebard's (1934) aerial photographs (Fig. 3.), the traces of fields or garden plots are visible in a CORONA satellite photograph dating from the 1960s (see the channelled plots around the castle on Fig. 6.). In the 1960s large seasonal *wadis* full of water were still visible in the region. Desertification has, however, taken over the area today.

Fig. 5. A CORONA satellite photograph from the 1960s showing the locations of Taibe and Qasr al-Hayr ash-Sharqi. Note the large number of fields, which have vanished today due to desertification.

Fig. 6. A CORONA satellite photograph from the 1960s showing the area of Qasr al-Hayr ash-Sharqi still surrounded by large seasonal wadis and fields. The canal systems, ancient field systems/ orchard areas and the so-called outer enclosure surrounding the castle are visible in the photo.

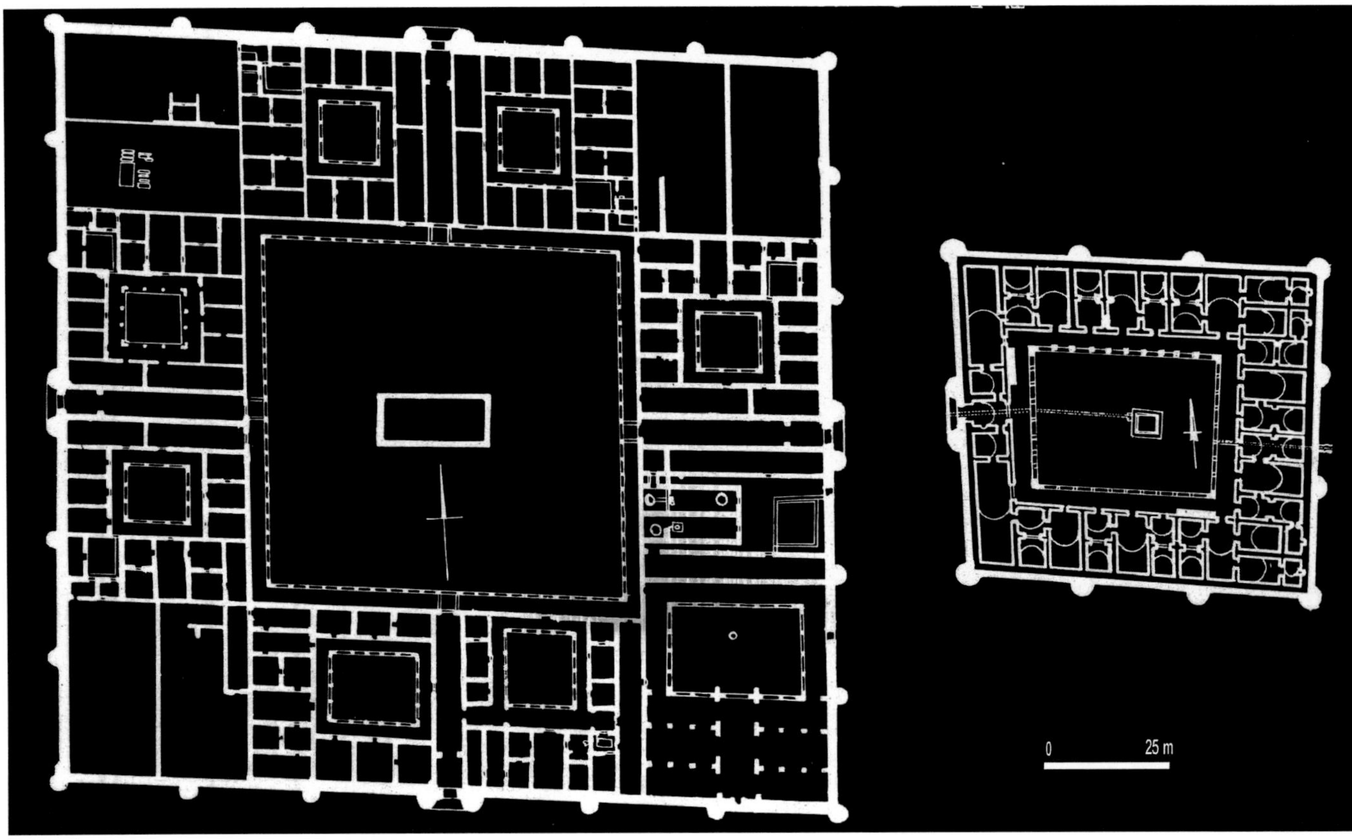

Fig. 7. The plan of the larger and smaller enclosure of Qasr al-Hayr ash-Sharqi after Grabar et al. 1978.

Fig.8. Side entrance to the main palace, Qasr al-Hayr ash-Sharqi. Photo: Minna Lönnqvist 2000.

Fig. 9. Walls of the larger enclosure and a minaret of the Ayyubid mosque to the right of the larger enclosure and the palace / caravanserai. Photo from SW: Minna Lönnqvist 2000.

Fig. 10. Inner structures at the ruins of Qasr al-Hayr. Photo: Margot Stout Whiting 2000.

According to sixteenth century descriptions, such as that of Tenreiro, the Bedouins of Taibe (the site of ancient Oriza) (Figs. 5. and 11.) sold skins and leather as well as alkaline for soap production. 500-600 inhabitants were living in the walled village at that time, and their sheikh had an armed force of Bedouins consisting of hundreds of men with camels. In the seventeenth century the site was still "enclosed" and consisted of as many as 250 houses built of sun-dried mud-brick with conical roofs. There was water and there were well-irrigated gardens of palm trees and fruit orchards. At that time the Abu Riche emirs, a family or a clan belonging to the Mawali, exercised power over the desert-steppe areas southeast of Aleppo. However, travellers such as Texeira, Pietro della Valle and Abbé Carré describe the gradual deterioration and the ceasing of the commercial activities along the caravan routes. (Grabar *et al.* 1978; Mundy and Musallam 2000; Lewis 1991: 67-78).

Taibe has seven springs, the waters of which had often not been sufficient for livelihood, and the site became

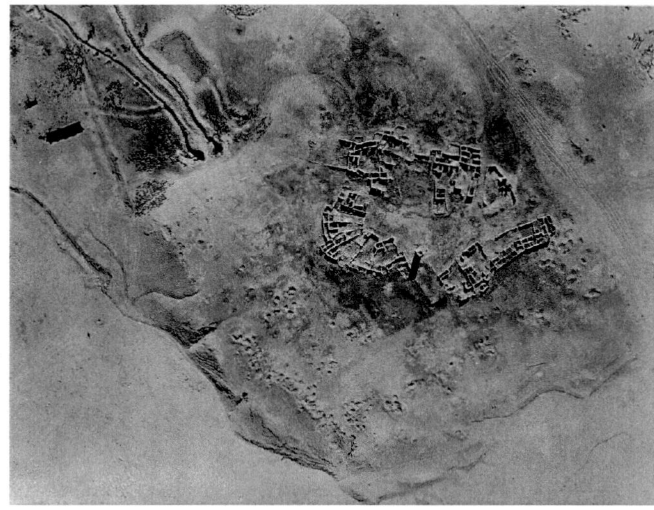

Fig. 11. An aerial photograph of the ruined and abandoned village of Taibe taken by Poidebard in the 1920s. Source: Poidebard 1934.

completely deserted in the eighteenth century. There are several explanations for this abandonment; some, like C. Niebuhr (1792), say that the site was destroyed, others tell that its inhabitants moved into the nearby village of Sukhne. It is said that in 1870 one of the members of the families that had moved received permission with a dozen other families to resettle the site again, whereas another group of families moved to the abandoned site of El Kowm. Gendarme stations were built in these resettled sites. (Mundy and Musallam 2000). However, in his economic description of the neighbourhood of Deir ez-Zor from the year 1881 Captain D.H. Stewart (*apud* Issawi 1988) tells that scattered villages such as Sukhne and Tadmor (probably confusing Tadmor with Taibe, as Palmyra has plenty of water) were poor and miserable places in a state of decay at the time. Their inhabitants were chiefly involved in extracting potash from desert weeds. The potash was sent, for example, to Orfa and Aleppo, where it was used for producing soap. It was informed that these villages had been considerably larger but the Bedouins had ruined them. A substantial number of deserted houses seemed to have supported this assumption. According to Stewart, the constant problem was the scarcity of water.

Bedouins plundered Taibe again in 1917, after which it was reoccupied in 1936. With the help of water channels the gardens of the oasis started flourishing again. In 1963 most of the inhabitants of the village were Bedouins, who watered their flocks from the village wells. (Mundy and Musallam 2000; Lewis 1991: 67-78; Wirth 1971: 442-447). Fields of shifting cultivation had been extended between Taibe and Qasr al-Hayr according to the

CORONA satellite photographs from the 1960s (see Fig. 5.).

The existence of caravan stations, *khans*, and gendarme stations is frequently reported by A. Musil (1927; 1928a) in the neighbourhood of Jebel Bishri. Musil (1928a: 176-177) notes that stone heaps had been piled up to guide the caravans in the desert-steppe at prominent points on Jebel Bishri. There are various kinds of guide posts for continuous travelling in deserts and steppes. (See Madoun 1996: 65-71). As previously mentioned (ch. 6. and 8.), we encountered several stone heaps, path finders or *rujm* markers (Fig. 12.), in the central areas of the mountain during our surveys.

Fig. 12. A rujm marker at Nadra on Jebel Bishri. Photo: Kenneth Lönnqvist 2004.

Such *rujm* markers also occur in Israel and Jordan and are associated with Bedouin movement. E. Gibson (2008: 108-110) has noticed that they were often visible from beneath, guiding the travellers to take a convenient path and not to make efforts in vain in climbing, which is more energy-consuming than descending. It is apparent that beside *rujm* markers, cairns/*tumuli* are sometimes called *rujms* - and as discussed in previous chapters - they are also generally located on hilltops or higher ridges on Jebel Bishri and have served as landmarks in the nomadic landscape.

14.2. Arab Fortresses on the Euphrates Side of the Mountain and Medieval Activities in the Central Areas of the Mountain

Our Finnish project SYGIS did not visit any Islamic castles or fortresses at the northern edge of Jebel Bishri following the Euphrates river. Justinian I had refortified Nicephorium-Callinicum on the other site of the Euphrates opposite Jebel Bishri in the sixth century A.D., but as earlier indicated, it fell to Islamic invasions in A.D. 633-639. The Khalif Mansûr fortified the city and designed it into a round plan following the design of Baghdad (Fig. 13.). The city became known as Rafiqah (the Comrade), i.e., modern Raqqa. It is especially famous for its Medieval Raqqa ware. The site, however, was apparently destroyed by Mongols when it was reported as being in ruins in the fourteenth century. G. Bell (1924: 54-60) has especially described the ancient Islamic monuments of Raqqa in her travel description *Amurath to Amurath.*

As previously mentioned (ch. 13.), Qala'at Safin is a castle hanging on the cliff called "Jabal Safin", forming a convenient guarding post on Jebel Bishri, on the southern side of the Deir ez-Zor–Aleppo -road ca. 16 km southeast of Raqqa at the site known as Siffin (Fig. 13.). Apart from the evidence of the Roman foundations, the site has also produced both Byzantine and Islamic sherds (Ohnuma and Al-Khabour 2007). In any event, Siffin is a known site to Arab sources. Hamza mentions that at Siffin there was a residence of a Ghassānid king, but the extent of the Ghassānid rule to the Euphrates has been later doubted. On the other hand, in the latest studies the rule is accepted, and according to the present view, the Ghassānids formed *foederati* on the Euphrates opposite to the Persians during the Byzantine era. (Shahîd 2002: 337). The archaeological evidence at Siffin also supports Byzantine activities at the site, and a fort could have been rebuilt against the Persian Lakhmides, who had their own fortress called Qal'at Ja'bar on the other side of the Euphrates. During the spread of Islam a famous battle took place at Siffin between Ali and Mu'awiya in A.D. 657 (Shahîd 2002: 337).

As previously mentioned (ch. 13.), Naqhila (Nukhailah) is an ancient fort situated ca. 23 km southeast of Raqqa, also situating on the northern ridge of Jebel Bishri (Fig. 13.). The fort has been mentioned by F. Sarre and E. Herzfeld (1911) as well as A. Musil (1927: 187), K. Kohlmeyer (1986: 58-60) and J. Napoli (2000). It appears on Poidebard's map of 1934 and on Mouterde's and Poidebard's map of 1945 bearing the name "Nheyla". We have not studied the site as it was recorded by the Syro-Japanese team in the year 2007 and also belongs to the Mouhafazat of Raqqa.

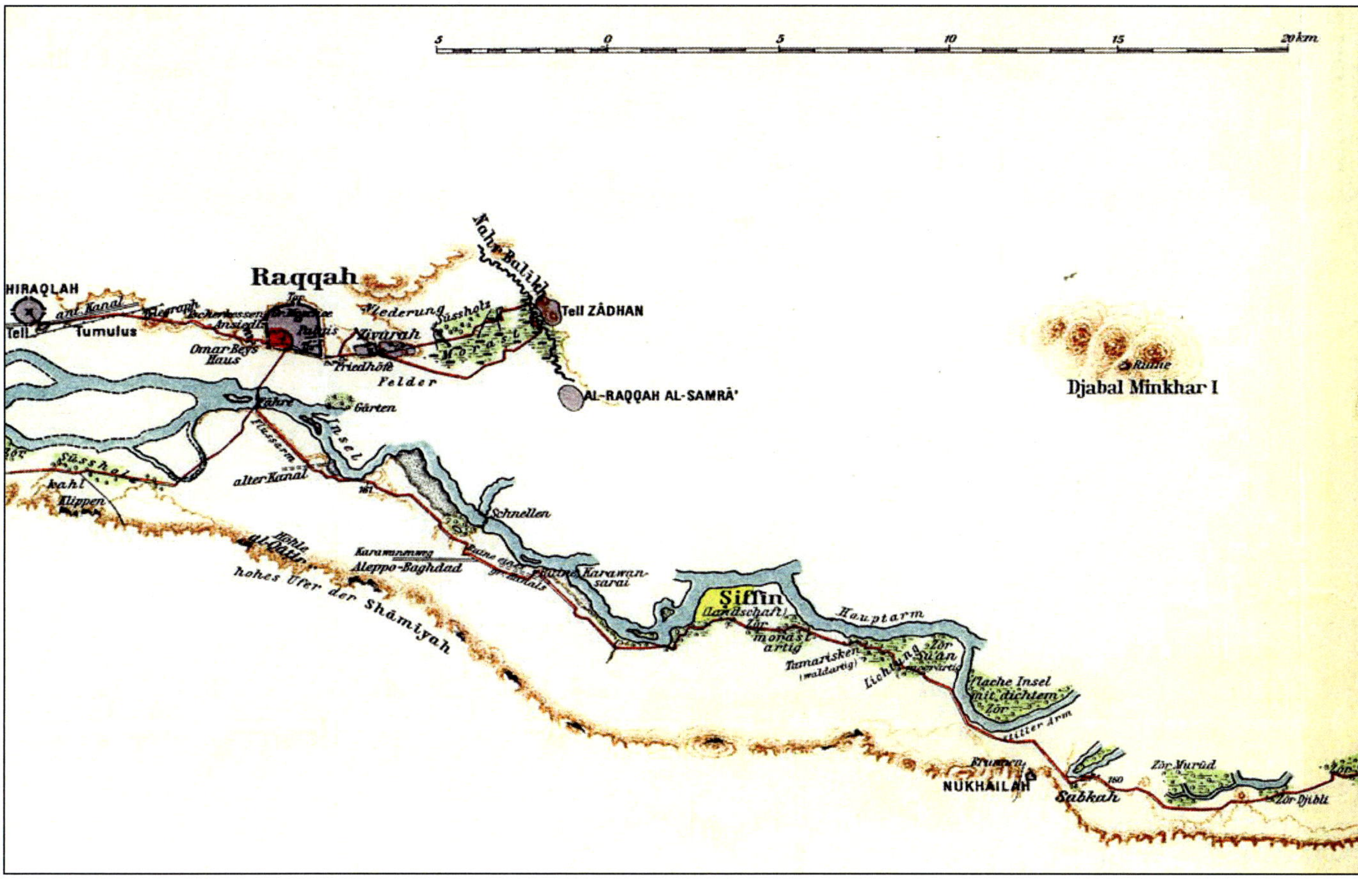

Fig. 13. The area of the Euphrates along Jebel Bishri from Raqqa to Sabkah. Source: Sarre and Herzfeld 1911.

Beneath the fort there is the village of Al-Rabt some 0.5 km south of the Deir ez-Zor–Aleppo -road. A wall on the southeast still stands to the height of 3 m. Like at the fort of Tabūs (see ch. 13.) the foundation is laid on large stone blocks, but here baked brick walls are surmounted on the blocks instead. There are two visible gates: one in the southeast and the other in the northeast. Associated scattered potsherds date from the Roman, Byzantine and Islamic period, which indicate that the foundations of the fortress, like that at Safin/Shiffin, date from an earlier period (Lauffray 1983: 70; Kohlmeyer 1986: 60; Napoli 2000). The fortress has quadrilateral corner towers which even in collapsed state still protrude from the ruins. The site clearly commands the surrounding area and was probably recognized as a good place for guarding in ancient times. (Ohnuma and Al-Khabour 2007; Al-Maqdissi and Ohnuma 2008: 120).

As previously mentioned (ch. 10.), while dealing with the Bronze and Iron Age settlements on the Euphrates banks, we recorded the coordinates of a medieval citadel on top of a tell at the village of Qasabi (UTM: 0569123, 3955156; ca. 226 m a.s.l.' see Fig. 14.). Qasabi is located on the peninsula of Halabiya northwest of the fort of Zenobia. The Arabic name "Qasabi" means "capital" and seems to refer to the citadel on the top of the tell. The site was visited by F. Sarre and E. Herzfeld (1911: 165) at the beginning of the twentieth century, and it is mentioned by J. Lauffray (1981: 69-70). K. Kohlmeyer (1986: 61-63) tells that at the site there was a *caravanserai* mentioned in the sixteenth century sources. The site is also mentioned in an unpublished report of the Syrian-Spanish mission led by J.L. Montero Fenollós in 2005. A lot of turquoise glazed pottery from the Parthian - Islamic period was scattered around, besides Early Bronze Age pieces. The site has a view to the Euphrates. The tell commands the river, where it enters the narrows known as the chanouca of Halabiya.

Fig. 14. Ruined rooms of the citadel at Qasabi with a view to the Euphrates. Photo from E: Minna Lönnqvist 2008.

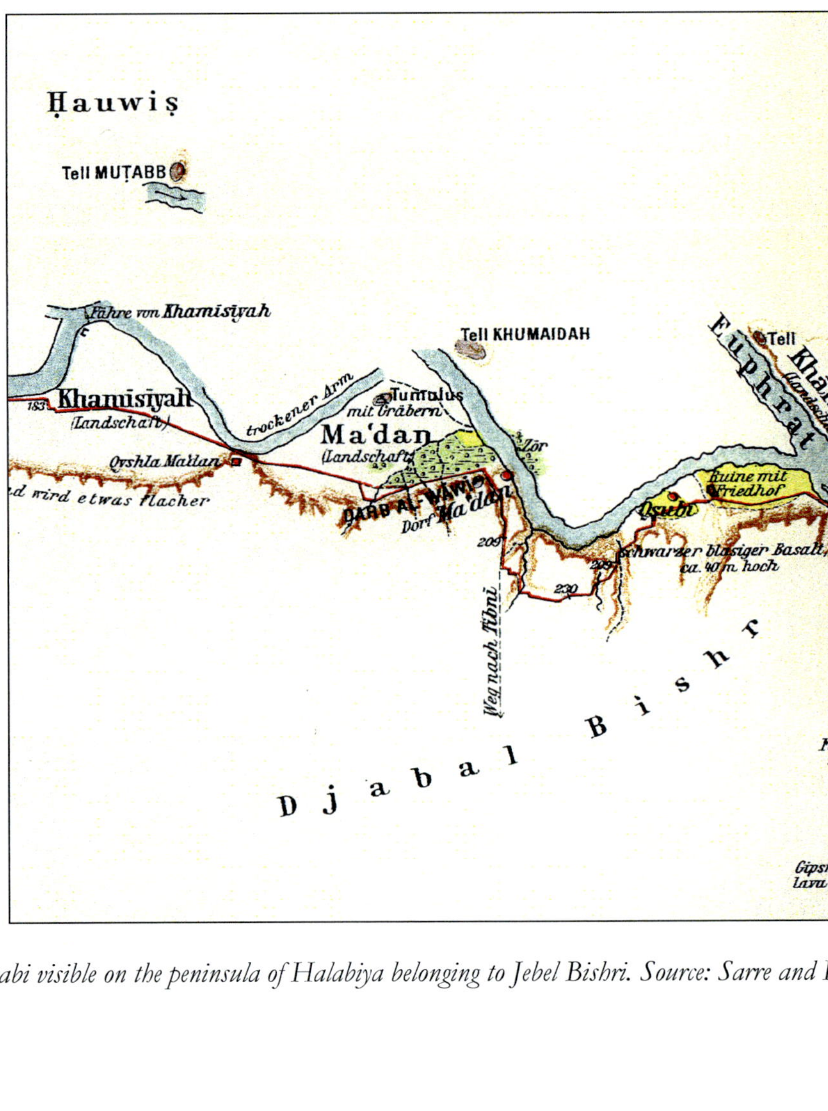

Fig. 15. Qasabi visible on the peninsula of Halabiya belonging to Jebel Bishri. Source: Sarre and Herzfeld 1911.

The stretch of the Euphrates from Halabiya to Deir ez-Zor produced Islamic pottery in various sites that we studied. For example, Tell Tibne (see the location in ch. 10) next to the peninsula of Halabiya provided sherds from the early Islamic period (Fig. 16.).

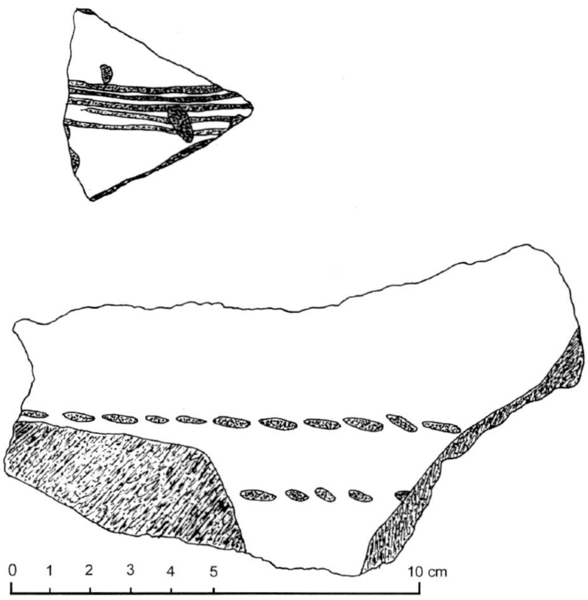

Fig. 16. Decorated Early Islamic pottery from Tell Tibne. Drawing: Michael Herles 2005.

Fig. 17. Northern slopes of Jebel Bishri on which there is an ancient gravefield of elongated cairns above the modern Islamic graveyard. Photo from E: Kenneth Lönnqvist.

Fig. 18. Badly weathered cairn/tumulus field H1a on the northern slope. Photo from SE: Kenneth Lönnqvist

Located directly from the plateau **H 1** (which is a Pleistocene terrace discussed in ch. 4., 5. and 10.; see the location information in paragraph 5.5., also see the map in ch. 13., Fig. 25.) way up to the south along the slope of Jebel Bishri there was a badly preserved graveyard **H 1a** (ca. 200 m x 50 m in size) consisting of dozens of elongated cairns which followed the slope aligned from the southeast to the northwest (Figs. 17. and 18.). The tombs were very hard to recognize because of the fact that several stones had rolled down from them. The alignment of the graves indicates that they could be Medieval Islamic graves (cf. Cauvin *et al.* 1985: 161, Fig. 2.). The dimensions of the graves were ca. 2 m x 0.6 m, constructed of stones ca. 0.2 – 0.3 m in diameter. On the western part of the slope there were a few round cairns. No associated pieces of pottery or flints to indicate dating were recognized at the graveyard, but Roman and Islamic pottery was recovered on the plateau beneath. On the southwestern part of the plateau there was a modern Islamic graveyard (Fig. 17.) which is currently in use. Otherwise the plateau had been largely used as a watering point for animals by local village herders as large pits had been dug here and there for collecting rainwater.

During our ground survey in the central eastern areas of Jebel Bishri several multi-period sites such as **C 10**, **I 8**, **I 15 a** and **I 19** (see the locations in ch. 5. and 6.) in association with the plain of Nadra provided evidence of Medieval activities and those of more recent Bedouins. The already described multi-period site **C 10** (ch. 6.) on the way from Shula to Ash-Shujiri produced Arabic inscriptions executed on the sandstone faces of a protruding rock-formation. The inscriptions date from various periods, showing that the site had obviously been popular and used by Bedouins and caravan travellers. One inscription (Fig. 18.) says: *'Ata Allah* with an Islamic date equivalent to 19th April 1771, and another mentions names *(Zamii) Ali Muhammad and 'Ata Allah al-*. (We owe the interpretation and translation to arabist Mikko Hakalin).

Fig. 19. An inscription at site C 10.
Photo from SW: Minna Lönnqvist 2006

Fig. 20. An inscription at site C 10.
Photo from SE: Minna Lönnqvist 2006

In 2000 the site C 10 also produced a fragment of a glass bracelet of black and coil twisted type (l. 2.98 cm, 0.74 cm in diam.). Different kinds of glass and enamel bracelets were common in Roman and Byzantine times, during the latter period also in the Islamic culture. In 2004 a similar bracelet fragment (Fig. 20.; size: l. 4.63 cm and round section 0.82 mm in diam., weight 2.7 g) to the one discovered at C 10 was found at site I 19 in the district of Nadra. This common type is comparable to those found at the Medieval cemetery of Cafer Höyük and Yumuktepe in Turkey. Several are known from Syria-Palestine, too (e.g., Site 45, al-Rujoum in Jordan in MacDonald 1992: 116, 242). The type is vaguely dated from the Roman to the Byzantine/early Islamic times being especially common in the latter period (see Cauvin *et al.* 1985: 146, Fig. 10: 37; also Köroğlu 1998, Fig. 7.). Furthermore, at sites I 8 and I 15a the finds included more colourful glass bracelets. One (l. 3.74 cm and round cross section 0.73 mm in diam., weight 4.0 g) from **I 8** (Fig. 21.) is multicoloured: white, yellow, reddish brown and green with special dots inserted with yellow and white colours in *millefiori* technique. The most elaborate fragment (l. 74.95 mm and flattened cross section 5.3 mm in diam., weight: 11.5 g) of a bracelet is from the site I 15a (Fig. 22.). It has silver thread inserted in glass. The alternating colours used in the bracelet are: turquoise, brown, blue, white and beige. Examples with flat cross-section have been attributed to later Islamic periods, like the Ottoman era (MacDonald 1992: 116). This item may, however, date from the Byzantine period. (See more on glass and enamelled bracelets, e.g., Köroğlu 1998).

Fig. 20. A fragment of a bracelet from site I 19 at Nadra.
Photo: Kenneth Lönnqvist 2008.

Fig. 21. A fragment of a bracelet from site I 8 at Nadra.
Photo: Kenneth Lönnqvist 2008.

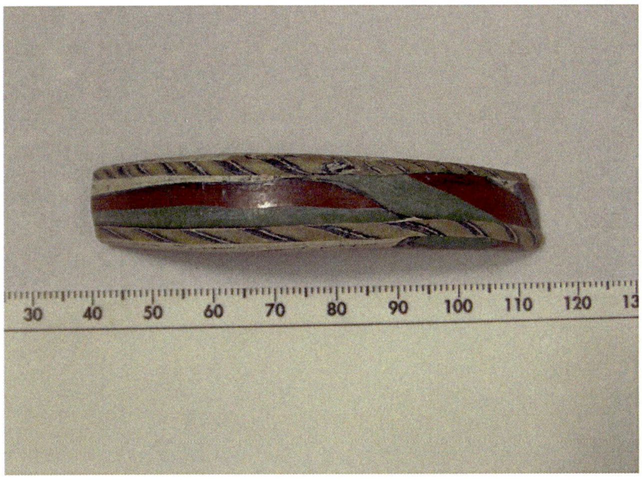

Fig. 22. A fragment of a bracelet from site I 15a at Nadra. Photo: Kenneth Lönnqvist 2008.

14.3. Bedouin Tribes Living in the Region

The Bedouin tribes currently occupying the mountain of Jebel Bishri and its surroundings originate from the camel nomads of the Arabian Peninsula and Iraq. However, many tribes have been living for centuries in Syria. Nowadays, the Bedouins met within the region of Jebel Bishri mainly consist of *Fid'ān, Sba'a* and *'Amarat tribes*. They all belong to *the 'Anazah federation* (see Fig. 23.), the history of which in Syria extends at least to the seventeenth century. As earlier indicated, our studies especially provided evidence of the tribal territories and grazing areas of the Fid'ān and Sba'a tribes, who are closely associated with vertical nomadism in the region. (Lewis 1987; Jabbur 1995; Nilhamn 2008). The shift northwards of the Shammar and 'Anazah tribes from Arabia and Iraq happened in the first half of the seventeenth century, and the tribal territories of these nomads were largely established at that time.

C. Niebuhr (1792: 179-180), who visited the areas of the Syrian Desert in the eighteenth century wrote about the Bedouins of the 'Anazah federation: *The tribe of Anæse are esteemed to be the greatest tribe in the desert of Syria. They have even spread into Nedsjed, where they are reckoned the most numerous tribe in the heart of Arabia. The caravans of Turkish pilgrims pay them a considerable duty for their free passage through the country. This tribe too, when dissatisfied, plunder the caravans. They often make war on the Pacha of Damascus. They lately routed and killed the Pacha of Ghassa in his own government… Within a day's journey of Damascus, the whole caravans were plundered by the tribe of Anæse. The trunk was opened. The Bedouins took what they chose, but left me my books, papers, a box of medals, and two watches. The camel driver collected the broken pieces of my trunk, and brought the whole honestly to Aleppo. Thus had I, at the same time, a proof of Arabian rapacity and Arabian integrity.*

The Hon. Charles Leonard Irby and James Mangles (1844: 71-79) describe an itinerary through Sukhne to Aleppo in 1817-1818 and also the violent nature of the local 'Anazah Bedouins at that time: *December 22,* 1817 *We have been detained at Aleppo nearly a month waiting for a caravan which brings kali from Sukne to this place, a journey of five days. Palmyra is of seven days journey from Aleppo….*

December 24, A caravan from Sukne arrived this day.

December 29, 1817 We were to have set off this evening to Palmyra, by way of Sukne, when accidentally meeting a merchant from Bagdad, a friend of Mr. Barker, he strongly dissuaded us from a measure, and urged us to go by the way of Hamah or Homs, as the Annasee Arabs are in the neighbourhood of Palmyra. We had understood that the cold had driven them all to the southward towards the banks of the Euphrates; but as it appears there yet, there remain two tribes of them, Homs will be the best place to start from…. Homs, is distant only four days' journey from Palmyra, which we had sanguine expectations of being able to reach from either one or the other of these towns. However, we had two other strings to our bow; either push on to Cariateen, which is only one day from Tadmor, thence to steal to that city before the Arabs were aware of our intention; or to take Turkish post-horses with an escort from Damascus, and go inspite of the Arabs….

January 26, 1818 At one p.m., after nineteen days negotiation at Hamah and Homs, we started with our three camels and as many conductors, with two skin bottles, in which they had poured the melted butter bought with Lady Hester's present… All our baggage consisted of a sheep-skin coat, the woolly side inwards, and the other side coloured red with ochre and greased to keep out of rain. We rode for five hours, our guides nearly the whole the time singing a favourite Arab song. On arriving to a Bedouin camp, we had some scruples about entering a tent, expecting they would have many objections against receiving us. Instead of which, to our surprise we were welcomed by both men and women; the latter smiling, said. We were Frangi (Franks) and retired into their part to prepare supper.

In the summer of 1854, which seemed to have been a hard year for Bedouins, consul N.W. Weeks of Aleppo complained that men from 'Anazah with their sheikh had carried out incursions against the fields of peasants and their villages, sharing their crops, and another time he writes that the route between Aleppo and Damascus was in the hands of these Bedouins. They caused which a threat to caravans and meant that escorts were needed,

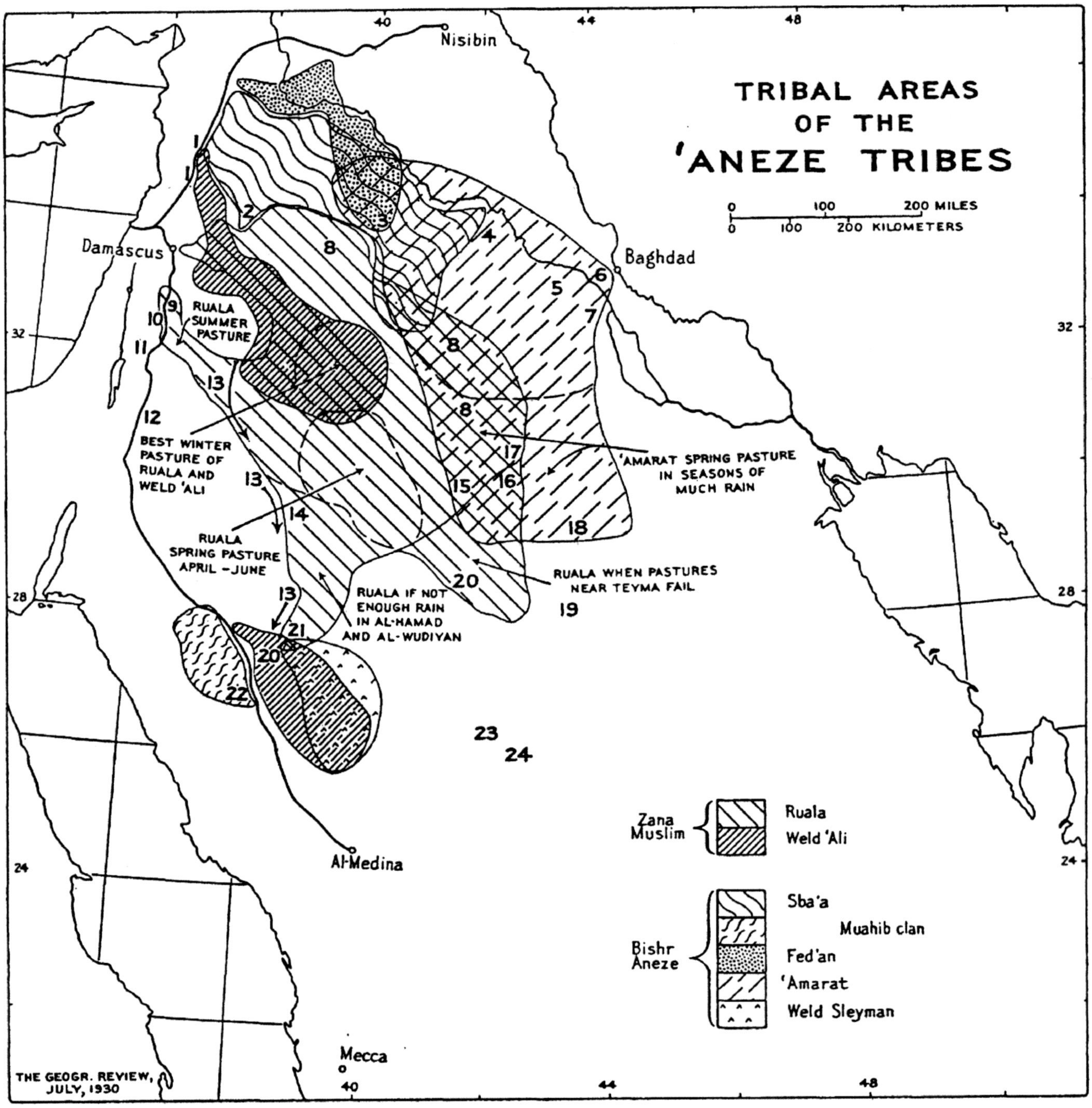

Fig. 23. Source: Raswan 1930.

varying from 20 to 58 cavalrymen. J. Jabbur (1995: 590) also notes that the Fid'ān and the Sba'a had participated in raiding and plundering activities in the past. A glimpse of a ravaging Bishri group was also provided by a Finnish arabist G.A. Wallin in a letter to his mentor Geitling, dated 12th June 1849, from a desert journey. 1870 meant the pacification of Palmyra and the surrounding desert-steppe, and as previously indicated, many gendarme stations were built to guard the areas. Walis of Damascus and Aleppo agreed to cooperate with the Ottoman rule in order to secure the areas (Lewis 1991: 73). However, A. Musil (1928a: 81) describes that at the beginning of the twentieth century there was still a possibility of violent clashes between travellers and the members of the 'Amarat tribe: *Suddenly ten 'Amârât came to us on camels and would have driven away our animals if we had let them. They asserted that we might be the spies of the enemy's troop whom their herdsmen had sighted on the previous day at al-Bišri.*

As there was a possibility to our being attacked not only by this troop but also by the marauding 'Amârât, we left the ruins (Qasr al-Hayr) in the evening and sought a sleeping place in the az-Zwêheč ridge.

In the nineteenth century R.D. Upton (1873: 159-162) observed that the 'Anazah tribes consisted of the Fid'ān tribe, the Jelas, the Rwala, the Seloot, the Erfuddi, the Shumlan, the Hayaza, the Ibu-Haddal and the Sba'a. The Fid'ān 'Anazah had nine subdivisions: Mehed, the people of Jedaan, Tooenis, Ajajara, Roas, Shmeilat, Griebat, Roaba and Hrisa. The Sba'a Anazah had seven subdivisions: Duam, Abideh, Ishoeieh, Mooeneh, Gomassa, Mooajeh and Beaiek. The Sba'a were then probably not the most numerous, but they were the wealthiest of all the tribes having huge herds and flocks, and they even possessed 12,000 horsemen. Among the tribes the Sba'a 'Anazah was the most famous for its horses, and the Gomassa had the reputation for the finest ones.

As earlier indicated, while the mountain of Jebel Bishri now chiefly belongs to the territory of the Fid'ān tribe, the western piedmont is largely occupied by the members of the Sba'a tribe. The life style of the members of the Sba'a and the 'Amarat tribes was especially scrutinized during our survey in the area of the western piedmont in 2000, while members of the Fid'ān and 'Agēdāt were encountered in the river zone and on the mountain in 2003-2008 (see more in ch. 15.). In autumn 2006 we encountered several members of the Fid'ān tribe, who were rearing large flocks of dromedaries on the mountain (Fig. 25.). As previously mentioned, Tar al-Sbai at the western edge of the mountain was a border zone that bears the name of the Sba'a tribe, and during our survey we noticed that the Fid'ān had erected tombs at the western edge, the upright capstones of which bore symbols of the tribe (Fig. 24.). The tombs also seemed to serve as territorial markers like discussed in ch. 8..

In the riverine zones the smaller tribes such as the 'Agēdāt, which does not belong to the 'Anazah federation, are involved in grazing along the Euphrates river. The tribe also has pastures on Jebel Bishri. The members of the 'Agēdāt are especially grazing in the area of the mountain during drier years. (Raswan 1930; Ashkenazi 1948; Wirth 1971, Karte 11; Jabbur 1995; D'Hont 1994: 210-211; our interviews among the local Bedouins). In the first half of the twentieth century the 'Agēdāt had stretched their territories to the Khabur, Maiadin and as far as to Abu Kemal at the Iraqi border (von Oppenheim 1939: 217).

Fig. 23. Grazing the flock of sheep at Nadra on Jebel Bishri in 2004. Photo: Gullög Nordquist 2004.

Fig. 24. Upright tomb stone as a territorial border sign, marked with a link-like symbol of the Fid'ān tribe, at the western edge of Jebel Bishri. Photo: Kirsi Lorentz 2000.

Fig. 25. Members of the Fid'ān tribe grazing dromedaries on Jebel Bishri in November. Photo: Minna Lönnqvist 2006.

14.4. The Changing Mode of Pastoral Nomadism in the Region

14.4.1 Nomadism of the Tribes in the Region of Jebel Bishri

Apart from some grazing of dromedaries the Bedouin tribes of the Jebel Bishri region have largely changed from camel nomadism to small livestock rearing in the neighbourhood of the river zone. C.R. Raswan (1930: 496) noted that in the Palmyride mountain belt to which Jebel Bishri belongs, sheep breeding nomads tend to rear their flocks north of the mountain chains and camel nomads to the south. The actualization of the change of the real Bedouin lifestyle identified with camel nomadism is visible in a symbiotic relationship with the Euphrates river valley, its sedentism and agricultural way of life as well as small livestock rearing. Those pastoralists closer to the desert still also concentrate on camels.

J.L. Burckhardt (1831: 32-33) considered the 'Anazah to be the only true Bedouin nation of Syria, at least as far as the originality of the desert manners were concerned. In his view at the time in the late eighteenth and early nineteenth century the 'Anazah were continuing the mobile life throughout the year. Their summer pastures were on the Syrian frontiers, while in the winter they stayed in the Syrian Desert or the Euphrates region. They usually stayed only three to four days in one spot and moved on after grazing the area. This reference to "true Bedouins" indicates that the degree of nomadism of these tribes has been different among various tribes, periods and areas.

R. Upton (1873) described the Great 'Anazah, or the 'Anazah proper, circuit. The tribes moved from Nejed to near Aleppo in the spring, and crossed the Euphrates, returning for winter to Nejed. They were the largest of the tribal groups and the most powerful one. The *dira*, the tribal road of the 'Amarat tribe, led its eastern wanderings as far as to Baghdad, Kerbela and An-Nejef. In the south the members of the tribe pastured their flocks in the region of Al-Hejera. (Raswan 1930: 494-495).

The reports of the years 1880-1881 (Issawi 1988: 60-61) noted that the nomadic tribes in the Middle Euphrates consisted of the Weldi, Affadli, Sba'a, Agedat and Baggara, which were totally under the Turkish authorities. The Weldi herded sheep and cattle, the Affadli buffaloes and camels; the Sba'a camels and sheep, the Agedat camels and horses and the Baggara only cattle. The people of the Middle Euphrates valley resided in mud hovels during the winter going into tents in the summer, but some others used tents all year round. Like Burchardt, these reports testify to that there were still nomads, who solely lived in tents in the region in the nineteenth century.

In the 1970s it was observed that true nomads were very rare in the Jezira, but on the southern side of the Euphrates valley the Fid'ān were closest to the kind of herding sheep and camels (van Dusseldorp 1971: 95).

Fig. 25. Tents at the foot of Jebel Bishri in February 2003. Photo: Minna Lönnqvist.

In February 2003 our project observed that on the Euphrates bank black tents were still located on the piedmont between the northwestern spurs of Jebel Bishri and the basalt peninsula of Halabiya (Fig. 25). The mountain was too cold, but these Bedouins still stayed in tents on the lower plateaus and did not live in the villages like the local transhumants during the winter season. This seemed to point to the fact that different degrees of nomadism are still met within the tribes on and around Jebel Bishri.

A traditional black tent, *beit sha'r*, the cover of which is woven from black goats' hair, is generally used among the pastoralists grazing in the region today. It protects the inner tent from the rain. On the southern piedmonts beige poplin tents are also used nowadays, some may be summer tents belonging to the members of the 'Agēdāt tribe (cf. D'Hont 1994: 193). J.L. Burckardt (1831) especially described the black tents and encampment types of the 'Anazah in the early nineteenth century. The encampments consisted of 1,000 to 800 tents. If they were just a few they were set in a circle, *dowár*, but if there were more of them they were pitched in a line or a row, for example, along a rivulet, sometimes behind three or four others. These are called *nezel*. In winter the tents were spread over the plains in three to four groups at distances of half an hour's journey. This shape of an encampment was called *fereik*. The Sheikhs' tents were situated in the west, because enemies and guests were expected to come from the west. A wealthy man was not supposed to pitch a tent in the east.

Fig. 26. A tent camp in the steppic area of Nadra on Jebel Bishri in May 2004. Note the cairns on the surrounding hill-tops. Photo: Gullög Nordquist.

14.4.2 The Mountain Herders and Transhumants at Jebel Bishri

In the 1930s Mijden ibn Mheyd, chief of the Fid'ān, owned several villages in Syria. Thousands of Bedouins were working in the fields, although the majority were still grazing flocks. (Raswan 1930: 49). J. Jabbur (1995: 527-533) notes that in 1956 the Al-Fid'ān consisted of 8,000 individuals with 598,480 dunums in Syria, Al-Sba'a Btaynat consisted of 6,000 individuals with the area of 19,450 dunums whereas Al-Sba'a al-'Abda comprised 12,000 individuals with the area of 100,965 dunums. The territorial space of the Fid'ān was larger compared to the Sba'a, when the number of individuals are taken into account. The Al-Rwala tribe had as many as 20,000 individuals but only 81,321 dunums. In northern Arabia the Rwala had as many as 12,000 tents. J. Jabbur (1995: 589) further mentions that six clans from the 'Anazah, including the Rwala and 'Amarat, frequent the villages of the province of Aleppo in the springs selling their milk and livestock products to buy grain. A. Musil (1928b) especially reports the manners and customs of the Rwala, and their use of camels. They also use falcons in their hunting activities. A. Musil (1930) reports the following: *The Rwala buy falcons from the settlers in Sheikh Miskin and Rheybe. They hold that the best hunting falcon is reddish-brown with many white spots on its tail, though dark brown and black falcons are also good for hunting.*

Musil (1928: 175-176) further described life of the Fid'ān and the impact of short-term fluctuations in weather conditions in the region of Jebel Bishri causing hardship for the Bedouin tribes in the year 1912: *After a while we were overtaken by two Fed'ân from a camp near Ab-al-Zir, who gave us a good description of the road we had to follow. Likewise our gendarme, a native of Dejr az-Zor, knew al-Bisri fairly well. Rotting sheep were seen everywhere. We were told that in the first days of January snow fell all the way from al-Bisri to ar-Resafa and remained on the ground forty-five days. The half-fellâhin and swâja (breeders of goats and sheep) who did not take their flocks to the Euphrates on time lost all their property, it was said. The animals died in cold and hunger...in the river bed of al-Gajri carcases were piled high. In one bend of the channel we also saw among the sheep three dead mules and even a gnawed skeleton of a small child. We were told that toward the end of February the snow had melted so fast that the tents of the poor half-fellahin, who had sought shelter against the freezing north winds in the deep gullies, floated away with the first rush of melted snow.*

Members of the Sba'a tribe, were also known to have sacked and destroyed the villages of Taibe and El Kowm in 1917 (Lewis 1991: 73); the Sba'a, however, have now partly settled in the area during the twentieth century along with a government-initiated sedentarization process (see ch. 15.). In February 2003 we visited a Bedouin family at the village of Tibne on the Euphrates side. This transhumant family stayed in winter in their house, but in the spring they left with their flocks for Jebel Bishri. E. Sachau (1900) notes that the Sba'a Bedouins had pitched their black tents all the way from Raqqa to Halabiya. A. Musil (1927: 188) also reports that in 1915 north of Meskéné, near the confluence of the Euphrates and Balikh area, down to Tibne (a village situated southeast of Halabiya), the camps of the Sba'a tribe, subdivided into al-Welde and al-Afâzle, were scattered on both sides of the Euphrates. In addition, M. von Oppenheim (1939: 211) reported that there were Bedouins at Tibne and Ma'dan at the beginning of the twentieth century, who belonged to the Sba'a tribe and spent the summers at the village and the winters on the other side of the Euphrates at Jebel Abd al-'Aziz.

In May 2004 we encountered several black tent camps of Bedouins rearing small livestock, such as goats and sheep, with donkeys in the neighbourhood of Nadra in the eastern central area of the mountain. The change in altitudes and environments from the Euphrates valley to the mountain of Jebel Bishri was pronounced. We visited the tent of a Sheikh from the Agedat tribe on the mountain. For him Jebel Bishri was also a seasonal grazing ground.

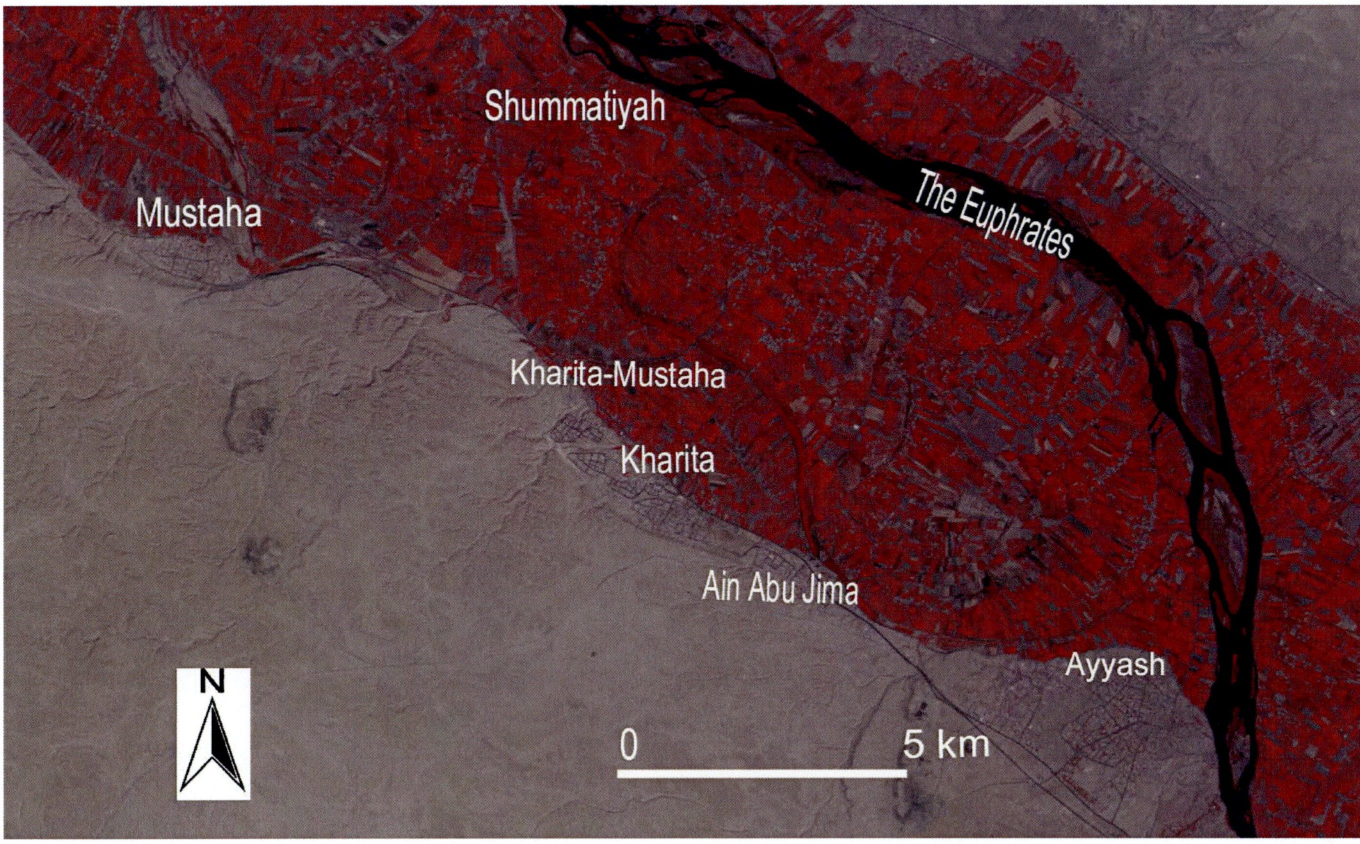

Fig. 27. Village locations, village layouts and fields in our research area on the Euphrates side visible on a SPOT image. Mapping by Minna Lönnqvist 2010.

The family of the Sheikh had a spacious house with several rooms and a kitchen in the village of Shummatiyah (see Fig. 27.) on the right bank of the Euphrates ca. 27 km from Deir ez-Zor towards Aleppo. We visited the house in 2004 and 2008 (see Fig. 28.). The eldest son of the family had got married between 2004 and 2008 and stayed now in his father's house but was building a house for himself next to that of his father's. That indicated patrilocal practices among these members of the tribe. The village habitations are based on kinship ties, basically on a *fakhad*, a patrilinear descent group, shared by the tribes living in the region nowadays. Marriages are still largely intra-family transactions. Usually, as described above, sons and their wives live for the first years of their marriages in their father's house. The tribe as a higher social unit has lost some of its significance. Sheikhs still carry out their duties as judges, but their powers as social and political leaders have diminished. (Van Dusseldorp 1971: 96). The base-village represents a semi-sedentary way of life in the river valley during the winter season, whereas tents are used as dwellings on the mountain in the spring and summer. In November 2008 we were told that the year had been so

Fig. 28. A breakfast in the Sheikh's house at the village of Shummatiyah on the Euphrates. The breakfast includes ground zatar herbs from Jebel Bishri with freshly baked bread, lebane yoghurt, falafel balls, scrambled eggs, hummus, marinated aubergines, tomatoes, olives, olive oil, lemon and apricot jam. Photo: Kenneth Lönnqvist 2008.

poor in rain that the family of the Agedat sheikh could not graze on the mountain but stayed in the village with its flocks all the year through. In this case the drought increased sedentism.

In any event, D. Eickelman (1981: 65) has emphasized that ecological or seasonal characterisations for nomadic groups are insufficient in themselves, like defining all the pastoral nomads in the Jebel Bishri region as village-based *transhumants* throughout the ages. Eickelman further points out that currently - as in the past - several tribes have both nomadic and settled components. Ethnopolitical questions of tribal identity in the region of Jebel Bishri are long-lived structures that we have already dealt with in the ch. 6.-10. dealing with the ancient nomadism in the region.

14.4.3 Cultivating Bedouins and Fellahin in the Euphrates River Valley

The settlement patterns of the Middle Euphrates region are made up of Bedouins and more sedentary agriculturalists know as *fellahin.* On the Euphrates bank the layouts of the villages on river terraces are more compact than in the valley (see Figs. 28., 31. and 32.) and on the southern and western piedmonts of Jebel Bishri. In the 1880s the extent of the fields in the district of Deir ez-Zor was quite modest. The chief crops, which were grown in the area in the 1880s, were wheat, millet, barley, rice and maize (Issawi 1988: 61-62). A growing number of fields can be observed in C. Héraud's (1922) map dating from the 1920s.

There was a fierce drought between 1958 and 1960 resulting in the loss of the herds of the Bedouins in the Middle Euphrates region, which in turn caused them to participate in growing cotton and carrying out large-scale irrigation projects. Therefore the area of cotton cultivation increased over ten times from 10,000 hectares to 120,000 hectares between 1952 and 1962. Subsistence farming changed from earlier cultivation of large fields of wheat and barley to more commercial cotton growing (Fig. 30.) which caused immigration into the area. (Van Dusseldorp 1971: 93-108). Nowadays cotton, wheat, maize and barley are mostly grown along the river.

The houses of pastoralists and *fellahin* on the Euphrates at the foot of Jebel Bishri follow the *Breitraum* layouts of simple rectangular buildings (Figs. 31.-34.), also comparable to Bedouin tents. In contrast to the western side of Jebel Bishri (see ch. 15.), on the Euphrates side the houses are built of concrete bricks or mud-brick. For example, on Halabiya, mud-brick (Fig. 35.) is still used, as in Mesopotamia in general. The houses of the river valley are more elaborately built and decorated than the houses on the mountain area and the western piedmont of Jebel Bishri. The houses along the river often have a porch which is supported by a screen of columns or pillars *in antis*, and they are also more often painted. The white colour is usually used on the walls, whereas iron doors are painted in purple or turquoise. The costumes worn by the women are strikingly brighter in colour on the Euphrates side compared to the women in the western part of the Jebel Bishri region.

Fig. 29. Winter grazing in the Euphrates valley at the foot of Jebel Bishri. Photo: Minna Lönnqvist, Feb. 2003.

Fig. 30. Herding with a donkey and dogs in the plateau of 'Ayyāsh on the Euphrates side of Jebel Bishri. Beyond the herders fields of maize and cotton are visible. Photo: Minna Lönnqvist, Oct. 2006.

Fig. 31. The villages of 'Ain Abu Jimā and Kharita on the northeastern edge of Jebel Bishri. Photo from E: Minna Lönnqvist 2006.

Fig. 32. The village of Mustaha on the terrace of the Euphrates valley at the foot of Jebel Bishri. Photo from SE: Minna Lönnqvist 2003.

Fig. 33. A Breitraum house with pillars in antis at the village of Mustaha on the Euphrates side at the foot of Jebel Bishri. Photo from N: Minna Lönnqvist Feb. 2003.

Fig. 34. A Breitraum house with columns in antis on the Euphrates side at the foot of Jebel Bishri. Photo from N: Minna Lönnqvist Feb. 2003.

Fig. 35. Drying of mud-brick at Halabiya, Nov. 2008. Photo from S: Minna Lönnqvist.

Fig. 36. Baking bread in an oven at a courtyard on Halabiya in early morning hours, Nov. 2008. Photo: Minna Lönnqvist.

14.5. Concluding Remarks

Two modes of pastoral nomadism, vertical and horizontal, have been observed within the area of Jebel Bishri throughout the past. As previously discussed (ch. 6.), the former mode of pastoral nomadism is best describes as *transhumant* pastoralism, which is based on semi-sedentary villages and movement between lowlands and highlands (see, e.g., Hole 1978, Eickelman 1981: 64). It is very typical of the movement between Jebel Bishri and the Euphrates river zone today. However, there still seems to be different degrees of nomadism among the pastoralists of the region today as some still sojourn in their tents during the winter season.

Many tribes that have formerly solely been camel nomads have engaged more on small livestock rearing on Jebel Bishri and near the Euphrates. The symbiosis with the agricultural life has become more pronounced near the river zone. The main tribes in the region, namely the Fidān and the Sba'a, belong to the powerful *'Anazah federation* and are engaged in vertical nomadism. The mountain proper belongs to the tribal territory of the Fidān who are more nomadic than the Sba'a and graze both dromedaries as well as goats/sheep. The Sba'a dominates the western piedmont and the Euphrates river zone. The members of the Agēdat tribe, who do not belong to the 'Anazah, are based on the Euphrates villages. They seem to take more part into the horizontal grazing beside the vertical nomadism – some of them have pastures on Jebel Bishri. Agriculture belongs to the seasonal economies of those living in the villages on the Euphrates side. In the river valley beside the semi-sedentary village herders, there are *fellahin*, whose economy is chiefly based on agriculture. Ethnographic analogies have been used for the Bronze Age nomads (see, e.g., Streck 2002: 159-168), who have lived on and around Jebel Bishri. However, it needs to be emphasized that even today different degrees of nomadism is met within the district of Jebel Bishri, and only a hundred years ago nomadism was more pronounced among the 'Anazah tribes than today. Sedentarization and the transhumant mode of pastoralism have become more frequent during the past century.

Bibliography

ANET = *Ancient Near Eastern Texts Relating to the Old Testament* (1969). Ed. by Pritchard, J. B. Princeton, N.J.: Princeton University Press.

Ashkenazi, T. (1948) The 'Anazah Tribes, in *Southwestern Journal of Anthropology*, Vol. 4, No. 2, pp. 222-239.

Al-Maqdissi, M. and Ohnuma. K. (2008) Preliminary Reports of the Syria-Japan Archaeological Joint Research in the Region of Ar-Raqqa, Syria, in *Al-Rāfidān*, Vol. XXIX, pp. 117-196.

Bell, G. (1924) *Amurath to Amurath.* Second edition. London: MacMillan & Co. Ltd.

Burchardt, J. L. (1831) *Notes on the Bedouins and Wahabys collected during his travels in the East* by the late John Lewis Burchardt. Vol. 1. London: Henry Colburn and Richard Bentley.

Cauvin, J., Chavane, M.-J. and Oziol, T. (1985) Le Cimitière Médiéval de Cafer Höyük (Malatya, Turquie), in *Cahiers de l'Euphrate* 4, pp. 135-173.

Cornu, G. (1985) La Passe d'El-Kūm–Taybé et les Géographes de l'Époque Classique, in *Cahiers de l'Euphrate* 4, pp. 207-211.

D'Hont, O. (1994) *Vie Quotidienne des 'Agēdāt: Techniques et Occupation de l'Espace sur le Moyen-Euphrate.* Préface de Cresswell, R.. Damas: Institut français de Damas.

Dignas, B. and Winter, E. (2007) *Rome and Persia in Late Antiquity: Neighbours and Rivals.* Cambridge: Cambridge University Press.

Eickelman, D.F. (1981) *The Middle East: An Anthropological Approach.* Englewood Cliffs, N.J.: Prentice–Hall, Inc.

Fairchild Ruggles, D. (2005) Gardens and Gardening, in *Medieval Islamic Civilization, an Encyclopedia,* Vol. 1, ed. by Meri, J.W., London and New York: Routledge, pp. 276-278.

Genequand, D. (2004-2005) Nouvelles recherches à Qasr al-Hayr al-Sharqi: La Mosque Ayyoubide et la nécropole, in *Les Annales Archéologiques Arabes Syriennes,* Vols. XLVII-XLVIII, pp. 271-293.

Genequand, D. (2005) From 'desert castle' to medieval town: Qasr al-Hayr al-Sharqi (Syria), in *Antiquity,* Vol. 79, pp. 350-361.

Genequand, D. (2008) The Sixth season of Excavations at Qar al-Hayr al-Sharqi (2007), in *CBRL 2008, Bulletin of the Council for British Research in the Levant 3*, pp. 96-100.

Genequand, D (2009) Syrian-Swiss Excavations and Research at Qasr al-Hayr al-Sharqi and Palmyra in 2008, in *CBRL 2009, Bulletin of the Council for British Research in the Levant 4*, pp. 41-45.

Gibson, E. (2008) Social Interaction in the Deserts of the Southern Levant, in *CBRL 2008, Bulletin of the Council for British Research in the Levant 3*, pp. 107-110.

Grabar, O., Holod, R., Knustad, J. and Trousdale, W. (1978) *City in the Desert: Qasr al-Hayr East.* An Account of the Excavations Carried out at Qasr al-Hayr East on Behalf of the Kelsey Museum of Archaeology at the University of Michigan, with the Help of Harvard University and the Oriental Institute of Chicago. Vols. I and II. Harvard Middle Eastern Monographs 23-24. Cambridge, Mass: Harvard University Press.

Héraud, C. (1922, repr.1988) *Une Mission de Reconnaissance de L'Euphrate en 1922.* Première Partie: les Cartes. Damas: Institut Français de Damas.

Hole, F. (1978) Pastoral Nomadism in Western Iran, in *Explorations in Ethnoarchaeology,* ed. by Gould, R., School of American Research Advanced, A Seminar Series, Albuquerque, New Mexico: University of New Mexico Press, pp. 127-167.

Irby, C.L. and Mangles, J. (1844) *Travels in Egypt and Nubia, Syria and the Holy Land. Including Journeys Round the Dead Sea and through the Country East of Jordan during the Years 1817 and 1818.* London: John Murray.

Issawi, C. (1988) *The Fertile Crescent, 1800-1914: A Documentary Economic History.* Studies in Middle Eastern History. USA: Oxford University Press.

Jabbur, J.S. (1995) *The Bedouins and the Desert: Aspects of Nomadic Life in the Arab East.* SUNY Series in Near Eastern Studies. Albany: State University of New York Press.

Kohlmeyer, K. (1986) Euphrat-Survey 1984, Zweiter Vorbericht über die Mitteln der Gerda Henkel Stiftung durchgeführte archäologische Geländebegehung im syrischen Euphrattal, in *Mitteilungen der Deutschen Orient-Gesellschaft,* Vol. 118, pp. 51-65.

Köroğlu, G. (1998) Selected Medieval Glass Artifacts from Yumuktepe Mound, in *Studia Anatolica,* Vol. 11, pp. 283-294.

Lauffray, J. (1983) *Halabiyya – Zenobia: Place Forte du Limes Oriental et La Haute-Mésopotamie au VIe Siècle.* Tome I: Les duchés frontaliers de Mésopotamie et les fortifications de Zenobia. Bibliothèque Archéologique et Historique T. CXIX. Institut français d'archéologie du Proche-Orient. Beyrouth – Damas – Amman. Paris: Paul Geuthner.

Lewis, N.N. (1987) *Nomads and Settlers in Syria and Jordan 1800-1980.* Cambridge Middle East Library. Cambridge: Cambridge University Press.

Lewis, N.N. (1991) Taïbe and El Kowm, 1600-1980, in *Cahiers de l'Euphrate* 5-6, pp. 67-78.

MacDonald, B. (1992) *The Southern Ghors and Northeast 'Arabah Archaeological Survey.* Sheffield Archaeological Monographs 5. Sheffield: J.R. Collins Publications.

Madoun, M.A. (1996) *The Silk Road and Palmyra: Bartering Cultures.* Damascus: Al Namir Publishing House.

Mouterde, R. and Poidebard, A. (1945) *Le Limes de Chalcis, Organisation de la Steppe en Haute Syrie Romaine,* Documents Aériens et Epigraphiques, Plans et Relevés de Lauffray, J. et Mazloium, S. Texte et Atlas. Bibliothèque Archéologique et Historique. Tome XXXVIII. Geuthner: Paris.

Mundy, M. and Musallam, B. (ed.) (2000) *The Transformation of Nomadic Society in the Arab East.* University of Cambridge Oriental Publications No. 58. Cambridge: Cambridge University press.

Musil, A. (1927) *The Middle Euphrates: A Topographical Itinerary.* American Geographical Society. Oriental Explorations and Studies No 3. New York: American Geographical Society.

Musil, A. (1928a) *Palmyrena: A Topographical Itinerary.* American Geographical Society, Oriental Explorations and Studies No 4. New York: American Geographical Society.

Musil, A. (1928b) *The Manners and Customs of the Rwala Bedouins.* American Geographical Society, Oriental Explorations and Studies No. 6. Edited by J.K. Wright. New York: American Geographical Society.

Musil, A. (1930) *In the Arabian Desert.* Edited by J.K. Wright. H. Liveright: New York.

Napoli, J. (2000) Les Remparts de la Fortresse de Djazla sur le Moyen-Euphrate, in *Syria,* Vol. 77, pp. 117-136.

Niebuhr, C. (1792) *Travels through Arabia and other Countries in the East,* performed by M. Niebuhr, translated into English by Heron, R., in Two Volumes, Vol. II. Edinburgh: Printed for R. Morison and Son.

Nilhamn, B. (2008) Nomadic Life in Central and Eastern Syria, A Perspective from the Present Life on Badiyah to the Amorite Nomadism in the Bronze Age, in *Jebel Bishri in Context, Introduction to the Archaeological Studies and the Neighbourhood of Jebel Bishri in Central Syria,* Proceedings of a Nordic Research Training Seminar in Syria, May 2004, ed. by Lönnqvist, M., BAR International Series, No. 1817, pp. 15-30.

Ohnuma, K. and Al-Khabour, A. (2007) Archaeological Research in the Bishri region – report of the second working season. Unpublished manuscript.

Oppenheim, M. von (1939) *Die Beduinen.* Band 1: Die Beduinenstämme in Mesopotamien und Syrien. Unter Mitbearbeitung von Erich Bräunlich und Werner Caskel. Leipzig: Otto Harrassowitz.

Poidebard, A. (1934) *La Trace de Rome dans le Désert de Syrie*: Le Limes de Trajan a la conquête arabe. Recherches aériennes (1925-1932). Texte. Atlas. Bibliothèque Archéologique et Historique. Tome XVIII. Paris: Geuthner.

Raswan, C.R. (1930) Tribal areas and migration lines of the North Arabian Bedouins, in *The Geographical Review,* Vol. 20, pp. 492-502.

Rostovtzeff, M. (1932, reprint 1971) *Caravan Cities.* New York: AMS Press.

Sachau, E. (1900) *Am Euphrat und Tigris: Reisenotizen aus dem Winter 1897-1898.* Leipzig: Hindrichs.

Sarre, F. and Herzfeld, E. (1911) *Archäologische Reise im Euphrat- und Tigris-Gebiet.* Forschungen zur islamischen Kunst I. Band I. Berlin: Dietrich Reimer (Ernst Vohsen).

Shahîd, I. (2002) *The Byzantium and the Arabs in the Sixth Century.* Vol. 2: 1. Toponymy, Monuments, Historical Geography and Frontier Studies. Dumbarton Oaks. Washington D.C: Harvard University Press.

Stewart, D.H. (1881) *apud* Issawi, C. (1988) *The Fertile Crescent 1800-1914: A Documentary Economic History.* Studies in Middle Eastern History. Oxford: Oxford University Press.

Streck, M.P. (2002) Zwischen Weide, Dorf und Stadt: Sozio-ökonomische Strukturen des amurritischen Nomadismus am Mittleren Euphrat, in *Baghdader Mitteilungen,* Vol. 33, pp. 155-209.

Upton, R. D. (1873) *New Market and Arabia.* An Examination of the Descent of Racers and Coursers. Original London. Repr. 1987 Hildesheim: Georg Olms Verlag.

Van Dusseldorp, D.B.W.M. (1971) *Planning of Service Centres in Rural Areas of Developing Countries.* Publication

15. Wagenigen, Netherlands: International Institute for Land Reclamation and Improvement.

Wirth, E. (1971) *Syrien: Eine Geografische Landeskunde.* Wissenschaftliche Landeskunde. Band 4/5. Darmstadt: Wissenschaftliche Buchgesellschaft.

15. A WINDOW TO "LIVING ARCHAEOLOGY": BEDOUIN CAMPS AND VILLAGES

Minna Lönnqvist, Markus Törmä and Kenneth Lönnqvist

Fig. 1. Ethnoarchaeological surveying at the village of Shanhas. Photo: Taija Turunen 2000.

15.1. Ethnoarchaeology among Pastoral Nomads

Ethnoarchaeological documentation and observations were further applied to our project design in order to scrutinize present-day pastoral nomadic lifestyle among the local Bedouins and its relation to the village life in the Jebel Bishri region. Our work has aimed to highlight and scrutinize further those factors that have shaped nomadic life in the region today, and perhaps, in the past as well. The studies of C. Kramer (1979), F. Hole (1978), Ø.S. LaBianca (1990), R. Cribb (1991a, b) and I. Finkelstein (1995) have generally provided important clues concerning the significance of ethnoarchaeology in studying and understanding nomadic sites and their development in the Near East.

In the 1970s the purpose of the research by F. Hole (1978) in Iran was to identify and categorize various forms of pastoral dwellings and known tribal territories. O. Aurenche, P. Desfarges (1983) and R. Jarno (1984) have also carried out ethnoarchaeological studies at an earlier point among the sedentarizing nomads at the semi-sedentary village of Qdeir (Fig. 9.) on the western piedmont of Jebel Bishri. (Interestingly, as mentioned in ch. 6., the archaeological evidence from Qdeir, like that of El Kowm-2 Caracol, among the oases of the western piedmont, has helped to clarify the earliest transition to nomadic life in the region during the Pre-Pottery Neolithic B phase.)

In 1988 S.R. Simms published his archaeological studies on Bedouin camp structures among the Bedul Bedouins of Petra in Jordan paying attention to the previous pioneering work by E.B. Banning and I. Köhler-Rollefson (1983) in the neighbourhood. Less well-known is an important study which N. Goren Inbar and I. Hershkovitz (1990) carried out in Southern Sinai in order to study Bedouin dwellings. Analogies were applied to the evidence concerning Pre-Pottery Neolithic B sites in the Sinai region. Special attention was paid to aspects of camp locations, their spatial organisation, and the number of tents in a camp, their intra-camp density and communal characters. As expressed earlier, C. Chang's (1992) approach to the pastoral landscape from an ethnoarchaeological point of view has also brought to light valuable points regarding the spatial arrangement of nomadic areas. Chang takes a view that the spatial organization of pastoral systems existing on the contemporary landscape provides a basis for investigating how pastoralism may be rendered visible in the archaeological record.

Understanding the *sedentarization of nomads* and possible *renomadization* process is a key issue regarding the changing pastoral life in the region of Jebel Bishri. With respect to our ethnographical observations we need to deal with the concept of sedentarization to some extent. It does not have to mean any final state, but can concern "any change in the long continuum from nomadism to sedentism" (Salzman 1980: 10-11). Ø.S. LaBianca (1990: 33) defines sedentarization from the subsistence point of view emphasising food system intensification, while nomadization is to be seen as food system abatement. In our study of the cycle from nomadism to sedentism and *vice versa*, special attention has been given to tracing the visible signs of *sedentarization*.

In our study of the sedentarization and renomadization *the phenomenon of abandonment* comes into focus. M. Schiffer (1987: 89-120), a developer of behavioural archaeology, has especially analysed the nature of different processes of abandonment and their influences on the formation of archaeological record, and his studies have provided stimulus to our inquires. Schiffer (1987: 89) states: *"Abandonment is the process whereby a place, an activity area, structure, or entire settlement is transformed to archaeological context."* C.M. Cameron (1993), S.A. Tomka (1993), L. Horne (1993) and G. Davis Stone (1993) among others have further studied the behaviour of abandonment especially applying ethnoarchaeological approaches to site and settlement analysis in nomadic areas in arid zones also observing the disappearance of agricultural areas. Apart from site-centred analyses, they have observed the *scales of abandonments from abandonment of structures and activity areas to settlements and regions* (Cameron 1993: 4). Our Finnish project SYGIS has also acknowledged that there is a need for understanding the dynamic forces behind changing habitation behaviour in marginal environments. The approach applied with ethnoarchaeology in nomadic and cyclic contexts not only reveals formation processes, but also shows the rationales for different kinds and scales of abandonments. A tent site abandonment usually leads to the continuation of camp movement or sedentarization in other areas.

Abandonments leave certain kinds of evidence on tent bases, house floors and courtyards depending on the nature of the abandonments, whether or not they are *planned seasonal short-term abandonments* or *planned final abandonments*. It is known that sudden catastrophes such as war, famine and environmental hazards leave most complete archaeological records of past activities on house floors. These are cases in which people do not have time to plan their escape, and if they are able to leave the site they may not return. Larger quantities of usable objects, the so-called *de facto refuse*, are then left behind. *Curate behaviour*, the removal of usable objects, takes place in the cases of planned and final abandonments. In mobile societies caching and storage at sites reflect planned departures and anticipated returns. (Passim Schiffer 1987). However, sometimes in the archaeological record the caching and storage of valuable items reflects that even if the departure was planned, nobody was able to return.

Answers to the site abandonment phenomenon were sought at the ground level by recording and documenting site structures and usable materials left behind by local Bedouins, with an anticipated return in mind or with the purpose of a final abandonment. Informants have been critically used where possible. We have concentrated on the study of layouts, spatial organisation and functions of camps and tents as well as houses, courtyards and villages. The types of usable materials and items were recorded. From recorded and documented data we have been trying to find patterns of sedentarization and reasons for either planned site abandonments or final site or regional abandonments in the area of Jebel Bishri.

The movement of the desert line affecting the subsistence economies of the local people in the district has been taken into account from a satellite perspective. Examining changes in agricultural behaviour has included the use of large-scale satellite image data, as well as GIS modelling of agricultural fields. Shifting cultivation and changes in seasonal agricultural fields in *wadis* bring about larger changes in landscapes and often lead to an environmental change (see Lönnqvist and Törmä 2005; Lönnqvist *et al.* 2010). The current situation provides means to understand ancient nomadic life and the choices in the context of the shared environment of the Levantine arid zone (cf. Hole 1978).

15.2. From Tents to Village Houses on the Western Piedmont

15.2.1 Abandoned Bedouin Tent Camps and Compounds

A. Musil (1928: 81) recorded in his travel accounts that at the beginning of the twentieth century there was a garden enclosed by a wall and the whole area of Qasr al-Hayr ash-Sharqi could have been cultivated – at least by irrigation. Interestingly, O. Grabar *et al.* (1978: 2), who conducted excavations at the Qasr in the 1960s and 1970s mention that a few Bedouin tents were surrounding the castle at the time, but there was only one well of 90 m in depth with brackish water to supply the Bedouins in the area. The CORONA satellite photographs from June 1967 reveal that at that time the western piedmont around Qasr al-Hayr was still a relatively good grazing area (see ch. 14: Fig. 6.) as a number of black tent, *beit shar,* camps and *wadis* with recent large torrents of water are visible. The photographs also clearly show that environmental conditions were more favourable to start shifting cultivation in the *wadi* areas in the 1960s compared to the present-day situation (see ch. 14: Fig. 5.).

Numerous camps of black tents visible in the CORONA photographs of the 1960s were not any more traceable on the ground in the neighbourhood of Qasr al-Hayr in the early spring (April) and summer (July) of 2000, when we worked in the area. However, the fact that spring is the grazing time at higher altitudes, has to be taken into account, and this fact could have been the reason for the tents being void in the area during our survey. On the other hand, the further study of satellite images from the 1990s revealed that this was not the only reason but instead the environment had drastically changed during the past decades making the neighbourhood too dry for being a good grazing ground any more (Lönnqvist and Törmä 2004; 2006).

On the western piedmont near Qasr al-Hayr and at Tar al-Sbai at the western edge of Jebel Bishri we identified *abandoned tent bases,* the locations of which were recorded in 2000. The resolution of our study did not aim to close documentation and analyses of the spatial distribution of left items or refuse on the tent floors. Earlier more detailed abandoned tent site documentations have been done, for example, by R. Cribb (1991) in Turkey, I.E. Banning and I. Köhler-Rollefson (1992) in Jordan and P. Goldberg with I. Whitbread (1993) in Israel. B.A. Saidel (2001) and C. Palmer with P. Daly (2006) also carried out somewhat comparable studies in Jordan contemporaneously with our studies in Syria. P. Goldberg's and I Whitbread's studies in Israel concentrated on "living floors" sampling hearths, storage areas producing small finds, a sleeping area and an abandoned animal pen. Micromorphological analyses were used to distinguish activity areas in soil. Chemical analyses of tent base sediments have also been executed by J.P. Grattan, S.I. Huxley and F.B. Pyatt (2003) in Jordan to determine the level of health risks caused by copper to which the Bedouins were exposed.

Sites A 2 and A 3
Tent Bases on the Western Piedmont

A 2: tent base and hearth
UTM: 0506062, 3881369
Size of the tent base: 13.7 m x 12.2 m;
hearth: 3 m in diam.

A 3: two tent bases and hearth
UTM: 0506078, 3881413; ca. 390 m a.s.l.
Sizes of the tent bases: a) 10.9 m x 5.2 m and
b) 8 m x 8 m; hearth: 3 m in diam.

Fig. 2. An abandoned tent base near Qasr al-Hayr ash-Sharqi. Photo from S: Minna Lönnqvist 2000.

In 2000 abandoned tent sites **A 2** and **A 3** were identified situating ca. 500 m north of Qasr al-Hayr. The area was silty sand. A 2 consisted of one rectangular feature of lighter colour marking the area of a tent base on the ground (size: 13.7 m x 12.2 m in plan). The longer side was facing north. Outside the northeastern corner there was a large round pit, 3.3 m in diameter, full of dung surrounded by yellow camomile vegetation. It seemed to have served as a place for a hearth. B. Saidel (2001: 155) has classified these kinds of hearths as "basin hearths" differing from the "surface hearths", because as pits they penetrate into sediments. Our hearths outside the tents could also have served as hearth refuse dump (cf. Simms 1988: 205). A shoe and a metal barrel had been left behind at site A 2.

A 3 consisted of two rectangular abandoned tent bases (a. 10.9 m x 5.2 m in size and b. 8 m x 8 m in size). The tents had been set in such a way as to form an L-shaped courtyard, so that the southern one was orientated on the south-north axis, and the northern one stretched east-west in width. Tent pole supports which consisted of piled tiles and pieces of large *pithoi* (see Fig. 2.) were found on the tent bases. A round pit filled with dung, 3 m in diameter, was also here set on the northern flank of the northern tent. Shoes, remains of a reed mattress and clothes had been left behind. Already S.R. Simms (1988: 208) noted that when a camp is abandoned larger and still usable objects had been carried away, and the remaining debris is shallow. The sizes of our camp sites A 2 and A 3 were comparable to the space to accommodate five to eight people – a family;

compounds like those studied by Simms (1988: 201) in Petra, Jordan.

As J. Yakar (2000: 202) has noted nomadic pastoralists prefer locations of historical sites for their camps like those located in the surroundings of Qasr al-Hayr. Nomads reuse materials from the ruins for construction. Also, already in the beginning of the twentieth century S. Guyer (*apud* Sarre and Herzfeld 1920: 1) observed that the Bedouins of the 'Anazah tribes (see more in the previous chapter) had pitched their tents in the midst of the Byzantine ruins of Resafa-Sergiopolis on the western piedmont of Jebel Bishri.

Cairns E 1 and Tent Base A 31

Site E 1: UTM: 0517927, 3907055
Size: unrecorded

Site A 31: UTM: 0518003, 3905149; ca. 704 m a.s.l.
Size: 6.1 m x 7.1 m

In 2000 locations of abandoned tent bases were also recorded at Tar al-Sbai (Jebel Sabah), at the western edge of the mountain (see Fig. 5.). Above at the mountain edge there were cairns at site **E 1**. According to Muhammed Taha, our Syrian archaeologist, the tombs belonged to the Sba'a tribe. From that location several abandoned tent floors were visible on the plateau beneath. Islamic graves were further encountered (see ch. 8, Table 1.) at the edge as well.

Site **A 31,** providing a tent base, was composed of stones set apart from each other. The stones (30-40 cm in diam.) had apparently kept the tent canvas with strings on the ground. There were an empty tin of *hummus* and a cloth that had been left behind.

Site C 2
Shepherd's or Hunter's Rock Shelters and A Tent Base

UTM: 0517971, 3906886
UTM: 0517925, 3907033

Sizes of the shelters: a) w. of the entrance 3.8 m, h. 1.4 m and depth 1.7 m and b) w. 4.5 m and h. 1.4 m; tent base unrecorded.

Fig. 3. Shepherd's or hunter's shelter at site C 2. Photo from W: Margot Stout Whiting 2000.

At site **C 2** we recorded apparent Bedouin rock shelters (Figs. 3. and 4.), naturally formed in porous limestone. Limestone boulders had been piled to form a protecting wall at the entrances of the both shelters. They appeared in association with a hunter's blind (see paragraph 5.2.5). A small tent base situated 20 m east of the shelters. It had large stones set apart to keep the canvas on the ground. Apart from the prehistoric flints, here no associated items were found.

Grazing of dromedaries (Fig. 4.) was observed beside small livestock rearing still in July (see a video clip: www.helsinki.fi/hum/arla/sygis) in these higher areas at the edge.

Fig. 4. Shepherd's or hunter's shelter at site C 2 and dromedaries grazing at Tar al-Sbai. Photo from S: Kirsi Lorentz 2000.

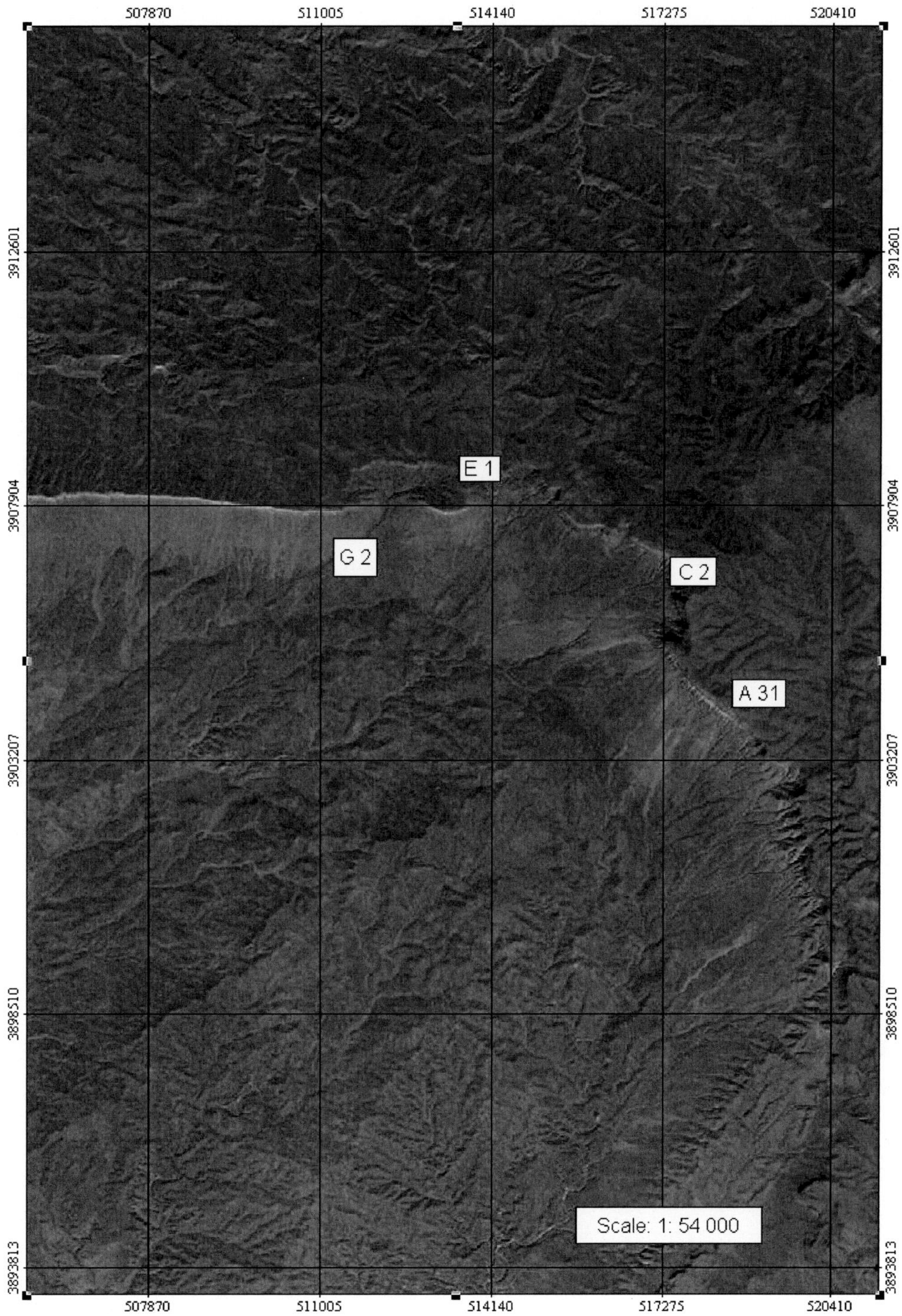

Fig. 5. Abandoned Bedouin sites at Wadi Sbai and Tar al-Sbai. Mapping by Minna Lönnqvist 2010.

Site E 5
Well and Bedouin Compound

UTM: 0507410, 3886173, ca. 500 m a.s.l.
Size: unrecorded

In 2000, among the sites on the western piedmont north of Qasr al-Hayr ash-Sharqi, a suspended well and several water cisterns were identified at the Bedouin compound called Shaharidz al-Wahab. The site was near the Roman/Byzantine tomb **B 1** (see ch. 11.). According to Mr. Muhammed al-Asa'ad, the representative of the Palmyra museum surveying with us, there were 20-25 water cisterns in the area. Each Bedouin family used one cistern.

Site G 2
Well and Bedouin House

UTM: 0512292, 3907189
Size: unrecorded

Below Tar al-Sbai on the western plateau in Wadi Sbai a well site called Be'er Sbai, site **G 2** (Fig. 5.), was also identified. The *wadi* and the well both carry the name of the Sba'a tribe, to whose territory the western piedmont belongs, as far as to the border zone of Tar al-Sbai (the Circle of the Sba'a tribe). Beside the well this site consisted of an abandoned L-shaped Bedouin house built of field stones and concrete, a square animal pen and a covered water cistern.

Fig. 6. The village of Shanhas. Photo from NW: Minna Lönnqvist 2000.

15.2.2 Shanhas - a Pastoral Village of Sedentarizing Nomads

The Bedouin village of Shanhas (the name meaning "Thirsty" in Arabic), site **A 6** (Fig. 6.), on the western piedmont was targeted for closer ethnoarchaeological study on the ground in 2000. The village is situated 1.4-1.6 km northeast of Qasr al-Hayr ash-Sharqi on a limestone outcrop (ca. 400 m a.s.l.). It turned out that the village was well applicable to study the processes of *sedentarization* and *abandonment*. We were able to observe the process of sedentarization through the layout of the village and its courtyard compounds. Because the pastoralists of the village were on their seasonal grazing grounds, we had an opportunity to study the process of *seasonal abandonment* and its effects on storage and site furniture. Effects of seasonal abandonment have earlier been studied, e.g., by S.A. Tomka (1993) among transhumant pastoralists. At the village of Shanhas our field studies and recording were also aimed at understanding the facilities that had been taking place and storage. S. Kent (1989) and A. Rapoport (1989), for example, have paid attention to the use of space in domestic architecture and variables which affect the chosen activities and house designs.

We have approached the degree of sedentarism at Shanhas by studying the CORONA satellite photographs from the 1960s and the village layout in Landsat-7 ETM as well as SPOT images from the 1990s and 2000. This information was associated with the empirical recording and documentation on the ground. The current layout of the village is visible in the Landsat images as well as in the SPOT images (ch. 5, Fig. 3; here Fig. 7.). The layout is open; fenceless courtyards have been set apart from each other. In the case of Shanhas there are no developed established dwellings within the closed courtyards. On the mountain instead, at the village of Ash-Shujiri, the houses often had closed courtyards, even if these courtyard houses had often also been set relatively far apart from each other.

A survey in the Fars province in Iran has distinguished the comparable open plan with a low settlement density, being closest to the nomadic camp type of plan in the development towards a compact village. (Cribb 1991a: 386). The plan of Shanhas still follows the plan of a nomadic camp (Fig. 7.), which can be seen as an initial settlement of sedentarizing nomads comparable to the first stage of development towards a sedentary lifestyle described by R. Cribb (1991a, see also our Fig. 8.). Cribb further states that campsites can develop into composite settlements and ultimately into villages. The process of infilling and joining up courtyards to form a dense agglomerative plan can be detected in the villages of sedentarizing nomads in Syria.

Only Mr. Hussein Ahmed Ruejah, the guard of the village, was present in the summer season of 2000 at Shanhas, and he served as our ethnographic informant. He was interviewed about the village and its life by Minna Lönnqvist on the 28th of July. Mr. Ruejah lived on the western edge of the village with his family and was originally from the village of Taibe (see more in ch. 14.). According to Mr. Ruejah, the people of Shanhas had left the village three months before in April/May heading to their seasonal grazing grounds on the Jezira. They were to turn in October, when the children's school was to be opened. This is the yearly cycle in which winter is spent in sojourn in the village in the piedmont that belongs to the desert-steppe currently under a heavy progress of desertification. The cycle differs from the neighbouring village of Qdeir which is occupied by pastoralists only for a few months of the year during the summer season (Jarno 1984: 191).

According to Mr. Ruejah, the village of Shanhas was founded in 1960. The foundation and the construction of houses are contemporaneous with the nearby village of Qdeir (cf. Jarno 1984: 191). In earlier times the inhabitants of Shanhas used to live solely in tents. Like at Qdeir the continuity of nomadism is still visible in the use of tents in the courtyards of the village (Fig. 9.). Today Shanhas consists of members from three Bedouin tribes: the Sba'a, the Shiat and the Amarat tribes, which belong to the 'Anazah federation (see more on the tribes in ch. 14.). These tribes currently enjoy good relations with each other. Like Shanhas, the inhabitants of Qdeir also consist of three Bedouin tribes, but beside the Sba'a at Qdeir there are members from the tribes of the 'Badat and the 'Fadli (Jarno 1984: 201). Therefore, the Sba'a is forming the only common tribal component in both villages. The origin of the people from Shanhas is directly traceable to Eastern Syria, and the members of the Amarat tribe come from the flooded Euphrates valley.

Fig. 7. The plan of Shanhas as seen on a SPOT image.

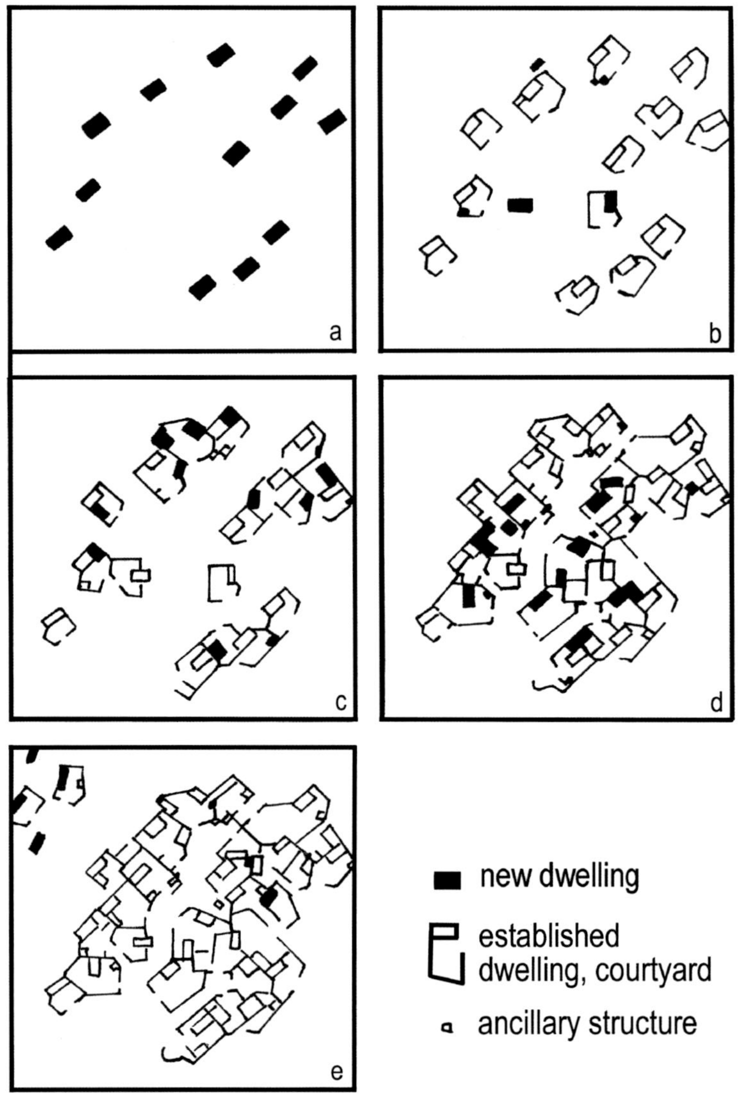

Fig. 8. R. Cribb's (1991a) scheme of the sequence in the development from a tent camp to a permanent village. In the scheme the stage a) applies to Shanhas. Redrawn by Minna Lönnqvist after Cribb 1991a: 387, Fig. 7.

The land of Shanhas is owned by the government. The water is brought in tanks from the neighbouring village of Taibe. The main economy is sheep and goat rearing in addition to which there is some wheat cultivation. This accords with the background and livelihood of the pastoralists in the nearby village of Qdeir as well (Jarno 1984: 201). In 2000 there lived 40-41 families at Shanhas headed by one to two chiefs. In a SPOT image there are approximately 40 open courtyards visible at Shanhas (Fig. 7.), which indicates that each family had its own courtyard. The layout of the courtyard compounds follows the local tent compounds, which are not set into a circle or lines but rather in the shape of an L. This courtyard plan was also observed among the abandoned tent compounds on the western piedmont and active tent compounds on the mountain. Therefore the camp layouts of the 'Anazah federation in the area of Jebel Bishri are nowadays quite simple compared to earlier times (cf. paragraph 14.4.1).

There was a school with 22-23 pupils at Shanhas. Both girls and boys went to school: the classes were mixed, and there were two teachers for them. There were no police or a doctor in the village. The men of the village were mainly rearing their flocks. There were 100-200 sheep in each family. The families also had donkeys and dogs which were used in rearing the flocks. Women milked the sheep, prepared *lebane* yoghurt, baked bread and collected truffles for sale. Some women took care of tomato orchards situated in the nearby village of El Kowm. Grown-up boys went into the army or to work as foreign labour in Lebanon.

In the year 2000 there had been a real shortage of rain, and that had actually also been the case during the previous two years. If the drought was to continue in 2000 the inhabitants of Shanhas were going to stay in the Jezira in the Habur region in their tents, rearing their flocks during the coming year. But when our project representative visited the village in June 2003, rain had returned and was plentiful. The inhabitants still stayed in their village, because the rain had allowed the grazing to continue in summer. In this case dry years opened a way to renomadization in other areas, and good rainy years for prolonged sedentism. The situation was clearly different from the case of the Agedat family living in the village of Shummatiyah on the Euphrates side (see ch. 14.) – in that case drought on Jebel Bishri led the family to stay in their village on the Euphrates side throughout the year 2008. It is interesting how drought and its impact on both movement and arrest are closely tied up with the availability of water in particular areas.

The degree of nomadism of the *transhumant* pastoralists at Shanhas is clearly semi-sedentary as the tribes occupy the village part of the year. However, according to the CORONA satellite photographs taken during the years 1967 and 1968, Mr. Ruejah's reference to the 1960 foundation year for the village seems to refer to a very early stage, as no permanent layout is visible. In any event, the information of the time of the establishment is in accordance with the general government policy of the 1960s in Syria and the contemporary establishment of Qdeir.

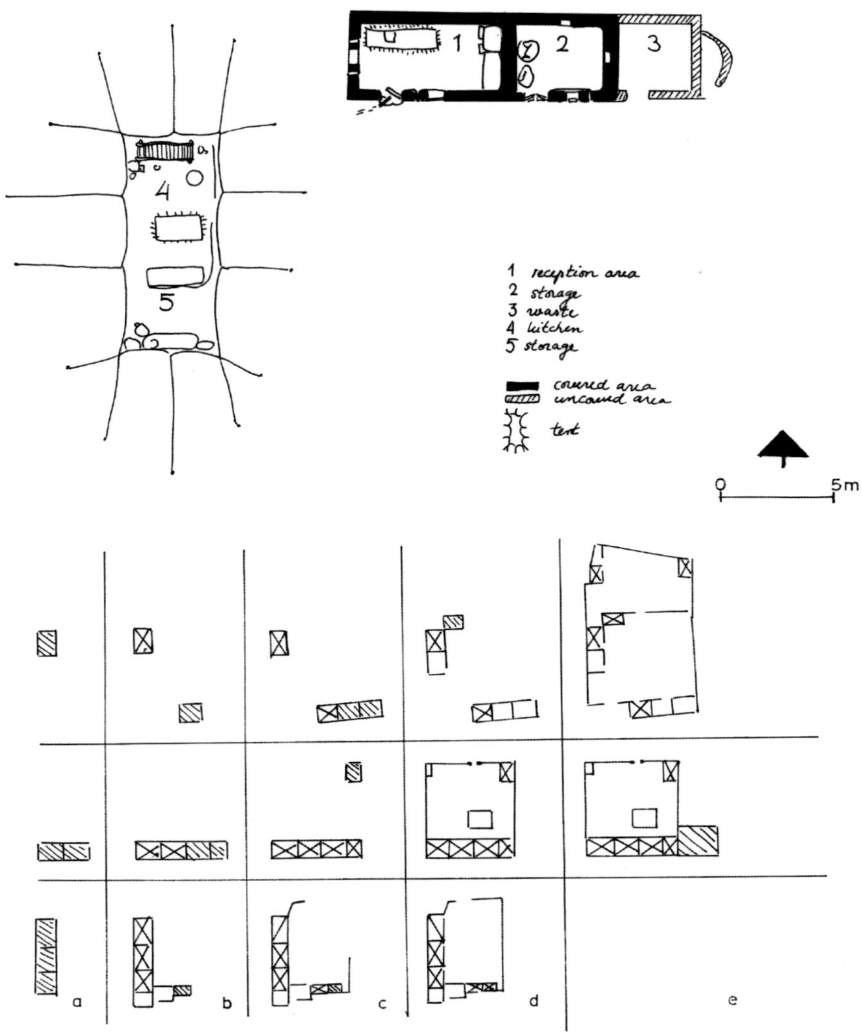

Fig. 9. The application of a rectangular tent layout into sedentary structures at the village of Qdeir on the western piedmont of Jebel Bishri. The courtyards of Shanhas represent the stages visible in sections without fences. Redrawn by Minna Lönnqvist after Aurenche and Desfarges 1983.

15.2.3 The Recorded and Documented Courtyard Compounds at Shanhas

We designated the village of Shanhas as site **A6**. Beside the recording and documentation of a few family courtyard compounds at the village we recorded (Fig. 10.) a school building and an unfinished possible mosque. The village is separated into the eastern and western parts by a *wadi*. Because of the time constraints typical courtyard compounds were chosen for recording and documentation.

The houses of Shanhas had been constructed of field limestones fitted together with *terre pisée* and roofs were made of wooden beams. The houses generally had one to two metal doors, in some cases with beautiful decorations (see Fig. 39.). Windows were rare. The plan is a rectangular *Breitraum*, doors set on the longer side. The simple rectangular plan of the houses follows the black tent as also regards the interior composition and furnishing of the houses. The nomadism of the inhabitants is therefore still clearly visible in the house designs. The houses did not generally have any painting (only in one case we saw a painted interior) or porches with screen of columns *in antis* like the houses of the Euphrates side (cf. ch. 14.). Only a few houses at one courtyard of Shanhas had more features in their façades, this distinction apparently signifying status, such as a chief's compound. For example, A. Rapoport (1989: 10) has stated that the built form can be related to family structure, clans, moieties, religious institutions, sex roles and status hierarchies.

The recorded features of the courtyards at Shanhas were as follows:

The Western Part of the Village
A 6a
Bedouin Compound

UTM: 0507533, 3882331
Size of the plan: max. ca. 70 m x 50 m.

The recorded units (see the plan in Fig. 11. and the structures as well as items Figs. 12.-22.) are:

1) Three rectangular houses (a. 9.1 m x 4.6 m; b. 8.6 m x 4.6 m; c. 10.4 m x 3.5 m)
2) Kitchen (2.9 m x 3.2 m)
3) Animal pen (13.2 m x 8.5 m)
4) Tent base (16.3 m x 6.5 m)
5) Two cylinder-shaped silos (1.8 m in diam.)
6) Two water tanks (2.1 m x 2.1 m each)
7) Three package blocks (ca. 3.5 m - 4.7 m in length)

Fig. 10. Recording and documenting the courtyard compound A6a in 2000. Photo from NE: Sanna Aro-Valjus 2000.

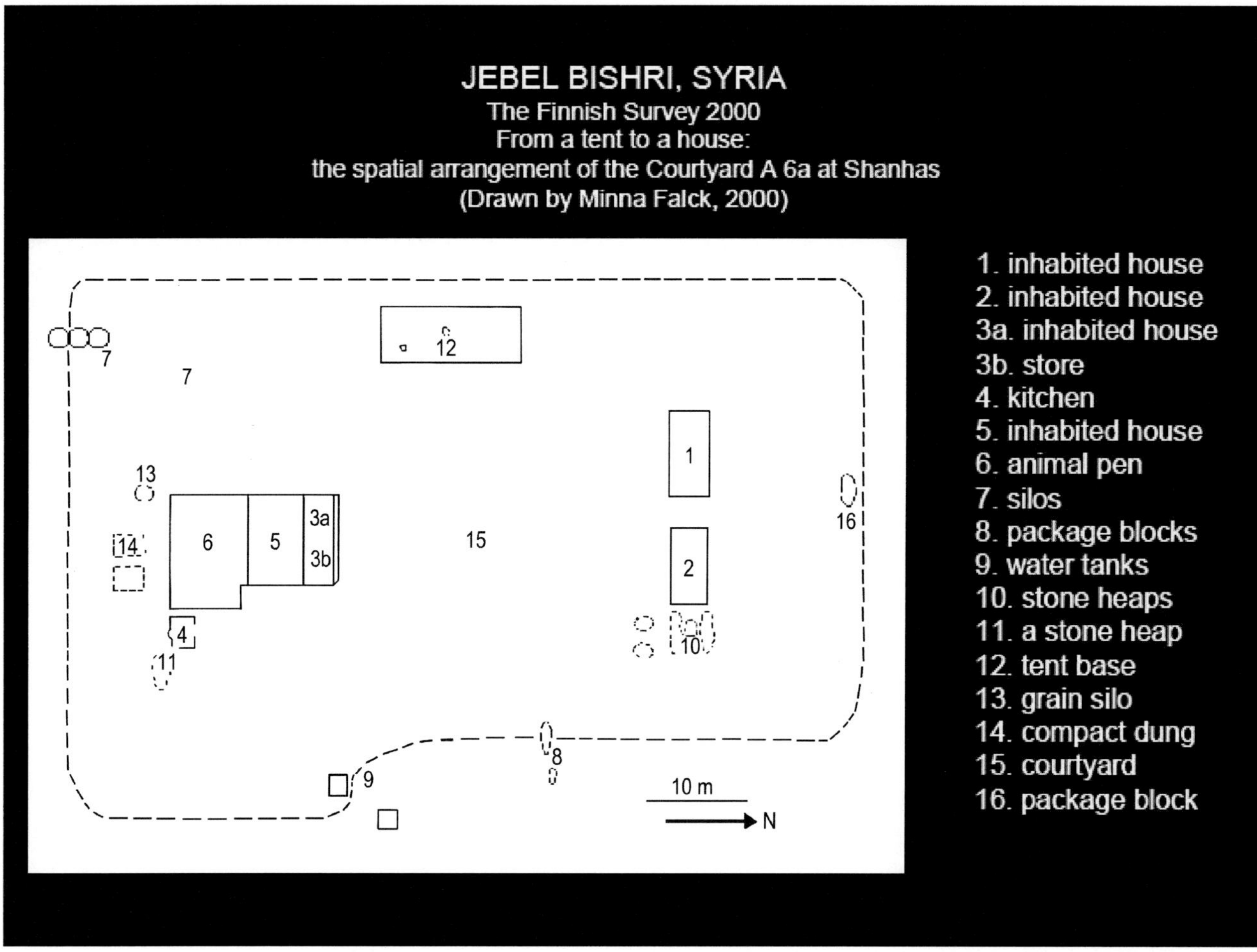

Fig.11.

Fig. 12. House No. 1 (plan 9.1 m x 4.6 m) at the courtyard compound of A6a. Photo from S: Sanna Aro-Valjus 2000.

Fig. 13. House No. 2 (plan 8.6 m x 4.6 m) at the courtyard compound of A6a. Photo from S: Sanna Aro-Valjus 2000.

Fig. 14. House No. 3 in the courtyard compound of A6a. Rolled carpets, blankets and pillows were visible inside the rooms. Photo from N: Sanna Aro-Valjus 2000.

Fig. 15. House No. 4 served as a kitchen at the courtyard compound A6a. Photo from N: Sanna Aro-Valjus 2000.

Fig. 16. Animal pens No.6 (overall plan 13.2 m x 8.5 m) at the courtyard compound A6a. Photo from NE: Sanna Aro-Valjus 2000.

Fig. 17. The silos of the courtyard compound A6a. See No. 7 in the plan. Photo from S: Sanna Aro-Valjus 2000.

Fig. 18. Collected bushes next to an oven at the courtyard compound A6a, No. 13 on the plan. Photo from SE: Sanna Aro-Valjus 2000.

Fig. 19. A tent-base at the courtyard compound A6a, No.12 on the plan. Package blocks and Qasr al-Hayr ash-Sharqi in the background. Photo from SE: Sanna Aro-Valjus 2000.

Fig. 20. Package blocks at the courtyard compound A6a. No. 8 on the plan. Photo from S: Sanna Aro-Valjus 2000.

A 6b
Bedouin Compound

UTM: 0507630, 3882663
Size of the plan: ca. 37 m x 60 m.

The recorded units (see Figs. 23.-30.; also designs of iron doors) are:

1) Three rectangular houses (a. 14.5 m x 5 m; b. 8 m x 4.5 m; c. 9.1 m x 4.7 m)
2) A circular kitchen hut (max. 2 m in diam.)
3) A sheep pen (ca. 8 m x 10 m)
4) Three tent bases (5 m x 14 m; 4.9 m x 15.9 m and 4.9 m x 11.5 m)
5) Storage blocks rolled and covered with sack canvas and robes
6) Four cylinder-shaped silos (size of each 1.3 m in diam.) sealed (with clay plaster from the top)

Fig. 21. Two cubic metal water tanks (sizes 2.1 m x 2.1 m) at the courtyard compound A6a. No. 9 on the plan. Photo from W: Sanna Aro-Valjus 2000.

Fig. 22. Building material at the courtyard compound A6a. No. 10 on the plan. Photo from W: Sanna Aro-Valjus 2000.

Fig. 23. House 1 with a green door from the courtyard compound A6b. Photo from S: Juhana Saukkonen 2000.

Fig. 24. House 2 (plan 14.5 m x 5 m) at the courtyard complex A6b. Photo from N: Juhana Saukkonen 2000.

Fig. 25. House 2 (plan 14.5 m x 4.5 m) from behind at the courtyard compound A6b.
Photo from S: Juhana Saukkonen 2000.

Fig. 28. The kitchen hut of the courtyard compound A6b from behind. Note a bread baking tray to the right on the ground.
Photo from S: Juhana Saukkonen 2000.

Fig. 26. House 3 at the courtyard compound A6b.
Photo from N: Juhana Saukkonen 2000.

Fig. 29. The animal pen of the courtyard compound A6b.
Photo from W: Juhana Saukkonen 2000.

Fig. 27. The kitchen hut of the courtyard compound A6b.
Photo from N: Juhana Saukkonen 2000.

Fig. 30. The silos of the courtyard compound A 6b.
Photo from E: Juhana Saukkonen 2000.

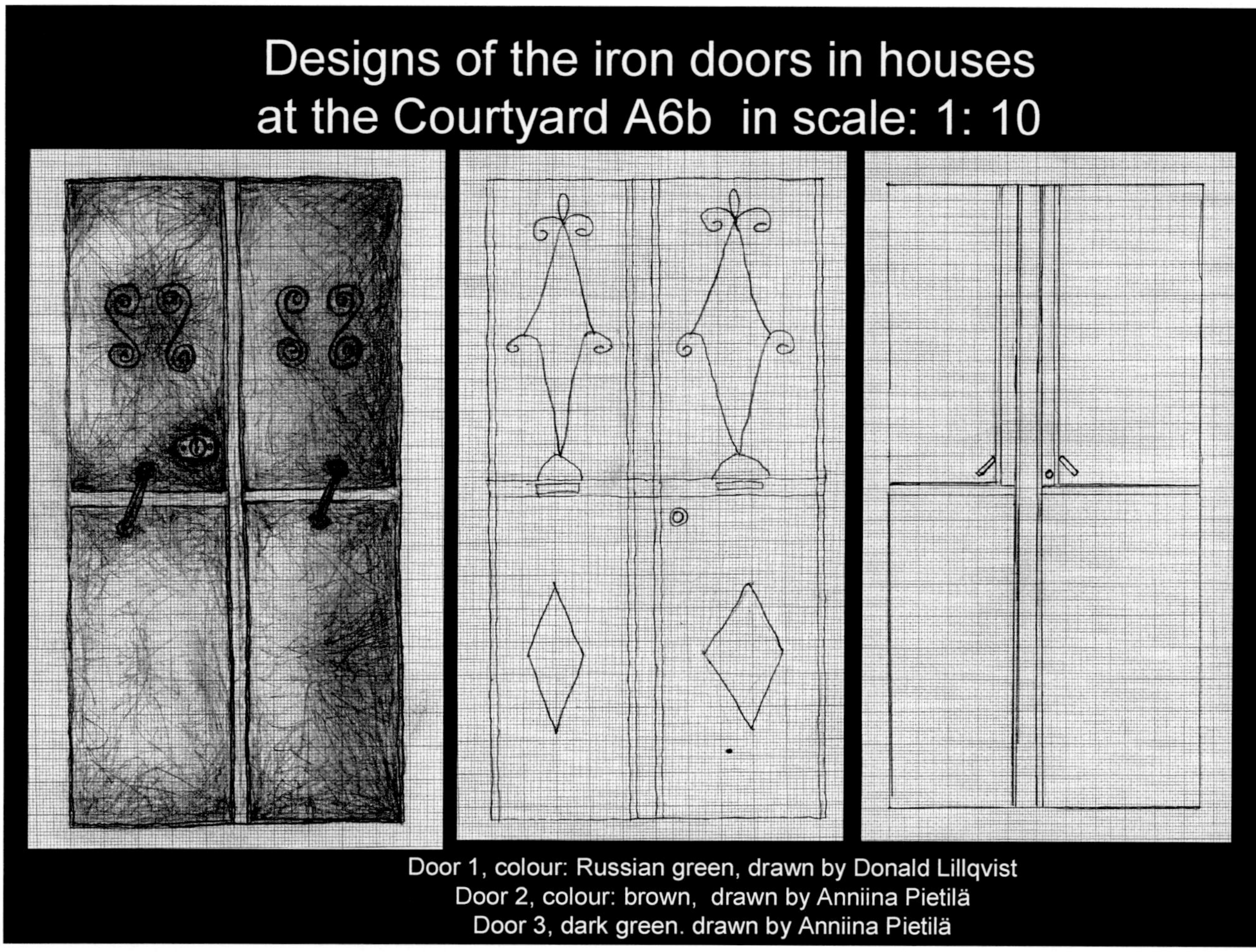

Fig. 31. Iron double doors of the houses at the courtyard compound A6b were documented by drawing to illustrate the designs of local craftsmen.

A 6c
Bedouin Compound

UTM: 0507724, 3882581
Size of the plan: unrecorded

The recorded units are:

1) A house (Fig. 32.; 4.65 m x 13.8 m, h. 3.4 m), with two doors, one wooden and the other of iron
2) A sheep pen (13.5 m x 6.6 m, h. 1.0 m) constructed from angular field stones fitted together with *terre pisée*, and nine black rubber *guphas* in the corner
3) Package blocks wrapped in brown poplin were set lengthwise on half-cut metal oil barrels, to keep the blocks out of the ground; stone weights had been set on the blocks to keep the material in place

Fig. 32. House at the courtyard compound A6c. Photo from S: Juhana Saukkonen 2000.

The Eastern Part of the Village

Site A 6d
Storage and Activity Area

UTM: 0508041, 3882461
Size: unrecorded

The recorded units are:

1) 17 blocks of tent packages
2) A heap (4 m x 5 m) of dry bushes for fuel
3) Four cylinder-shaped silos (max. 5.2 m in diam., h. 1.6 m)
4) A small house for poultry (1.25 m x 1.05 m, h. 0.6 m) built of angular field stones fitted together with clay
5) A cooking hut (11.7 m in diam.), cylinder-shaped, narrowing upwards with a small oven and zinc trays for baking bread left behind
6) Building material (rectangular stones) in a heap (21 m x 13 m)

Site A 6e
Unfinished Mosque?

UTM: 0508041, 3802467
Size of the plan: 21 m x 13 m

We also encountered a building **A6e** under construction (Fig. 33.) in the central area of the village. It is apparently a mosque. It had a rectangular plan (21 m x 13 m), and it had been constructed of evenly dressed limestone. It had eight windows 1.00 x 1.05 m in size. A pair of iron doors with iron ring-handles on the longer wall entered from the north. The surroundings of the door and the windows were made of concrete blocks. Opposite the door inside the building, there was a niche (*mihrab*?) which could be approached by two symmetric staircases going up on both sides of the niche. Above the niche there was a window.

Site A6f
School

Coordinates: unrecorded
Size of the plan: 6.20 m x 19.26 m.

A school **A6f**, the layout of which was rectangular (6.20 m x 19.26 m in plan, Fig. 34.), situated in the centre of the village. The building was plastered all over with cement and was probably built with blocks of concrete. There was a pair of iron doors on the longer side and a platform of cement in front.

Fig. 33. An apparent unfinished mosque at Shanhas. Photo from S: Margot Stout Whiting 2000.

Fig. 34. The school of Shanhas. Photo from NW: Margot Stout Whiting 2000.

As presented here, a tent was still a vital mode of shelter for the people of Shanhas. Like at the village of Qdeir (see Fig. 9.), the tents had been surmounted next to the houses to provide additional sheltered space. Hearths were identified in several tent bases at the courtyards, and it is apparent that tents also served as cooking facilities. R. Cribb (1991a: 386) has noted that sedentarizing nomads may often still use tents in their courtyards so that the village consists of both houses and tents in the stage of a developing village. As described above, tents had been collected and stored in material blocks covered by sack canvases and tied with robes to wait for the return of the inhabitants from the grazing period. They were set on half-cut metal barrels to protect them from water currents. So, some usable tents had been left behind while others obviously had been taken to the seasonal grazing grounds on the Jezira.

Based on the documentation it is of special interest here to pay attention to facilities and storage areas which reflect the semi-sedentary lifestyle and the planned seasonal abandonment with an anticipated return in mind. Facilities or activities in general consisted of those associated with animal husbandry, agriculture and cooking. At Shanhas, beside houses and tents, the presence of the animal-based economy of the village is reflected in architectural remains such as pens and poultry houses (cf. Kramer 1979: 144). In one animal pen there were 9 black rubber *guphas* that had been left behind. They had been used to distribute fodder for animals.

The additional source of living was shifting cultivation. Agricultural activities were architecturally represented in cylinder-shaped silos at the site. They were meticulously sealed with clay. The storage of cereals was a precaution taken by the pastoralists, who had planned for the seasonal abandonment with an anticipated return in mind. Ovens in some huts imply their use as kitchens (cf. Kramer 1979: 147), and some of them had additional implements such as baking trays made of zinc or tin found *in situ*. There seem to have been special storage sites for each compound surrounding the courtyards, but then the courtyard compound **A6 d** also served as a larger area allocated especially for common storage. Dried bushes collected in heaps were probably used for fuel, although they may have also been used for producing potash to make soap in the area.

Apart from the mentioned recorded and usable items, rolled mattresses, carpets and pillows had been left in the houses in a way that suggests that return was anticipated. This kind of storage of bedding materials was also observed by S.R. Simms (1988: 206) among the Bedul Bedouins of Petra in Jordan. Material blocks with tents obviously included other items as well. However, it seemed that not all the usable items had been stored but some had been taken to the grazing grounds. This differs from the cases in which tranhumants cache entire assemblages (cf. Tomka 1993: 15-16).

F. Hole (1978: 152-153) has noted that when nomads carry out some farming, they have special constructions for storage, and grain is stored until the people return from their seasonal grazing areas like at Shanhas. In her studies S. Kent (1991: 39) has further pointed out that in the structure of the mobile hunter-gatherer sites, formal storage areas appear in the sites that have an anticipated medium- or long-term occupation. In her studies no storage areas appeared at any sites with short-term occupations. So, the mobility strategies, and degree of movement, clearly affect the will to construct and provide extra space with more storage facilities.

Fig. 35. A semi-sedentary extended family compound of Bedouins living between Tar al-Sbai and Tell Tletuat on the western part of the mountain, where our project camped in 2000. Photo from W: Jean-Pierre Matar 2000.

15.3. From Sedentarization to Regional Abandonment on the Mountain

15.3.1 An Extended Family Compound at Tar al-Sbai

In the year 2000 we stayed at a Bedouin compound near Tar al-Sbai, situating at the western edge of Jebel Bishri. We only mapped the location of this extended family compound on a satellite image (see the fields in Fig. 46.) not recording the coordinates *in situ*, and no measurements of the houses were taken in this living community (Lönnqvist *et al.* 2010). We, however, observed that some houses had been built of limestones fitted together with *terre pisée*, others had been built of cement bricks and covered with cement plaster. Women stayed at the compound in July while most of the men had left for grazing their flocks on the other side of the Euphrates on the Jezira. The spring and summer grazing grounds of this extended family were comparable to those of Shanhas. We were informed that these semi-sedentary pastoral villagers near Tar al-Sbai on the mountain had been told that the government had stopped the water supply for their tomato orchards, which had severed any agricultural activities in the region. This was obviously tied up with the year 1995 ban on agriculture in the vulnerable steppe areas by the government.

15.3.2 The Partly Abandoned Bedouin Village of Ash-Shujiri

Apart from the compound of Tar al-Sbai, most of the semi-sedentary village house compounds that we visited on the mountain in 2000-2008 had been permanently abandoned while tents were seen in abundance, especially in the eastern, southern and northern steppe high plateaus during the early and late spring of 2004.

As described in previous chapters, Ash-Shujiri is a village in a hilly central region (see ch. 12, Fig. 1.) of the mountain near the highest point of the area known as Jebel Al-Bishri. It is a former village of sedentarized Bedouins, now largely abandoned. The village generally consists of closed courtyard compounds, which we did not record or document but observed in our QuickBird satellite images and generally by "window screening" from cars on the ground. As already mentioned, the closed courtyards are scattered and separate. The distances between them can vary from a few dozens, to several hundreds of metres. As the name of Ash-Shujiri indicates (in Arabic meaning "tree"), one would expect trees, but there are hardly any natural trees except for a few acacias in the neighbourhood (see ch. 4, Fig. 16.). Recent plantations belong to the ACSAD program to prevent further desertification.

Fig. 36. An abandoned Bedouin houses near the village of Ash-Shujiri on Jebel Bishri. Photo: Minna Lönnqvist 2006.

Site G 3
Abandoned and Ruining Bedouin Compound

UTM: 0533438, 3915325, ca. 809 m a.s.l.
Size of the plan: unrecorded

About 3 km east of the Shula - Ash-Shujiri-crossroads on the mountain, there runs a road heading in the direction of Ash-Shujiri. On the way to the village we recorded the location of an abandoned Bedouin house compound registered as site **G 3,** consisting of three rectangular houses (Fig. 36.) built of irregular limestones fitted together with mud/*terre pisée*. Originally the houses seemed to have been covered with mud-plaster. Wooden posts held roofs filled with reeds inside (Fig. 37.) and covered by mud-rubble outside. The houses stood side by side oriented north-south, the entrances being towards the east. Doors were missing, and small round windows brought some light into the interior areas. Square animal enclosures were attached to each house at the northern end, and in the middle with one at the southern end. The most western houses had additional separate square animal enclosures as well. No usable items had been left behind indicating that the place had been permanently abandoned. A ruined *tannur* oven (Fig. 38.) was found serving as a fixture at the site (Fig. 38., see more on the ovens in Mulder-Heymans 2002 and Parker and Uzel 2007).

Fig. 37. One of the abandoned Bedouin houses at Ash-Shujiri from inside. Note the small windows and reed used in constructing the roof. Photo: Minna Lönnqvist 2006.

Fig. 38. A tannur oven of an abandoned house near the village of Ash-Shujiri. Photo: Gullög Nordquist 2004.

The process of gradual decay at the site had already set in. It is a generally known fact that when a house is not occupied and maintained, it is vulnerable to processes of decay and ruining. Collapsed roofs or missing doors and windows are clear signs of that the house is not occupied for continuous living any more, like in the case of our site G 3. Mud-brick/adobe walls and stonewalls seem to deteriorate and crack in different ways. In mud-brick a stepped crack takes place, while stone walls crack in a vertical direction. Doors or window frames tend to move from their original position. (Rotchild *et al.* 1993: 126).

15.3.3 An Abandoned and Ruined Village at Nadra

According to ACSAD, permanent livelihood is very difficult on the mountain, not only because of the lack of rain but also because wells and cisterns are not easily built in gypsum-based terrain. As described in ch. 4.-6., the valley of Nadra, the site of a palaeolake, on the eastern central high plateau has large seasonal *wadis,* and its landscape is more steppic than in the western parts of the mountain. Therefore the grazing potential of this area is clearly higher than at the western edge of the mountain which is under the progressive desertification.

Site I 1a-c
Abandoned and Ruined Bedouin Village

UTM: 0546295, 3911090; ca. 595 m a.s.l.
Size of the plan: ca. 60 m x 80 m.

Despite the steppe-type landscape an abandoned village compound was encountered at Nadra, the location of which was recorded at cross-roads where two desert tracks intersected at Nadra (see the plan in Fig. 40.). The size of the whole village was comparable to one courtyard compound of Shanhas, A 6. This site in the central eastern area was completely ruined (see Fig. 39.) differing from more recently abandoned houses of Ash-Shujiri. (The tracks crossing this Bedouin village may date from the Roman period as discussed in previous chapters.)

This small pastoral village consisted of altogether eight houses. The houses here were in a more deteriorated and ruined state than those at the Bedouin compound G 3 near Ash-Shujiri. Here, four houses were grouped together in the west, and four in the east, built in a rectangular, square and L-shaped plan. The houses had been built from angular limestones fitted together with mud or *terre pisée.* The number of rooms in houses varied from one to five. The compound area to the west of the crossroads is linked to a *wadi* that offers agricultural terraces and corrals, used as pens for animals. There were signs that at some point rain-fed agriculture may have been possible on seasonal basis at this village. There was a well (1 m in diam.) witnessing to the past availability of water (Fig. 41.).

Fig. 39. Ruined village houses at an abandoned Bedouin village at Nadra. Photo: Gullög Nordquist 2004.

A mill-stone (Fig. 42.) and an oven were discovered documenting bread baking in this small compound. 8 m north of the cistern there was an Arabic inscription on a limestone slab (see Figs. 40., 43.; 45 cm x 30 cm in size, with 5-7 cm high letters). An Islamic graveyard (Figs. 40., 44.) on the southern slope of a *wadi* consisted of ca. 100 tombs of elongated stone heaps and cairns (2.5-3 m x 1.5 m in size) allocated in parallel rows with small round cairns (0.75 m in diam.) indicating the size of the group that had lived in the area. The larger graves were set higher, with the smaller ones lower on the slope. It was clear that this Bedouin compound had been left for good a long time ago with no anticipated return in mind. Raiding or planned curate behaviour seemed to have taken place as no usable portable items had been left behind, only the fixtures such as a mill-stone and an inscription were discovered.

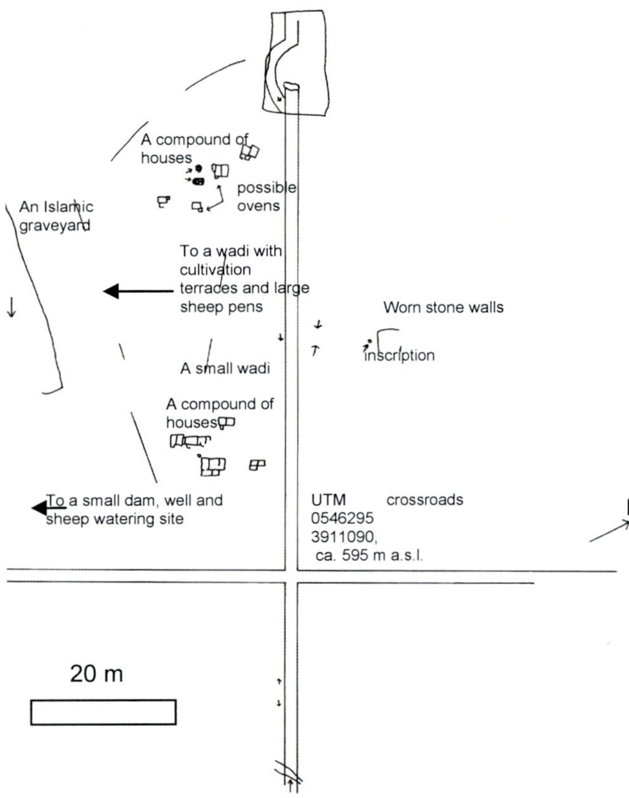

Fig. 40. A sketch plan of the village I 1. Drawing by Eivind Seland 2004.

Fig. 41. A well near some rectangular structures. Photo: Christian Meyer 2004.

Fig. 42. A mill-stone at the ruined village in Nadra. Photo: Gullög Nordquist 2004.

Fig. 43. An inscription at the ruined village in Nadra. Photo: Gullög Nordquist 200.

Fig. 44. An Islamic graveyard of the ruined village at Nadra. Photo: Christian Meyer 2004

15.4. Government Policies and Desertification Monitoring

In 1961 a major change occurred in the governmental policy of the Syrian Ba'ath party that caused the nomads to settle. The old land tenure system had broken down in the 1950s in Syria when all the uncultivated steppe areas became public property. N.N. Lewis (1987: 186-187) notes that between 1950s and 1980s the most large-scale building of houses took place in the 1960s among tribesmen as the consequence of droughts and the beneficiaries of the Land Reform. The government was involved in the sedentarizing process of these pastoral nomads. (See also Van Dusseldorp 1971: 95). Similar speeded processes of sedentarization have been encountered among the 'Azazma Bedouins of Jordan (Prag 1992). But still among many Bedouin tribes in Syria there were purely nomadic groups living in tents without any houses in the 1970s – in some groups consisting of over 30%. As mentioned in the previous chapter, we had recorded some comparable groups in the district of Jebel Bishri in 2003.

The drought and government initiated sedentarization led to the digging of new wells and redistribution of water in other ways to encourage and stimulate fresh agricultural activity in the steppe. These policies have also affected Jebel Bishri, and they must be taken into account, when considering the emergence of the semi-sedentary Bedouin villages in the region at the time. It is apparent that the emergence of the semi-sedentary Bedouin villages of Shanhas and Qdeir in the 1960s is linked with these historical changes. According to ACSAD, the year 1982 was the major year of increasing cultivation on Jebel Bishri. The alarming progress of rapidly expanding desert line is immediately followed by the increase in cultivation and is especially detectable during a period of about a dozen years from 1986 to 2000 in the Landsat images analysed by the ACSAD, a German project and our project (Hoffman and Geerken 1997: 71-73; Hoffman and Geerken 1998; Lönnqvist and Törmä 2004). The rehabilitation program was under way, but the efforts to harness the progress of the sand cover had not yet yielded visible results in the satellite image data in the year of 2000.

We also made an analysis of the emergence and development of field areas on the western piedmont and the mountain proper comparing CORONA satellite photographs from the 1960s to the Landsat images of the 1970s up until 2001. The impact of desertification and the changes in cultivation were analysed by comparing the size and position of the cultivated land over the time-span (see Lönnqvist and Törmä 2006; Lönnqvist *et al.* 2010). Three areas that we knew from experimental ground visits or studies were chosen for remote sensing analyses: 1) The semi-sedentary Bedouin village of Shanhas (Fig. 45.) with its fields on the western piedmont of Jebel Bishri in the neighbourhood of Qasr al-Hayr ash-Sharqi; 2) a semi-sedentary Bedouin compound between Tar al-Sbai and Tell Tletuat (Fig. 46.) near the western edge of the mountain, where we had camped in 2000; 3) The semi-sedentary Bedouin village of Ash-Shujiri (Fig. 47.) on the mountain which we visited in the years 2003-2008.

The following satellite images were used in studying the fields of Bedouin sites:

*CORONA satellite photographs taken 28.6.1966 and 22.1.1967
* Landsat MSS-images 186/36 taken 14.9.1972, 185/35 and 185/36 27.6.1975, 172/35 and 172/36 12.7.1983
* Landsat TM-images 172/35 and 172/36 taken 1.9.1990
* Landsat ETM-images 172/36 taken 11.8.2000 and 172/35 26.5.2001

The Landsat images used are part of the large collection of georeferenced satellite images available free of charge (Tucker *et al.* 2004). They are called 'GeoCover-images' and are available from Global Land Cover Facility-web server (URL http://glcf.umiacs.umd.edu/).

The shifting agricultural fields of Shanhas are situated in *wadis* southeast from the village. The satellite images reveal the movement of the field areas according to the availability of water and the dramatic changes of the desert line (Fig. 45.). The same seems to apply to the fields of Tar al-Sbai (Fig. 46.) and Ash-Shujiri (Fig. 47.) during the past decades. Agriculture has also affected the loss of natural rangeland, and as previously mentioned, in the year 1995 agriculture had generally been banned in the Syrian steppe zone. Since then, the aim has been to rehabilitate the rangeland. But as C. Chang (1992: 77-78) has noted, the multi-resource nature of village agro-pastoralism, from both spatial and temporal perspectives, present a particular problem for interpreting the landscape. Abandoned agricultural land in upland regions in her studies had also opened up new areas for pastoral land use. Accordingly, we noticed that even if village compounds had been abandoned and left to ruin, like at the area of Nadra, the district was continuously used as grazing grounds from tents. Our satellite image monitoring further revealed that this area and the district of Ash-Shujiri (see ch. 7.) were not affected by the most large-scale desertification either and offered the best grazing grounds on Jebel Bishri.

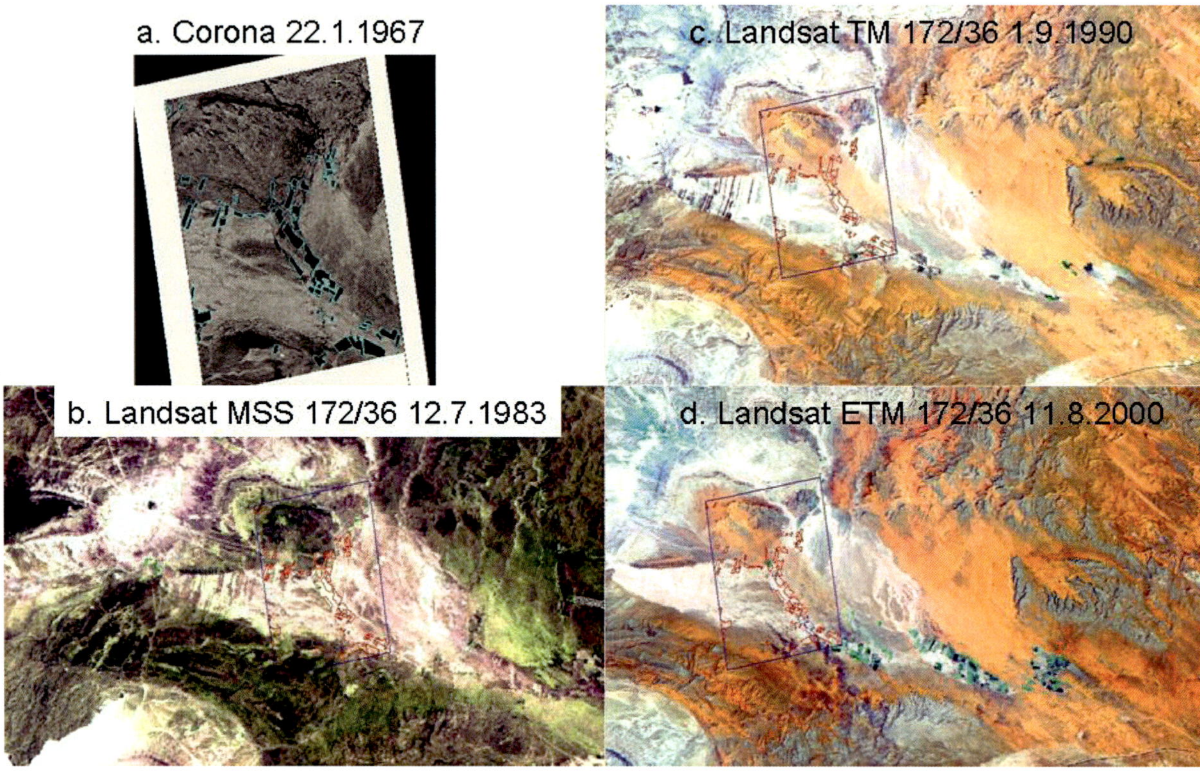

Fig. 45. The disappearance of the fields and the conquering desert line at the semi-sedentary Bedouin village of Shanhas. GIS modelling with satellite images by Markus Törmä 2005.

Fig. 46. The disappearance of the fields and the conquering desert line at a semi-sedentary Bedouin compound between Tar al-Sbai and Tell Tletuat. GIS modelling with satellite images by Markus Törmä 2005.

Fig. 47. The disappearance of the fields and the conquering desert line at the semi-sedentary Bedouin village of Ash-Shujiri. GIS modelling with satellite images by Markus Törmä 2005.

There have been periods of drought which had affected these areas. However, the reasons for the abandonment of the village at Nadra could have differed from that of Ash-Shujiri, which had been apparently deserted not only because of the shortage of rain but also because of the 1995 ban on agriculture in the steppic areas. In both cases, however, the question is of the planned and final abandonment as *curate behaviour* is visible: apart from the fixtures, no usable items had been left behind. The sites had been left to ruin. The more ruined state of the village at Nadra indicated, however, that it had been abandoned a longer time ago, and thus its abandonment was not apparently the result of the most recent environmental changes or government policies that had affected the region. According to historical accounts, there was a large scale desertion of Bedouin villages already in the late eighteenth century in Syria. There was no protection, and some could not pay protection money to sheikhs. (Lewis 1987: 12-15). This took especially place in the eastern and northern parts of the country for several reasons. As mentioned in the previous chapter, the village of Taibe had been raided by Bedouins also at the beginning of the twentieth century. Apart from the environmental reasons, raiding could also have been the reason to leave the village at Nadra.

According to the ESCWA UN report of 2007, the rehabilitation of the Jebel Bishri region started in 1995 and the grazing was limited into the areas where there was enough vegetation. Combating desertification included tree planting to prevent wind erosion, and plantations of rangeland vegetation in different patterns to avoid run-off mechanisms. In the rehabilitation of the vegetation cover thousands of rangeland seedlings had been carried out using mainly indigenous plants. The Ruta plant was used in most of the cases, when the minority of cases consisted of various types of Atriplex. Water harvesting was also developed by building collection points. The small reservoirs (up to 300 m^3) were aimed for household use, the medium-sized (5000 m^3 or less) ones for animals (see Fig. 48.) and the large ones (up to 10, 000 m^3) for irrigation purposes. This combination of plants and water has been successful in recent years. Vegetable productivity increased from 95 kg/ha to 929 kg/ha from 1995 to 2005. In this period rangeland productivity increased from 45 kg/ha to 464 kg/h and rangeland capacity increased from 0.17 head/ha to 1.7 head/ha.

Fig. 48. A Bedouin camp near a watering site at Ash-Shujiri in the spring time after the rains in May 2004. Photo: Kenneth Lönnqvist 2004.

15.5. Summarizing Remarks

"The living archaeology" of the presented Bedouin sites and their satellite image analyses have offered us a loupe for scrutinizing and visualizing possibilities of ancient village life and to reconsider the environmental constraints and socio-political circumstances in the area of Jebel Bishri which has been combating desertification. Residential instability is clearly visible in this marginal zone.

During seasonal grazing, tents are still used as shelters in the region of Jebel Bishri. As discussed in ch. 14, the northern areas have indicated that there still were pastoralists, who used tents the whole year around in 2003. In the satellite photographs of the 1960s, tents could be seen in June surrounding Qasr al-Hayr ash-Sharqi on the western piedmont. We did not see any tent camps around the Qasr in the spring and summer of 2000 any more. However, abandoned tent bases and Bedouin compounds were recorded on the western piedmont. The sites had been left for good, marking a *final abandonment*; and no usable items had been left behind.

According to our ethnographic interviews and historical studies, government initiated larger-scale sedentarization process of local Bedouins had started in the 1960s. The Bedouin village of Shanhas occupied by sedentarizing pastoralists was founded at that time near the Qasr. The plan of the village is open and follows the layout of nomadic tent camps. The plans and interiors of the houses also followed tents reflecting the nomadism of the people. The Bedouins of the village belong to the 'Anazah federation and had been solely living in tents before. Tents still serve as accommodation during seasonal grazing as well as an extra space for living in the village courtyards among these sedentarizing pastoralists. *Transhumance* had, however, become a common mode of pastoralism of the people occupying Shanhas. Shifting cultivation had become widespread among these semi-sedentary pastoralists, especially among the Sba'a tribe, but this has speeded up the spread of the desert line. *The seasonal abandonment*, such as recorded at the village of Shanhas, had caused packing of tents and storage of grain that was witnessed at the site allocating storage in particular areas. There also were usable items that had been left behind such as mattresses, *guphas* and baking trays. These items reflected an anticipated return.

Today the vertical seasonal movement of the sheep and goat pastoralists takes place from the base-villages of the western desert piedmont and western edge of Jebel Bishri, back and forth to the grazing from tents on the Jezira in the spring and summer season, and in turn from the Euphrates valley villages to grazing from tents on Jebel Bishri during the spring and summer. The rationale, as to why the people of Shanhas did not use the upper mountains of Jebel Bishri for their grazing lands during the spring and summer, seemed to be both tribal and territorial, depending on *dirah*, the tribal road. The members of the Sba'a tribe headed towards the Jezira with small livestock. In contrast, the Fida'an tribe, still grazing dromedaries in the region, has had a more nomadic way of living up till recent years.

Despite the environmental hazards in the Jebel Bishri region the mobility of the nomads offers a flexible choice in adapting to changing situations. But compared to the piedmonts and the Euphrates, most of the semi-sedentary village house compounds that we visited on the mountain in 2000-2008 had been *permanently abandoned*. However, tents were seen in abundance especially on the eastern, southern and northern high plateaus of the mountain in the spring. Exceptions in the semi-sedentary village compounds on the mountain consist of a few active households similar to the one where our research team camped at Tar al-Sbai and the partly abandoned village of Ash-Shujiri. In the vicinity of Ash-Shujiri we recorded an abandoned Bedouin compound in the process of ruining. A village compound at Nadra that we studied in the central eastern area was clearly completely abandoned and was already in a ruined state. (See also Lönnqvist *et al.* 2006).

Archaeologically there was a distinctive difference in site inventories at the seasonally abandoned sites compared to the permanently abandoned sites, which all had been left followed by intentional planning of the move. Totally abandoned village compounds of Ash-Shujiri and Nadra with no plan for returning afforded no storage facilities and usable portable utensils. Only heavy fixtures had been left behind at these sites with no anticipated return in mind.

The mountain, that nowadays is mainly the tribal area of the Fid'ān tribe, but also used by the Agedat tribe, has faced *a larger scale regional abandonment* of its village house compounds. This is also visible in the disappearance of agricultural fields, which is coeval with the process of desertification detectable through satellite images. Ploughing, as well as the government's ban on agriculture on the steppe, *badiyah*, in 1995 had clearly irrevocably changed the equilibrium, causing the regional and final abandonments of house compounds. Climate change and desertification also cause regional abandonment, in which case the nomads or agro-pastoralists choose to move to better grazing grounds.

Historical hazards like insecure political conditions and raids could have earlier affected the abandonment of the steppic site of Nadra which had been left in ruins a much longer time ago than some house compounds near and at Ash-Shujiri. A living semi-sedentary compound at Tar al-Sbai still tried to manage without the government's water supply on the mountain in 2000. Even if there seems to be organized water harvesting and springs at Nadra, the water in the dry years has not been sufficient for agricultural purposes any more. However, pastoral nomadism has not ceased in the area, and the rehabilitation program has apparently had its good effect on the range land during the first decade of the twenty-first century. So, the revival of the more nomadic life on Jebel Bishri is under way, but the possibility of continued grazing is still limited to certain areas because of heavy sand cover that has affected the areas of the western parts.

According to the interviews, *seasonal abandonment* can be prolonged and stretched to total abandonment in the areas under desertification. In those cases droughts in one area can cause renomadization towards the Jezira, where the rainfall is more sufficient and predictable. However, in our previous study (ch. 14.) in the villages of the Euphrates drought on the mountain increased sedentarism among those Bedouins who had base-villages along the Euphrates. According to the ethnographic documentation and our change detection carried out with the satellite image analyses, apart from instability caused by raiding activities or government policies, desertification plays a significant part in the processes of *regional final abandonment* on the mountain.

Bibliography

ACSAD = The Arab Centre for the Studies of Arid Zones and Drylands, Damascus.

Aurenche, O. and Desfarges, P. (1983) Travaux d' ethnoarchéologie en Syrie et en Jordanie, Rapports préliminaires, in *Syria*, Vol. LX, pp. 147-185.

Banning, E.B. and Köhler-Rollefson, I. (1983) Ethnoarcaheological survey in the Beidha area, southern Jordan, in *Annals of the Department of Antiquities of Jordan*, Vol. 27, pp. 375-384.

Banning, E.B. and Köhler-Rollefson, I. (1992) Ethnographic Lessons for the Pastoral Past: Camp Locations and Material Remains near Beidha, Southern Jordan, in *Pastoralism in the Levant: Archaeological Materials in Anthropological Perspectives*, ed. by Bar-Yosef, O. and Khazanov, A.. Monographs in World Archaeology, No. 10, Madison, Wisconsin: Prehistory Press, pp. 181-204.

Burchardt, J.L. (1831) *Notes on the Bedouins and Wahabys collected During his travels in the East* by the late John Lewis Burchardt. Vol. 1. London: Henry Colburn and Richard Bentley.

Cameron, C.M. (1993) Abandonment and Archaeological Interpretation, in *Abandonment of settlements and regions: Ethnoarchaeological and archaeological approaches,* ed. by Cameron, C.A. and. Tomka, S.A., New directions in archaeology, Cambridge: Cambridge University Press, pp. 3-7.

Chang, C. (1992) Archaeological Landscapes, The Ethnoarchaeology of Pastoral Land Use in the Grevena Province of Greece, in *Space, Time, and Archeological Landscapes*, Interdisciplinary Contributions to Archaeology, ed. by Rossignol, J. & Wandsnider, L., New York and London: Plenum Press, pp. 65-89.

Cribb, R. (1991a) Mobile villagers: the structure and organization of nomadic pastoral campsites in the Near east, in *Ethnoarchaeological Approaches to Mobile Campsites, Hunter-Gatherer and Pastoralist Case Studies*, ed. by Gamble, C.S. and Bosmier, W.A., International Monographs in Prehistory, Ethnoarchaeological Series 1, Ann Arbor, Michigan: International Monographs in Prehistory, pp. 371-393.

Cribb, R. (1991b) *Nomads in Archaeology.* New Studies in Archaeology. Cambridge: Cambridge University Press.

Davis Stone, G. (1993) Agricultural Abandonment: A Comparative Study in Historical Ecology, in *Abandonment of settlements and regions: Ethnoarchaeological and archaeological approaches,* ed. by Cameron, C.A. and. Tomka, S.A., New directions in archaeology, Cambridge: Cambridge University Press, pp. 74-84.

ESCWA 2007 = *Economic and Social Commission for Western Asian, Land Degradation Assessment and Prevention*: Selected Case Studies from the ESCWA Region. New York: United Nations.

Finkelstein, I. (1995) *Living on the Fringe: The Archaeology and History of the Negev, Sinai and Neighbouring Regions in the Bronze and Iron Ages.* Monographs in Mediterranean Archaeology 6. Midsomer Norton, Somerset: Sheffield Academic Press.

Goldberg, P. and Whitbread, I. (1993) Micromorphological Study of a Bedouin Tent Floor, in *Formation Processes in Archaeological Context*, ed. by Goldberg, P., Nash, D.T. and Petraglia, M.D., Madison: Prehistory Press, pp. 165-188.

Goren Inbar, N. and Hershkovitz, I. (1990) Ethnoarchaeology: The Southern Sinai Bedouin as a Case Study, in *The Biblical Archaeology Today*, The II International Congress on Biblical Archaeology, Jerusalem, 24 June – 4 July 1990, Abstracts, p. 113.

Grabar, O., Holod, H., Knustad, J. and Trousdale, W., *et al.* (1978) *City in the Desert: Qasr al-Hayr East.* An Account of the Excavations Carried out at Qasr al-Hayr East on Behalf of the Kelsey Museum of Archaeology at the University of Michigan, with the Help of Harvard University and the Oriental Institute of Chicago. Harvard Middle Eastern Monographs 23-24. Cambridge, Mass: Harvard University Press.

Grattan, J.P., Huxley, S.I. and Pyatt, F.B. (2003) Modern Bedouin Exposures to Copper Contaminations: an Imperial Legacy? in *Ecotoxicology and Environmental Safety,* 55, pp. 108-115.

Hoffman, K. and Geerken, R. (1997) Degredations- und Desertifikationsprozesse in den Bishri-Bergen Syriens, in *Spektrum der Wissenschaft,* No. 2, pp, 71-73.

Hoffman, K. and Geerken, R. (1998) Multisensor analysis of desertification processes in the Bishri Mountains (Syria), in *Future Trends in Remote Sensing*, ed. by Gundmandsen, P., Rotterdam: Balkema, pp. 73 - 76.

Hole, F. (1978) Pastoral Nomadism in Western Iran, in *Explorations in Ethnoarchaeology*, ed. by Gould, R., School of American Research Advanced Seminar Series, Albuquerque: University of New Mexico Press, pp. 127-167.

Horne, L. (1993) Occupational and locational instability in arid land settlement, in *Abandonment of settlements and regions: Ethnoarchaeological and archaeological approaches,* ed. by Cameron C.A. and. Tomka, S.A, New directions in archaeology, Cambridge: Cambridge University Press, pp. 43-53.

Jarno, R. (1984) Tente et Maison: Le Jeu Annuel de la Sédentarisation a Qdeir (Syrie), in *Nomades et Sédentaires: Perspectives Ethnoarchéologiques*, Textes rassemblés par Aurenche, O., Centre Jean Palerne, Mémoires IV, Éditions Recherches sur les Civilisations, Mémoire no. 40, Paris: Éditions Recherches sur les Civilisations, pp. 191-229.

Kent, S. (1989) Activity areas and Architecture: an interdisciplinary view of the relationship between use of space and domestic built environments, in *Domestic architecture and the use of space: An interdisciplinary cross-cultural study,* ed. by Kent, S., New Directions in Archaeology, Cambridge: Cambridge University Press, pp. 1-8.

Kent, S. (1991) The Relationship between Mobility Strategies and Site Structure, in *The Interpretation of Archaeological Spatial Patterning*, ed. by Kroll, E.M., Price, T.D., Interdisciplinary Contributions to Archaeology, New York: Springer, pp. 33-59.

Kramer, C. (1979) An Archaeological View of a Contemporary Kurdish Village: Domestic Architecture, Household Size and Wealth in *Ethnoarchaeology: Implications of Ethnoarchaeology for Archaeology*, ed. Kramer, C., New York: Columbia University Press, pp. 139-163.

La Bianca, Ø.S. (1990) *Sedentarization and Nomadization: Food System Cycles at Hesban and Vicinity in Transjordan. Hesban 1.* Berrien Springs, MI: The Institute of Archaeology, St. Andrews University Press.

Lönnqvist, M. and Törmä, M. (2004) Different Implications of a Spatial Boundary, Jebel Bishri between the Desert and the Sown in Syria, in *The ISPRS XXth Congress Proceedings*, ed. by Altan, M.O., *The ISPRS International Archives of the Photogrammetry, Remote Sensing and Spatial Information Sciences,* XXXVB, Istanbul, pp. 897-902.

Lönnqvist, M. and Törmä, M. (2006) Observing Changes and their Effects in Nomadic Environment, Remote-sensing and Archaeology for Sustainable Development on Jebel Bishri in Syria, in *Environmental Geoinformatics and Modeling: Proceedings of ICEM '05: International Conference on Environmental Management (28-30 Oct. 2005)*, ed. by Reddy, M.A., Vol. IV, Hyderabad.

Lönnqvist, M. with the contributions by Törmä, M., Nuñez, M., Lönnqvist, K., Stout Whiting, M., Okkonen, J., Riihiaho, H., Nissinen M. (2006) Archaeological Surveys of Jebel Bishri 2000-2004, Preliminary Reports of the Finnish Mission to Syria, in *KASKAL*, Vol. 3, pp. 203-240.

Lönnqvist, M., Törmä, M., Nuñez, M., Okkonen, J., Stout Whiting, M., Riihiaho, H., Nissinen, M. (2010) Desertification and Ethnoarchaeology: Studying Hazards in the Nomadic Environment of Jebel Bishri in Syria, in *Proceedings of the 6th International Congress on the Archaeology of the Ancient Near East, May 5th-10th 2008, "Sapienza"-Università di Roma*, Vol. 1, Near Eastern Archaeology in the Past, Present and Future: Heritage and Identity, Ethnoarchaeological and Interdisciplinary Approach: Results and Perspectives, Visual Expression and Craft Production in the Definition of Social Relations and Status, ed. by Matthiae, P., Pinnock, F., Nigro, L., Marchetti, N. with the collaboration of Romano, L., Wiesbaden: Harrassowitz Verlag, pp. 369-390.

Mulder-Heymans, N. (2002) Archaeology, experimental archaeology and ethnoarchaeology on bread ovens in Syria, in *Civilisations*, Vol. 49, pp. 197-221.

Musil, A. (1928) Musil, A. (1928) *Palmyrena: A Topographical Itinerary*. American Geographical Society. Oriental Explorations and Studies No 4. New York: American Geographical Society.

Palmer, C. and Daly, P. (2006) Jouma's Tent: Bedouin and digital archaeology, in *Digital Archaeology: Bridging Method and Theory*, ed. by Evans, T.L., Daly, P., London: Routledge, pp. 97-127.

Parker, B.J. and Uzel, M.B. (2007) The Tradition of Tandir Cooking in Southeastern Anatolia: An Ethnoarchaeological Perspective, in *Ethnoarchaeological Investigations in Rural Anatolia*, ed. by Takaoğlu, T., Vol. 4, pp. 7-43.

Prag, K. (1992) Bronze Age Settlement Patterns in the South Jordan Valley: Archaeology, Environment and Ethnology, in *Studies in the History and Archaeology of Jordan*, Vol. IV, pp. 155-160.

Rapoport, A. (1989) Systems of activities and systems of settings, in *Domestic architecture and the use of space: An interdisciplinary cross-cultural study*, ed. by Kent, S., New Directions in Archaeology. Cambridge: Cambridge University Press, pp. 9-20.

Rotchild, N.A., Mills, B.J., Ferguson, T.J. and Dublin, S. (1993) Abandonment at Zuni farming villages, in *Abandonment of settlements and regions: Ethnoarchaeological and archaeological approaches*, ed. by, Cameron, C.M. and Tomka, S.A., Cambridge: Cambridge University Press, pp. 123-137.

Saidel, B.A. (2001) Abandoned Tent Camps in Southern Jordan, in *Near Eastern Archaeology*, 64/3, pp. 150-157.

Salzman, C.P. (1980) Introduction: Processes of Sedentarisation as Adaptation and Response, in *When Nomads Settle, Processes of Sedentarisation as Adaptation and Response*, ed. by Salzman, C.P. with the assistance of Sadala, E., Praeger Special Studies New York: Praeger Publishers, pp. 1-20.

Sarre, F. and Herzfeld, E. (1911-1920) *Archäologische Reise im Euphrat- und Tigris-Gebiet*. Band I-II. Forschungen zur islamischen Kunst I. Berlin: Dietrich Reimer (Ernst Vohsen).

Schiffer, M.B. (1987) *Formation Processes of the Archaeological Record*. Albuquerque: University of New Mexico Press.

Simms, S.R. (1988) The Archaeological Structure of a Bedouin Camp, in *Journal of Archaeological Science*, Vol. 15, pp. 197-211.

Tomka, S.A. (1993) Site abandonment behavior among transhumant agro-pastoralists: the effects of delayed curation on assemblage composition, in *Abandonment of settlements and regions: Ethnoarchaeological and archaeological approaches*, ed. by Cameron, C.A. and Tomka, S.A., New directions in archaeology, Cambridge: Cambridge University Press, pp. 11-24.

Tucker, C.J, Grant, D.M. and Dykstra, J.D. (2004) NASA's Global Orthorectified Landsat Data Set, in *Photogrammetric Engineering and Remote Sensing*, Vol. 70, pp. 313-322.

Van Dusseldorp, D.B.W.M. (1971) *Planning of Service Centres in Rural Areas of Developing Countries*. Publication 15. Wageningen, the Netherlands: International Institute of Land Reclamation and Improvement.

Yakar, J. (2000) *Ethnoarchaeology of Anatolia: Rural Socio-economy in the Bronze and Iron Ages*. Tel Aviv University, Sonia and Marco Nadler Institute of Archaeology Monograph Series No. 17. Tel Aviv: Tel Aviv University Press.

16. CONCLUDING REMARKS

Minna Lönnqvist

Jebel Bishri, a mountain in Central Syria, has been the focus of the archaeological studies of the Finnish project SYGIS. The aim of the project has been to carry out prospecting, to survey and map the sites of visible human activities by remote sensing and field walking on the ground. The UTM coordinates have served as basic parameters in site identification. This has been the basis on which also to manage the sites in the view of preventing looting or disappearance of the site information by natural hazards like desertification and flooding. No preference has been given to certain periods, and the information was collected evenly from prehistoric to recently abandoned Bedouin sites. The mountain of Jebel Bishri forms a triangle which had previously been regarded as an empty space in archaeological maps. This, however, is an illusion created by the fact that the area was largely unexplored by archaeological means before the Finnish project started its work in the area in 1999/2000. It should be noted, though, that Jebel Bishri has been known by Assyriologists for decades as an important area for the Mesopotamian people.

The Finnish project has been following in the footsteps of Father A. Poidebard, who began archaeological remote sensing by aerial studies in the neighbourhood as early as in the 1920s and 1930s. Poidebard was searching for the Roman remains of the Eastern *limes* of the empire. With the benefit of modern technology, the Finnish project has been able to utilize satellite images as the major data source in our remote sensing studies and mapping with GIS (Geographic Information Systems). The satellite images have served as a valuable aid in *macro-scale* in order to grasp the large area covering ca. 1 million hectares as well as recognizing the differences in its environment and landscape. The evolving spatial differences and contrasts in the environment can be detected not only in the desert zone and the steppe of the mountain limited to the meandering Euphrates river valley, but also in altitudes from the lowlands of the piedmonts and the Euphrates river valley to the highlands of the mountain of Jebel Bishri itself. The movement of the desert line and its impact have been studied through satellite images. Landscape models and panoramic views have been constructed to visualize the landscape and visibility from ancient sites dating from various periods.

From the perspective of the Landsat satellite images, best applicable for environmental studies, the area seemed, at first, to be void of archaeological remains. This is, however, due to the low spatial resolution of the images, which only reveal monumental remains. The QuickBird images appeared as the best data sources for prospecting even the remains of mobile groups on Jebel Bishri. In the archaeological data traced and collected from the space and ground-level we can detect the difference between highland and lowland cultures. The ground data provides *medium* and *micro scale* perspectives. The ground data is important as empirical evidence and a chronological source of reference – even providing data for further studies in *minute scale.* Three distinct environmental zones were targeted for the chosen plots of the systematic and intensive fieldwork: desert-steppe, steppe and alluvial river zone. Geographically they correspond to: the Qasr al-Hayr – Tar al Sbai region, the Nadra and Ash-Shujiri region and the 'Ayyāsh – Tabūs river zone.

Our findings reveal that the lowland is the area of sedentary people, while the material remains left by the people living on the mountain represent more or less mobile groups: hunter-gatherers or pastoral nomads. The availability of water in the Euphrates valley must have been a major attraction of the Lower Palaeolithic hunters and gatherers representing Achaeulian culture living on the escarpments facing the river. On the other hand their impact is not very visible on Jebel Bishri but reappears on the western piedmont in the oasis area of El Kowm. The palaeolake of Nadra in the southeastern area of the mountain and nearer to the Euphrates also attracted hunters and gatherers, if not Lower Palaeolithic, then Middle Palaeolithic groups. Not with standing, the search for prey and fish was the primary reason for the round-movement of the groups. The highland-lowland interaction seems to have started already in the Middle Palaeolithic era. Apparently the groups producing the Middle Palaeolithic Hummalian industries identified on Jebel Bishri had their base in the El Kowm basin.

As far as the Stone Age is concerned, the Middle Palaeolithic is best attested in each studied environmental district varying from desert-steppe, steppe to the alluvial river zone. The clearest evidence of the highland-lowland interaction during the Stone Age can be detected at Tar al-Sbai at the western edge of Jebel Bishri. Tar al-Sbai apparently served as an offshoot site of the people living in the El Kowm basin. Especially Upper Palaeolithic and Epipalaeolithic hunters with apparent bases on the western piedmont used *abris*, rock shelters, in the area. The locations of these sites were apparently chosen because of their extraordinary good visibility and impressive landscapes. The rock shelters and hunting blinds appear on the route of the prey such as gazelles. However, the flint finds from the sites indicate that apart from serving as bases for hunting the sites were used more for processing the carcasses of the hunted animals. Kebaran impact is particularly visible in the finds. Geometric Kebaran was the phase which expanded and exploited the marginal zones in the Levant; some resemblance in the finds of Jebel Bishri at the time can be observed in the implements of the so-called Madamaghan industries in Jordan. The Madamaghan industries, related to the Kebaran, had especially adapted into arid areas.

Further there is continuity from the Epipalaeolithic Kebaran and Natufian occupation to the Neolithic one in the steppe and desert fringe indicating more systematic exploitation of the available resources in the region of Jebel Bishri. The Epipalaeolithic and Neolithic periods were not as visible on the Euphrates side, although some indications of sedentarization could be detected. On the other hand, the Pre-Pottery Neolithic period is well attested throughout the desert-steppe and steppic areas of Jebel Bishri. The earliest stone enclosure sites on the mountain may already date far as back as the Pre-Pottery Neolithic B period; there is, by any means, continuity from the Pre-Pottery Neolithic to the Chalcolithic and Bronze Age in the material remains of a few sites. The pastoral technocomplex with cairns/*tumuli* and stone enclosures such as corrals, animal pens, was in use already in the Chalcolithic and Bronze Age, while some of these structures may date from fairly late periods. In addition, there is evidence of their use in the Roman period, not to mention that ethnographic parallels indicate that such structures have been used by recent Bedouins as well.

The role of the interaction between Jebel Bishri and the Euphrates becomes most visible in the Chalcolithic and Early Bronze Age. This is the time of emergence of urban way of living, although some villages may have emerged on the Euphrates side of Jebel Bishri during the Neolithic period. The mountain served as the seasonal grazing arena of semi-sedentary people living in the lower regions. This is particularly typical of the *tranhumant* mode of life. The mountain has been - and still is - the living grounds of small-scale compounds with non-monumental remains, while the piedmonts and the river valley are the area of sedentary sites and more monumental remains. The small finds furthermore differ in nature between the mountain and the surrounding areas: flint, for instance, seems to be the most frequently found material on the mountain and its western piedmont, whereas pottery is more plentiful in sedentary sites in the river valley. The collected data and conclusions based on that data are, nevertheless, only strictly applicable to the systematically studied transects on the ground. Moreover, animals such as gazelles for hunters and goats/sheep for pastoral nomads are reflected in the archaeological remains as rock shelters are represented by hunter-gatherers and animal pens like corrals were constructed by pastoralists. Animals were central for the economy of the people moving on the mountain.

The mountain appears as the nomad's realm of ancestors and animals. The burial grounds, represented by cairns/*tumuli* form the most frequent architectural features met within the landscape. Cairns/*tumuli* as well as stone enclosures are allocated to non-arable lands and on higher ridges. Some sites are composed both of these structures. This is a typical feature of the desert-steppe and steppe environments of the mountain. Tombs are visible from afar to travellers serving also as path finders, some even with more elevated stone heaps on top, the so-called *rujm* markers. Beside tombs other ritual types of sites appear in areas with good visibility and in striking landscapes. The sites are intermingled with burial grounds such as the more visible sites of the pastoralists. Animal pens, i.e., corrals concentrate on the steppic highland and plateau regions, where the most significant grazing potential is clearly detectable in our cluster analysis of the satellite images. Pastoral sites have been generally set in one to two hour's hiking distance to seasonal *wadis*. There is one stone enclosure site more or less in every space of one kilometre on average on the highest area of the mountain at Ash-Shujiri and Nadra.

The impact of the Greater Mesopotamian culture and civilization is visible in the sedentary sites such as the tells on the Euphrates side of Jebel Bishri starting with impact of the Al-Ubaid and Uruk cultures. The cuneiform references to the area date from the third to the first millennium B.C. Jebel Bishri is a deified mountain to which sacrifices were made; an observation which may also explain it as a large graveyard attested in archaeological remains. In the cuneiform sources of the third millennium B.C. the mountain is mentioned as an Amorite mountain, or later in the first millennium B.C. sources, it is known as an Aramaean mountain. Ethnically the area was therefore especially occupied by pastoral nomads of these West Semitic and tribal people.

The tells beneath Jebel Bishri on the Euphrates side are, on the other hand, fairly small and indicate their involvement in agro-pastoralist economy or service as transit stations of trade comparable to the villages and *transhumant* pastoralism in the area today. The Early Bronze Age occupation is especially pronounced in the material at several tells and gravefields on the Euphrates side of the mountain. These small settled sites have visible components of finds related to pastoral nomadic cultures. In the area surveyed by our teams there were cairns/*tumuli* and cists, and possibly some shaft graves dating from the Early Bronze Age between Deir ez-Zor and Halabiya. Some produced evidence from the Middle Bronze Age as well. Tell Kharita ca. 22 km northwest of Deir ez-Zor appears as an early site most closely integrated with the foot of the mountain and producing pottery types from EBIII and EBIV periods typical of the pastoral nomads of the periods. The evidence of the tell seems to actualize the sedentarization of pastoral nomadic groups from the mountain.

Tell Tibne some 43 km northwest of Deir ez-Zor has produced evidence which could make it an ancient transit place for caravans and boats due to its topographic location. The ancient river channel reached the tell during the ancient times, but now it has silted up and the channel has changed further to the north. The site could have served as an ancient harbour for transporting stone and salt to Lower Mesopotamia. The pottery repertoire offers parallels to Ebla, Emar and Mari, major Syro-Mesopotamian sites in the Early and Middle Bronze Age both inland country and on the river. Tell Tibne could also have served Dûr Yahdun-Lim's port during the heyday of Mari in the Middle Bronze Age II period. In the Mari sources, in particular, Jebel Bishri appears as the Sutean and the Benjaminite tribal region. In these tribal territories Tell Tibne seems to fall into the territory of Benjaminites. Later on Tell Kharita, if not Tell Tibne, could also be connected to the Assyrian and Aramaean conflicts described in the cuneiform sources and taking place at the foot of Jebel Bishri. The number of identified tells between Deir ez-Zor and Raqqa conforms to the number of sites mentioned as having been destroyed at the foot of Jebel Bishri by the Assyrian rulers. Inscriptions discovered at the site possibly are applicable to the Aramaean era. The Aramaean kingdom of Laqē probably was more closely associated with the mountain of Jebel Bishri than several previous studies have suggested.

The Hellenistic and Persian periods are quite invisible in our material collected so far, but the impact of the Roman and Byzantine realms is well represented in the region. The Roman sites are clearly linked with the military presence of troops in the eastern border zone of the empire. From that time on there are monumental remains such as military posts on the desert border, but particularly forts and fortresses at the outer ridges of Jebel Bishri to protect the Euphrates *limes* against Parthians and later against Persians. Our evidence shows that most of the Roman sites in the inner area of Jebel Bishri and its outer fringes originally date from the second and third century A.D., from the time of the establishment of a more permanent Roman presence on the Euphrates ca. A.D. 160. It is further possible that the city of Palmyra also held the command of several Roman sites on the desert and Euphrates borders at some point.

A military camp and two forts such as Qebaqeb and Qseybe were recorded on the desert-steppe piedmonts of the mountain. Possible evidence of local Palmyrene Aramaean impact at Qseybe was recovered among the Roman pottery finds. Water harvesting was well organised in the desert areas with *qanats* apparently dating from the Roman period; and the evidence of a large Roman reservoir with a tower in the central summit region reinstates the systematic use of the "Bishri corridor" for logistics over the mountain. A line of small sites with Roman pottery bordered a way from a camp on the top to Qseybe. The Severan expansion is clearly visible on the Euphrates, and the organisation of sites to serve logistics for the fort such as of Qreiye and fortress city of Dura Europos seem to be as credible as the earlier suggestions that the itinerary over Jebel Bishri served Circesium.

However, the Roman sites in the inner areas of the mountain seem to have been abandoned along with Dura's fall ca. A.D. 256. The refortification of the sites by Diocletian in the *Strata Diocletiana* on the western piedmont of the mountain has been earlier studied by other expeditions. Some sites, e.g. tombs, which we studied in the region, can be possibly to be associated with early Christian pilgrimage to the fortress city of Resafa-Sergiopolis. Two sites which we studied on the Euphrates *limes*, namely Tibne (the fortress of Mambri?) and the fort of Tabūs, were also apparently occupied already in the second century A.D., the latter even during the first centuries around the common era, but were refortified during Diocletian and Justinian I. Tabūs probably served as a relay station in military reconnaissance, and its connection with Zenobia's fortress on Halabiya in the Justinianic era is obvious. The roads which we recorded on the Euphrates seem to have been established between the sites during the Later Roman period.

Arabs were populating the area already in the Assyrian and Roman times, and the early Christian Arabs known as Ghassānids were occupying Resafa-Sergiopolis during the Late Antiquity. The famous battle of Siffin related to the Islamic conquerors took place on the Euphrates side of Jebel Bishri. Some sites on the mountain have produced early Islamic evidence in ceramic types and glass bracelets, but later Bedouin presence is most visible. The Bedouin tribes in the region provided us with good ethnographic evidence of pastoral nomadism and its ways of life in the neighbourhood. The relations between highland and lowland were experienced in *transhumance* on a seasonal basis.

The seasonally abandoned Bedouin village of Shanhas on the western piedmont served as a laboratory of living archaeology to us while its inhabitants were on their grazing grounds on the Jezira. We were able to study the process of sedentarization and abandonment in different degrees at Shanhas like at other local Bedouin sites. We utilized satellite image data to compare the impact of desertification on these processes. Knowledge of political situations and governments initiation in the sedentarization in the 1960s is essential in order to understand the processes. The expanding desert line followed the government's ban for agricultural activities on the steppe zone to revive the grazing grounds have obviously caused the regional abandonment of sites on the mountain. Earlier rivalry between different tribal groups had apparently ruined some sites before that.

Clearly seasonal abandonment with anticipated return in mind, like that observed at the village of Shanhas on the western piedmont, provided evidence of storage and more usable items to record in the Bedouin compounds compared to the Bedouin sites which had been left for good in the mountain area. Final and more large-scale abandonments were detected at semi-sedentary house compounds on the mountain due to environmental hazards, while some were also caused by the government's ban on agriculture in the vulnerable steppe areas. One Bedouin site may already have been abandoned during the tribal clashes in the 19th century. Only fixtures, such as inscriptions or heavier grinding stones were found in the totally abandoned sites. However, still these areas were used by local Bedouins as seasonal grazing grounds from tents.

Spatial boundaries of different environment and landscapes between zones of desert, steppe and irrigated river valley have affected the mode of life in these zones in which Jebel Bishri is enclosed. Mobility is the typical mode of life visible in the archaeological remains in the desert-steppe landscape of the mountain, while oases and river banks are testified by sedentism. The mountain people often appear to have included the same groups of people who also were affecting the life on the piedmonts and sedentarizing in the river valley. This mutual relationship of the areas opens the understanding of closer interaction between the highlands and lowlands. It needs, however, to be emphasized that the degrees of mobility has varied as much as degrees of sedentism. Among the pastoral nomads the change from the camel nomadism of the desert areas to small livestock economy nowadays increases on the mountain of Jebel Bishri, where the relationship with the Euphrates zone is affecting the economy. In any event, this is small-scale society, and buildings as well as villages are constructed following pastoral and tribal models. The sizes of tells beneath Jebel Bishri also indicates that the same models were used in the past as well. During the Roman and Islamic periods steppe and desert areas were more systematically employed by irrigation to serve more dense and larger settlements or even monumental sites.

INDEX OF SITES AND COORDINATES

Site	Type	Coordinates
A 1	open air site	UTM 0505934, 3881500
A 2	tent base	UTM 0506062, 3881369
A 3	tent base	UTM 0506078, 3881413
A6a	open air site, Bedouin compound	UTM 0507533, 3882331
A6b	Bedouin compound	UTM 0507630, 3882663
A6c	Bedouin compound	UTM 0507724, 3882581
A6d	storage and activity area	UTM 0508041, 3882461
A 8	open air site	UTM 0507929, 3886671
A 9	open air site	UTM 0508606, 3986349
A 10	open air site	UTM 0509253, 3886067
A 11	open air site	UTM 0508554, 3886464
A 12	cairn/*tumulus*	UTM 0518060, 3905850
A 13	stone circles	UTM 0518060, 3905850
A 14	cairn/*tumulus*	UTM 0518039, 3905940
A 15	hunting blind	UTM 0518015, 3905919
A 16	cairn/*tumulus*	UTM 0528064, 3905893
A 17	cairn/*tumulus*	UTM 0518065, 3905826
A 18	cairn/*tumulus*	UTM 0518065, 3905812
A 19	house, stone enclosures	UTM 0519050, 3905748
A 20	cairn/*tumulus*	UTM 0518021, 3905688
A 21	hunting blind	UTM 0518020, 3905640
A 22	cairns/*tumuli*	UTM 0517994, 3905479
A 23	cairn/*tumulus*	UTM 0518080, 3905418
A 24	Islamic grave	UTM 0517827, 3905445
A 25	hunting blind	UTM 0517827, 3905445
A 26	cairn/*tumulus*, stone circle	UTM 0517717, 3905479
A 27	megalithic cairns, stone enclosure	UTM 0517793, 3905377
A 28	hunting blind	UTM 0517939, 3905199
A 30	cairns/*tumuli*, hunting blind	UTM 0517951, 3905178
A 31	Bedouin tent base	UTM 0518003, 3905149
A 32	cairn/*tumulus*	UTM 0518002, 3905073
A 33	hunting blind	UTM 0518002, 3905073
A 34	hunting blind	UTM 0518052, 3904927
A 35	cairns/*tumuli*	UTM 0518003, 3905149
A 36	cairn/*tumulus*	UTM 0518358, 3904687
B 1	tomb	UTM 0507596, 3886103
B 2	open air site, tomb	UTM 0504815, 3884501
B 3	cairn/*tumulus*	UTM 0518039, 3906573
B 4	complex of cairns, stone circle	UTM 0517997, 3906520
B 5	rock shelter	UTM 0517950, 3906502
B 6	cairns/*tumuli*, stone circles, hunting blind	UTM 0518016, 3906375
B 7	cairns/*tumuli*	UTM 0517974, 3906208
B 8	undefined structures	UTM 0517819, 3906239
B 9	cairn/*tumulus*	UTM 0517953, 3900172
B 10	cairns/*tumuli*	UTM 0517983, 3906206
B 11	Islamic grave	UTM 0518016, 3906062

B 12	cairn/*tumulus*	UTM 0518001, 3906086
B 13	Islamic graves	UTM 0518001, 3906092
B 14	stone circle	UTM 0517997, 3906112
B 15	boat-like stone structure, cairn, stone wall	UTM 0517993, 3906120
B 16	cairn/*tumulus*	UTM 0517991, 3906157
B 17	cairn/*tumulus*	UTM 0517978, 3906179
C 1	rock shelter	UTM 0517906, 3907065
C 2	Bedouin tent base, rock shelter,	UTM 0517927, 3907055
	hunting blind	UTM 0517971, 3906886
C 3	cairn/*tumulus*	UTM 0518003, 3906919
C 4	Islamic grave	UTM 0518028, 3906813
C 5	rock shelter	UTM 0518067, 3906770
C 6	huts	UTM 0518087, 3906742
C 7	rock shelter	UTM 0518077, 3906718
C 8	cairns, ring-tumulus, stone lines	UTM 0518033, 3906677
C 9	cairns/*tumuli*	UTM 0518072, 3906810
C 10	corrals, stone enclosures	UTM 0545940, 3908253
C 11	cairn/*tumulus*	UTM 0518246, 3904739
C 12	cairns/*tumuli*, stone walls	UTM 0518358, 3904633
C 13	cairns/*tumuli*, stone lines	UTM 0518358, 3904627
C 14	stone complex of cairns/*tumuli*	UTM 0518724, 3904434
C 15	cairns/*tumuli*, stone circle, hunting blind	UTM 0520095, 3910264
D 1	cairns/*tumuli*, hunting blind	UTM 0520981, 3912099
D 2	cairns/*tumuli*, hunting blind, stone circles	UTM 0521037, 3912210
D 4	graves	UTM 0520209, 3903263
E 1	cairn/*tumulus*	UTM 0517927, 3907055
E 2	open air site	UTM 0517946, 3907031
E 3	stray find?	UTM 0520138, 3903162
E 4	stray find?	UTM 0517686, 3901743
E 5	well, Bedouin compound	UTM 0507410, 3886173
F 1	stone enclosure	UTM 0540300, 3922350
F 2	stone enclosure	UTM 0539800, 3922200
F 3	stone enclosure	UTM 0539850, 3921950
F 4	stone enclosure	UTM 0540770, 3922400
F 5	stone enclosure	UTM 0540700, 3922200
F 6	stone enclosure	UTM 0540400, 3920300
F 7	stone enclosure	UTM 0540500, 3920120
F 8	stone enclosure	UTM 0540700, 3920300
F 9	stone enclosure	UTM 0539920, 3916700
F 10	stone enclosure	UTM 0540080, 3916800
F 11	stone enclosure	UTM 0540520, 3916850
F 12	stone enclosure	UTM 0540576, 3916775
F 13	stone enclosure	UTM 0542300, 3921800
F 14	stone enclosure	UTM 0542500, 3921800
F 15	stone enclosure	UTM 0543200, 3921950
F 16	stone enclosure	UTM 0541650, 3917200
F 17	stone enclosure	UTM 0543800, 3921700
F 18	stone enclosure	UTM 0545420, 3921200
F 19	stone enclosure	UTM 0544000, 3919900

F 20	stone enclosure	UTM 0544100, 3916900
F 21	stone enclosure	UTM 0544500, 3916700
F 22	stone enclosure	UTM 0547080, 3924300
F 23	stone enclosure	UTM 0546650, 3924350
F 24	stone enclosure	UTM 0547530, 3922400
F 25	stone enclosure	UTM 0546500, 3921180
F 26	stone enclosure	UTM 0546700, 3920900
F 27	stone enclosure	UTM 0546500, 3921220
F 28	stone enclosure	UTM 0547200, 3920220
F 29	stone enclosure	UTM 0547230, 3920250
F 30	stone enclosure	UTM 0546800, 3919930
F 31	stone enclosure	UTM 0546700, 3919780
F 32	stone enclosure	UTM 0547600, 3918520
F 33	stone enclosure	UTM 0547400, 3918450
F 34	stone enclosure	UTM 0547450, 3918600
F 35	stone enclosure	UTM 0547280, 3918480
F 36	stone enclosure	UTM 0546400, 3918350
F 37	stone enclosure	UTM 0546264, 3918376
F 38	stone enclosure	UTM 0546600, 3917200
F 39	stone enclosure	UTM 0546600, 3917300
F 40	stone enclosure	UTM 0546600, 3917180
F 41	stone enclosure	UTM 0546550, 3917180
G 1	tell, settlement	UTM 0521037, 3912210
G 2	Bedouin house, well	UTM 0512292, 3907189
G 3	Bedouin compound	UTM 0533438, 3915325
G 4	terrace	UTM 0580307, 3930531
G 5	terrace	UTM 0580484, 3929221
H 1	plateau/open accumulated site	UTM 0587782, 3924814
		UTM 0588385, 3924630
		UTM 0588017, 3924508
		UTM 0587641, 3924709
H 2	cist tomb	UTM 0587937, 3924488
H 3a	cist tomb	UTM 0588061, 3924426
H 3b	cist tomb	UTM 0588166, 3924423
H 4	cist tomb	UTM 0588097, 3924407
H 5	cist tomb	UTM 0588343, 3924358
H 6	cist tomb	UTM 0588678, 3923973
H 7	cist tomb	UTM 0588709, 3923923
H 8	cist tomb	UTM 0588709, 3923923
H 9	cist tomb	UTM 0588728, 3923984
H 10	cist tomb	UTM 0588761, 3923953
H 11	cist tomb	UTM 0588796, 3923980
H 12	cist tomb	UTM 0588729, 3923848
H 13	road	UTM 0587800, 3924989
		UTM 0588873, 3924616
H 14	bridge	UTM 0587835, 3925027
H 15	road	UTM 0587778, 3925051
		UTM 0587276, 3925041
H 16	plateau	UTM 0588791, 3924304

H 17	graveyard	UTM 0586819, 3925363
H 18	graveyard	UTM 0586868, 3925336
H 19	fort	UTM 0586948, 3925414
H 21	tower	UTM 0586457, 3925596
H 23	tower	UTM 0586109, 3925467
H 24	water harvesting site	UTM 0587192, 3925188
I 1	ruined Bedouin village, road	UTM 0546295, 3911090
I 2	cairn/*tumulus* field	UTM 0545921, 3909111
I 3	cairn/*tumulus*	UTM 0546817, 3909626
I 4	cairn/*tumulus*	UTM 0546948, 3909304
I 5	cairn/*tumulus* field	UTM 0546966, 3909161
I 6	cairn/*tumulus*	UTM 0547122, 3909010
I 7	cairn/*tumulus*, house base, stone structure	UTM 0551944, 3910910 UTM 0551964, 3910907 UTM 0552037, 3910796 UTM 0552092, 3910598
I 8	open air site	UTM 0551925, 3910802
I 9	stone enclosures	UTM 0551776, 3910907
I 10	cairns/*tumuli*, stone enclosures	UTM 0551639, 3911226 UTM 0551653, 3911253
I 11	cairns/*tumuli*, stone enclosures	UTM 0551461, 3912985 UTM 0551521, 3912895
I 12	cairn/*tumulus*, stone enclosures, walls	UTM 0552185, 3912482
I 13	open air site, cairn/*tumulus*	UTM 0546557, 3910187
I 14	water harvesting site	UTM 0546723, 3910264
I 15	stone enclosure, water harvesting site, bridge/dam	UTM 0545414, 3911581 UTM 0545504, 3911420
I 16	settlement	UTM 0545733, 3912088
I 17	ring-tumulus	UTM 0545933, 3912115
I 18	open accumulated site	UTM 0545479, 3911838
I 19	open air site	UTM 0537957, 3912474
I 20	stone enclosures	UTM 0544927, 3912428
I 21	cairn/*tumulus*, stone enclosures	UTM 0545502, 3912971 UTM 0545511, 3913001
I 21b	water harvesting site	UTM 0545513, 3912930
I 22	cairn/*tumulus*, stone ring	UTM 0540078, 3911932
I 23	stone enclosure	UTM 0537819, 3912408
I 24	cairns/*tumuli*, stone enclosure	UTM 0538241, 3911973
I 25	disturbed graveyard	UTM 0543949, 3913372
I 26	ring-tumulus	UTM 0543994, 3913239
I 27	open air site	UTM 0544260, 3913041
I 28	water source	UTM 0543776, 3913955
I 29/30	stone enclosures	UTM 0543775, 3913502
I 31	ring tumulus, Islamic grave, ruim marker	UTM 0543686, 3913508
I 32a-d	stone enclosures	UTM 0543561, 3913420 UTM 0543562, 3913446
I 33	cairns/*tumuli*, stone enclosures	UTM 0538333, 3912899
I 34	water harvesting site	UTM 0546819, 3910057

I 35	stone enclosures	UTM 0543344, 3913613
I 36	stone enclosures	UTM 0546150, 3908053
I 37	rock shelter	UTM 0537782, 3913311
I 38	stone enclosure	UTM 0537085, 3913155
J 1	open air site	UTM 0595032, 3919265
J 2	open air site	UTM 0594872, 3919389
J 3	open air site	UTM 0594649, 3919584
J 4	open air site	UTM 0594526, 3919679
J 5	open air site, plateau	UTM 0593827, 3920245
J 6	road	UTM 0593719, 3920320
J 7	bridge	UTM 0593354, 3920769
J 8	cairns/tumuli	UTM 0593324, 3920353
J 9	open air site	UTM 0594113, 3920045
J 10	open air site, road	UTM 0593109, 3921225
J 11	open air site, open accumulated site	UTM 0593185, 3921140
J 12	open air site, cairns/*tumuli*	UTM 0593991, 3920688
J 13	open accumulated site	UTM 0594016, 3920882
J 14	open air site, cairn/*tumulus*	UTM 0593793, 3920928
J 15	cairn/*tumulus*	UTM 0593887, 3920922
J 16	cairn/*tumulus*	UTM 0593927, 3920971
J 17	cairn/*tumulus*	UTM 0593936, 3920984
J 18	cairn/*tumulus*	UTM 0594046, 3921018
J 19	open accumulated site	UTM 0593221, 3920934
J 20	open accumulated site	UTM 0592654, 3821375
J 21	open accumulated site	UTM 0592577, 3921370
J 22	graveyard	UTM 0586904, 3924695
K 1	military camp	UTM 0540793, 3917317
K 2	ruined houses	UTM 0539429, 3919109
K 3	water harvesting site with tower	UTM 0543591, 3920667
K 4	open air site	UTM 0568664, 3911075
K 5	settlement	UTM 0568815, 3911211
K 6	fort, military post	UTM 0568664, 3911075
K 8	fort	UTM 0558858, 3878879
K 9	tower	UTM 0558255, 3878241
K 10	well and water channel	UTM 0563635, 3894940
L 1	road	UTM 0592625, 3921364
L 2	open air site	UTM 0590597, 3922084
L 3	open air site	UTM 0590486, 3922129
L 4	open air site	UTM 0590480, 3922232
L 5	open air site	UTM 0589867, 3922154
L 6	open air site	UTM 0589806, 3922076
L 7	open air site	UTM 0589678, 3921809
L 8	open air site	UTM 0589482, 3921729
L 9	open air site	UTM 0589678, 3921809
L 10	open air site	UTM 0589343, 3921728
L 11	open air site	UTM 0589220, 3921826
L 12	open air site	UTM 0588721, 3921601
L 13	open air site	UTM 0588505, 3921850
L 14	tell, settlement	UTM 0589436, 3923932

INDEX OF MUNSELL READINGS OF ILLUSTRATED POTTERY

Chapter 6
Nadra

Fig. 16: Munsell Ware 5YR 7/2 pinkish gray
outside 10YR 4/1 dark gray
inside not available

Chapter 10
Tell Tibne

Fig. 14.1: Munsell Ware 10R 8/3 pink
outside 2.5Y 8/4 pale yellow
inside 2.5Y 7/4 pale yellow

Fig. 14.2: Munsell Ware 5 YR 6/6 reddish yellow
outside 5Y 8/3 pale yellow
inside 5YR 7/4 pink

Fig. 14.3: Munsell Ware 5YR 6/6 reddish yellow
outside 2.5Y 8/4 pale yellow
inside 2.5Y 7/3 pale yellow

Fig. 14.4: Munsell Ware 10YR 7/3 very pale brown
outside 2.5Y 8/3 pale yellow
inside 2.5Y 8/3 pale yellow

Fig. 14.5: Munsell Ware 7.5YR 8/4 pink
outside 2.5Y 8/3 pale yellow
inside 2.5Y 8/3 pale yellow

Fig. 14.6: Munsell Ware 2.5Y 8/4 pale yellow
outside 10YR 8/3 very pale brown
inside 10YR 8/3 very pale brown

Fig. 14.7: Munsell Ware 10YR 8/3 very pale brown
outside 2.5Y 8/3 pale yellow
inside 10 YR 7/3 very pale brown

Fig. 14.8: Munsell Ware 7.5YR 6/6 reddish yellow
outside 7.5YR 7/4 pink
inside 5YR 6/4 light reddish brown

Fig. 14.9: Munsell Ware 10YR 7/3 very pale brown
outside 2.5Y 8/3 pale yellow
inside 10YR 8/4 very pale brown

Fig. 15.1: Munsell Ware 5YR 8/3 pink
outside 10YR 8/3 very pale brown
inside 10YR 8/3 very pale brown

Fig. 15.2: Munsell Ware 2.5Y 8/2 pale yellow
outside 5Y 8/3 pale yellow
inside 5Y 8/3 pale yellow

Fig. 15.3: Munsell Ware 2.5YR 8/3 pink and 2.5Y 7/4 pale yellow
outside 10R 7/6 light red
inside 10R 6/6 light red

Fig. 15.4: Munsell Ware 7.5YR 7/3 pink
outside 5Y 8/3 pale yellow
inside 5Y 8/3 pale yellow

Fig. 15.5: Munsell Ware 2.5Y 5/3 light olive brown
outside 2.5Y 6/3 light yellowish brown
inside 2.5Y 6/3 light yellowish brown

Fig. 15.6: Munsell Ware 5YR 5/4 reddish brown
outside 10YR 8/3 very pale brown
inside 2.5YR 6/6 light red

Fig. 15.7: Munsell Ware 10YR 7/3 very pale brown
outside 2.5Y 7/3 pale yellow
inside 2.5Y 7/3 pale yellow

Fig. 15.8: Munsell Ware 7.5YR 6/4 light brown
outside 2.5Y 7/3 pale yellow
inside 2.5Y 7/4 pale yellow

Fig. 15.9: Munsell Ware 10YR 8/3 very pale brown
outside 5Y 8/3 pale yellow
inside 5Y 8/3 pale yellow

Fig. 16.1: Munsell Ware 2.5YR 6/8 light red
outside 10YR 8/4 very pale brown
inside 5YR 7/6 reddish yellow

Fig. 16.2: Munsell Ware 2.5Y 8/4 pale yellow
outside 2.5Y 8/3 pale yellow
inside 2.5Y 8/3 pale yellow

Fig. 17.1: Munsell Ware 5YR 6/6 reddish yellow
outside 10YR 8/3 very pale brown
inside 2.5YR 7/6 light red

Fig. 17.2: Munsell Ware 5YR 7/4 pink
outside 2.5Y 7/3 pale yellow
inside 2.5Y 8/3 pale yellow

Fig. 17.3: Munsell Ware 5YR 6/6 reddish yellow
outside 2.5Y 8/4 pale yellow
inside 2.5Y 8/4 pale yellow

H 1

Fig. 20.1: Munsell Ware 10YR 8/4 very pale brown
outside 10YR 8/4 very pale brown
inside 2.5Y 7/3 pale yellow

Fig. 20.2: Munsell Ware 10R 7/3 pale red
outside 2.5Y 8/4 pale yellow
inside 10YR 7/3 very pale brown

H 2

Fig. 24.1: Munsell Ware 7.5YR 8/4 pink
outside 2.5YR 8/4 pink
inside 2.5Y 8/3 pale yellow

Fig. 24.2: Munsell Ware 7.5YR 8/4 pink
outside 2.5Y 8/4 pale yellow
inside 2.5Y 8/3 pale yellow

Ayyash

Fig. 36.1: Munsell Ware 5YR 6/4 light reddish brown
outside 5YR 7/3 pink
inside 5YR 6/4 light reddish brown

Fig. 36.2: Munsell Ware 10R 5/6 red
outside 10YR 6/4 light yellowish brown
inside 2.5YR 5/6 red

Tell El Kharita

Fig. 42: Munsell Ware blackish
outside 5YR 6/4 light reddish brown
inside 5YR 6/4 light reddish brown

Fig. 46: Munsell Ware Gley 1, 3/N very dark gray
outside 10YR 8/4 very pale brown
inside 10YR 8/4 very pale brown

Fig. 48: Munsell Ware 5YR 7/3 pink
outside 5YR 6/3 light reddish brown
inside 5YR 6/4 light reddish brown

Fig. 49: Munsell Ware 5YR 8/3 pink
outside 10YR 8/3 very pale brown
inside 10YR 8/3 very pale brown

Fig. 50.1: Munsell Ware 10YR 7/4 very pale brown
outside 10YR 8/2 very pale brown
inside 10YR 7/4 very pale brown

Fig. 50.2: Munsell Ware pinkish clay
outside 10YR 7/4 very pale brown
inside 10YR 7/4 very pale brown

Fig. 50.3: Munsell Ware 10YR 8/2 very pale brown
outside 10YR 8/3 very pale brown
inside 10YR 8/3 very pale brown

Fig. 50.4: Munsell Ware 5YR 7/4 pink
outside 5YR 7/4 pink
inside 5YR 7/4 pink

Fig. 50.5: Munsell Ware pinkish coarse clay, greenish white slip
outside 10YR 7/4 very pale brown
inside 10YR 7/4 very pale brown

Fig. 51.1: Munsell Ware 10YR 8/4 very pale brown
outside 10YR 8/4 very pale brown
inside 10YR 8/4 very pale brown

Fig. 51.2: Munsell Ware not available
outside 2.5Y 7/4 pale yellow
inside 10YR 7/4 very pale brown

Fig. 52.1: Munsell Ware 10YR 8/3 very pale brown
outside 10YR 8/3 very pale brown
inside 10YR 8/3 very pale brown

Fig. 52.2: Munsell Ware pinkish clay, greenish white slip
outside wheeled
inside not available

Fig. 52.3: Munsell Ware 2.5Y 7/4 pale yellow, greenish white slip
outside 2.5Y 7/4 pale yellow
inside 2.5Y 7/4 pale yellow

Fig. 52.4: Munsell Ware brownish clay, greenish-white slip
with brown stripes on the both sides of mouth
outside wheeled
inside not available

Fig. 52.5: Munsell Ware not available
outside 7.5.YR 8/4 pink
inside 7.5.YR 6/8 reddish yellow

Fig. 53: Munsell Ware Gley2 6/1 bluish gray
outside 2.5Y 7/4 pale yellow
inside black stripe inside

Fig. 54.1: Munsell Ware 10YR 7/3 very pale brown
outside 2.5Y 8/3 pale yellow
inside 2.5Y 6/3 light yellowish brown

Fig. 54.2: Munsell Ware course greyish pink clay
outside 10YR 7/2 light gray
inside 10YR 7/2 light gray

Fig. 54.3: Munsell Ware reddish pink clay
outside greenish white slip, wheeled
inside not available

Fig. 54.4: Munsell Ware pinkish clay
outside yellowish white slip
inside not available

Ayyash

Fig. 56: Munsell Ware Gley 1 3N very dark gray
outside 10R 6/4 pale red
inside 10R 6/4 pale red

Fig. 57: Munsell Ware 10YR 8/1 white or 8/2 very pale brown
outside 10YR 8/3 very pale brown
inside 10YR 8/1 white

Fig. 58.1: Munsell Ware 5YR 7/4 pink
outside 2.5YR 7/4 light reddish brown
inside 5YR 6/4 light reddish brown

Fig. 58.2: Munsell Ware 5Y 8/2 pale yellow
outside 5Y 7/3 pale yellow
inside 10YR 7/3 very pale brown

Fig. 59.1: Munsell Ware 5YR 5/6 yellowish red
outside 10YR 7/2 light gray
inside 10YR 7/2 light gray

Fig. 59.2: Munsell Ware 7.5YR 6/4 light brown
outside 5Y 8/3 pale yellow
inside 7.5YR 7/3 pink

Fig. 59.3: Munsell Ware 5YR 5/6 yellowish red
outside 5YR 6/4 light reddish brown
inside 5YR 6/4 light reddish brown

Fig. 61.1: Munsell Ware 5Y 6/3 pale olive
outside 2.5Y 8/3 pale yellow
inside 2.5Y 6/4 light yellowish brown

Fig. 61.2: Munsell Ware 10YR 7/4 very pale brown
outside 10YR 7/4 very pale brown
inside 10YR 7/4 very pale brown

Fig. 61.3: Munsell Ware 5Y 6/3 pale olive
outside 5Y 7/3 pale yellow
inside 5Y 7/1 light gray

Fig. 63: Munsell Ware 2.5YR 6/6 light red
outside 10YR 8/3 very pale brown
inside 5YR 7/4 pink

Fig. 65: Munsell Ware 2.5YR 5/4 reddish brown
outside 10YR 7/3 very pale brown
inside 7.5YR 6/4 light brown

Fig. 67: Munsell Ware 5Y 7/4 pale yellow
outside 5Y 7/4 pale yellow
inside 5Y 7/4 pale yellow

Fig. 69: Munsell Ware 5YR 7/4 pink
outside 10YR 7/4 very pale brown
inside 10YR 7/4 very pale brown

Fig. 70.1: Munsell Ware 7.5YR 6/4 light brown
outside 7.5YR 7/3 pink
inside 7.5YR 7/3 pink

Fig. 70.2: Munsell Ware 5Y 8/2 pale yellow
outside 5Y 7/3 pale yellow
inside 2.5Y 7/4 pale yellow

Chapter 11
B 1

Fig. 10.1: Munsell Ware reddish brittle ware
outside 5YR 4/1 dark gray
inside not available

Fig. 10.2: Munsell Ware not available
outside 5YR 8/2 pinkish white
inside not available

Chapter 12
K 1

Fig. 19.1: Munsell Ware 5YR 6/4 light reddish brown
outside 5YR 5/3 reddish brown
inside 5YR 6/4 light reddish brown

Fig. 19.2: Munsell Ware 10YR 6/6 brownish yellow
outside 10YR 7/6 yellow
inside 10YR 6/6 brownish yellow

Fig. 19.3: Munsell Ware Gley1 4/N dark gray
outside Gley1 4/Ndark gray
inside Gley1 4/N dark gray

Fig. 19.4: Munsell Ware Gley1 4/N dark gray
outside Gley1 4/N dark gray
inside Gley1 4/N dark gray

Fig. 19.5: Munsell Ware 2.5YR 6/6 light red
outside 10R 6/6 light red
inside 2.5YR 6/6 light red

Fig. 19.6: Munsell Ware Gley1 4/N dark gray
outside Gley1 4/N dark gray
inside Gley1 4/N dark gray

Fig. 19.7: Munsell Ware Gley1 4/N dark gray
outside Gley1 4/N dark gray
inside Gley1 4/N dark gray

Nadra

Fig. 28.1: Munsell Ware 10YR 8/4 very pale brown
outside 2.5Y 8/4 pale yellow
inside 10YR 8/4 very pale brown

Fig. 28.2: Munsell Ware 10YR 8/4 very pale brown
outside 2.5Y 8/4 pale yellow
inside 10YR 8/4 very pale brown

Chapter 13
Tell Tibne

Fig. 12.1: Munsell Ware 7.5YR 6/6 reddish yellow
outside 2.5Y 8/4 pale yellow
inside 2.5Y 8/3 pale yellow

Fig. 12.2: Munsell Ware 7.5YR 7/3 pink
outside 10YR 7/3 very pale brown
inside 10YR 8/4 very pale brown

Fig. 12.3: Munsell Ware 5Y 8/2 pale yellow and 10YR 5/4 yellowish brown
outside 2.5Y 7/3 pale yellow
inside 2.5Y 7/3 pale yellow

Fig. 12.4: Munsell Ware 2.5YR 7/4 light reddish brown
outside 2.5YR 8/3 pink
inside 7.5YR 7/3 pink

Tell Tabūs

Fig. 35.1: Munsell Ware not available
outside 5YR 8/2 pinkish white
inside 5YR 7/8 reddish yellow

Fig. 35.2: Munsell Ware red and grey brittle ware
outside 10R 4/3 weak red
inside 10R 4/8 red

Fig. 35.3: Munsell Ware not available
outside 5YR 6/8 reddish yellow
inside 5YR 6/8 reddish yellow

Fig. 35.4: Munsell Ware not available
outside 5YR 7/4 pink
inside 10R 6/6 light red

Fig. 36: Munsell Ware black brittle ware
outside 10YR 4/1 dark gray
inside 10YR 3/3 dark brown

Fig. 37: Munsell Ware 10YR 7/4 very pale brown
outside 10YR 7/2 light grey
inside 10YR 7/2 light grey

Ayyash

Fig. 68.1: Munsell Ware 2.5YR 6/4 light reddish brown
outside 2.5YR 6/4 light reddish brown
inside 2.5YR 6/4 light reddish brown

Fig. 68.2: Munsell Ware 5Y 7/3 pale yellow
outside 5Y 7/3 pale yellow
inside 5Y 7/3 pale yellow

Fig. 73.1: Munsell Ware 2.5Y 8/2 pale yellow
outside 2.5Y 8/4 pale yellow
inside 2.5Y 8/4 pale yellow

Fig. 73.2: Munsell Ware 2.5Y 7/4 pale yellow
outside 10YR 8/4 very pale brown
inside 2.5Y 7/4 pale yellow

Fig. 73.3: Munsell Ware 10R 4/6 red
outside 10YR 2/1 black
inside 10R 4/6 red

Fig. 74: Munsell Ware 5YR 6/6 reddish yellow
outside 2.5Y 8/3 pale yellow
inside 2.5Y 7/3 pale yellow

Chapter 14
Tell Tibne

Fig. 16.1: Munsell Ware 5YR 6/6 reddish yellow
outside 2.5Y 8/3 pale yellow
inside 5YR 7/4 pink

Fig. 16.2: Munsell Ware 5Y 4/1 dark gray
outside 5Y 7/3 pale yellow
inside 5Y 7/3 pale yellow

INDEX OF AUTHORS, GEOGRAPHICAL PLACES AND SUBJECTS OF INTEREST